Hands-On Introduction to
LabVIEW™
for Scientists and Engineers

Hands-On Introduction to
LabVIEW™
for Scientists and Engineers

Fourth Edition

John Essick

Reed College

New York   Oxford
OXFORD UNIVERSITY PRESS

Library of Congress Cataloging-in-Publication Data

Names: Essick, John, author.
Title: Hands-on introduction to LabVIEW for scientists and engineers.
Description: Fourth edition. | New York, NY : Oxford University Press, [2019] | Includes index.
Identifiers: LCCN 2018008842 | ISBN 9780190853068 (Paperback) | ISBN 9780190853082 (Ebook)
Subjects: LCSH: Scientific apparatus and instruments—Computer simulation. | Electronic apparatus
    and appliances—Computer simulation. | LabVIEW. | Science—Experiments—Data
    processing. | Computer graphics. | Computer programming.
Classification: LCC Q185 .E69 2018 | DDC 502.85/53—dc23
LC record available at https://lccn.loc.gov/2018008842

Printing number: 9 8 7 6 5

Printed by LSC Communications, United States of America

To my wife, Katie

Contents

Preface

Hands-On Introduction to LabVIEW for Scientists and Engineers provides a learn-by-doing approach to acquiring the computer-based skills used daily in experimental work. This book is not a manual-like presentation of LabVIEW. Rather, *Hands-On Introduction to LabVIEW* leads its readers to mastery of LabVIEW through the process of using this powerful laboratory tool to carry out interesting and relevant projects. Readers, who are assumed to have no prior computer programming or LabVIEW background, begin writing meaningful programs in the first few pages.

Hands-On Introduction to LabVIEW can be used as a text in an instructional lab course or for self-study by individual researchers. The book is designed for flexible use so that readers can easily choose the desired depth of coverage. The first six chapters, which form the foundation appropriate for all readers, focus on the fundamentals of LabVIEW programming as well as the basics of computer-based experimentation using a National Instruments data acquisition (DAQ) device. These opening chapters can be used as the basis of a three- or four-week introduction to LabVIEW-based data acquisition. Subsequent chapters have been written as independently as possible so that an instructor or self-learner can fill out their course of study as desired. Those who work through most of the text's chapters will attain an intermediate skill level in computer-based data acquisition and analysis.

The progression of topics in *Hands-On Introduction to LabVIEW* is as follows:

Chapters 1–4: Fundamentals of the LabVIEW Graphical Programming Language. Central features of LabVIEW including its programming environment, control loop structures, graphing modes, mathematical functions, and text-based MathScript (and Formula Node) commands are learned in the course of writing digitized waveform simulation programs.

Chapter 5: Introduction to Data Acquisition Devices Using MAX. Features of National Instruments DAQ devices are presented, along with concepts of digitized data such as resolution, sampling frequency, and aliasing. Then, using the Measurement & Automation Explorer (MAX), readers interactively control the full functionality (analog-to-digital, digital-to-analog, digital input/output, and pulse counting) of a National Instruments DAQ device.

Chapter 6: Data Acquisition Using DAQ Assistant. Using the high-level DAQ Assistant Express VI, readers write LabVIEW programs that execute analog-to-digital, digital-to-analog, and digital input/output tasks on a National Instruments DAQ device. Computer-based instruments constructed include a DC voltmeter, digital oscilloscope, DC voltage source, waveform generator, and blinking LED array.

Chapters 7–10: More LabVIEW Programming Fundamentals. Implementation of data file input/output, local memory, and conditional branching in LabVIEW is investigated while writing several useful programs (e.g., spreadsheet data storage, moving averager) and learning the powerful state machine program architecture. Additionally, LabVIEW's control flow approach to computer programming is studied.

Chapters 11 and 12: Data Analysis. Proper use of LabVIEW's curve fitting and fast Fourier transform (FFT) functions is investigated. Using Express VIs to control a DAQ device, two computer-based instruments—a digital thermometer and a spectrum analyzer—are constructed.

Chapter 13: Data Acquisition Using DAQmx. Programs are written to carry out analog-to-digital, digital-to-analog, and digital counter tasks on a DAQ device using the conventions of DAQmx. This lower-level approach (in comparison to the high-level Express VIs) allows utilization of the full available range of DAQ device features. A DC voltmeter, DC voltage source, waveform generator, and frequency meter are constructed, as well as a sophisticated digital oscilloscope based on the state machine architecture.

Chapter 14: Control of Stand-Alone Instruments. Using LabVIEW's VISA communication driver, control of a stand-alone instrument over the General Purpose Interface Bus (GPIB) as well as the Universal Serial Bus (USB) is studied. A Keysight/Agilent 34410A Multimeter is used to demonstrate the central concepts of interface bus communication between a PC and stand-alone instrument.

Appendix A: Formula Node Supplement. After a brief introduction to the Formula Node, instructions are given for carrying out Chapter 4 exercises using the Formula Node (rather than the MathScript Node).

Appendix B: FFT Supplement. A mathematical description of the leakage and windowing effects associated with fast Fourier transform analysis is presented.

Appendix C: Temperature Control Project. The LabVIEW skills acquired throughout the book are used to construct a Proportional-Integral-Derivative (PID) temperature control system. A design for the hardware required for this project is included.

Key features of *Hands-On Introduction to LabVIEW* include its emphasis on real-world problem solving, its early introduction and routine use of data acquisition hardware, its Do It Yourself projects and Use It! examples at the end of each chapter, and its healthy offering of back-of-the-chapter homework problems.

Real-World Problem Solving: Chapter topics and exercises provide examples of how commonly encountered problems are solved by scientists and engineers in the lab. LabVIEW features, along with relevant mathematical background, are introduced in the course of solving these problems. The "best practice" strategies presented (such as modularity and data dependency) equip readers to optimize their use of LabVIEW.

Data Acquisition Usage Throughout: LabVIEW's Express VIs allow exercises involving DAQ hardware to appear early and then routinely in *Hands-On Introduction to LabVIEW*. Express VIs package common measurement tasks into a single graphical icon and so allow the user to write a program with minimal effort. Of particular note, following the book's first four software-only chapters that teach the fundamentals of the LabVIEW programming language, data acquisition using a DAQ device is covered in Chapters 5 and 6. For a professor or self-learner who wishes to devote only three or four weeks to instruction in computer-based data acquisition, Chapters 1 through 6 will provide the needed instructional materials. For those planning a more comprehensive study of LabVIEW, the Express VIs allow construction of a state-machine digital oscilloscope, digital thermometer, and spectrum analyzer in Chapters 9, 11, and 12, respectively. In Chapter 13, the control of a DAQ device via the more advanced programming DAQmx icons is covered. In contrast to the Express VIs, the DAQmx icons enable a user to utilize the full available range of the DAQ-device features. In Chapter 14, data are acquired remotely from a stand-alone instrument using the GPIB and/or USB interface bus and, in Appendix C, interested readers can use a DAQ device to precisely control the temperature of an aluminum block. Additionally, commonly used interfacing circuits consisting of low-cost integrated circuits are presented. Circuits include an anti-aliasing filter, thermocouple signal conditioner, and digital potentiometer that communicates via the Serial Peripheral Interface (SPI).

Do It Yourself Projects: To allow readers to gauge their understanding of the presented material, each chapter of *Hands-On Introduction to LabVIEW* concludes with a Do It Yourself project. Each of these projects poses an interesting problem and (loosely) directs readers in applying the chapter's material to find a solution. In some chapters, this project involves writing a program that functions as a stopwatch (Chapter 2) or determines a person's reaction time (Chapter 10); in other chapters the reader constructs a computer-based instrument including a digital thermometer (Chapter 11), a spectrum analyzer (Chapter 12), and a frequency meter (Chapter 13).

Use It! Examples: Ready-to-use example programs, which carry out common tasks encountered in laboratory work, are presented at the end of each chapter. Some of these examples involve programming solutions, for example, showing how to input parameters at the beginning of a data run, save and plot data during runtime, and apply a criterion to a sequence of values to selectively build a data array. Others examples are low-cost hardware solutions, including anti-aliasing through the use of an eighth-order Butterworth low-pass filter, amplification and cold-junction compensation for a thermocouple temperature measurement, control of integrated circuits using SPI communication, and construction of an Arduino-based voltmeter and digital oscilloscope.

Back-of-the-Chapter Homework Problems: A selection of homework-style problems is included at the end of each chapter so that interested readers can further develop their LabVIEW-based skills. In some of these problems, readers test their understanding by applying the chapter topics to new applications (e.g., Bode magnitude plot); in others, readers use programs written within the chapter to explore important experimental issues (e.g., frequency resolution of a fast Fourier transform). Finally, a number of problems introduce readers to features of LabVIEW relevant to, but not included in, the chapter's text (e.g., data storage in binary format).

Improvements to the Fourth Edition: This new edition includes the following improvements:

- New chapter interactively introduces all features of National Instruments DAQ devices using the Measurement & Automation Explorer (MAX). [Chapter 5]
- New **Use It!** examples at the end of each chapter present ready-to-use programs that carry out common tasks encountered in laboratory work.
- Commonly used, low-cost integrated circuits (for example, eighth-order Butterworth low-pass filter, thermocouple signal conditioner) highlighted in end-of-the-chapter problems and **Use It!** examples.
- LabVIEW control of an Arduino is demonstrated through construction of Arduino-based voltmeter and digital oscilloscope. [Chapter 14]
- All chapters are fully updated to the latest version of LabVIEW. DAQ hardware now commonly used in instructional laboratories and self-learning is highlighted.
- 14 new end-of-the-chapter problems appear throughout the book.

Hands-On Introduction to LabVIEW is fully compatible with the Full Development System, Professional Development System, and Student Edition of LabVIEW. In addition, all chapters may be carried out by Base Development System owners, with the exception of Chapters 11 and 12 (since the Base Development System does not include curve fitting and fast Fourier transform functionality). An instructor might consider having students purchase personal copies of the low-cost Student Edition software (the Student Edition can now be purchased by itself at a very affordable price; that is, it is no longer necessary to buy an expensive bundled book/software package). With their own LabVIEW software, students can perform non-hardware-related chapter sections and/or back-of-the-chapter problems as homework on their own computers.

To aid readers in creating their LabVIEW programs, the following conventions are used throughout the book: **Bold** text designates the features such as graphical icons, palettes, pull-down menus, and menu selections that are to be manipulated in the course of constructing a program. The descriptive names that label controls, indicators, custom-made icons, programs, disk files, and directories (or folders) are given the **straight** font. *Italic* text highlights character strings that the programmer must enter using the keyboard and also signals the first-time use of important terms and concepts.

Any suggestions or corrections are gladly welcomed and can be sent to John Essick, Reed College, 3203 SE Woodstock Boulevard, Portland, OR 97202, USA, or jessick@reed.edu.

Updates, answers to frequently asked questions, and ancillary materials for *Hands-On Introduction to LabVIEW* are available at http://academic.reed.edu/physics/faculty/essick.

Additionally, solutions to the even-numbered back-of-the-chapter problems can be downloaded at www.oup.com/us/essick. Instructors who adopt this book for a course can obtain a password-protected link to the solution set for every problem from Oxford University Press.

For their advice and assistance in preparing this revision of *Hands-On Introduction to LabVIEW*, I thank Dan Kaveney, Megan Carlson, and Claudia Dukeshire of Oxford University Press. For their helpful comments and suggestions, I express my appreciation to the reviewers.

- Prathap Basappa, Norfolk State University
- Armando Carrasco, Austin Community College
- James Doyle, Macalester College
- Hector Gutierrez, Florida Institute of Technology
- Aubri Hanson, Chipola College
- Robert Haring-Kaye, Ohio Wesleyan University
- Saliman Isa, South Carolina State University
- Robert Muratore, Hofstra University
- Robert Polak, Loyola University Chicago
- John Viator, Duquesne University
- Zifeng Yang, Wright State University

Finally, to my family: Thank you for your love and support while I worked on this project.

John Essick
Portland, Oregon

About the Author

John Essick is a professor at Reed College with research interests in the optoelectronic properties of semiconductors. Since 1993, he has taught computer-based experimentation using LabVIEW as part of Reed's junior-level Advanced Laboratory and used LabVIEW to carry out many research projects.

CHAPTER 1

LabVIEW Program Development

1.1 LABVIEW PROGRAMMING ENVIRONMENT

Welcome to the world of *LabVIEW*, an innovative graphical programming system designed to facilitate computer-controlled data acquisition and analysis. In this world, you—the LabVIEW user—will be operating in a programming environment that is different from that offered by most other programming systems (for example, the C and Python computer languages). Rather than creating programs by writing lines of text-based statements, in LabVIEW, you will code programming ideas by selecting and then properly patterning a collection of graphical icons.

A LabVIEW program consists of two windows, the *front panel* and the *block diagram*. Once you have completed a program, the front panel appears as the face of a laboratory instrument with your own design of knobs, switches, meters, graphs, and/or strip charts. The front panel is the program's user interface; that is, it facilitates the interaction of supplying inputs to and observing outputs from the program as it runs.

The block diagram is the actual LabVIEW programming code. Here reside the graphical images that you have appropriately selected from LabVIEW's well-stocked libraries of icons. Each icon represents a block of underlying executable code that performs a particular useful function. Your programming task is to make the proper connections between these icons using a process called *wiring* so that data flow amongst the graphical images to accomplish a desired purpose. Because the icon libraries are designed specifically with the needs of scientists and engineers in mind, LabVIEW enables you—the modern-day experimentalist—to write programs that perform all of the laboratory tasks required for your state-of-the-art research and industrial applications, including instrument control, data acquisition, data analysis, data presentation, and data storage.

To begin developing your skill in LabVIEW graphical programming, in this first chapter, I will guide you through the steps of creating a LabVIEW program. Together, we will explore key features of the LabVIEW programming environment as we write a program that detects the *parity* of a given integer, that is, whether it is even or odd. The parity of an integer is defined by the following test: Using long division, divide the integer by 2. If the remainder is 0 (that is, it "divides evenly"), the integer is even; if the remainder is 1, the integer is odd (by convention, the remainder is assumed to be positive when dividing a negative integer by a positive integer). According to this definition, for example, the integers 0, 4, and −4 are even, while 3 and −3 are odd. To write a program that carries out this parity test, we begin with a blank slate.

1.2   BLANK VI

Quite commonly, a LabVIEW program is written to create a computer-based system that has the same functionality as a traditional stand-alone laboratory instrument. The input and output objects on such a program's front panel (called *controls* and *indicators*, respectively) mimic the appearance of the physical instrument, while code contained on the block diagram directs the computer in carrying out the intended instrumental operations. Given that LabVIEW was chiefly developed for this purpose—that is, as a tool for morphing a computing system (consisting of a PC and data acquisition device) into the form and function of a physical instrument—a LabVIEW program is called a *virtual instrument* (or *VI*, for short), even those programs designed for tasks much smaller in scope than a full computer-based instrument.

To begin writing our parity-detecting program (or, in LabVIEWspeak, our parity-detecting VI), find the LabVIEW application on your computer (which may have a desktop icon or may be in a folder named **National Instruments, LabVIEW,** or **LabVIEW Student Edition**, depending on your particular computing system) and launch it by double-clicking on its icon or name.

After a few moments, a *Getting Started Window* labeled **LabVIEW** (in older LabVIEW versions, this window was labeled **Getting Started**) will appear. Place your mouse cursor over the **Blank VI** selection (depending on your version of LabVIEW, you may have to select **Create Project** first) and then click. Alternatively, you can select **New VI** in the Getting Started Window's **File** pull-down menu. An untitled front panel will appear in the foreground of your screen with a slightly offset block diagram in the background. In addition, a menu bar for an array of pull-down menus that contain LabVIEW editing items will appear at the top of the window.

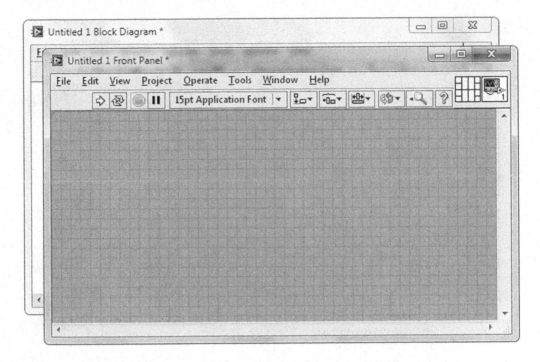

There are three ways to toggle the block diagram between the background and foreground:

- Select **Show Block Diagram** (when it is in the background) or **Show Front Panel** (when the block diagram is in the foreground) from the **Window** pull-down menu.
- Click your mouse cursor on a visible region of the window (block diagram or front panel) that is currently in the background.
- Use the keyboard shortcut by typing *<Ctrl+E>*.

   Practice toggling the block diagram between the foreground and background using each of the three available methods.

1.3 FRONT-PANEL EDITING

To begin adding programming elements to your VI, situate the blank front panel in the foreground. The objects to be placed on this front panel as you write your LabVIEW program are found in a repository called the *Controls Palette*. If a Controls Palette is not already visible, activate one by selecting **Controls Palette** from the **View** menu.

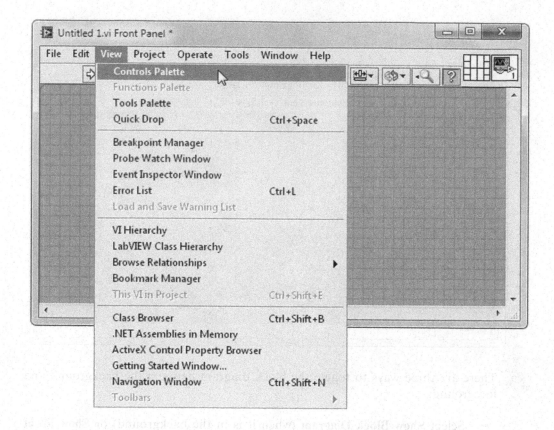

A "floating" Controls Palette will appear on your screen, which can be placed in a convenient spot by "clicking and dragging" on its title bar.

The Controls Palette is an organized collection of *categories* (such as *Modern*, *Silver*, and *Express*), where each category contains a palette of related programming objects. The Controls Palette may be configured in a multitude of ways, and the particular configuration that you see will depend on choices made by past users (if any) of your LabVIEW system.

For our work, configure the appearance of the Controls Palette as follows:

- Click on the **Customize** button [Customize▾] near the top of the Controls Palette (for older LabVIEW versions, it is called the **View** button). In the menu that appears, select the **Change Visible Palettes…** option. In the dialog window that follows, click on **Select All** and then press the **OK** button.

• Click on the **Customize** button again. This time, in the menu, click on **Options....** In the dialog window that appears, select **Controls/Functions Palettes**, and then, under **Formatting**, choose **Category (Icons and Text)** for the **Palette** option. Finally, press the **OK** button.

The Controls Palette is now configured. A particular category's palette can be toggled between visible ("open") and hidden ("closed") by clicking on its name. Open the **Modern** palette and close all others so that your Controls Palette appears as follows (where I've resized the Controls Palette's window so that it better fits on the page). The categories listed on your Controls Palette may differ slightly from this illustration if your system has some optional LabVIEW add-ons.

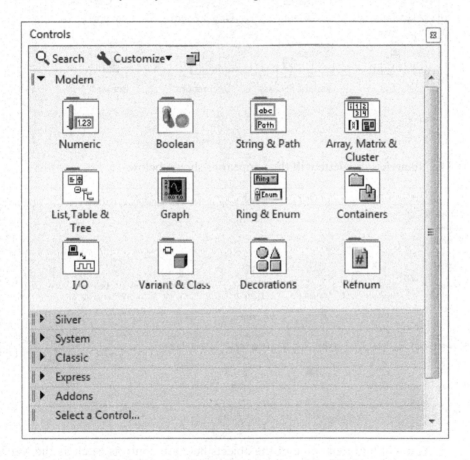

Briefly familiarize yourself with the contents of the **Modern** palette. Then place the cursor over the **Numeric** button and click.

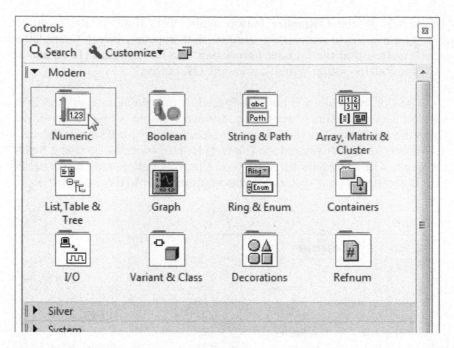

The **Numeric** subpalette will then appear as shown below.

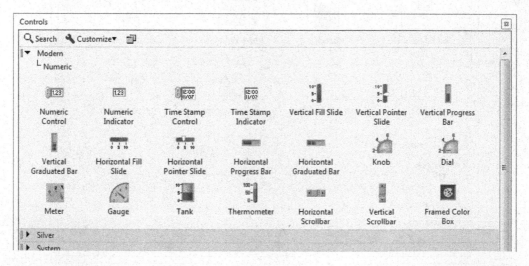

You will find that some of the objects here are controls (such as the **Numeric Control** and **Horizontal Pointer Slide**), which offer a user the opportunity to input numeric data to the VI. Other objects are indicators (such as the **Numeric Indicator** and **Meter**), which output results from the VI for viewing by the user. To close this subpalette and return to the **Modern** palette, simply click on **Numeric**.

Those with one of the latest versions of LabVIEW will have the choice of obtaining needed front-panel programming objects from either the **Modern** or **Silver** palettes. The analogous controls and indicators found in these two palettes are functionally the same; they differ only in appearance. The **Modern** icons have been available in many versions of LabVIEW, while the **Silver** icons were introduced to commemorate the 25th anniversary of LabVIEW and have a contemporary look with rounded edges and sleek shadowing. Because not all readers have the latest LabVIEW update, I will use the **Modern** controls and indicators throughout the book. You may, however, enjoy using the equivalent **Silver** objects instead if you have them available.

On the front panel of our parity-detecting VI, we will need one control (for a user to input the integer of interest) and one indicator (to alert the user whether the integer is odd or not). For the required control, we will use a **Numeric Control**. Here's how to place this icon on the front panel: First, select the **Numeric Control** icon from the Controls Palette. To find this icon, open the **Modern** palette by clicking on the **Modern** category and then open the **Numeric** subpalette. From now on, such a sequence of choices will be indicated as follows: **Controls>>Modern>>Numeric**. Then select the **Numeric Control** by placing the mouse cursor over this object and clicking.

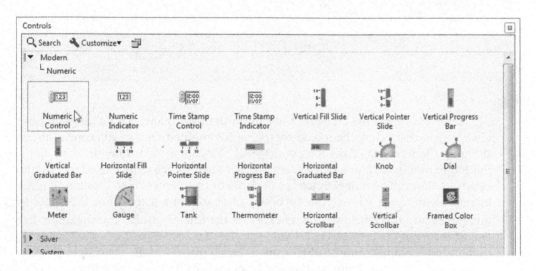

Now that you've selected the **Numeric Control** icon, place the cursor at the location on your front panel where you wish the icon to reside.

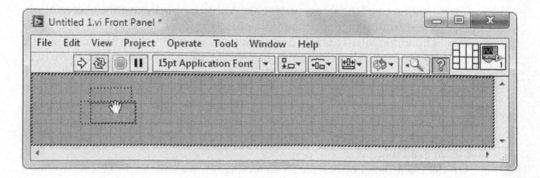

Then click the mouse and—Whoomp!—there the icon is.

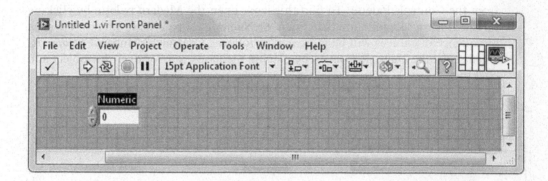

After you place the **Numeric Control** on the front panel with a single mouse click, if you don't click the mouse further, a highlighted region will appear above the upper left corner of the icon containing the default text **Numeric**. This highlighted region is the Numeric Control's *owned label*, which can be used to give the control a descriptive name. Since the purpose of this control is to input an integer of interest into our VI, use the keyboard to change the text within this label to **Integer**. Secure this label by either clicking on the Enter button ☑ at the upper left end of the front panel, pressing *<Enter>* on the numeric keypad of your keyboard, or simply clicking the mouse cursor on an empty region of the front panel. If you accidentally click the mouse so that the label loses its highlighting before you have a chance to enter a name, keep the default text **Numeric** for now; you'll learn how to restore the highlighting in a minute.

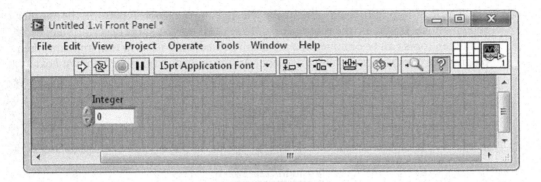

If you decide that you would like to move this icon to some other location on the front panel, you can do so through use of the *Positioning Tool* , whose job is to select, move, and resize objects. The is one of the several available LabVIEW editing tools displayed in the *Tools Palette*, which may or may not be visible at the moment.

Activate the Tools Palette by selecting **Tools Palette** in the **View** pull-down menu. In our shorthand convention, this menu selection is **View>>Tools Palette**.

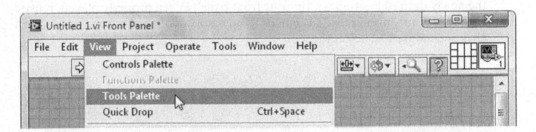

The "floating" Tools Palette will appear as shown next and, just like the Controls Palette, can be relocated to a convenient location by "clicking and dragging" its title bar.

The Positioning Tool may already be selected on your Tools Palette, in the manner shown above. If not, use the mouse cursor to select it now. A short description of each of the ten tools can be obtained from a small box called a *tip strip* that appears when placing the cursor over each tool's button. An example of a tip strip is shown next.

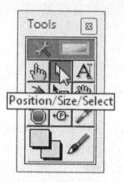

To move the **Numeric Control**, place the ⤢ over this icon and click. When selected in this way, the icon will become highlighted by a moving, dashed border called a *marquee*. With the mouse button depressed, drag the highlighted icon to the newly desired location within your front panel. When properly placed, release the mouse button and move the cursor to an empty spot on the front panel. Then click the mouse to deselect the icon.

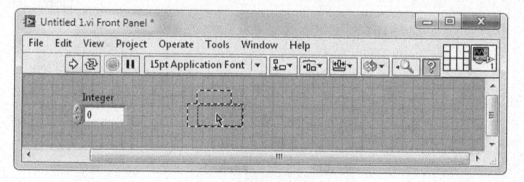

Once an object is highlighted with a marquee, there are a couple of other handy movement techniques. If you hold down the <*Shift*> key and then drag the object, LabVIEW will only allow purely horizontal or vertical motion. Also, you can move the selected object in small, precise increments by pressing the keyboard's <*Arrow*> keys, rather than dragging with the mouse. Try moving the **Numeric Control** icon using both of these techniques.

A few moments ago, if you lost highlighting in the Numeric Control's label before you had a chance to enter the text **Integer**, use the *Labeling Tool* [A] found in the Tools Palette to rehighlight the label region, then enter the text.

Next, let's place the indicator that will report the result of the parity test (i.e., whether the integer is found to be odd or not) on the front panel. For this program feature, we will use a **Round LED**. This icon is found in **Controls>>Modern>>Boolean**, a palette that contains switches, pushbuttons, and LEDs useful for entering and displaying Boolean (TRUE or FALSE) values. Open **Controls>>Modern>>Boolean** and select a **Round LED** icon by clicking on it.

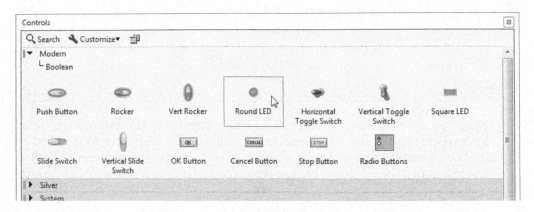

Then place this icon on the front panel, labeling it **Odd?**. The finished front panel for our VI is shown next.

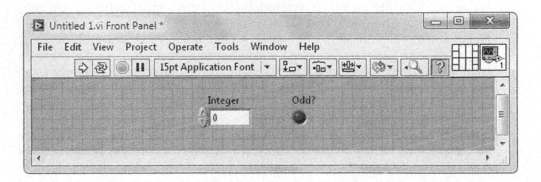

Now that the front panel is complete, it's time to move to the block diagram. There, we will build the code to accept an integer from the **Integer** control, determine its parity, and then light the **Odd?** LED, if the integer is found to be odd.

1.4 BLOCK-DIAGRAM EDITING

Switch to block diagram by selecting **Window>>Show Block Diagram** or using the keyboard shortcut <*Ctrl+E*>.

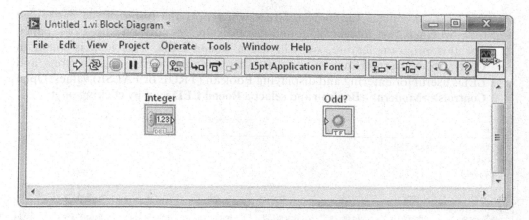

Here, you will find the block diagram is already populated with two objects, which have appeared automatically in association with the two objects that were placed on the front panel. These two *icon terminals* function as block-diagram portals to their associated front-panel icons. That is, the numeric data input at the **Integer** control on the front panel is available on the block diagram at , while the Boolean data produced on the block diagram are passed to the front-panel **Odd?** indicator via . Note that the **Integer** icon terminal contains a small picture of numbers, along with an outward-directed arrow, indicating it delivers numeric data from a Numeric Control. Also, this icon has an orange border with text **DBL** denoting that its delivered data are formatted as double-precision floating-point numbers, a computer-adapted version of scientific notation for representing both integers and non-integers (more on this later in the chapter). The icon is further identified by the **Integer** label that you entered on the front panel. Similarly, the **Odd?** icon terminal contains a small picture of a lit light-emitting diode, along with an inward-directed arrow, indicating it accepts data for a Round LED. Additionally, the icon has a green border with text **TF** denoting that its accepted data are formatted as Boolean values (i.e., TRUE or FALSE). This icon terminal is further identified by the **Odd?** label. If you need to reposition these icons on the block diagram, you can do so using the .

Okay, hopefully the details of what we need to accomplish on the block diagram are coming into focus. Here is our task: Construct block-diagram code that accepts the integer delivered from , determines this integer's parity, and then sends a TRUE Boolean value to if the integer is odd.

The programming objects we will use to build the required LabVIEW code are found in an organized collection of categories called the *Functions Palette*. If a

Functions Palette is not already visible, activate one by selecting **View>>Functions Palette**. Once activated, the Functions Palette will appear whenever the block diagram is brought to the foreground. It will disappear and be replaced by the Controls Palette when the front panel is subsequently toggled into the foreground. If only a few categories are visible in your Functions Palette, click the **Customize** button near its upper right corner and then select **Change Visible Palettes...>>Select All**.

A particular category's palette can be toggled between visible ("open") and hidden ("closed") by clicking on its name. Open the **Programming** palette and close all others so that your Functions Palette appears as follows (again, I've resized the Functions Palette to better fit the page). The categories listed on your Functions Palette may differ slightly from this illustration if your system has some optional LabVIEW add-ons.

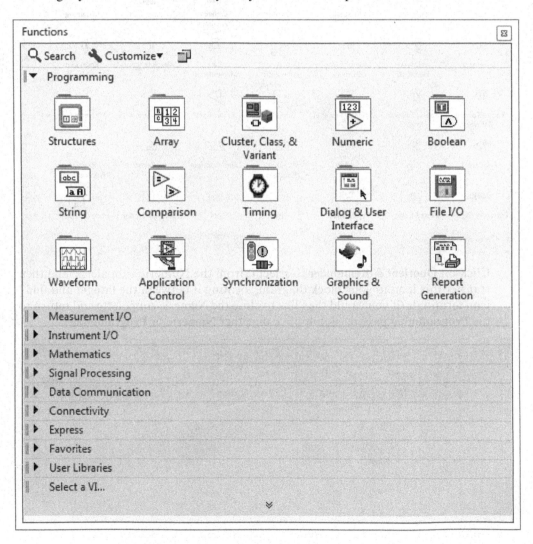

We will use an icon called **Quotient & Remainder**, which is found in **Functions>> Programming>>Numeric**, to determine whether an integer is odd or even. Place this icon on your block diagram as follows: With the **Programming** palette open, place the cursor over the **Numeric** button and click. The **Numeric** subpalette will then appear as shown below, where you can find **Quotient & Remainder**.

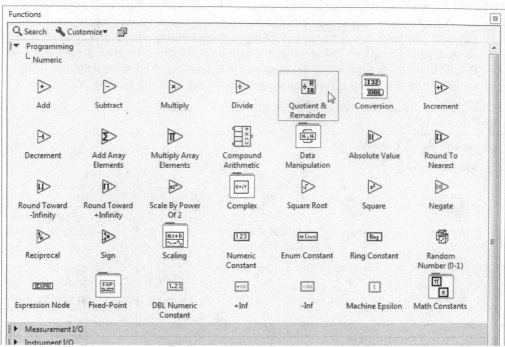

Click on **Quotient & Remainder** to select it from the Numeric subpalette, and then transfer this icon to your block diagram, locating it between the Integer and Odd? icon terminals (if you would then like to close the **Numeric** subpalette and return to the **Programming** palette, simply click on either **Numeric** or **Programming**).

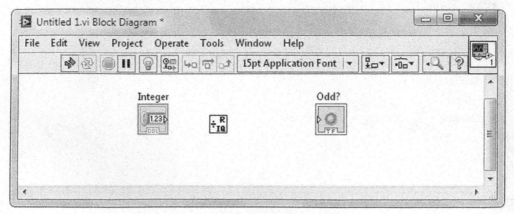

The function of the **Quotient & Remainder** icon is described in the following *Context Help Window.*

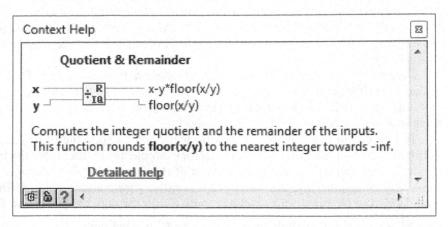

In a Context Help Window, the icon's input *terminals* are shown on the left and output terminals on the right. Thus, we see that the **Quotient & Remainder** icon accepts input values at its **x** and **y** terminals and returns output values at its **remainder (R)** and **integer quotient (IQ)** terminals. The value at the **R** terminal is the remainder R that results when x is divided by y. This quantity, which corresponds to the modulo function *mod(x,y)* of other programming languages, is calculated using the following algorithm: $R(x, y) = x - y \times floor(x/y)$, where *floor* truncates its argument to the next lowest integer. Thus, if x is an integer and $y = 2$, then R equals 0 and 1 when the integer x is even and odd, respectively. For example, if the integer is 4, $R(4,2) = 4 - 2 \times floor(4/2) = 4 - 2 \times (2) = 0$, while if the integer is 3, $R(3,2) = 3 - 2 \times floor(3/2) = 3 - 2 \times (1) = 1$. For the negative integer –3, $R(-3,2) = -3 - 2 \times floor(-3/2) = -3 - 2 \times (-2) = 1$.

You may see this Context Help Window for yourself by selecting **Help>>Show Context Help.**

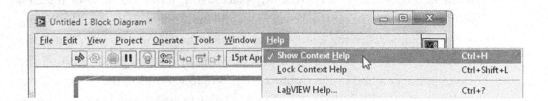

After the Context Help Window appears, place it in a convenient position using the ⬉. Then place the ⬉ over the **Quotient & Remainder** icon to view its description. When you have no further need for the Context Help Window, it can be toggled off by deselecting **Help>>Show Context Help**. As an alternative to mouse control, try toggling the Context Help Window on and off using the following keyboard shortcut: *<Ctrl+H>*.

Here is our scheme for how to use the **Quotient & Remainder** icon: Deliver the integer from **Integer** to this icon's **x** input terminal and make the value at its **y** input terminal equal to 2. The **R** output terminal of the icon will then tell us the parity of **Integer**: $R = 0$ means even, $R = 1$ means odd.

To implement this scheme, we need to learn how to code the transfer of data between two block-diagram icons so that, for example, the integer from the **Integer** icon terminal can be moved to the **x** input of the **Quotient & Remainder** icon. In LabVIEW programming, this coding task is called *wiring* and is accomplished by connecting the terminals of two objects with a *wire* through the use of the *Wiring Tool* ✲. Go to the **Tools Palette** and select the ✲ by clicking on it.

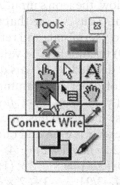

Here is a method for producing proper wiring between the **Integer** icon terminal and the **x** input of **Quotient & Remainder**. First, activate the Context Help Window by the keyboard short cut *<Ctrl+H>*, and then position the Context Help Window in a convenient, out-of-the-way place. Now, move the ✲ over the ⬛ icon. *Wire stubs* will appear from each input and output terminal when the Wiring Tool closely approaches the icon. As you position the ✲ over each terminal, the terminal's identity will be displayed in a tip strip, and it will blink both on the block-diagram icon and in the Context Help Window. Position the ✲ over the **x** input terminal, then click the mouse. You have now tacked down one end of a wire to the **Quotient & Remainder** icon.

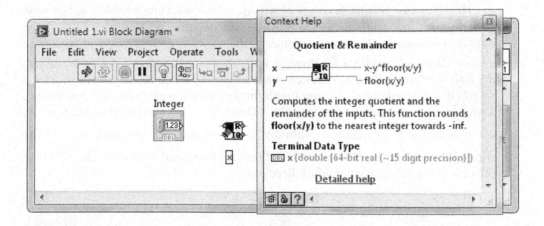

Next, move the ❦ smoothly to the left until it is over the **Integer** icon terminal

and the ⬚1.23 is blinking (in the region near the small outward-directed arrow). Click the mouse to complete the wiring. If done correctly, an orange wire will con-

nect ⬚1.23 to the **x** input terminal of ⬚R·IQ. The wire's orange color is LabVIEW's way of signifying that the data being passed between these icons are formatted as floating-point numbers.

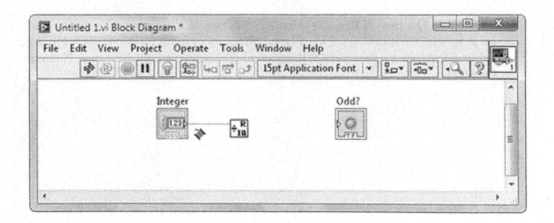

Here are two further features of wiring, which are demonstrated in the next illustration: First, while using the 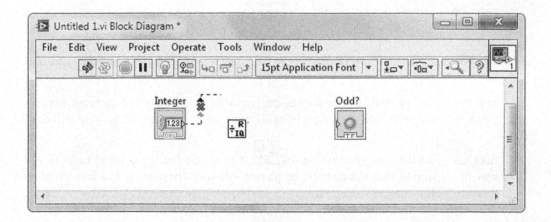, if you make a mistake and try to form an improper connection between two objects, the resulting *broken wire* will appear as a black dashed line with a ✖ through it. Such a mistake can be erased by selecting **Edit >>Remove Broken Wires** or, more simply, by using the keyboard shortcut: *<Ctrl+B>*. Alternatively, you can highlight the broken wire using the Positioning Tool and then erase it by pressing the *<Delete>* key on your keyboard. Second, while using the ✎, if at some location you wish to make a right-angle bend in the wire, click the mouse when you arrive at that location, and then move the ✎ off in the newly desired (perpendicular) direction.

For the **Quotient & Remainder** icon to function as a parity checker of the integer at its **x** input, the integer 2 must be supplied at its **y** input terminal. To fulfill this requirement, go to **Functions>>Programming>>Numeric** and obtain a **Numeric Constant**.

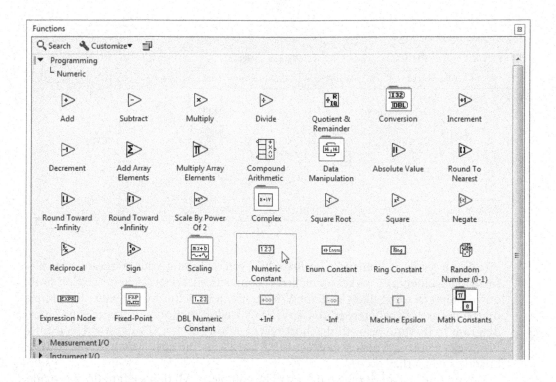

Place the **Numeric Constant** on your block diagram in the vicinity of Quotient & Remainder's **y** input terminal and click the mouse once. The **Numeric Constant** will appear as a blue-bordered rectangular box with its interior highlighted, ready to receive the numeric value of your choosing. If the mouse is inadvertently clicked a second time, the icon's interior highlighting will be lost, but can be restored again using the (or the not-yet discussed). Using the keyboard, enter the integer *2* into **Numeric Constant**'s highlighted interior, then secure this choice in the usual way (by either clicking on the Enter button , pressing *<Enter>* on the numeric keypad of your keyboard, or simply clicking the mouse cursor on an empty region of the block diagram). Next, wire the to the **y** input terminal of . You may find that an effort-saving feature of LabVIEW called *autowiring* has already made this connection automatically when you initially brought the **Numeric Constant** nearby the **Quotient & Remainder** icon. If so, then you're in good shape; if not, use the to make the connection. Note that and the wire connecting it to are colored blue, LabVIEW's code for integer-formatted data.

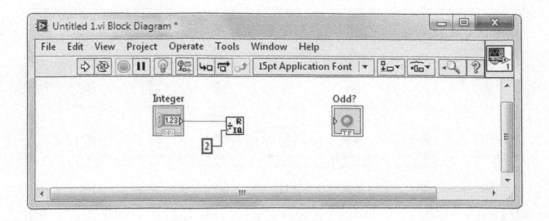

As an aside, if a mistake occurs while you're coding your program (say, autowiring accidentally makes an incorrect connection), select **Undo** in the **Edit** pull-down menu (a word like **Create** or **Wiring** will follow the word **Undo**, depending on the nature of the latest edit). LabVIEW's **Undo** function will revert your block diagram to its state prior to the most recent editing action. The **Edit>>Undo** editing trick, whose keyboard shortcut is *<Ctrl+Z>*, is the easy way to erase editing mistakes when they occur.

With one last addition, your VI will be complete. All that's left to do is provide a method for checking whether the **R** output of **Quotient & Remainder** is 0 or 1 and then sending the correct Boolean response (FALSE for **R** = 0, TRUE for **R** = 1) to the **Odd?** Round LED's icon terminal.

Open the **Functions>>Programming>>Comparison** subpalette. There, you will find a collection of icons that offer several options for carrying out the required task. **Not Equal To 0?** is especially promising as it provides a one-icon solution to our problem. Select **Not Equal To 0?**.

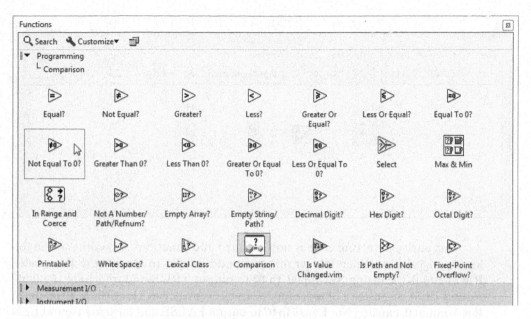

The Context Help Window for **Not Equal To 0?** is shown next. Here, we find that, when given a number at its **x** input, the icon outputs a Boolean value of TRUE and FALSE if $x \neq 0$ and $x = 0$, respectively.

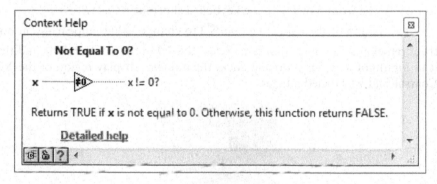

On your block diagram, place the **Not Equal To 0?** icon between **Quotient & Remainder** and the **Odd?** icon terminal (the ⟨⟩ will come in handy, if you need to move icons to open up blank space). Use the ⟨⟩ first to connect Quotient & Remainder's **R** output terminal to the **x** input of ⟨⟩, then second to connect the output terminal of ⟨⟩ to ⟨⟩ as shown below. Note that the first and second resulting wires are orange and green, indicating that they carry floating-point numbers and Boolean values, respectively.

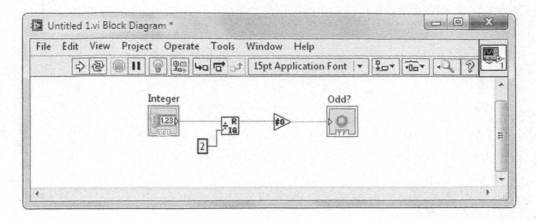

Our parity-detecting code is now finished and functions according to the following algorithm: When Integer inputs an odd integer to **Quotient & Remainder**, **R** will equal 1, causing **Not Equal To 0?** to output TRUE and thus light the Odd? Round LED. If, instead, Integer inputs an even integer to **Quotient & Remainder**, **R** will equal 0, causing **Not Equal To 0?** to output FALSE and the Odd? Round LED will be unlit.

1.5 PROGRAM EXECUTION

Return to the front panel with the keyboard shortcut <*Ctrl+E*>.

It is the job of the *Operating Tool* 🖑 to change values appearing on both the front panel and the block diagram. Select the 🖑 in the Tools Palette and then use it to highlight (e.g., via a double-click) the numeric display region of the **Numeric Control** that we labeled Integer.

Next, use the keyboard to enter the integer whose parity is to be tested—let's try *3*—then secure this choice in the usual way.

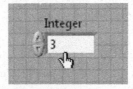

Alternatively, the desired integer value can be entered by using the 🖑 to click on the increment/decrement buttons on the left-hand side of the Numeric Control.

You now are ready to run your program to test the parity of the integer 3 with the help of the toolbar shown below.

The leftmost button on the toolbar is the **Run** button ⇨. To start your program, simply click on the ⇨. If your VI is programmed correctly, the **Odd?** Round LED will light, indicating that the integer 3 is odd.

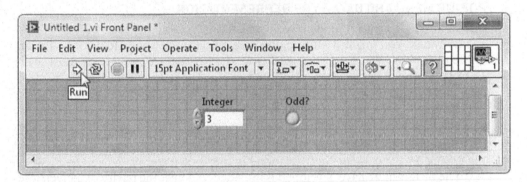

Verify that your VI gives valid results for a few other integers such as 4, 5, –3, and –4.

Now for some bad news. Try entering a non-integer value such as 3.6 into the Integer control, then run your VI. As shown below, you will find that your program incorrectly identifies this number as odd, a designation that only applies to integers.

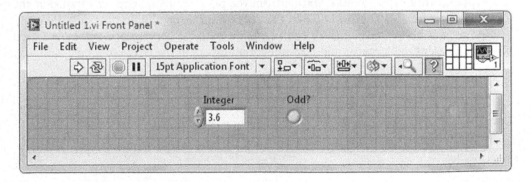

There is obviously a bug in our program, and it is not too hard to trace it to our use of ⊨ for checking the value of **R** to determine parity. In the above example, when the non-integer value of 3.6 is supplied to the **x** input of **Quotient & Remainder**, the **R** output value will be determined by the following calculation:

$$R(3.6, 2) = 3.6 - 2 \times floor(3.6/2) = 3.6 - 2 \times (1) = 1.6.$$

Thus, since 1.6 is not equal to zero, the **Not Equal To 0?** icon will output a TRUE value, lighting the **Odd?** LED.

Since our VI works perfectly as long as only integers are used as input, one solution to the bug we have uncovered is to simply include a cautionary (and confidence-sapping) message on the front panel, such as "WARNING: Program Produces Incorrect Results for Non-Integer Input." However, we can do better than that. Let's fix the bug.

1.6   POP-UP MENU AND DATA-TYPE REPRESENTATION

You are about to enter a hidden world and ascend to a higher level as a LabVIEW programmer. You might wish to pause and reflect upon your life prior to the enlightenment that you are about to attain. Here is the deep secret: Almost every LabVIEW object has its own associated *pop-up menu* that is accessed by right-clicking the mouse on the object. By gaining access to its pop-up menu, you—the programmer—are empowered to control all of the many available features of the associated object. To *pop up* on an object, it is simplest to just right-click on its icon. However, you can also pop up on an object using the *Pop-Up Tool* (officially called the *Object Shortcut Menu Tool*) 🖳 found in the Tools Palette.

The program we have developed is intended to detect the parity of a given integer and thus the front-panel **Integer** control should rightly be restricted to accepting only integers as input. However, as we have seen, **Integer** presently is formatted to accept both integers and non-integers formatted as double-precision floating-point numbers (the default format for a **Numeric Control**). It is this formatting choice that has introduced a bug into our VI.

To remedy this problem, pop up ("right-click") on the front-panel **Integer** control and then select **Representation** from the pop-up menu. You will find that the *data type* for this control is **DBL**, the abbreviation for double-precision floating-point numbers. Under this formatting, eight bytes of computer memory are used to represent real (integer and non-integer) numbers in the range of $\pm 1.79 \times 10^{308}$.

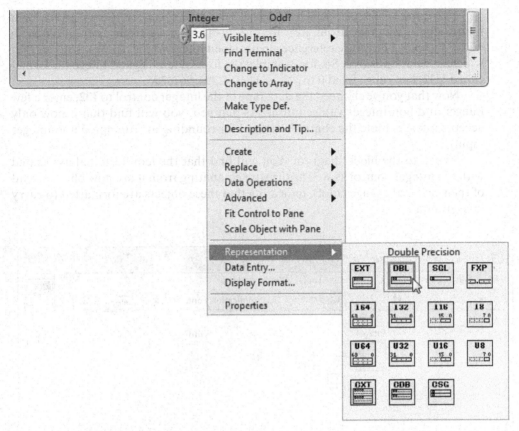

Change the data type of the **Integer** control to a four-byte (32-bit) signed integer (also called *long signed integer*) by selecting **I32** from the **Representation** palette.

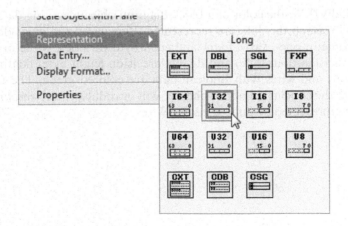

This type of integer, which sacrifices one of its bits for use as a plus or a minus sign, can range from -2^{31} to $+\left(2^{31}-1\right)$, that is, from –2,147,483,648 to +2,147,483,647. For future reference, an unsigned integer data type is also available, which is always positive. So, for example, the four-byte unsigned integer, a format called **U32**, can range from 0 to $\left(2^{32}-1\right)=4,294,967,295$.

Now that you've changed the data type of the Integer control to **I32**, enter a few integer and non-integer values into it. As planned, you will find that it now only accepts integers (note the control's method for rounding an attempted non-integer input).

Switch to the block diagram. You will find that the icon terminal associated with the Integer control as well as the wire emanating from it are now blue (instead of their original orange color), indicating that these objects are formatted to carry integer data.

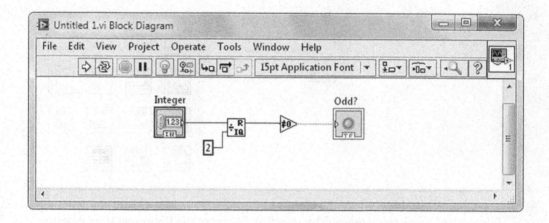

In LabVIEW, the color of a block-diagram object carrying data indicates the manner in which the data are represented. Blue wires and icons denote integers, which come in one-, two-, four-, and eight-byte varieties, while orange indicates single-precision (four-byte) and double-precision (eight-byte) floating-point numbers. Icons and wires with Boolean data are colored green. The Do It Yourself project at the conclusion of this chapter will introduce you to the *string* data type, which is represented by pink icons and wires.

1.7 PROGRAM STORAGE

Your program detects the parity of an integer, so let's save this VI under the descriptive name **Integer Parity Detector.** In the **File** pull-down menu, select **Save**, which will activate the **Name the VI** dialog window shown next. Because you will be writing and saving many VIs in your study of LabVIEW, let's first create a folder in which to store this work. Use the **Save in:** box to navigate to a desired storage place on your computing system (e.g., **Desktop**, **Documents** folder, flash drive), and then click on the **Create New Folder** button.

In the highlighted box that appears next to the folder icon, entitle the new folder **YourName** and then click **Open** to open this folder.

The VI you have written is associated with the first chapter of this book. Thus, within the **YourName** folder, create a subfolder named **Chapter 1** and save the VI there as follows: With **YourName** open, click on the **Create New Folder** button. In the highlighted box that appears next to the folder icon, name the new folder **Chapter 1** and click **Open** to open this folder. Finally in the **File name:** box, type **Integer Parity Detector** and then press **OK**. Your VI will be saved in the **Chapter 1** folder, which is within the **YourName** folder (a location we will denote as **YourName\Chapter 1**) with the extension **.vi** appended to its name.

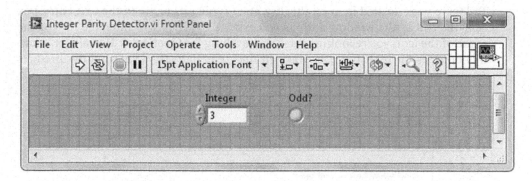

1.8 QUICK DROP

To develop your skill as a LabVIEW programmer, take some time to examine the various palettes and subpalettes (and sub-subpalettes) of the Controls Palette and, particularly, the Functions Palette. I think you'll be impressed by the hundreds of data acquisition and analysis tools available for your use. While writing your programs, though, it can be time-consuming to find a particular icon, especially if it is located in a sub-subpalette of the Functions Palette. To aid in such a search, LabVIEW provides an extremely helpful editing trick called *Quick Drop*.

Let's demonstrate the use of Quick Drop using two of the block-diagram objects we became familiar with during our work in this chapter. Switch to the block diagram. To activate Quick Drop, select **View>>Quick Drop** or, better yet, use the keyboard shortcut <*Ctrl+Space*>. The Quick Drop dialog window will appear, loaded with the list of all of the block-diagram objects available in the Functions Palette.

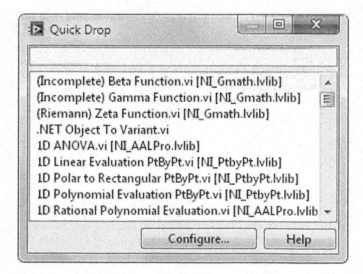

To locate your desired item on this list, start typing its name in the text entry box near the top of the window and, as you type, LabVIEW will filter the list for the name of the icon you're looking for. Try searching for **Quotient & Remainder**. Since this name is fairly distinctive, you will find that LabVIEW successfully predicts what you're looking for after inputting just a few characters.

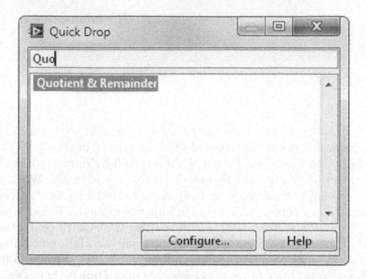

Now, simply use the mouse to click on the text **Quotient & Remainder** within the dialog window.

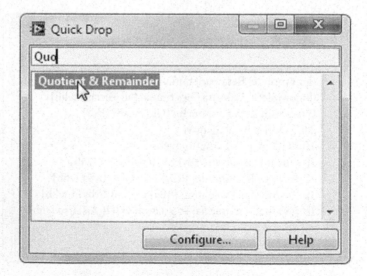

Then move the cursor over to the desired location on the block diagram. When you click the mouse button again, the **Quotient & Remainder** will appear there, ready for wiring.

Let's try Quick Drop with another familiar icon—**Not Equal To 0?**. After typing several of the first characters of the desired icon's name, LabVIEW has narrowed down the search to just a few possibilities.

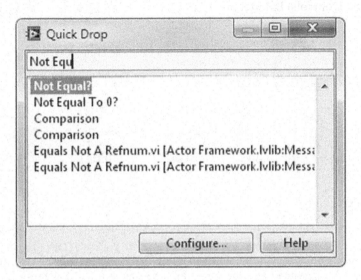

At that point, one can usually stop typing, scan the short list for the icon of interest, and then select it with a mouse click.

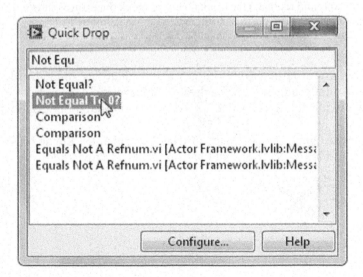

With this selection, move the cursor over to the desired location on the block diagram and click to place the icon there.

Quick Drop lets you rapidly find and place front-panel and block-diagram objects without having to navigate through palettes. It is activated most simply with the keyboard shortcut: *<Ctrl+Space>*. You may wish to explore the use of a similar tool that is available by clicking on the **Search Button** at the top of the Functions (and Controls) Palette.

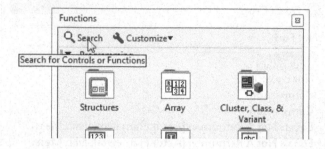

DO IT YOURSELF

Each of the book's chapters concludes with a Do It Yourself (DIY) project that poses an interesting problem and directs you in finding a solution by applying the chapter's material. A suggested name for the solution VI that you will write is given first; icons used in constructing this VI can be found with the aid of **Quick Drop** (icons' palette locations will not be given explicitly in the DIY project description). For this first DIY project, a completed solution VI is given and you are asked to simply find the required icons and replicate the front panel and block diagram as shown. In subsequent chapters, the DIY projects will be much more loosely directed.

Palindromic Word Detector.vi Write a VI that identifies single-word palindromes (i.e., words that read the same backward as forward). This project will introduce you to the string data type, which is used to carry alphanumeric data via pink icons and wires.

First, use Quick Drop to place a **String Control** and a **Round LED** labeled Word and **Palindrome?**, respectively, on the front panel as shown here.

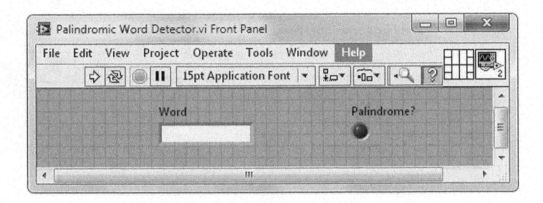

Next, code the block diagram shown below, which includes the string-related icons **To Lower Case** and **Reverse String** (the function of each of these icons can be found in their Context Help Windows). The **To Lower Case** icon allows proper nouns (with their first letter capitalized) to be tested.

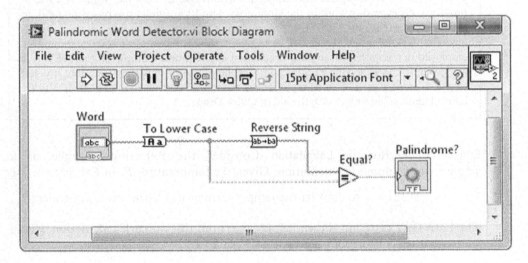

When completed, test **Palindromic Word Detector** with the following inputs: *radar, rotator, Otto, Hannah.*

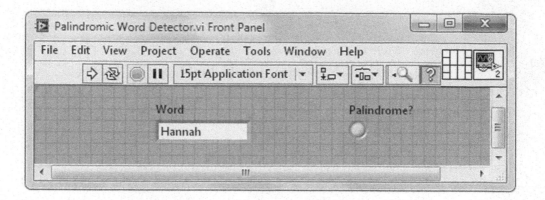

USE IT!

> Each of the book's chapters also concludes with a Use It! example, which is a useful tool for your everyday work in the laboratory. Use It! examples are based on (and, in some cases, extend) the LabVIEW programming techniques presented in their associated chapters. Many of the Use It! examples highlight hardware circuitry commonly employed in computer-based data acquisition applications. Completed LabVIEW VIs are given in Use It! examples, along with explanations of how they function; it is left to you to employ your skills to replicate the examples. Icons used in the Use It! VIs are labeled and can be found with the aid of **Quick Drop**.

Single-Variable Numeric Calculation Consider the following example of a single-variable numeric calculation: Given the temperature T_F in Fahrenheit, use $T_C = \left(T_F - 32\right) \times \dfrac{5}{9}$ to convert the temperature to its Celsius scale equivalent T_C.

The LabVIEW program shown below carries out this numeric calculation, where we see that the VI correctly converts 68°F to 20°C .

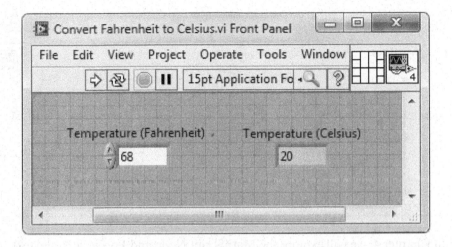

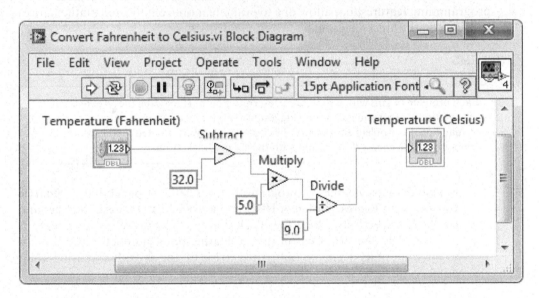

In most cases, the LabVIEW graphical programming language greatly simplifies the coding of desired computer operations. However, in the case of encoding mathematical relations, graphical programming is typically more cumbersome than text-based languages. For this reason, LabVIEW offers the *Expression Node*, an easy-to-use icon that can evaluate a text-based single-variable formula directly on the block diagram. Using an **Expression Node**, which is found in **Functions>> Programming>>Numeric**, the block diagram for our temperature-conversion program is as shown below.

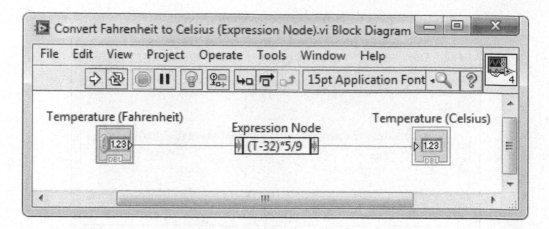

In Chapter 4, we will explore the MathScript Node and Formula Node, text-based programming features that allow one to evaluate more complicated mathematical relations on the block diagram.

PROBLEMS

> Each problem begins with a suggested descriptive name (including the **.vi** extension) for the solution VI that you will write. Suggested icons for use in the VI are given at the end of many of the problem statements. The palette locations of cited icons are not given explicitly; these icons can be found with the aid of **Quick Drop**.

1. **Integer Detector.vi** Create a program that, on its front panel, has a **Numeric Control** (with **Representation** of **DBL**) and a **Round LED** labeled **Number** and **Integer?**, respectively. Code the block diagram so that when an integer is input via the **Numeric Control** (i.e., a floating-point number with all zeros to the right of its decimal point), the LED is lit; for any other floating-point number, the LED is unlit. Suggested icon: **Round To Nearest**.

2. **Convert Celsius to Fahrenheit.vi** Write a program that, given a temperature T_C in Celsius (input via a **Numeric Control**), converts the temperature to its Fahrenheit scale equivalent T_F, and then displays this value on a front-panel **Numeric Indicator**. The formula to convert between the two temperature scales is $T_F = \dfrac{9}{5} T_C + 32$.

3. **Parallel Resistors.vi** For two resistors R_1 and R_2 in parallel, the equivalent resistance R is given by $R = \dfrac{1}{\dfrac{1}{R_1} + \dfrac{1}{R_2}}$. Write a VI that, given the resistance values R_1 and R_2 in two **Numeric Controls**, calculates R, and then displays this value in a front-panel **Numeric Indicator** labeled **Equivalent Resistance**. Suggested icon: **Reciprocal**.

4. **Sine Function.vi** Write a VI that, on its front panel, accepts an angle in degrees in a **Numeric Control** and displays the sine of this angle in a **Numeric Indicator**. On the block diagram, evaluate the sine function using the **Sine** icon. From the Context Help Window for this icon, you will find that its argument **x** input must be in radians, so you must first convert the angle input by the **Numeric Control** from degrees to radians. Suggested icon: **Degree to Radian** (older versions of LabVIEW do not have this icon, in which case you will have to write your own angle-conversion code; the icon **Pi** will come in handy).

5. **Account Balance.vi** If an investor deposits an initial amount P ("principal") into a bank account with a fixed annual interest rate I, the accrued amount B ("balance") in the account after N years is given by $B = P \times (1 + I)^N$. Write a VI that has three **Numeric Controls** on its front panel to input the values of P, I, and N. Given these three input values, calculate the account balance on the block diagram, and then display the resulting value in a front-panel **Numeric Indicator**. Suggested icon: **Power of X**.

6. **Integer Parity Detector (String Display).vi** Write an integer parity-detecting program similar to **Integer Parity Detector**, except that this new VI displays the result of the test as a string (that says either *Even* or *Odd*) within a front-panel **String Indicator**. On the block diagram, use the **Select** icon as shown below to decide which of two strings should be sent to the front-panel **String Indicator**. Each of the two strings (*Even* and *Odd*) is contained within a **String Constant**.

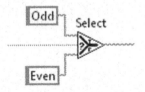

7. **Coin Toss.vi** Build a VI that simulates the toss of a coin. On your block diagram, use **Random Number (0–1)** to generate a random floating-point number x in the range from 0 up to (but not including) 1. When run, if $x \geq 0.5$, assign the result of the coin toss to be *Heads*; otherwise, the result is *Tails*. Then use a **Select** icon as shown below to decide which of two strings should be sent to a front-panel **String Indicator** to display the toss result. Each of the two strings (one *Heads*, the other *Tails*) is contained within a **String Constant**.

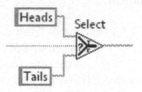

8. **AND Gate.vi** On the front panel of this VI, place two **Push Buttons**, labeled A and B, respectively, and also a **Round LED** named A AND B. Then code the block diagram using an **And** icon so that when the VI is run with both **Push Buttons** pressed, the **Round LED** lights, while for any other combinations of inputs (e.g., A pressed, B not pressed), the LED is unlit.

CHAPTER 2

The While Loop and Waveform Chart

Throughout most of the remainder of the book, only the code to be written on block diagrams will be illustrated, rather than the entire block-diagram windows. Front-panel windows, however, will be shown in their entirety.

Also, the **Modern** palette will be used for front-panel controls. If you prefer to use the **Silver** palette, select **File>>VI Properties>>Category>>Editor Options>>Control Style for Create Control/Indicator>>Silver Style** so that **Silver** (rather than **Modern**) controls are produced when you implement Lab-VIEW's automatic creation feature (see Section 2.10).

Finally, in order to familiarize you with the contents of the Functions Palette and the Controls Palette, the palette location of icons used within the chapter texts will be given. However, you may wish to access these icons using **Quick Drop**.

2.1 PROGRAMMING STRUCTURES AND GRAPHING MODES

LabVIEW is a programming language. As such, it provides the means for creating a set of instructions to carry out a desired computer-related task. By design, LabVIEW is especially well suited for programming the laboratory and industrial tasks of concern to scientists and engineers, including computer-based instrument control, data acquisition, analysis, and reporting. Over the years, LabVIEW has grown into a full-featured programming language so that (if one wishes) it can also be used to carry out computationally intensive tasks such as simulations.

In a typical LabVIEW program, tasks are programmed using a combination of *structures,* each of which carries out one of the following functions: *looping, branching, sequencing.* A looping structure executes a given task repeatedly until a certain condition is met. LabVIEW provides two types of looping structures—the *While Loop* and the *For Loop.* The While Loop, which is explored in this chapter, repeatedly executes a task until an associated Boolean quantity changes its value (e.g., from FALSE to TRUE). In contrast, a For Loop, the subject of the following chapter, repeatedly executes a task for an initially specified number of times. A branching structure allows a program to follow alternative paths of execution, depending on the value of a (Boolean or numeric) selector. In Chapter 9, we will learn how to implement branching in LabVIEW using the *Case Structure.* Sequencing refers to controlling the ordered execution of instructions, that is, assuring that the correct instructions are executed in the correct order. Unlike many other programming languages, LabVIEW is capable of executing multiple threads of a program in parallel, making the concept of sequencing a somewhat involved topic that we will discuss as we develop VIs in this and in subsequent chapters. LabVIEW does provide an explicit method of imposing sequencing, which is (naturally enough) called the *Sequence Structure.* In Chapter 10, we will study the Sequence Structure and find that, quite commonly, it is more elegant to impose sequencing in a LabVIEW program using a concept called *data dependency.* Interestingly, we will also discover (in Chapter 9) that a nested combination of two structures—a Case Structure within a While Loop—provides a powerful method for sequencing tasks through a programming architecture called the *state machine.*

Data display is an essential requirement of many programs. To meet this requirement, LabVIEW offers the programmer several easy-to-implement, yet powerful graphing modes, including the commonly used *Waveform Chart, Waveform Graph,* and *XY Graph.* The Waveform Chart behaves like a laboratory strip chart, producing a real-time plot as each new data point is generated. This graphing mode pairs nicely with the While Loop and so will be used as part of the VI that we build in this chapter. In contrast, the **Waveform Graph** and **XY Graph** (which we will study in Chapters 3 and 4, respectively) display an array of data points that was produced at an earlier time.

2.2   WHILE LOOP BASICS

The While Loop structure is used to control repetitive operations. By default, it will repeatedly execute the subprogram written within its borders (called the *subdiagram*) until a specified Boolean value is no longer FALSE. Thus, the While Loop is equivalent to the following text-based code:

> Do
>> Execute subprogram (which sets *condition*)
>
> While *condition* is FALSE

This loop structure is found in **Functions>>Programming>>Structures** and appears on your block diagram as shown below.

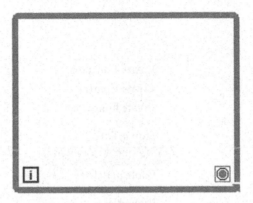

Within the While Loop, you will find the *iteration terminal* as well as the *conditional terminal* , which is set to its **Stop if True** state by default. At the end of each loop iteration, LabVIEW checks the value of . If it is FALSE, the value of is incremented by 1, and the loop begins another execution; if TRUE, the loop ceases execution. The initial value of (during the first iteration of the While loop) is 0. As an example, in a While Loop where is continuously FALSE until becoming TRUE during the 10th iteration, the loop will execute exactly ten times and the final value of will be 9.

The setting of the While Loop's conditional terminal can be found by popping up (right-clicking) on the as shown below.

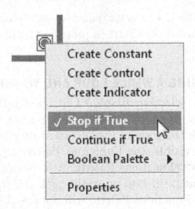

Here, the check mark indicates that the 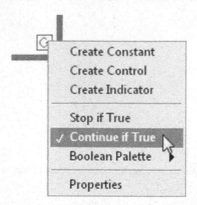 is set to its default value of **Stop if True**. As shown next, the alternate setting **Continue if True** can be selected in the pop-up menu.

When set to **Stop if True** and **Continue if True**, the conditional terminal appears as and , respectively. As an alternative to using the pop-up menu, one can toggle between these two settings by simply clicking on the conditional terminal with the .

In this chapter, we will cause a While Loop to execute numerous iterations by connecting its to a Boolean control terminal that is usually FALSE. As the loop iterates, we will use the to systematically increment the argument of the sine function and use a Waveform Chart to plot the resulting sine-wave values as time progresses.

2.3 SINE-WAVE PLOT USING A WHILE LOOP AND WAVEFORM CHART

Create a fresh VI by selecting **Blank VI** in the Getting Started Window or, if a program is already open, **File>>New VI**. Switch to the block diagram via **Window>>Show Block Diagram** or the keyboard shortcut *<Ctrl+E>*.

If it isn't already visible, the Functions Palette can be activated in the usual way of selecting **View>>Functions Palette**. Alternatively, here's a handy shortcut for accomplishing that task: Pop up (right-click the mouse button) on the blank block diagram and a Functions Palette will appear. This palette is temporary, in that it will disappear as soon as you release the mouse button. However, you may secure the palette by placing the mouse cursor over the thumbtack located in the upper left corner of its window and then releasing the mouse button. This "tacked-down" Functions Palette can then be placed in a convenient location by "clicking and dragging" on its title bar. The same method works for activating the Controls Palette on the front panel. Additionally, the "thumbtacking" procedure can be employed to make any frequently used subpalette of the Functions Palette or Controls Palette continuously visible.

Open **Functions>>Programming>>Structures** and select a **While Loop**. Once you've made this selection, the cursor will appear as a miniature of the While Loop structure ⌐ⓖ when it is placed within the block-diagram window. To place a While Loop within the diagram, click where you want the upper left corner of the loop to be. Then, while holding down the mouse button, drag the cursor to define the size of your loop.

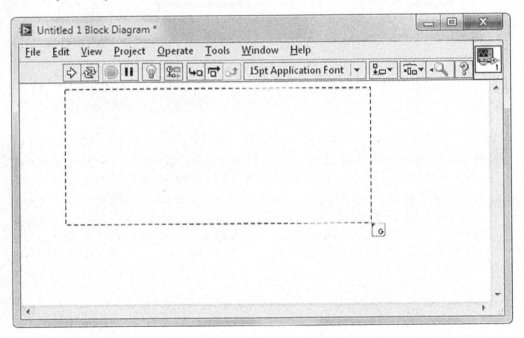

When you release the mouse button, the While Loop will appear as shown here.

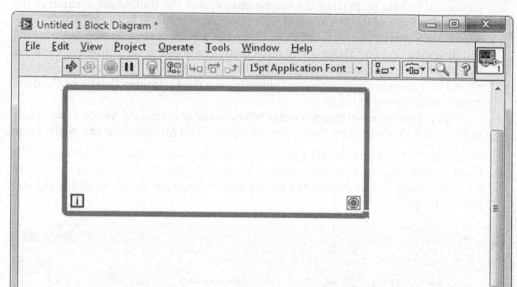

If you are dissatisfied with the dimensions of your While Loop, this can be remedied using the ⬉ (to activate the Tools Palette, select **View>>Tools Palette**). The Positioning Tool is used to resize your While Loop by the following procedure. Place the ⬉ at one of the loop's corners. At the corner, the ⬉ will transform into a *Resizing Cursor* ⬉. Click and drag this cursor to redefine the dimensions of your While Loop. When you release the mouse button, the While Loop of desired size will appear.

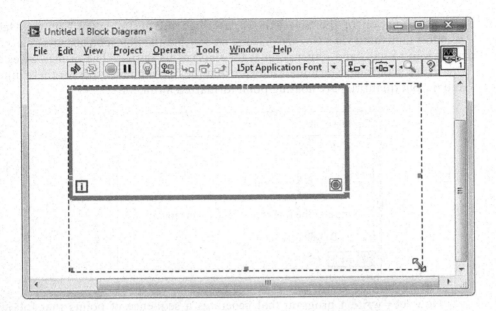

Now, let's write a program that will generate and plot a sine wave. With a While Loop already on your block diagram, select the **Sine** icon from **Functions>> Mathematics>>Elementary & Special Functions>>Trigonometric Functions** and place it on your block diagram.

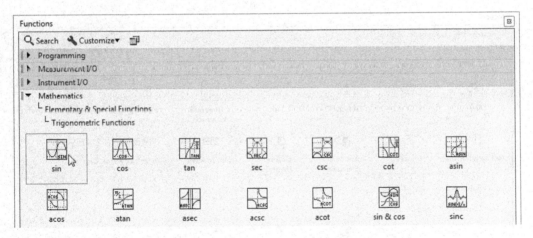

The function of the **Sine** icon is described in the following Context Help Window (activated via *<Ctrl+H>*). Here, we see that the **Sine** icon accepts an argument **x** in radians and returns values for **sin(x)** at its output. The orange wires at this icon's **x** and **sin(x)** terminals indicate that the default data types for these input and output quantities are floating-point formatted numbers.

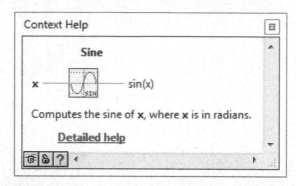

Now let's write a program that generates a sequence of points that follows the sine function. First, we need to configure the While Loop so that it will repetitively perform the operation defined within its borders. One (crude) method of accomplishing this goal is the following: Select a **False Constant** from **Functions>> Programming>>Boolean**.

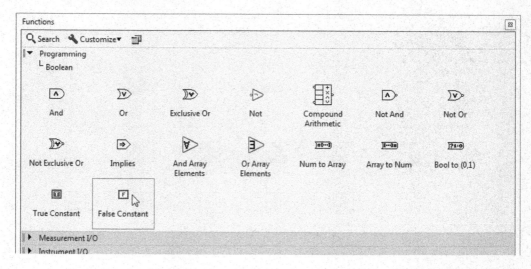

Place the **False Constant** (in older LabVIEW versions, this icon appears as 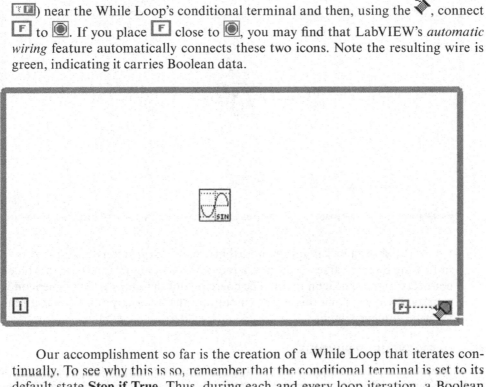) near the While Loop's conditional terminal and then, using the ▼, connect to . If you place close to , you may find that LabVIEW's *automatic wiring* feature automatically connects these two icons. Note the resulting wire is green, indicating it carries Boolean data.

Our accomplishment so far is the creation of a While Loop that iterates continually. To see why this is so, remember that the conditional terminal is set to its default state **Stop if True**. Thus, during each and every loop iteration, a Boolean FALSE value will flow through the wire from to , setting the conditional terminal to FALSE. When LabVIEW checks the value of at the end of the iteration, the While Loop will be instructed to re-execute (rather than stop).

We will now use the While Loop's iteration terminal as a source of ever-increasing argument **x** for our sine function. Use the ↖ to position in the neighborhood of the **Sine** icon, and then wire the to the Sine's argument **x** input. Remember, if you make a mistake, the illegal wiring will result in a black dashed line that can be most easily erased by the **Remove Broken Wires** command's keyboard shortcut: *<Ctrl+B>*.

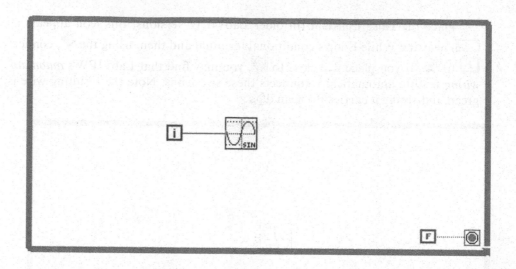

In the above block diagram, note that and wires emanating from it are blue, indicating integer values. However, a red *coercion dot* appears where this blue wire connects to the Sine icon input. The coercion dot denotes that the Sine icon is automatically converting this integer input into the floating-point format it requires for its argument **x** input.

2.4 LABVIEW HELP WINDOW

To obtain more detailed information about the inner workings of a particular icon, the online reference resource *LabVIEW Help* can be accessed by clicking on the blue **Detailed Help** hypertext or the Question Mark [?] at the bottom of the icon's Context Help Window. Alternatively, you can select **Help>>LabVIEW Help...** and then search for the LabVIEW Help Window describing your icon of interest.

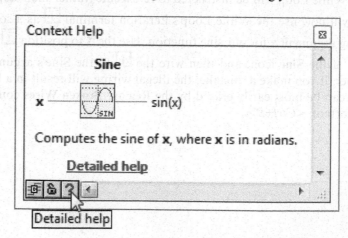

The LabVIEW Help Window for the Sine icon is shown next.

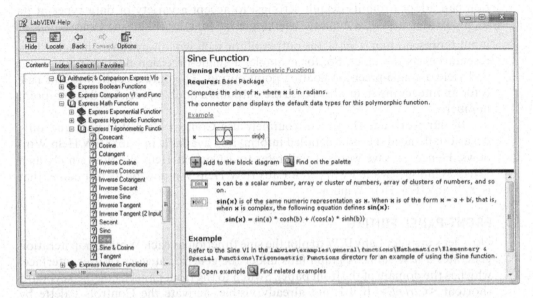

The window informs us that the default data type for the **x** input (i.e., its representation) is 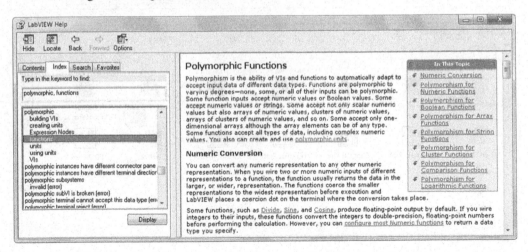, that is, a double-precision floating-point number. However, we also are told that Sine is a *polymorphic function*. The definition of polymorphic can be found by clicking on the **Index** tab, typing the keyword **polymorphic**, and then selecting the subtopic **functions**.

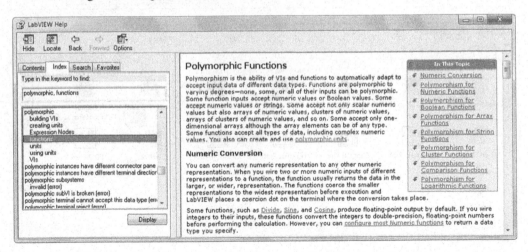

Reading the relevant sections of this window, we find that a polymorphic function can adapt (from its default settings) to accept a variety of data types at its input. In the case of a numeric trigonometric function like **Sine** (whose default input representation is 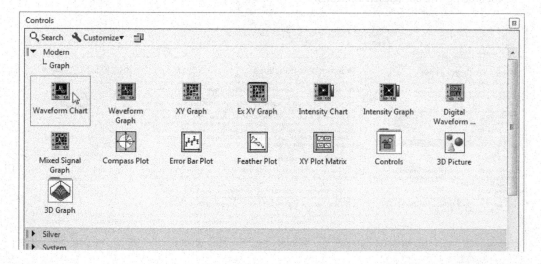), the output of the icon will always have the same representation as the input. So, for example, a single-precision floating-point input will yield a single-precision floating-point output. The only exception to this rule is for an integer input, in which case the output is a double-precision floating-point number.

In our work ahead, we will routinely use Context Help Windows and only occasionally need the more detailed information available in LabVIEW Help Windows. Hence, to save words, I will refer to a Context Help Window simply as a "Help Window," while I will call a LabVIEW Help Window by its full name, that is, a "LabVIEW Help Window."

2.5   FRONT-PANEL EDITING

Now, let's instruct LabVIEW to plot the sine function as each While Loop iteration generates a new value. A graph falls under the generic category of user interface, which is the domain of the front panel. Switch to the front panel using the keyboard shortcut *<Ctrl+E>*. If it is not already visible, activate the Controls Palette by either selecting **View>>Controls Palette** or right-clicking on a blank region of the front panel.

In **Controls>>Modern>>Graph**, select a **Waveform Chart**. The Waveform Chart, one of three commonly used graphical modes in LabVIEW, behaves like a laboratory strip chart, producing a real-time plot as each new data point is generated.

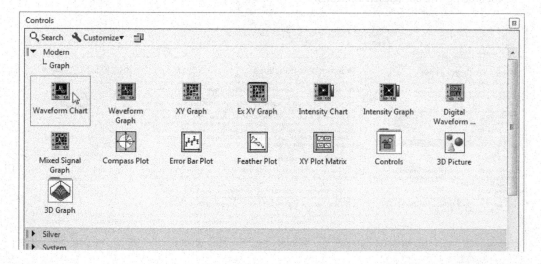

After you place the **Waveform Chart** on the front panel with a single mouse click, if you don't click the mouse further, a highlighted region will appear above the upper left corner of the plot containing the default text **Waveform Chart**.

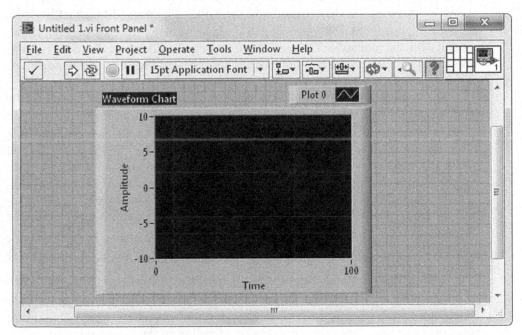

This highlighted region is the Chart's *Label*. You can use the keyboard to give the plot a descriptive name, but let's just keep the default **Waveform Chart** for this program. Secure this label by either clicking on the Enter button ☑ at the upper left end of the front panel, pressing *<Enter>* on the numeric keypad of your keyboard, or simply clicking the mouse cursor on an empty region of the front panel.

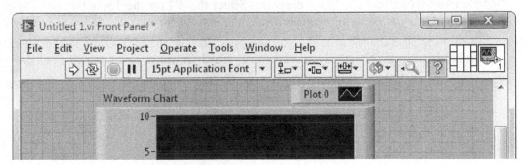

In the future, if you accidentally click the mouse so that the label loses its highlighting before you have a chance to enter a name, the label region can be rehighlighted using the *Labeling Tool* $\boxed{\text{A}}$ found in the Tools Palette.

In addition to the chart region and the Label, the Waveform Chart includes the *Plot Legend* $\boxed{\text{Plot 0}\ \text{\tiny $\sim$}}$. As we will discover shortly, the Plot Legend allows control over the plot style. Through it, one can choose such plot characteristics as whether data will be plotted as points or as interpolated lines and the shape, if any, of the data points.

Let's explore how to reposition and resize the Waveform Chart. First, using the $\text{\tiny $\uparrow$}$, highlight the entire Waveform Chart (including the chart region, Label, and Plot Legend) with a marquee by placing the $\text{\tiny $\uparrow$}$ over the chart region and clicking. With the mouse button depressed, drag the highlighted object to the newly desired position. Once properly placed, release the mouse button and move the cursor to an empty spot on the block diagram. Then click the mouse to deselect the object. In addition, the Label and Plot Legend can each be moved independently. To demonstrate this feature, highlight just the Plot Legend with a marquee by placing the $\text{\tiny $\uparrow$}$ directly over this object and clicking. Then you will be able to drag this single object to a convenient location. Finally, to resize the chart region, position the $\text{\tiny $\uparrow$}$ at one of the Waveform Chart's corners. At the corner, the $\text{\tiny $\uparrow$}$ will transform into a Resizing Cursor $\text{\tiny $\nwarrow$}$. Click and drag this cursor to redefine the dimensions of your chart region. Experiment with this resizing for a few moments. You will find that you can resize both the actual chart region as well as its background frame.

A more important adjustment is the manner in which the axes should be scaled. Soon, we will be plotting sine-wave values on the Waveform Chart's y-axis, so this axis must be prepared to chart data in the range of -1.0 to $+1.0$. Note that the default setting for the Waveform Chart's y-axis data range is -10.0 to $+10.0$.

It is the job of the Operating Tool $\text{\tiny ☝}$ to change values appearing on both the front panel and the block diagram. Select the $\text{\tiny ☝}$ on the Tools Palette. Then change the y-axis data range from its default setting through the following procedure: First, use the $\text{\tiny ☝}$ to highlight the upper limit of the y-axis data range. Once highlighted, enter the newly desired value, which in our case is *1.0*, and then click on $\boxed{\checkmark}$ (or click the mouse on an open area of the front panel).

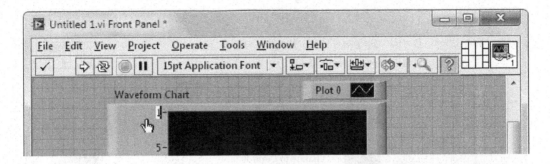

Then, in a similar manner, define the lower limit of the *y*-axis data range to be *–1.0*. LabVIEW will automatically redefine the intermediate *y*-axis labeling. To save the redefined *y*-axis labeling scheme, simply select **Edit>>Make Current Values Default**.

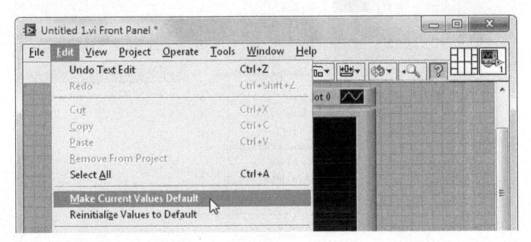

2.6 WAVEFORM CHART POP-UP MENU

Pop up (i.e., right-click) on the chart region of the **Waveform Chart**. As a first example of controlling the features of this object using its pop-up menu, toggle the Label on and off by selecting **Visible Items>>Label**. You will find that when a feature such as Label is activated ("on"), a check mark appears by its name in the pop-up menu.

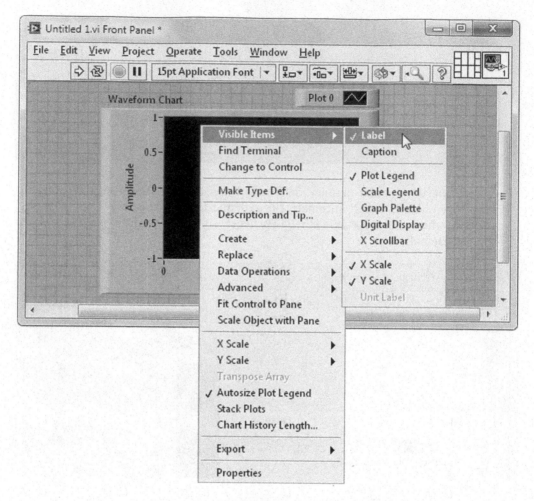

Now, let's use the Waveform Chart's pop-up menu to assist us in appropriately scaling its *y*-axis. A few moments ago, we manually chose proper scaling for this axis using the Operating Tool. A simpler method to accomplish this same goal is via activation of the Waveform Chart's autoscaling feature: Pop up on the chart region of the Waveform Chart and inspect the **Y Scale>>Autoscale Y** option. You will find it checked, meaning it is already activated. By default, the autoscaling on the Waveform Chart's *y*-axis (but not its *x*-axis) is activated.

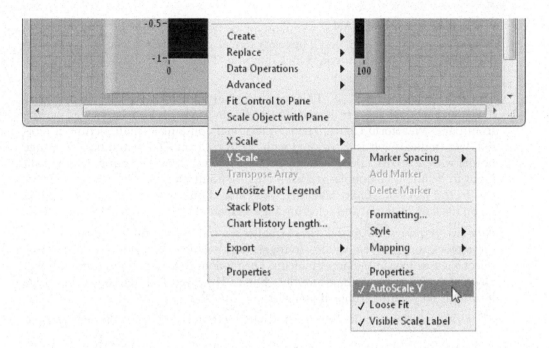

Finally, let's use the pop-up menu to activate one of the Waveform Chart's as-yet-unseen features. Pop up on the chart region and select **Visible Items>>Scale Legend**. The *Scale Legend*, which allows access to several useful functions that determine the scaling and labeling of the *x*- and *y*-axes, will then appear below the Waveform Chart.

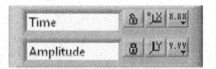

Here, we see that, by default, the *x*- and *y*-axes are labeled **Time** and **Amplitude**, respectively. We, of course, will be plotting *sin(x)* vs. *x*, so these default labels are inappropriate choices for our plot. Use the 🖑 to highlight **Time**, replace it with **x (radians)**, and press *<Enter>*. Similarly, replace **Amplitude** with **sin(x)**.

Additionally, the Scale Legend provides a simple method for activating autoscaling of axes. Simply use the 🖑 to click on the **Scale Lock** button of the desired axis. When the Scale Lock is open 🔓, autoscaling is OFF. When it is closed 🔒, autoscaling is continuously ON. If you just want to autoscale, say, the *x*-axis once (i.e., not continuously), click on the *Autoscale* button. The small green indicator in the Autoscale button lights when autoscaling is activated.

2.7   FINISHING THE PROGRAM

With one last editing step, you'll be ready to run your VI. Use the keyboard short-cut *<Ctrl+E>* to return to your block diagram. There you will find the Waveform

Chart's icon terminal ![icon]. This icon terminal is the Waveform Chart's block-diagram portal; that is, it accepts the block-diagram data to be plotted on the front-panel Waveform Chart. The icon terminal contains a small picture of plotted data (indicating it is associated with the Waveform Chart) and has an orange border with text **DBL**, along with an inward-directed arrow, denoting that the data input to it should be double-precision floating-point numbers. It is further identi-fied by the **Waveform Chart** label that you entered on the front panel. By popping up on the icon terminal and selecting **View as Icon**, the terminal will morph into its

data-type terminal ![DBL] guise, whose smaller size can be helpful if you need to con-serve block-diagram real estate. To toggle back to the icon terminal form, simply select **View as Icon** in the pop-up menu. Throughout this book, icon terminals will always be used in block-diagram illustrations. However, feel free to use data-type terminals on your diagrams, if you wish.

Using the ![cursor], place the Waveform Chart's icon terminal near the **sin(x)** output

of the **Sine** icon, and then use the ![wire] to connect the **sin(x)** output to ![icon]. Remem-ber that the Help Window can be activated with *<Ctrl+H>* to aid in this wiring operation.

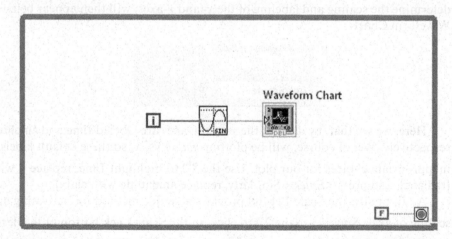

2.8 PROGRAM EXECUTION

Return to the front panel. Your program is now complete and ready to run. Start your VI by clicking on the **Run** button ⬦ in the toolbar. As your program executes, you will observe a rather jagged-looking sine wave produced on your Waveform Chart in strip chart fashion.

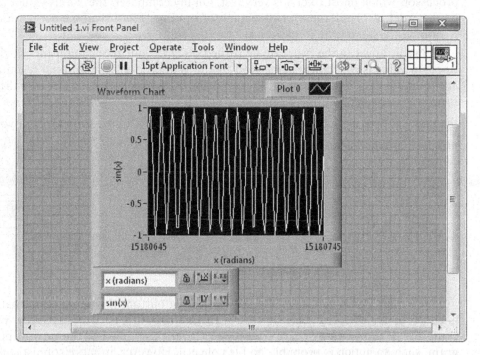

After watching the sine wave move across the Waveform Chart for a while, you may start to wonder about a few things. For example, the Waveform Chart's block-diagram terminal is supplied only with y-axis (sine-wave) values, so how is the Waveform Chart producing the associated x-axis values? What determines the speed at which the sine wave appears to move? And, now that the program is running, how do we turn it off?

Let's answer the first question first. Concisely stated, the x-axis denotes the *index* of each plotted data point. That is, the Waveform Chart keeps track of the sequence of data values it has received and associates an integer-count index with each datum. For the ith data value in the sequence, the plot displays an (x, y) point, where $x = i$ and $y =$ the actual data value. In its default setting, the Waveform Chart's x-axis values are the indices of the last 101 y-axis (in our case, sine-wave) values supplied to its block-diagram terminal.

In regard to the second question, the speed at which the sine wave moves across the chart region is determined by the calculational time delay necessary to produce each new data point. So this speed simply reflects the time it takes for each iteration of your block diagram's While Loop. Because we are allowing the program to run freely, this time per iteration is determined by the speed of your computer's processor, which (most likely) is very fast. On my computer, the x-axis values are on the order of 70 million after 10 seconds of runtime. Thus, the time per iteration is approximately $10 \text{ s}/70 \times 10^6 \approx 1.4 \times 10^{-7}\text{s} = 0.14\,\mu\text{s}$.

Finally, how can the program be turned off? On your block diagram, a False Boolean value is constantly being read by the While Loop's conditional terminal at the end of each loop iteration, so your program will run forever. Because there is no way provided within the program to halt execution, your only recourse at the moment is to click on the **Abort Execution** button ⬤ in the toolbar. Halt your program by clicking on the ⬤.

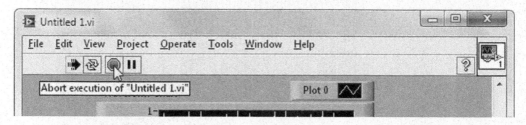

Now that you have used the **Abort Execution** button, let me mention that it is bad practice to do so. This button, at the moment it is activated, causes your computer to immediately cease operation of your program. For the present program, such an action is probably no big concern. However, in more sophisticated programs, pressing **Abort Execution** at the wrong time could halt execution while reading data from a file or during communication with data acquisition devices connected to your computer. These situations can lead to data corruption and other undesirable consequences. Thus, it is always best to code a built-in stopping mechanism into your program.

After halting the program, if you would like to clear your plot, pop up on the Waveform Chart and select **Data Operations>>Clear Chart** (this pop-up menu is altered while the program runs, but the "runtime" menu also allows the chart to be cleared).

2.9 PROGRAM IMPROVEMENTS

Based on the preceding observations, let's upgrade your program in the following three ways: (1) furnish a front-panel button that, when pressed, allows the program to complete its current While Loop iteration and then halt its operation; (2) provide a control over the speed at which the While Loop iterates; and (3) improve the resolution of the sine wave being produced so that it has a smoother appearance.

Front-Panel Switch. LabVIEW provides a multitude of front-panel switches that can be used to bring your programs to a graceful stop. Halt your program so that you may edit its front panel. In **Controls>>Modern>>Boolean**, select a **Stop Button**.

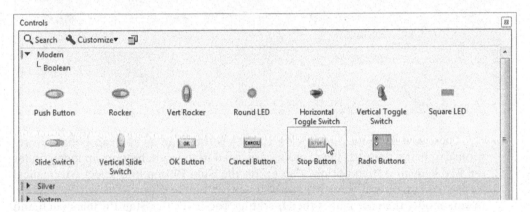

Using the ⬉, put this button in a convenient location on your front panel. The default Boolean value for the Stop Button is FALSE.

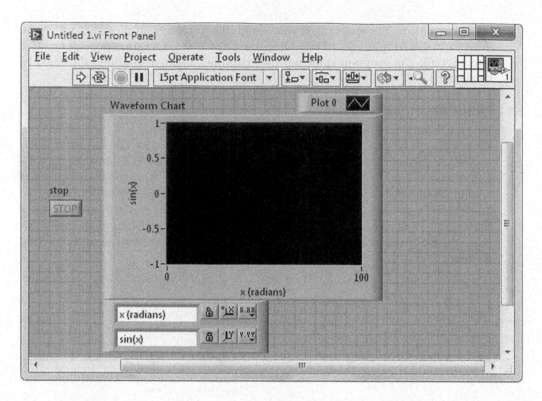

There are six modes in which a LabVIEW Boolean switch can behave in response to being pressed. These modes are listed in the switch's pop-up menu under the **Mechanical Action** option. Pop up on the **Stop Button** and select **Mechanical Action**. You will find that this switch is set to **Latch When Released** by default. In this mode, the user places the Operating Tool over the **Stop Button** switch and depresses the mouse button. Then, at the later time when the user releases the mouse button, the switch changes from its default value to the opposite Boolean value. The switch retains ("latches") this new value until the program reads it once, at which time the switch returns to its default setting. You can find detailed explanations of each mode by selecting **Help>>LabVIEW Help...** and then searching for **Mechanical Action**.

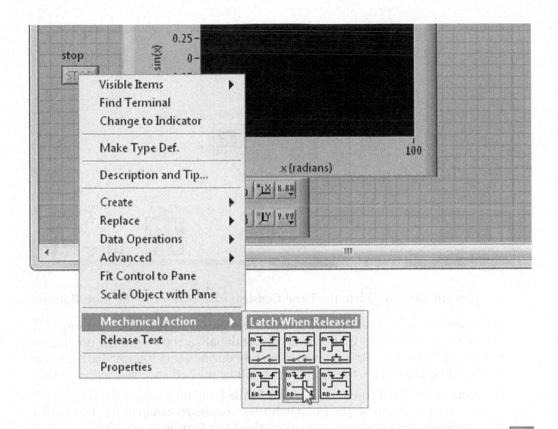

Now toggle to the block diagram. There, you will find the icon terminal that feeds the value of the front panel's **Stop Button** to the block diagram. Our plan is to connect to the While Loop's conditional terminal. To accomplish this task, select the wire connecting the **False Constant** to the conditional terminal by clicking on it with the ⬉. Once it becomes highlighted with a marquee, erase this wire by pressing *<Delete>* on your keyboard.

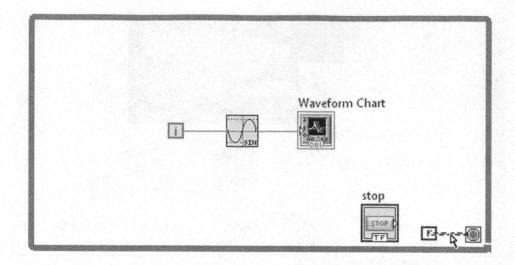

In a similar way, delete the **False Constant** ⬚. Now drag the Stop Button's icon terminal ⬚ near to the ⬚ and wire these two objects together. Because ⬚ is FALSE by default, the While Loop will continually iterate when the program is started. At some later time, almost assuredly during some intermediate portion of the While Loop cycle, the Stop Button will be pressed and then released, thus changing to the TRUE state. Because the While Loop only checks the value of ⬚ at the end of an iteration, the Loop will fully execute its calculations during that final iteration, before ceasing operation. Once the Stop Button's TRUE value is read, it will return to its default value of FALSE so the program is ready for the next time it is run.

Controlling Iteration Rate. While we are on the block diagram, let's also provide a method for controlling the rate at which the While Loop iterates. Through an icon called **Wait (ms)**, LabVIEW provides a means of delaying subsequent program operations for a specified time period. In **Wait (ms)**'s Help Window shown next, we see that by specifying a numeric value at the icon's input, the program will wait for that specified number of milliseconds.

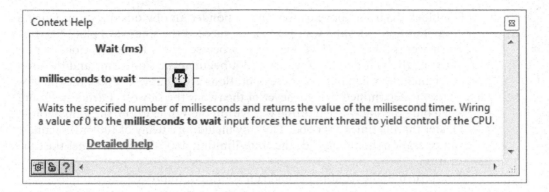

In **Functions>>Programming>>Timing**, select the **Wait (ms)** icon and position it within the While Loop. In the Help Window above, the blue wire emanating from the **milliseconds to wait** input indicates that this terminal should be wired to an integer value. Select a **Numeric Constant** from **Functions>>Programming>> Numeric**. When the **Numeric Constant** is first selected, its interior is highlighted and ready for you to enter a number. If you click the mouse, this highlighting will disappear. To restore the highlighting, use the 🖑. Type the integer *100* into the **Numeric Constant**, press *<Enter>* (or click on a blank region of the block diagram), and then wire this icon to the input of **Wait (ms)**.

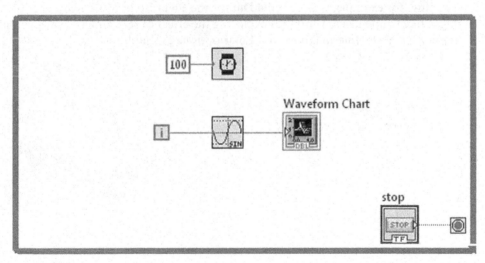

The block diagram above causes one to ponder an obvious question. During each loop iteration, does the wait precede or succeed the sine-wave plot? The surprising answer is that LabVIEW effectively processes these two operations in parallel. That is, the wait and the sine-wave plot begin at the same time and the loop iteration concludes when both of these operations are complete. Thus, the time for an iteration is determined by whichever of the two is the slowest. Earlier we found that when only the sine-wave plotting was within the While Loop, the loop iterated much faster than 10 times a second. Thus, by including a delay of 100 milliseconds, this wait operation should provide the "rate-limiting step" that determines the time per iteration.

Now return to the front panel and run your new, improved program by pressing ⇨. Because of the **Wait (ms)** icon, you will find the sine wave is now generated in an easy-to-observe manner. Use the front panel's **Stop Button** to stop your program gracefully.

Improving Sine-Wave Resolution. To see why our sine wave has a jagged appearance, pop up on the black region of the Waveform Chart's **Plot Legend** Plot 0 ⟋. Under **Interpolation**, we discover that, by default, the Waveform Chart presents data in a "connect-the-dots" fashion. That is, a straight line is drawn from each data point to its neighbor and the entire collection of these lines represents the waveform. Also, the default **Point Style** is **None**, which means that no symbol is placed at each data point. However, we can make the actual data points visible by selecting, for example, a large **Solid Dot** for the **Point Style**. While you're at it, you might like to spend a few minutes exploring the effect of selecting the various **Point Style**, **Line Style**, **Interpolation**, and **Color** options available.

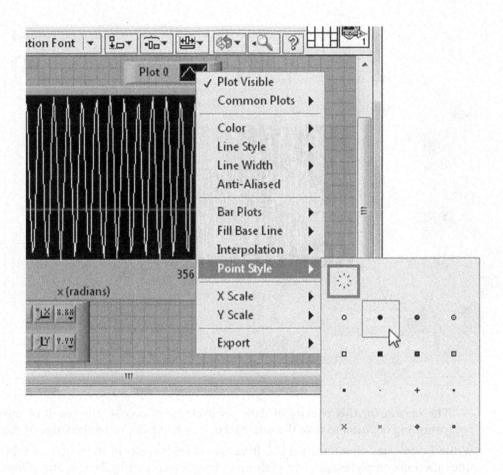

With the data points now visible, we see that each cycle of the sine wave is only represented by a few sampled locations. It is this sparse sampling that leads to the jagged waveform appearance when lines are interpolated between adjacent data points.

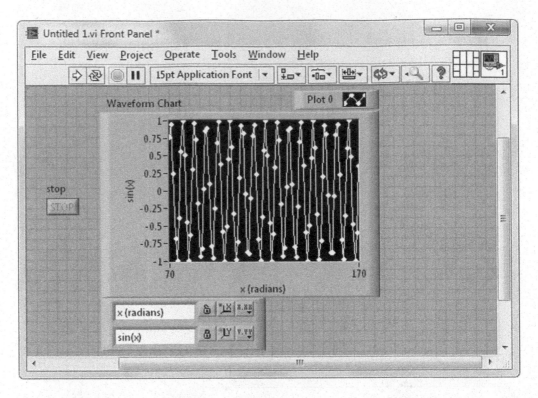

The reason for this paucity of data per cycle is, of course, the result of our programming decision to take the sine-function argument **x** to be the value of the While Loop's iteration terminal ⊡. Because ⊡ increments in steps of 1 and the sine function completes each new cycle every time **x** increases by $2\pi \approx 6$, only about six locations are sampled during each sine-wave cycle.

To sample the sine function with higher resolution, let's take **x** to be one-fifth of ⊡, rather than simply ⊡. Then it will take five times more iterations of the While Loop for **x** to increase by 2π, resulting in five times more sampled points per cycle. To accomplish this feat, select the **Divide** icon (whose Help Window is shown next) from **Functions>>Programming>>Numeric** and place it on your block diagram.

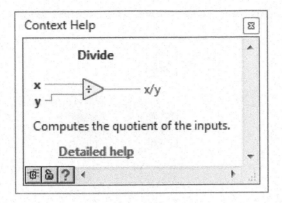

Then, also from **Functions>>Programming>>Numeric**, obtain a **Numeric Constant**. When you first position the **Numeric Constant** on your diagram (before the next mouse click), it is a blue-bordered rectangular box with its interior high-lighted, ready to receive the numeric value of your choosing. Enter the number *5.0*. The Numeric Constant's border will change from blue to orange, indicating that it contains a floating-point (as opposed to integer) number.

If instead you had programmed the **Numeric Constant** with the number *5* (rather than *5.0*), the icon would remain blue, indicating that it contains an integer. On your diagram, although you programmed the **Numeric Constant** with *5.0*, it may appear as an orange 5 rather than an orange 5.0 because an option called **Hide tailing zeros** is activated. To deactivate this option, pop up on the **Numeric Constant** and select **Display Format...**. In the dialog window that appears, choose **Default editing mode**, and then make **Digits** and **Precision Type** equal to **1** and **Digits of precision**, respectively, uncheck the **Hide trailing zeros** box, and finally press **OK**. In this book, I will always take the trouble to uncheck **Hide trailing zeros** so that it is apparent when a **Numeric Constant** contains a floating-point number. On your block diagrams, you may or may not wish to go to this trouble. The icon's blue or orange border will tell you whether it is programmed with an integer or floating-point number, respectively.

Position the **Divide** and 5.0 icons conveniently and wire them so that one-fifth of i is input to the Sine's **x** input. Note the red coercion dot at the iteration terminal's connection to the **Divide**. This dot denotes that the integer from i is converted to a floating-point number. Thus, the Divide icon performs a floating-point divide operation.

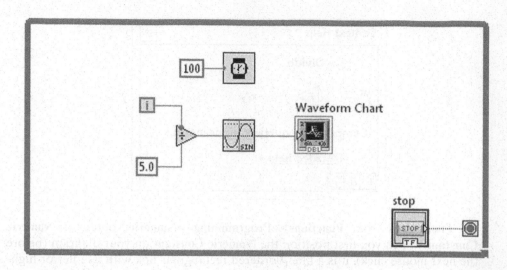

Return to the front panel. There is one thing we still need to fix—the *x*-axis calibration. If we make no further changes, the quantity plotted on the *x*-axis will be the index associated with each sine-wave value on the *y*-axis. Remembering that we have modified our program so that it takes the sine function argument for a particular data point to be the index value *i* divided by 5, we see that the index is no longer equal to the radian argument **x**, and so the wrong quantity is being plotted on the *x*-axis. However, this axis can be calibrated as **x** in radians by simply multiplying each index by 0.2. Such calibration by a multiplicative constant is a common need when using the Waveform Chart, and thus LabVIEW provides a mechanism for implementing it. Pop up on the **Waveform Chart** and select **Properties**. In the **Chart Properties** dialog window that appears, click on the **Scales** tab. Then under the **x (radians) (X-Axis)** menu, set **Multiplier** equal to *0.2* and then press the **OK** button.

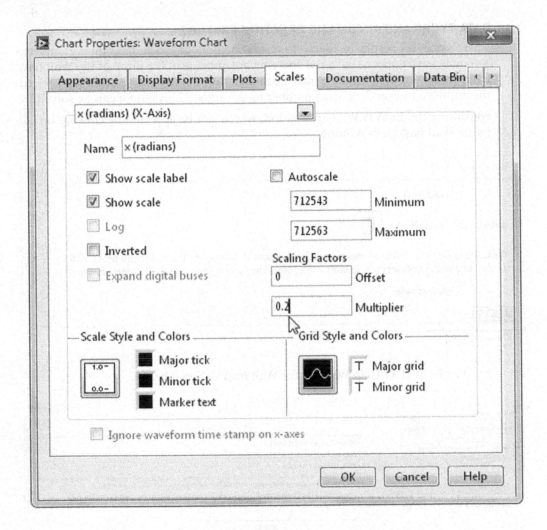

Run this final program and enjoy the beautifully realized sine wave being generated. While your program is running, imagine writing the mountains of code necessary to replicate this real-time sine-wave plot, complete with the attractive user interface, using a text-based language such as C. I think you'll agree that the graphical-based LabVIEW programming language is simple, yet very powerful.

2.10 DATA TYPES AND AUTOMATIC CREATION FEATURE

There's one last perplexing detail to resolve before our block diagram is complete. You may have noticed that a red coercion dot appears at the input of the **Wait (ms)** icon, indicating a numeric-format mismatch, although we have seemingly wired the requisite integer-formatted number to this input. This puzzle can be solved by consulting the LabVIEW Help Window, which can be accessed by clicking on [?] in the **Wait (ms)** Help Window.

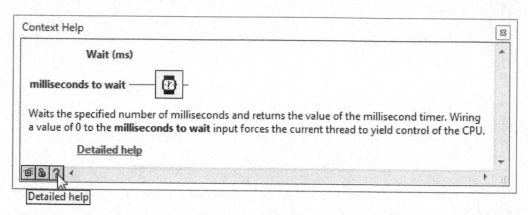

The LabVIEW Help Window for **Wait (ms)** is shown below.

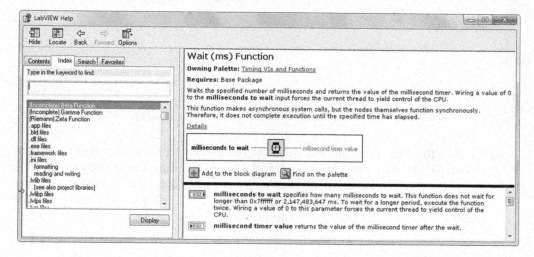

Here, we find that the **milliseconds to wait** input is configured to accept an unsigned four-byte (32-bit) integer, a numeric data type whose shorthand name is **U32**. The unsigned integer data type is in contrast to the signed integer data type (such as **I32**), which sacrifices one of its bits for use as a plus or a minus sign.

To discover the data type of the integer on your block diagram, pop up on 100,
and then select **Representation** from the pop-up menu. You will find that this integer is formatted as **I32**, the default integer data type for a **Numeric Constant**.

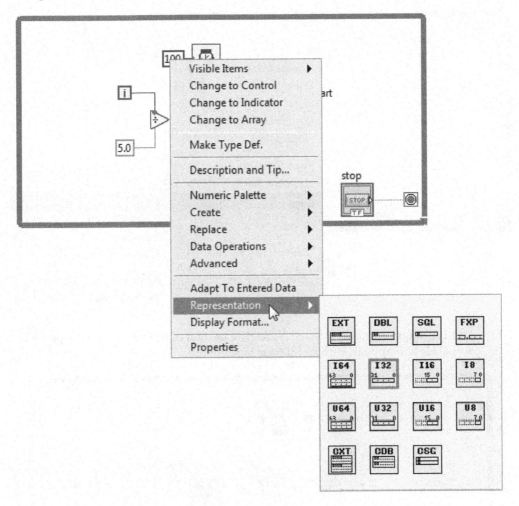

Change the data type of 100 to an unsigned integer by selecting **Representation>> U32** in its pop-up menu. You will then find that the coercion dot disappears.
Another mystery solved!

Now that you understand some of the subtleties involved in wiring up constants to icon terminals, here's a time-saving shortcut—LabVIEW's *automatic creation* feature—I think you'll appreciate. First, delete the Numeric Constant 100 and its wire from your block diagram. Next, place the 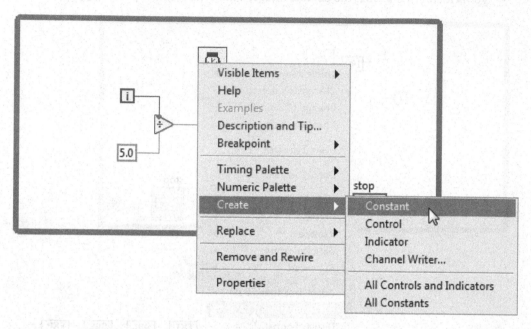 over the **milliseconds to wait** input of **Wait (ms)**, then pop up and select **Create>>Constant** from the menu that appears.

Like magic, an already wired **Numeric Constant** of the correct data type (in this case, **U32**) will appear, with its interior highlighted.

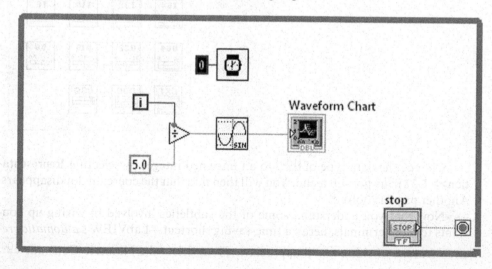

All you need to do is type in the desired integer value of *100* from the keyboard, press *<Enter>*, and you're finished.

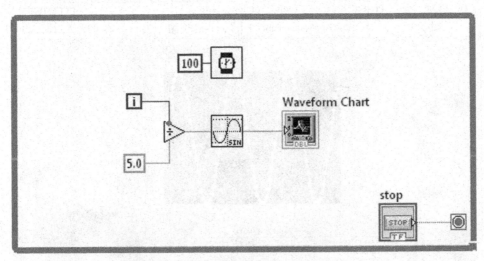

Automatic creation of "terminal-appropriate" objects is a universal feature of LabVIEW icons. Use of these time-saving "creation" options, available in the pop-up menus at the input and output terminals of all block diagram icons, will speed up your program development time.

Using **File>>Save**, create a folder named **Chapter 2** within the **YourName** folder, and then save this VI under the name **Sine Wave Chart (While Loop)** in the **YourName\ Chapter 2** folder.

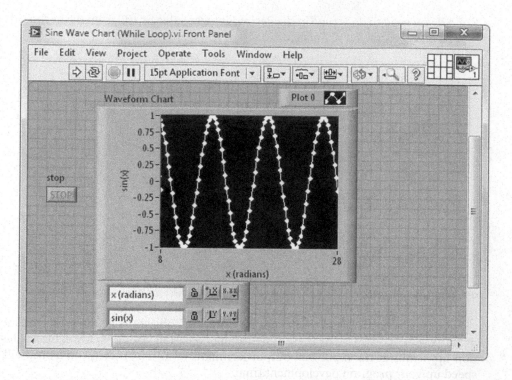

DO IT YOURSELF

Stopwatch.vi Write a VI that functions as a stopwatch with a precision of 0.01 second. Design the VI so that it continuously displays the elapsed time from when its **Run** button is pressed until a **Stop Button** on its front panel is pressed. The front panel should appear as shown below. Here, the elapsed time is displayed on a **Numeric Indicator** labeled **Elapsed Time (second)**. Choose the Stop Button's **Mechanical Action** appropriately.

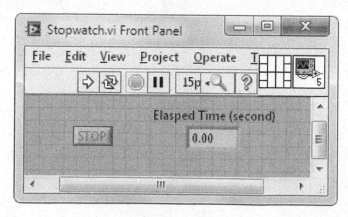

USE IT!

Configuring an Experiment at the Beginning of a Run Data-taking programs commonly prompt users to enter experimental parameters (e.g., the range of voltages to be applied during the data run) at the beginning of runtime. The following LabVIEW program, which is a modified version of **Sine Wave Chart (While Loop)**, demonstrates one method for providing such functionality. Here, rather than

"hardwiring" 5.0 as the divisor of i, the user provides a value for the divisor when the VI first starts its run.

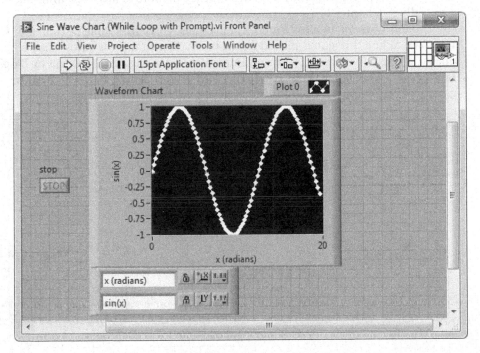

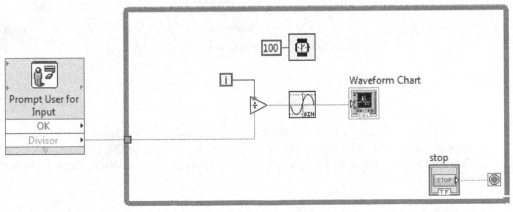

In Section 2.9, we learned of the following LabVIEW feature: two block-diagram objects that are not connected by wires execute in parallel. This present program exploits another LabVIEW feature called *data dependency*, which applies to two block-diagram objects that are connected by wires—in this case, the **Prompt User for Input** icon and the **While Loop**. When the **Prompt User for Input** icon executes, it outputs a numeric quantity named Divisor. Divisor is an input for the While Loop and so this loop must wait to receive the value of Divisor before it can begin to execute. Thus, when the VI is run, **Prompt User for Input** executes first, followed by the While Loop. We will study data dependency in more detail in coming chapters, especially in Chapter 10.

Prompt User for Input is found in **Functions>>Programming>>Dialog & User Interface**. When this icon is placed on the block diagram, the window shown below opens, where the icon can be programmed with the desired **Message to Display**, **Input Name**, and **Input Data Type**. Once the programming is finished, close this window by pressing the **OK** button.

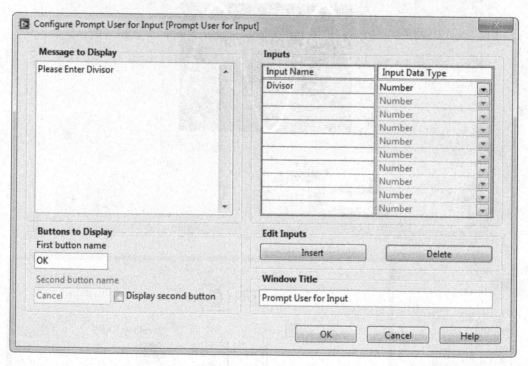

Then, when the VI is run, the following **Prompt User for Input** dialog window appears, where the user enters the value for Divisor. After the **OK** button is pressed, **Prompt User for Input** will output the value of Divisor and the While Loop can then begin to execute.

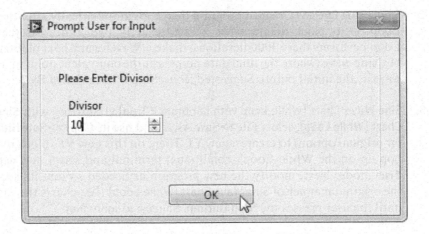

To better understand this VI's sequenced execution due to data dependency, try implementing *Highlight Execution* (see Section 10.4). In the block-diagram toolbar, enable **Highlight Execution** by clicking on the button containing the light bulb. Then click on the **Run** button. The VI will execute in slow-motion animation, with the passage of data on the block diagram marked by bubbles moving along the wires and important data values given in automatic pop-up tip strips.

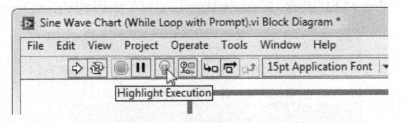

PROBLEMS

For a problem that involves writing a new program, the problem statement begins with a suggested descriptive name (including the **.vi** extension) for the VI that you will write; icons needed for the VI may be found with the aid of **Quick Drop** (activated most simply by the keyboard shortcut *<Ctrl+Space>*). For a problem that involves use of a VI already written as part of the chapter text, the problem's topic is given in bold at the beginning of the problem statement.

1. **Greater Than Ten.vi** Write a VI that lights a **Round LED** indicator on its front
    panel when the iteration terminal □ of a While Loop on its block diagram has
    a value greater than 10. Suggested icons: **Select**, **True Constant**, **False Constant**.

2. **Single Sine Cycle.vi** Create a program that stops automatically after the While Loop on its block diagram has iterated exactly 1000 times. As the While Loop performs these 1000 iterations, make a Waveform Chart plot one cycle of a sine wave (where the final data point is at the equivalent point on the sine wave as the initial point). Suggested icons: **Sine, Pi Multiplied By 2**.

3. **Sine Wave Chart (While Loop with Continue if True).vi** Starting with **Sine Wave Chart (While Loop)**, select **File>>Save As...** (and use its **Copy>>Substitute copy for original** option) to create a new VI. Then, on this new VI's block diagram, pop up on the While Loop's conditional terminal and select its **Continue if True** mode. Next, modify the new program as needed so that it executes in the original manner of **Sine Wave Chart (While Loop)** (i.e., charts the sine wave until the user presses the **Stop Button**). Suggested icon: **Not**.

4. **Odd Integer Search.vi** The odd integers are given by $2i + 1$, where $i = 0, 1, 2, \ldots$ Write a program that answers the following question: What is the smallest odd integer that is divisible by 3 and also, when cubed, yields a value greater than 4000? Display the answer to this question in a front-panel **Numeric Indicator**. Suggested icons: **Quotient & Remainder, Power Of X, And**.

5. **LED Metronome.vi** Write a VI whose front panel includes a **Numeric Control** and **Round LED** labeled **Blinks Per Minute** and **LED**, respectively, as shown below.

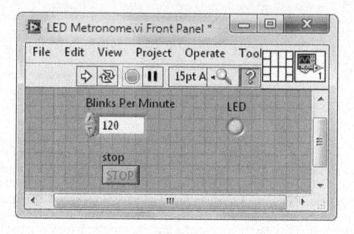

Construct the program's block diagram so that, when run, the front-panel LED blinks N times every minute until the **Stop Button** is pressed, where N is the value entered in the **Blinks Per Minute** control. For each blink, make the LED's on-time and off-time equal. Suggested icons: **Quotient & Remainder, Select, True Constant, False Constant**.

6. **EvenOdd.vi** Construct a VI whose front panel has two **Round LED** indicators and a **Stop Button**. Label one indicator **Even** and the other **Odd** as shown below. Place a While Loop on the block diagram, which iterates once every 0.5 seconds, and construct a subdiagram within it so that the **Even** indicator is lit (and **Odd** is unlit) during iterations for which the value of the iteration terminal ⬜ is even, while the **Odd** indicator is lit (and **Even** is unlit) during iteration for which the value of ⬜ is odd. Suggested icons: **Quotient & Remainder, Select, True Constant, False Constant**.

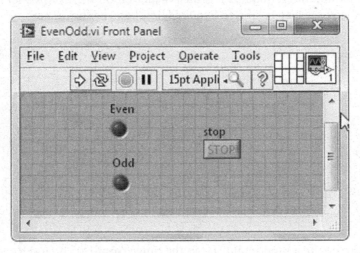

7. **CPU Hogging** Explore the impact of While Loop execution on central processing unit (CPU) usage. Open the **Windows Task Manager** by right-clicking an empty area on the taskbar at the bottom of your monitor and then selecting **Start Task Manager** or by pressing <*Ctrl+Shift+Esc*>. You will find that the **Task Manager** displays CPU usage (expressed as a percentage of your computer's maximum value) at the bottom of its window.

   (a) Open **Sine Wave Chart (While Loop)** and run this VI with **Wait (ms)** programmed to produce one While Loop iteration every 100 ms. What is the approximate increase in CPU usage that results when the program is running?

   (b) Next, program **Wait (ms)** to produce one While Loop iteration every 1 ms, and then run the VI. What is the approximate increase in CPU usage that results when the program is now running?

   (c) Finally, delete **Wait (ms)** from the VI's block diagram so that the While Loop will iterate as fast as possible on your computing system (e.g., on the order of a microsecond per iteration), and then run the program. What is the approximate increase in CPU usage that results when the program is now run? [Note that such "CPU hogging" should be avoided unless your program is executing a high-speed task (i.e., not simply updating a plot as in this case).]

8. **Iterations Until Integer Equals Five.vi** Write a program that randomly generates an integer in the range from 1 to 10 and iterates this random process until the integer equals 5. The number of iterations required until the randomly generated integer equals 5 is then displayed on the front panel in a **Numeric Indicator**, which is labeled **Required Iterations** as shown next.

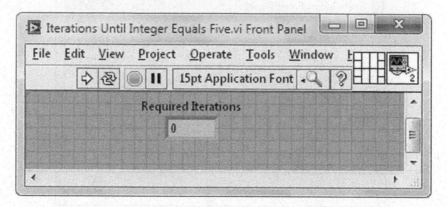

To create a random integer in the range from 1 to 10, use **Random Number (0–1)** to generate a random floating-point number in the range from 0 up to (but not including) 1, multiply this number by 10, and then use **Round Toward + Infinity** to round the floating-point number to the next highest integer (note **Round Toward + Infinity** rounds $x.000$ to x) as shown here.

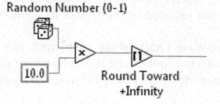

To check whether this integer equals 5, use the **Equal?** icon. Finally, think carefully about how to use the While Loop-related icons to determine the value of **Required Iterations**.

Knowing that the integers are created randomly (i.e., equal probability for producing integers 1 through 10 each iteration), what do you expect the value of **Required Iterations** to be, on average? Run your VI 20 times and record the value of **Required Iterations** resulting from each run. Is the average of these 20 values close to the value that you expected?

9. **Sine Wave Chart (While Loop With Property Nodes).vi** A characteristic of the Waveform Chart can be controlled within a program using a **Property Node**.

   (a) **Chart Multiplier** In this chapter, you set a Waveform Chart's **X-Axis Multiplier** equal to 0.2 using the **Properties** dialog window. Alternatively, this property of the Waveform Chart can be set on the block diagram via a Property Node as follows: With **Sine Wave Chart (While Loop)** open, select **File>>Save As...** (and use its **Copy** option) to create a new VI. On the front panel of this new VI, pop up on the Waveform Chart and select **Properties**. In the Chart Properties window, under the **Scales** tab, set **X-Axis Multiplier** equal to *1*, then press the **OK** button. Next, on the block diagram, pop up on the Waveform Chart's icon terminal and select **Create>>Property Node>>X Scale>>Offset and Multiplier>>Multiplier**, and then place the resulting Property Node on the block diagram. This Property Node is created in its read mode; hence, the small outward-directed black arrow on its right side indicating that the icon is configured to report the current value of the **Multiplier** property. Switch this icon to its write mode by popping up on its lower section and selecting **Change To Write**. The black arrow will now be inward directed on the icon's left side, indicating that data may be input there to set the **Multiplier** value. Using **Create>>Constant**, wire a value of *0.2* to the **XScale.Multiplier** as shown below.

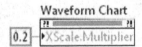

   Run your VI and assure yourself that the Property Node is indeed setting the **X-Axis Multiplier** to 0.2.

   (b) **Chart Clear** Program your VI so that its Waveform Chart is cleared at the start of each run. The appropriate Property Node is made by popping up on the Waveform Chart's icon terminal and selecting **Create>>Property Node>>History Data**. After changing this Property Node to its write mode by selecting **Change to Write** in its pop-up menu, **Create>>Constant** will produce the needed input called an **Empty Array** as shown below.

Waveform Chart

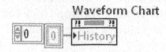

    To clear the Chart at the start of each program execution (i.e., immediately after the Run button is pressed), should this Property Node be placed inside or outside of the While Loop? Run your VI several successive times and demonstrate that the Property Node performs as intended.

(c) **Chart Background Color** Change the background color of your VI's Waveform Chart during runtime as follows: Place a Property Node within the While Loop by popping up on the Waveform Chart's icon terminal and using **Create>>Property Node>>Plot Area>>Colors>>BG Color**, and then change this Property Node to its write mode by selecting **Change To Write** in its pop-up menu. On the front panel, place a **Framed Color Box**, and then on the block diagram wire its terminal to the Property Node. Run your VI and demonstrate that the Waveform Chart's background color can be controlled using the Operating Tool.

CHAPTER 3

The For Loop and Waveform Graph

3.1 FOR LOOP BASICS

The LabVIEW programming language provides two possible loop structures to control repetitive operations in a program. In the previous chapter, we explored the While Loop, which (by default) executes the subdiagram within its borders until the Boolean value wired to its conditional terminal is TRUE. That is, the While Loop executes until a specified condition is no longer FALSE. Now we will focus our attention on the *For Loop*, LabVIEW's other available loop structure. In contrast to the While Loop, which is controlled by the value of a specified condition, the For Loop simply executes the subdiagram within its borders an initially specified number of times. This loop structure is found in **Functions>>Programming>> Structures** and appears on your block diagram as shown below.

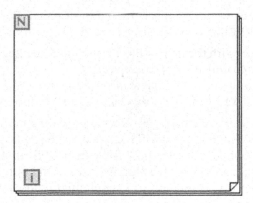

In this graphical structure, your "to-be-iterated" subprogram is written in the currently blank region within the borders. Once this subdiagram is written, the value of the *count terminal* determines the total number of times N that the

For Loop will iterate. The value of the count terminal is set by wiring a **Numeric Constant** (located outside the loop) to the ▣. The For Loop's other internal icon is familiar from your work in the previous chapter. As in the While Loop, the iteration terminal ▣ contains the current number of completed loop iterations. The value of ▣ is 0 during the first loop iteration, 1 during the second,..., and $N - 1$ during the last iteration. Thus, the For Loop is equivalent to the following text-based code:

$$\text{For } i = 0 \text{ to } N - 1$$
$$\text{Execute Subprogram (which increments } i\text{)}$$

In this chapter, you will write a sine-wave plotting program based on a For Loop. In the process of writing this VI, you will explore LabVIEW's method of storing numeric data in the form of an array and learn to operate a *Waveform Graph*, the second in LabVIEW's triad of commonly used graphing options. Additionally, you will further hone your LabVIEW editing skills.

3.2 SINE-WAVE PLOT USING A FOR LOOP AND WAVEFORM GRAPH

Create a fresh VI by selecting **Blank VI** in the Getting Started Window or, if a program is already open, **File>>New VI**. If it is not already visible, activate the Controls Palette by either selecting **View>>Controls Palette** or right-clicking on a blank region of the front panel.

Place a **Waveform Graph** on the front panel from **Controls>>Modern>>Waveform Graph**. Secure its default label **Waveform Graph** by pressing <*Enter*> and use the ▹ to position the entire graph nicely on the front panel. Pop up on the Waveform Graph and select **Visible Items>>Scale Legend**. You will be creating a program to plot *sin(x)* vs. *x*, so (with the help of the ▹) use the Scale Legend to label the *x*- and *y*-axes as **x (radians)** and **sin(x)**, respectively.

By default, autoscaling is activated on both the *x*-axis and the *y*-axis for a Waveform Graph. You can verify this fact yourself by noting that the **Scale Lock** buttons for both axes are closed 🔒. Keep autoscaling activated on both axes by changing nothing and then toggle off (i.e., hide) the Scale Legend by popping up on the Waveform Graph and deselecting **Visible Items>>Scale Legend**.

If, as in this present case, labeling the axes is the only required task, rather than using the Scale Legend, you can simply highlight each axis label using the Labeling Tool 🅰 from the Tools Palette and type in the desired text directly. Try this time-saving tip.

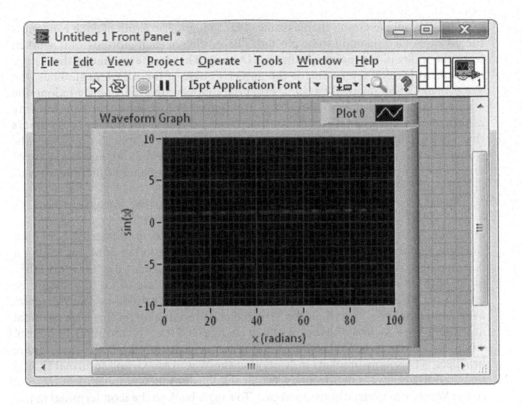

3.3 WAVEFORM GRAPH

What is the Waveform Graph and how does it differ from the Waveform Chart? In our previous work, we saw that a Waveform Chart acts as a real-time data plotter. As each new data point is generated on the block diagram, this single numeric value is passed to the Waveform Chart's terminal and then immediately displayed in the front-panel plot. The new data point is appended to the already existing data plot, so you can view the current value in the context of the previous values.

In contrast to the Waveform Chart's interactive method of plotting data, the Waveform Graph (as well as the yet unstudied XY Graph) accepts an entire block of data that has been generated previously. The Waveform Graph's terminal accepts this data block as a one-dimensional (1D) array of N numeric elements with the order of the elements indexed by integers in the range of 0 to $N - 1$. For example, a 10-element array would have the following form:

Index	0	1	2	3	4	5	6	7	8	9
Array	1.20	1.30	1.40	1.50	1.60	1.70	1.80	1.90	2.00	2.10

The Waveform Graph then assumes that the numeric value of each element is the y-value of a data point to be plotted. The x-value of each plotted point is taken to be the index of that point within the 1D array. That is, the x-value of the array's ith element is i and the y-value is the numeric value of the ith element. Thus, the Waveform Graph implicitly assumes that data points are evenly spaced along the x-axis, a situation that is often realized in practice. For example, the time-varying nature of an analog signal is commonly represented by a set of discrete data points that were digitized ("sampled") during a sequence of equally spaced time intervals. You will find later that the XY Graph allows one to plot the more general case of an array of unequally spaced data points (as well as multivalued functions such as circular shapes).

Switch to the block diagram. There, you will find the Waveform Graph's terminal, in its icon terminal ![icon] guise, with its label located nearby. Note that the icon terminal contains a distinctive small picture of plotted data (indicating it is associated with the Waveform Graph) and has an orange border with text **DBL**, along with an inward-directed arrow, denoting that the data input to it should be double-precision floating-point numbers. It is further identified by the **Waveform Graph** label that you entered on the front panel. If you pop up on this icon terminal and select **View As Icon**, the terminal will change its data-type terminal ![DBL] form, where we note that the DBL data-type specifier is enclosed within a pair of square brackets. The square brackets indicate that the Waveform Graph terminal should be wired to accept a numeric array, rather than just a single number as in the case of the Waveform Chart discussed above. To toggle back to the icon terminal manifestation, pop up and select **View As Icon**.

3.4   OWNED AND FREE LABELS

Let me digress for a moment and discuss LabVIEW labels. The Waveform Graph's label currently on your block diagram (which contains the text **Waveform Graph**) is what is known as an *owned label*. An owned label belongs to a particular object (in this case, the Waveform Graph's terminal) and moves with that object. To demonstrate this fact, use the ↳ to highlight the terminal, and then drag it to a new location on the block diagram. Note that the label accompanies the icon to its new location. Deselect the icon by clicking on an empty spot of the block diagram, and then click on top of the label itself. A marquee will then appear solely around the label, and you will be able to drag just the label to some newly desired location relative to its owning icon. An owned label can be made visible or invisible using the **Visible Items>>Label** option in the object's pop-up menu. Experiment a bit on the block diagram until you understand the functioning of an owned label.

An alternate form of LabVIEW annotation is called the *free label*. Free labels are not attached to any object and can be created, moved, or disposed of independently. They can be used as explanatory comments on your front panels and block diagrams. To create a free label, select the Labeling Tool [A] from the Tools Palette and click on the desired location for the label. A small, bordered box appears, ready to accept text input. After you type the text message, press *<Enter>*, and then the free label is complete. Try producing a free label on your block diagram. After you succeed in creating a free label, click on it with the ⬧ and delete it.

3.5 CREATION OF SINE WAVE USING A FOR LOOP

Now select a **For Loop** from **Functions>>Programming>>Structures** and add it to your block diagram (the Functions Palette can be activated by popping up on a blank region of the block diagram). If you are dissatisfied with your initial choice of dimensions, you can place the ⬧ at a corner (or at the midpoint of a side) of the For Loop, where it will morph into the Resizing Cursor and allow you to reshape the loop borders.

Our programming strategy is this: Use the For Loop to create a one-dimensional array of data points that represent a few cycles of the sine function, and then pass this array to the Waveform Graph's icon terminal for plotting on the front panel. Because we desire data to be passed to the Waveform Graph's terminal only after the For Loop has completed its final iteration, we must place the terminal in a region outside the For Loop's boundary. Use the ⬧ to build the following block diagram.

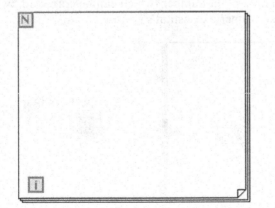

Waveform Graph

Now let's write the code for creating the sine-wave values within the For Loop. To achieve this goal, obtain a **Sine** icon from **Functions>>Mathematics>>Elementary & Special Functions>>Trigonometric Functions** and place it inside the For Loop.

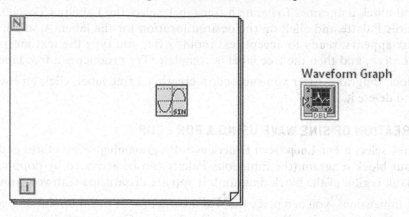

Waveform Graph

Next use **Divide**, from **Functions>>Programming>>Numeric**, to write the sine-wave creation code shown below. After placing **Divide** on your diagram, pop up on its bottom input and select **Create>>Constant** to automatically create the **Numeric Constant** 5.0 (if the constant appears as an orange 5, you can change it to 5.0 using **Display Format...** in the pop-up menu, if desired). Next, wire i to the top input of **Divide**, and then wire Divide's output to the **x** input of Sine. If you reverse the order in which 5.0 and i are wired to **Divide**, the **Numeric Constant** will be autocreated as an integer rather than a floating-point number (which can be corrected by selecting **Representation>>DBL** in the Numeric Constant's pop-up menu).

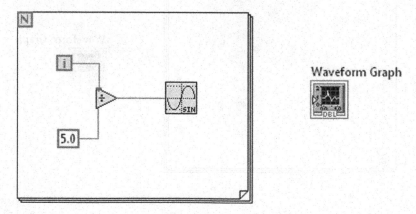

Waveform Graph

Ready for another editing shortcut? When coding a block diagram, you will find that the most commonly used tools are ⬆ and ▼. Rather than having to go to the Tools Palette each time to switch back and forth between these tools, simply press <*Space*> on your keyboard. With each depression of the <*Space*> key, the mouse cursor will toggle back and forth between the Positioning and Wiring Tools.

Alternatively, by clicking on the **Automatic Tool Selection** button [🔧 ▬] near the top of the Tools Palette, its green LED indicator will illuminate, denoting that the *automatic tool selection* option is activated. Then, when you move the cursor to a position on the block diagram (or front panel), LabVIEW will automatically select the tool that is most appropriate for that location. When this option is activated (indicated by the illuminated green LED), it can be toggled off by clicking on the **Automatic Tool Selection** button. Finally, after you have toggled off Automatic Tool Selection, your keyboard's <*Tab*> key can be used to manually select tools. As you press <*Tab*> repeatedly, the mouse cursor will cycle from ✋ to ⬆ to [A] to ▼ in the Tools Palette.

Now we need to tell the For Loop the number of iterations *N* we would like it to perform. We know from our experience of writing similar block-diagram code in the last chapter that about 30 loop iterations will be required to generate one cycle of the sine function on this block diagram. Let's set $N = 100$ so that we generate (a little more than) three sine-wave cycles. To accomplish this setting, wire a **Numeric Constant** (defined to be *100*) to the [N]. You already know two methods for acquiring the required **Numeric Constant** icon: the "hard" way, by obtaining it directly from **Functions>>Programming>>Numeric**, and the "easy" way, by simply popping up on the [N] and selecting **Create>>Constant**. Be good to yourself—use the "easy" way.

3.6 CLONING BLOCK-DIAGRAM ICONS

Here's one final editing option that you can add to your bag of tricks. It's called *cloning* and can be used in the present situation to produce the required Numeric Constant (to be wired to the [N]) because an equivalent icon already exists on the block diagram. To clone the existing **Numeric Constant** [5.0] on your block diagram, place the Positioning Tool over this icon.

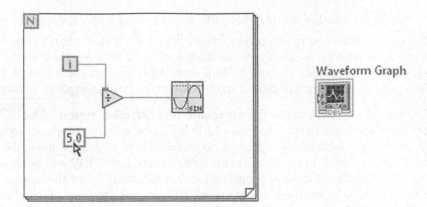

Now click the mouse button while depressing *<Ctrl>* on your keyboard. Then, by moving the mouse while still holding down its button, you will drag a copy of the icon with you, while the original stays in place. Once you arrive at the newly desired location, release the mouse button and the cloned icon will reside there (if you place within the autowiring range of the , a wire will automatically connect the two).

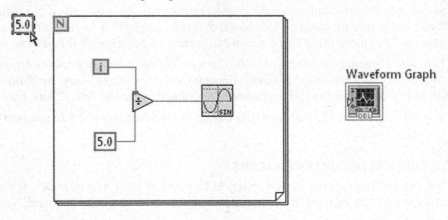

Use the Operating Tool to change the value of this new **Numeric Constant** to *100* and then wire it to the 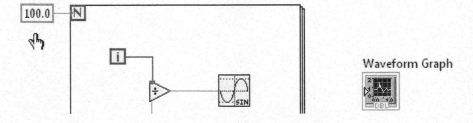.

The For Loop's count terminal is configured to accept the numeric format **I32** (i.e., a four-byte signed integer, also called a *long signed integer*). Depending on how its **Display Format…** option is configured, the Numeric Constant may or may not automatically adjust its data type to **I32**. If this adjustment is not made automatically, a red coercion dot, indicating a data-type mismatch, will appear as shown in the above diagram. This coercion dot can then be erased by popping up on the **Numeric Constant** and selecting **Representation>>I32**.

3.7 AUTO-INDEXING FEATURE

Finally, we need to store the sequence of sine-function values produced by the For Loop in a 100-element array so that we can pass this array to the Waveform Graph's terminal for plotting. Does it sound like there's some hard programming ahead? Surprisingly, there's not! Array storage of the sequence of values generated by the multiple iterations of a loop structure is such a commonly required operation that LabVIEW provides it as a built-in optional feature of both the For Loop and the While Loop. Perhaps it's easiest if we first implement this feature and then decipher what we've done.

Using ✎, simply wire the **sin(x)** output of the **Sine** icon to the Waveform Graph's terminal. You're finished!

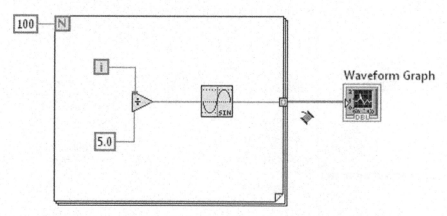

Here, we've accomplished our goal by implementing the *auto-indexing* capability of the For Loop. With this feature, as the For Loop creates a new value with each loop iteration, this new value is indexed and appended to an array at the loop boundary. After the loop completes its final iteration, the array is passed out of the loop to the Waveform Graph's terminal. Note that, in the region within the For Loop, the wire emanating from the Sine output is thin, denoting the fact that this icon produces just a single numeric value with each loop iteration. However, when this wire passes through the black-bordered square ▣ at the loop's boundary (called a *tunnel*), the wire becomes much thicker. This thick wire is LabVIEW's way of denoting a one-dimensional array of data values.

Let's explore this auto-indexing feature a little further. First, note the pair of square brackets contained within the tunnel icon 🔲. Such a bracket pair is Lab-VIEW code for an array. Thus, these brackets are the graphical indicator that an array is being created at the tunnel as the For Loop iterates (i.e., auto-indexing is activated).

Place the Positioning Tool over the tunnel at the For Loop's boundary.

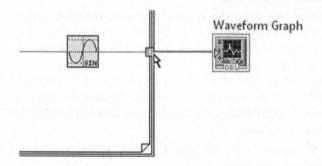

Then pop up on this tunnel to reveal its pop-up menu.

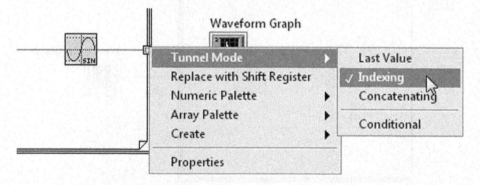

This menu offers you the ability of toggling the loop's auto-indexing feature on and off with the **Tunnel Mode** option. As we surmised earlier, a For Loop's default setting is **Indexing**. In older versions of LabVIEW, an option to **Disable Indexing** is offered (rather than Tunnel Mode), implying that auto-indexing is enabled. Toggle auto-indexing off by selecting **Tunnel Mode>>Last Value** (or **Disable Indexing**, if using an older version of LabVIEW).

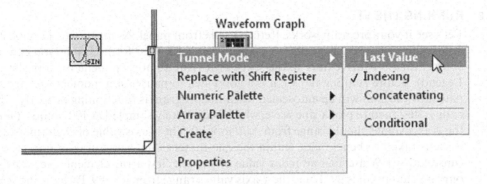

Now, rather than accumulating all the loop's iterated values into an array, the tunnel will only pass out a single value, the value of the sine function determined during the last loop iteration. To reflect this functional change for the tunnel, the bracket pair is no longer present within its icon, replaced instead by a solid color. As shown below, with the auto-indexing disabled, a broken wire now appears connected to the Waveform Graph's terminal because this icon expects to receive an entire array, not just a single numeric value.

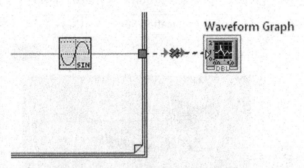

To restore the proper programming, pop up on the tunnel and toggle the auto-indexing back on by selecting **Tunnel Mode>>Indexing** (or **Enable Indexing** in older LabVIEW).

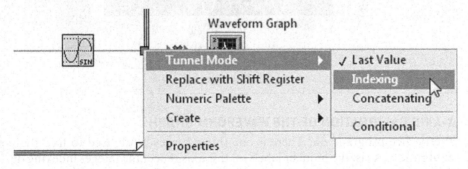

3.8   RUNNING THE VI

Let's see if your program works. Return to the front panel. As mentioned previously, autoscaling is activated by default on both axes of a Waveform Graph (you can, of course, turn off this feature in the Waveform Graph's pop-up menu or in the Scale Legend). Before running the VI, it is a good idea to pause for a moment and anticipate how the axes will be autoscaled when the program is functioning properly. The code is designed to plot a sine wave, which is discretely sampled at 100 points. Thus, the y-axis values should range from -1.0 to $+1.0$. The x-axis value of each sine-wave value is taken to be its index within the one-dimensional array of data points. Because LabVIEW ascribes an index value of 0 to the first array element, the index of our last data point is 99. Thus, the x-axis values range from 0 to 99. Perhaps the scaling will be more pleasing if LabVIEW's autoscaling feature chooses 100 as the x-axis upper limit (which is the default value for the Waveform Graph).

Okay, run your program. Hopefully, it will produce a plot as shown next, where the axes are scaled as we anticipated.

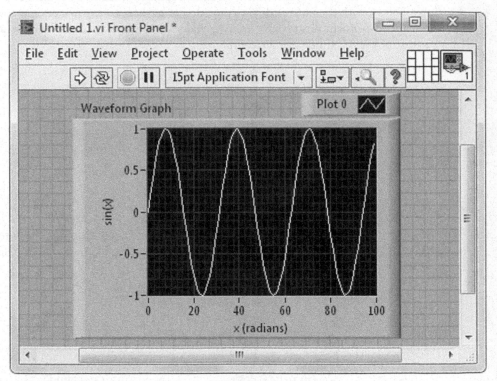

3.9   *X*-AXIS CALIBRATION OF THE WAVEFORM GRAPH

Pretty nice program, eh? There is one thing, however, we need to fix—the x-axis calibration. As is, the number labels on this axis reflect the integer indexing derived from the array storage of your data, while the text label indicates that this axis is intended to plot the radian argument of the sine function being graphed on the *y*-axis. Remembering that our program takes the sine function argument for a particular data point to be the array index value *i* divided by 5, we see that the *x*-axis can be calibrated in radians by simply multiplying each number label by 0.2. Alternatively, we can re-express this fact by saying that, by default, the Waveform Graph assumes the *x*-axis spacing between adjacent data points Δx is 1. Calibration is then accomplished by defining Δx to be 0.2.

Such calibration by a multiplicative constant is a common need when using the Waveform Graph, and therefore LabVIEW provides mechanisms for this procedure. In the previous chapter, we accomplished *x*-axis calibration of a Waveform Chart by defining a **Multiplier** in the **Chart Properties** dialog window. This mechanism can be implemented on the Waveform Graph also. Here, we will introduce an alternative calibration mechanism, which appears explicitly on the block diagram. This calibration mechanism involves a new LabVIEW process—bundling together several objects into a *cluster*. The resulting cluster is not a mathematical object, but rather a LabVIEW convenience. By grouping a number of objects together into a single cluster, one is able to transport a large quantity of data across a block diagram using a single cluster wire. Thus, clustering greatly simplifies the appearance and readability of your block diagrams.

Return to the block diagram, place the mouse cursor over the Waveform Graph's terminal, and activate its Help Window. As shown next, this Help Window is divided into a top and bottom portion.

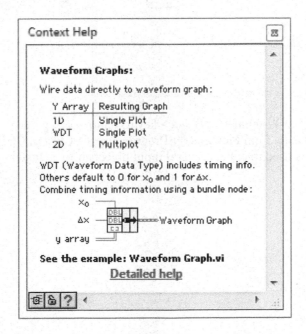

In the top portion, we find that a Waveform Graph with default data-point spacing $\Delta x = 1$ is created by simply connecting an array wire to the Waveform Graph terminal. This is the manner in which we've used the Waveform Graph thus far. We produced a single plot by feeding a one-dimensional data array to the Waveform Graph's terminal. Note that two (or more) waveforms can be displayed simultaneously on a Waveform Graph by connecting a two-dimensional (2D) array to the terminal; we will discuss 2D arrays in later chapters. The bottom portion of the Help Window indicates that, when given an appropriate cluster, the Waveform Graph produces a plot on the front panel with its x-axis properly calibrated. The appropriate cluster is formed by bundling together the following sequence of objects: a value for the x-axis origin x_0, a value for the x-axis spacing Δx, and the array of numeric data to be used as the y-axis values of the waveform's points.

Here's how to include the appropriate cluster in your code. First, remove the array wire connected to the Waveform Graph terminal. Using the Positioning Tool, click on this wire to produce a marquee.

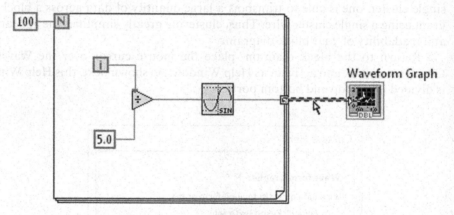

Then press your keyboard's <*Delete*> key to remove this wire.

Select **Bundle** from **Functions>>Programming>>Cluster, Class, & Variant**. The **Bundle** icon is used to collect several input objects into a cluster.

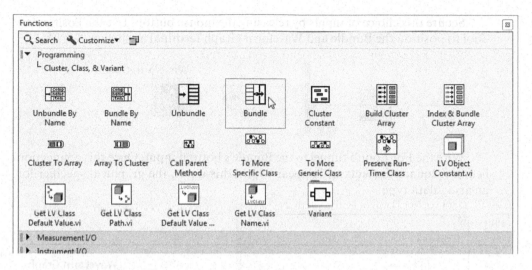

Place the **Bundle** icon on your block diagram. You will find that its leftmost section consists of two inputs by default. We require three inputs, so you must resize the icon.

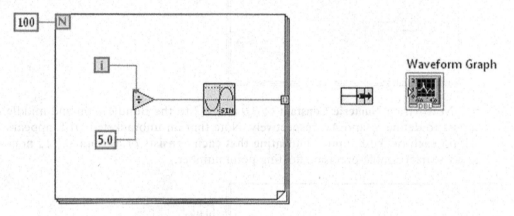

To resize the **Bundle** icon so that it has three inputs, place the ⬉ at the icon's bottom center (or top center) so that the tool morphs into a Resizing Cursor ↕. Then drag the cursor until an additional input appears.

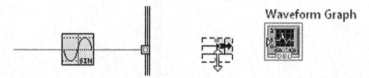

Secure this choice of inputs by releasing the mouse button. Use the Positioning Tool to position the Bundle and Waveform Graph terminal as shown.

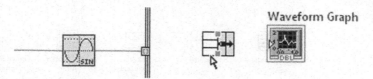

Wire the For Loop's tunnel to the Bundle's bottom input. Once this connection is made, square brackets will appear within this input, the graphical specifier for an array data type.

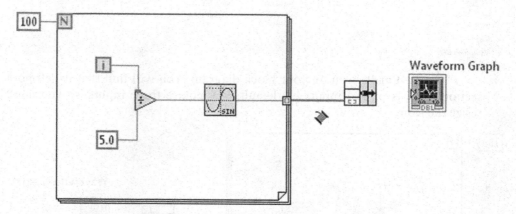

Now wire a **Numeric Constant** of *0.0* and *0.2* to the Bundle's top and middle inputs to define x_0 and Δx, respectively. Note that an unbracketed **DBL** appears within each of these inputs, indicating that each consists of a solitary (i.e., non-array scalar) double-precision floating-point number.

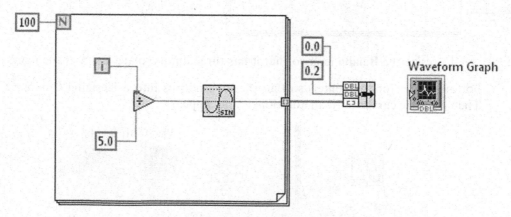

The **Bundle** icon receives these three inputs on its left side, bundles them together, and outputs the resulting cluster at its rightmost section. Wire this cluster output to the Waveform Graph's terminal.

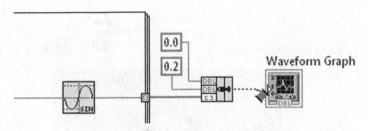

You will find that as soon as this connection is made, the icon terminal's data-type specifier morphs from an orange floating-point numeric specifier into a pink cluster specifier. In addition, the wire connecting the Bundle output to the Waveform Graph's terminal has a pink banded appearance, denoting cluster data.

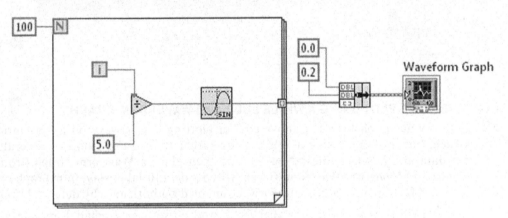

Now return to the front panel and run your final program. Note the *x*-axis is calibrated in radians.

Using **File>>Save**, create a folder named **Chapter 3** within the **YourName** folder, and then save this VI under the name **Sine Wave Graph (For Loop)** in **YourName\ Chapter 3**.

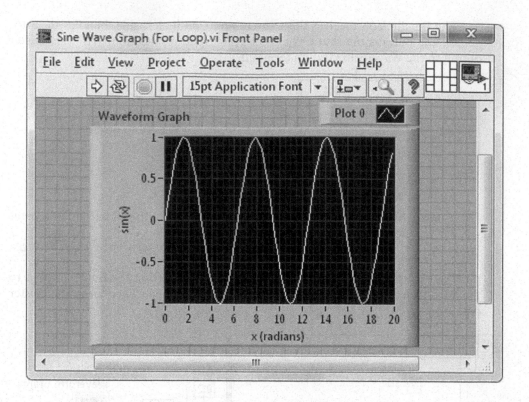

3.10 SINE-WAVE PLOT USING A WHILE LOOP AND WAVEFORM GRAPH

Let's try accomplishing the above goal of plotting a sine wave on a **Waveform Graph**, but this time using a While Loop rather than a For Loop to generate the data array. Select **File>>New VI**, and then place a **Waveform Graph** (from **Controls>>Modern>>Waveform Graph**) with its default label Waveform Graph on the fresh front panel. Autoscaling will be enabled on both axes by default. Using the A̲ (or the Scale Legend), label the *x*- and *y*-axes as x (radians) and sin(x), respectively.

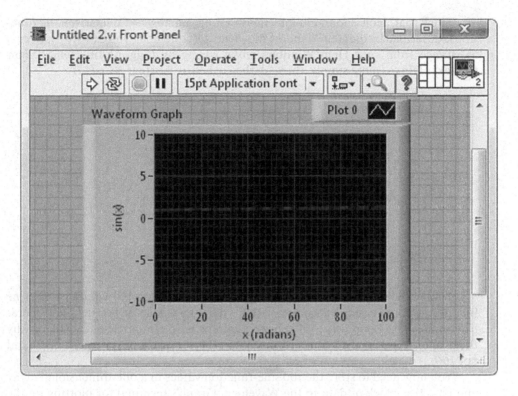

Switch to the block diagram and write the following code.

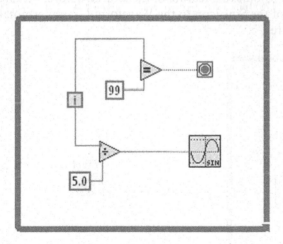

Waveform Graph

The **Equal?** icon is found in **Functions>>Programming>>Comparison**. Its function is described by the Help Window shown next. Remember, you can always access assistance from such Help Windows while wiring by toggling on **Help>>Show Context Help** or, more simply, through the keyboard shortcut *<Ctrl+H>*.

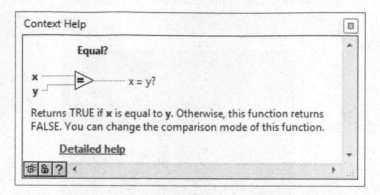

The While Loop on your block diagram will generate 100 values of the sine function before ceasing execution. Using your knowledge of the detailed operation of a While Loop, can you explain why this is so? In particular, explain why the proper value for the **Numeric Constant** in the comparison statement is *99* (and not *100*).

You now need to store the 100 sine-function values in a one-dimensional array and pass this block of data to the Waveform Graph's terminal for plotting on the front panel. Making use of the auto-indexing feature of the loop structure, we need simply to wire the **sin(x)** output of the Sine icon to the Waveform Graph terminal. Perform this wiring as shown here.

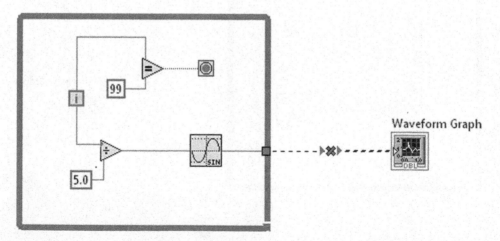

You will note that this seemingly simple connection produces a broken wire. To find out why, pop up on the tunnel at the While Loop's border.

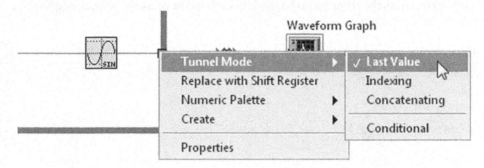

In the tunnel's pop-up menu, you will find that **Tunnel Mode>>Last Value** is activated (or, in an older version of LabVIEW, you will be presented with the option of enabling the loop's auto-indexing feature). Thus, we see that, in contrast to a For Loop, the auto-indexing feature of a While Loop is toggled off by default. Turn the auto-indexing on by selecting **Tunnel Mode>>Indexing** (or **Enable Indexing**, in older LabVIEW). Also include a **Bundle** icon to scale the *x*-axis in radians, as shown below.

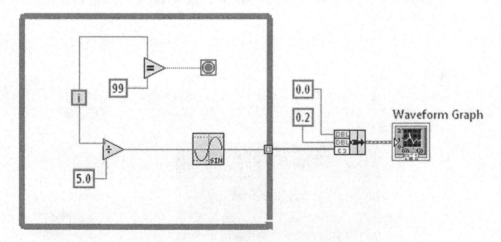

Your program is now complete. Return to the front panel and run it. You might try using the keyboard shortcut for the **Run** command: *<Ctrl+R>*.

3.11 FRONT-PANEL ARRAY INDICATOR

Are you unconvinced that there are actually 100 elements in the array that you produced? You can easily check this fact by inspecting the array using an *Array Indicator*. An Array Indictor is constructed as follows. Select an **Array** shell from

Controls>>Modern>>Array, Matrix & Cluster. As described in the next paragraph, the **Array** shell can be made to operate in either of two modes—as a *control* to input an array from the front panel to the block diagram or as an *indicator* to output an array from the block diagram to the front panel.

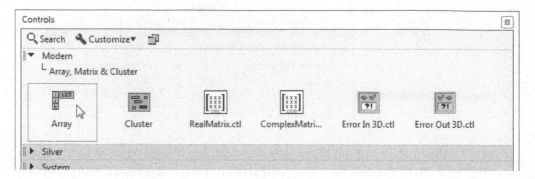

Place the **Array** shell on your front panel and then enter the label **Array**. You will find that the shell consists of an *index display* on the left and a large blank region on the right. By filling the blank region with an appropriate object (called the *element display*), one may choose whether the **Array** shell will function as an input or output device.

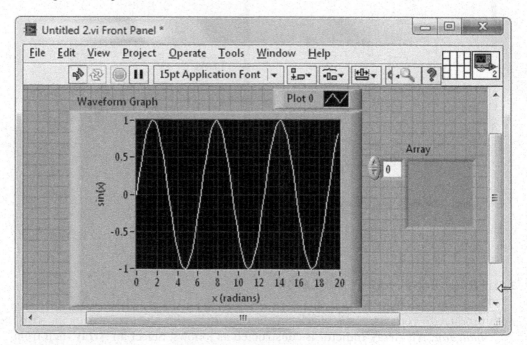

In the present situation, we wish to construct an *Array Indicator* that outputs an array from the block diagram, so we will choose our element display to be a **Numeric Indicator**. If instead we wished to construct an *Array Control* that inputs an array to the block diagram, we would use a **Numeric Control** for the element display.

To place a **Numeric Indicator** within the **Array** shell, do the following. Select a **Numeric Indicator** from **Controls>>Modern>>Numeric**.

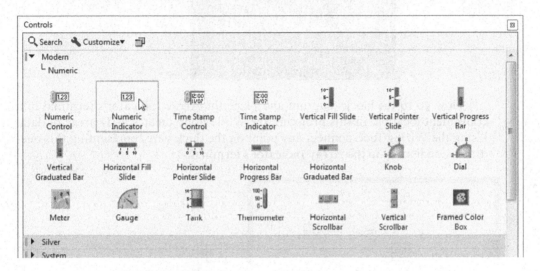

Then place the cursor in the large blank region of the **Array** shell.

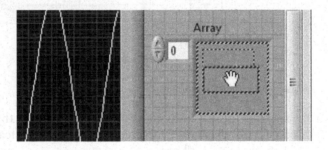

When you click the mouse button, the Array Indicator will be complete, as shown below.

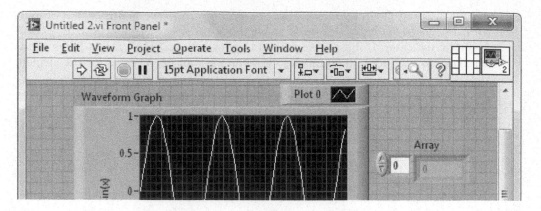

Now go to the block diagram and place the Array Indicator's terminal in a convenient location. In this program, the While Loop creates a 1D array of data. Using the Wiring Tool, connect any point on the thick wire representing this one-dimensional array to the Array Indicator's terminal.

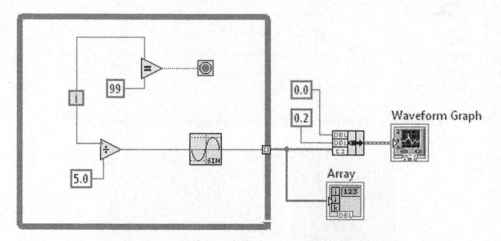

Return to the front panel and run your program. After the VI completes execution, you can inspect the array generated via the Array Indicator (if you need to resize it, use the ⬁). Using the ✋, look at various elements in the array. To scan the array, you may increment or decrement the index number in steps of 1 by repeatedly clicking on the index display's up/down arrows. Alternatively, to view a particular array element, you can use the ✋ to highlight the index display, then enter the desired element's index value from the keyboard, and press *<Enter>* or click on ✓.

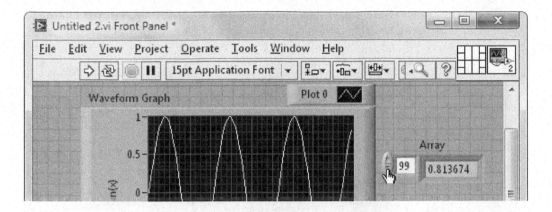

Through such exploration, you should find that your array consists of 100 elements, with indices running from 0 to 99.

Finally, you might try producing an Array Indicator the "painless" way. For example, pop up on the thick orange wire (or the tunnel) that carries the sine-wave array output from the While Loop, and then select **Create>>Indicator**.

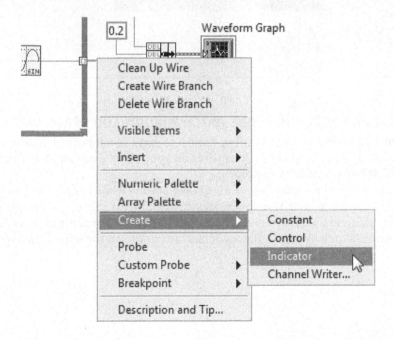

Now check the front panel. You will find an automatically produced Array Indicator there. Run your program and discover whether it works as expected.

Using **File>>Save**, save this VI in the **YourName\Chapter 3** (i.e., the **Chapter 3** folder within the **YourName** folder) under the name **Sine Wave Graph (While Loop)**.

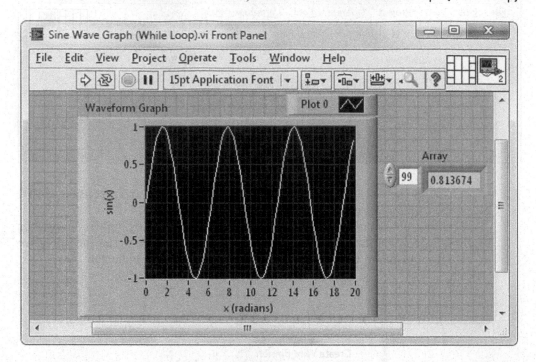

3.12 DEBUGGING WITH THE PROBE WATCH WINDOW AND ERROR LIST

When debugging a VI, it is often handy to have a temporary tool available for examining the contents of various wires on your block diagram. LabVIEW's *Probe Watch Window* is this handy tool.

Open the block diagram of **Sine Wave Graph (While Loop)**. Let's demonstrate how the Probe Watch Window works by using it to inspect the same sine-wave array's elements we examined a moment ago via an Array Indicator. To activate the Probe Watch Window, place the mouse cursor on the wire carrying the array data.

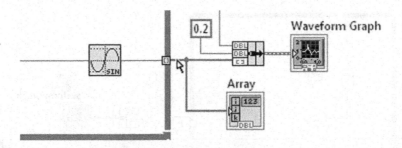

Then pop up on this wire and select **Probe**.

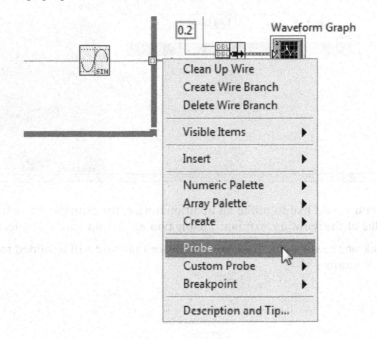

A Probe Watch Window will appear containing a probe for the selected wire, along with an associated numbered rectangle on the block diagram marking the wire location being probed. The Probe Watch Window can also be activated by clicking on the wire of interest with the *Probe Tool* found in the Tools Palette. The probe generated in this case is an Array Indicator, which displays the array element whose index is selected in the index display; the next six array elements are displayed as well.

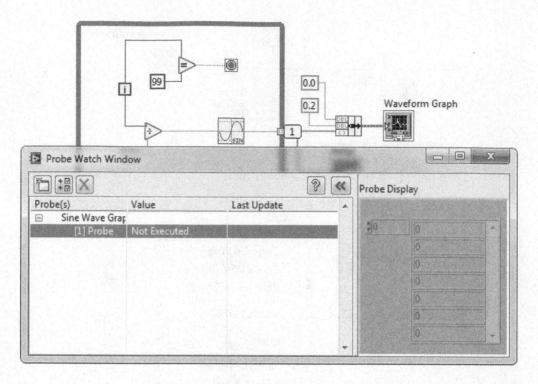

If you would like to probe an additional wire, for example, the wire carrying the value of the iteration terminal, simply pop up on that wire and select **Probe** (or else click on the wire with), and a probe for that wire will be added to the Probe Watch Window.

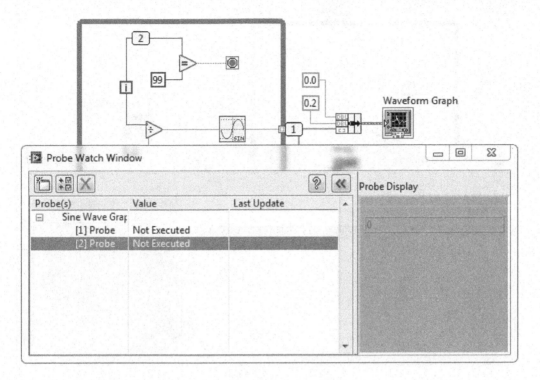

Run your program. Then, within the Probe Watch Window, click on **[1] Probe** to select the probe associated with the wire carrying the elements of the sine-wave array. In the **Probe Display** region, that probe's Array Indicator will be made available to inspect the elements of the sine-wave array. In a similar way, you can inspect the other probe that, during runtime, reports the current value of the iteration terminal and, when the VI has completed its run, gives the final iteration terminal value.

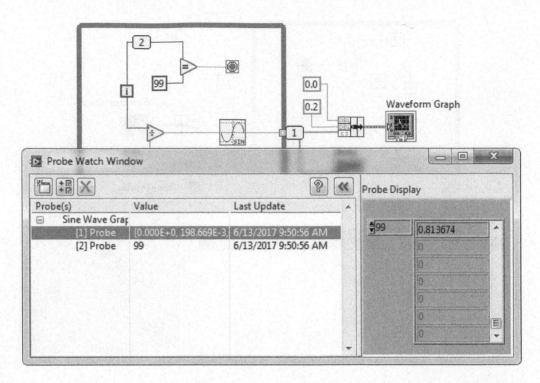

Within the Probe Watch Window, pop up on, for example, **[1] Probe** and explore the options available in a probe's pop-up menu such as deleting the probe or opening its own stand-alone window. These options are also available in the Probe Watch Window's toolbar.

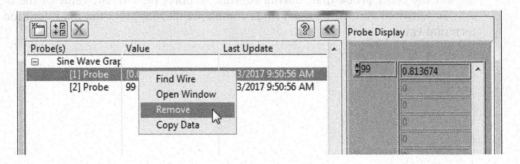

Close the Probe Watch Window when you're done, and it will disappear along with its associated numbered rectangles on the block diagram. The Probe Watch Window is an invaluable tool for debugging LabVIEW programs that execute but are producing questionable or unexpected results.

Finally, let's explore another of LabVIEW's handy program-debugging features, the *Error List*. Intentionally put an error on the block diagram of **Sine Wave Graph (While Loop)** by popping up on the tunnel at the While Loop's border and selecting **Tunnel Mode>>Last Value** (or **Disable Indexing**, if using an older version of LabVIEW). As shown below, with the auto-indexing disabled, a broken wire now appears connected to the Bundle's bottom terminal because this icon expects to receive an entire array, not just a single numeric value.

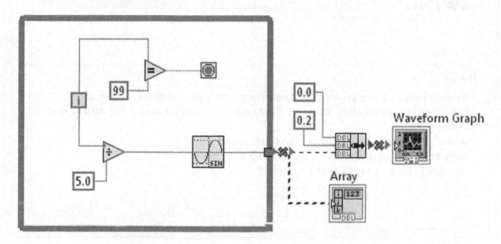

Scrutinize the toolbar. There, you will find that the **Run** button appears broken ![broken run button], which is LabVIEW's way of communicating that an error exists within your program. But this indicator isn't the only help available.

Click on the ![broken run button] (a tip strip alerts you that this action will produce a list of program errors). A dialog window will appear, listing all of the errors in your program. In the present case, two errors exist, both related to a data-type mismatch, namely, a scalar wire connected to an input formatted to accept an array.

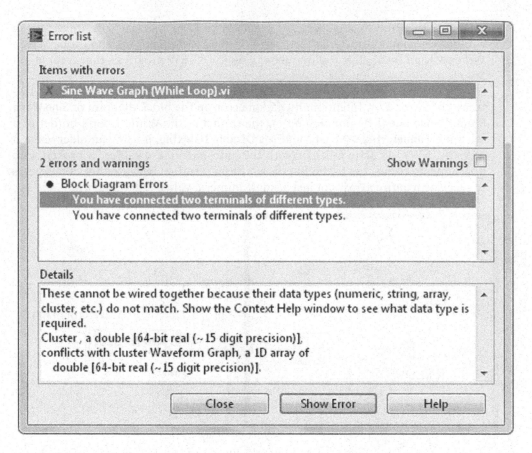

In the **Details** box, a detailed description is given of the error that is high-lighted in the center box. To highlight a different error, simply click on that error in the center box. Clicking on the **Show Error** button will take you to the error's location (via a marquee) on the block diagram, where you can then make the required repair. With all of this interactive help, quite commonly a LabVIEW program can be debugged without having to consult the more detailed online reference resource **LabVIEW Help**.

Repair the error in your VI by popping up on the tunnel on the While Loop's border and selecting **Tunnel Mode>>Indexing** (or **Enable Indexing**, in older Lab-VIEW). You will note that the **Run** button ⇨ becomes whole again, signaling that the program is now executable. Save your VI as you close it.

DO IT YOURSELF

Four Flipping Coins.vi A flipped coin can land in one of two states: *Heads* or *Tails*. Write a VI that simulates simultaneously flipping four coins and determines how many of these coins land in the Heads state.

(a) First, build the front panel as shown here, where the state of each coin is represented by a **Round LED** Boolean indicator. A lit (unlit) LED corresponds to Heads (Tails). Also include a **Numeric Indicator** labeled **Number of Heads** that gives the number of coins in the Heads state.

Using the block-diagram code shown next, model the flip of each of the four coins using a **Random Number (0–1)** icon. When this code is run, if the **Random Number (0–1)**'s output is greater than or equal to 0.5, the corresponding coin is assigned an output value of 1 (corresponding to the Heads state); otherwise, that coin is assigned an output value of 0 (corresponding to the Tails state).

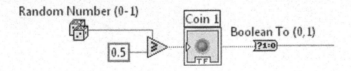

Use a **Compound Arithmetic** icon to tally the output values of all four coins and so determine the number of coins in the Heads state. Output this integer to the **Number of Heads** front-panel indicator. Note that there are five possible outcomes: 0, 1, 2, 3, and 4 Heads. Run your VI to verify that it is working correctly.

(b) The present code on your block diagram simulates one *trial* of simultaneously flipping the four coins. Now simulate repeating such a trial over and over again. Place this code within a **For Loop** so that you can execute N trials, where N is given by the value in a front-panel **Numeric Control** labeled **Total Number of Trials**. Store the value of **Number of Heads** from each trial in an array using the For Loop's auto-indexing feature. Then, after the N trials are complete, plot a histogram on a **Waveform Graph** that displays the number of trials whose outcome was 0, 1, 2, 3, and 4 Heads. To create a histogram, use the **Histogram.vi** icon as shown below with its **intervals** input wired to an integer *5* (because there are five possible outcomes for **Number of Heads**). Alternatively, the value for **intervals** can be calculated via the relation *intervals = max value − min value* +1 using the **Array Max & Min**, **Subtract**, and **Increment** icons.

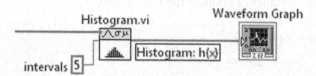

The completed front panel of your VI will then appear as shown next.

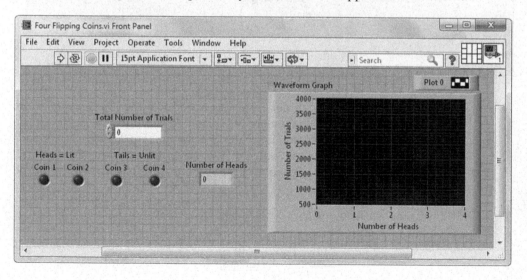

Run **Four Flipping Coins** with several choices of **Total Number of Trials**. In the Waveform Graph's **Plot Legend**, you may wish to select a **Bar Plots** option in its pop-up menu. From statistical theory, it is expected that when the number of trials N is large, the number of trials with an outcome of 0 or 4 Heads should equal $N \left(\dfrac{1}{2}\right)^4 = \dfrac{1}{16} N$ and the number of trials with an outcome of 2 Heads should be

$N \dfrac{4!}{2!2!}\left(\dfrac{1}{2}\right)^4 = \dfrac{3}{8}\,N$. Above what (approximate) value of N are your results consistent with this expectation?

USE IT!

Using Auto-Indexing to Input an Array to a Loop In addition to its ability to build an array at a loop's output, auto-indexing can be used to sequence through an array at a loop's input. The following LabVIEW program illustrates this feature. When this VI is run, two cycles of a sine wave are plotted over a time span of about 10 seconds.

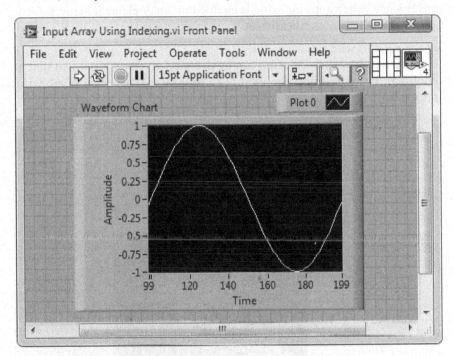

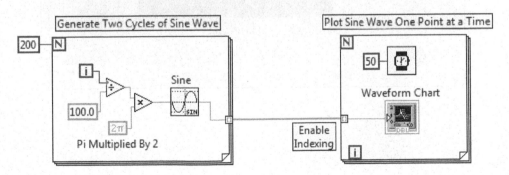

On the block diagram, the left-hand For Loop builds an array containing two cycles of a sine wave. For the right-hand For Loop, note that the array wire is thick (denoting an array) outside the loop and then becomes thin (denoting a scalar) inside the loop. Why is this so? Auto-indexing is activated in a For Loop by default, so upon execution, the loop will sequentially input one element from the array each time it iterates. That is, on the first loop iteration when 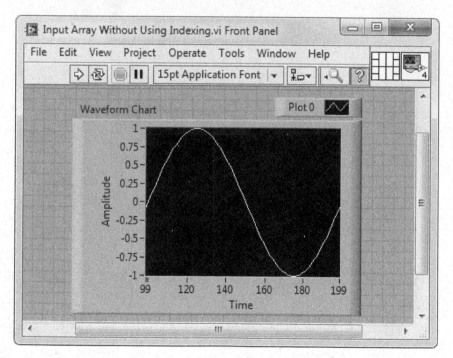 equals 0, the element of the array with index 0 will be input and plotted by the Waveform Chart; on the second iteration when equals 1, the element of the array with index 1 will be input and plotted; and so on until the end of the array is reached. As an added convenience, when auto-indexing is enabled on an N-element array entering a For Loop, LabVIEW automatically sets the loop's count terminal to N, thus eliminating the need to wire a value to .

Alternatively, auto-indexing can be disabled at a loop's input so that an entire array may be passed into the loop. The program shown next demonstrates this approach.

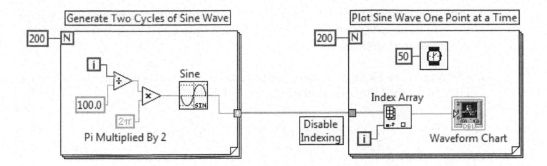

Here, by popping up on the tunnel of the right-hand For Loop and selecting **Disable Indexing**, auto-indexing has been switched off. Now the entire array is passed into the For Loop (with auto-indexing disabled, the count terminal needs to be wired). Within this loop, the **Index Array** icon (consult its Help Window for an explanation of its function) is used to sequence the array elements for plotting each element one at a time on the Waveform Chart as the For Loop iterates. When run, this VI performs in the same manner as **Input Array Using Indexing.vi**.

PROBLEMS

Each problem statement begins with a suggested descriptive name (including the **.vi** extension) for the program you will write. Suggested icons for use in the VI can be found with the aid of **Quick Drop**.

1. **Month's Total Pay.vi** I have agreed to hire you for a 30-day month. I pay you 1 penny on the first day, 2 pennies on the second day, and continue to double your daily pay on each subsequent day up to (and including) Day 30. How many total dollars do you earn by the end of the month? Write a For Loop based program that determines the answer to this question. (One dollar equals 100 pennies.) Suggested icons: **Power Of 2**, **Add Array Elements**.

2. **Finite Geometric Series.vi** A finite geometric series of N terms is given by $y_N = 1 + x + x^2 + x^3 + \dots + x^{N-1}$, where $|x| < 1$. It is well known that when $N = \infty$, $y_\infty = \dfrac{1}{1-x}$.

   (a) Write a For Loop-based program that, given N and x as inputs via front-panel **Numeric Controls**, determines y_N, calculates its percentage deviation from y_∞, and displays these two results in **Numeric Indicators**. Suggested icons: **Power Of X**, **Add Array Elements**, **Reciprocal**.

(b) Use your VI to answer the following question: If $x = 0.5$ and one wishes y_N to deviate from $y_\infty = 2.000\ldots$ by less than 0.1%, what is the minimum required value of N?

3. **Sum of Three Sines.vi** A unity-amplitude square wave $y(t)$ can be approximated by the following sum of three sine waves:

$$y(t) \approx \frac{4}{\pi}\left[\sin(x) + \frac{1}{3}\sin(3x) + \frac{1}{5}\sin(5x)\right]$$

where x is in radians. Write a VI that evaluates this sum at 300 equally spaced x-values in the range from $x = 0$ to $x = 6\pi$ (i.e., for three cycles of the first sine function in the sum) and then plots the resulting y vs. x on a Waveform Graph, whose x-axis is calibrated in radians. When the VI is run, it should produce the front-panel plot shown next. Suggested icons: **Sine, Compound Arithmetic, Pi.**

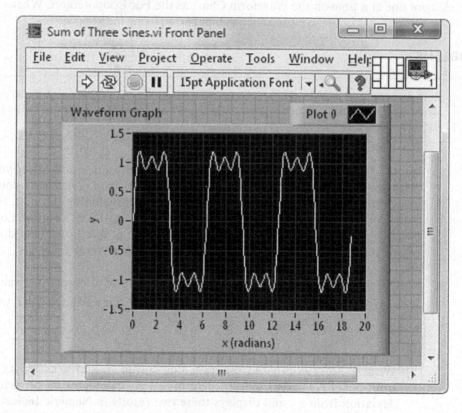

4. **Wait Test.vi** Write a program that iterates the following task $N = 10,000$ times: During the ith iteration, execute the **Wait (ms)** icon with **milliseconds to wait** equal to I, and then store the resultant value of **millisecond timer value** (which is the number of milliseconds that have elapsed since your computer was turned on) as the ith element of an array. After this sequence of N iterations, make your program plot the resultant N-element array of **millisecond timer value** values, after first subtracting off the initial "baseline" value (i.e., the first array element value). Also, display this resultant array in an Array

   Indicator called **Elasped Time (ms)** (this indicator can be resized using the ↖). The block diagram of your VI should appear as shown below. The **Index Array** icon is used to determine the initial "baseline" value of **millisecond timer value** (for an explanation of how this icon functions, consult its Help Window).

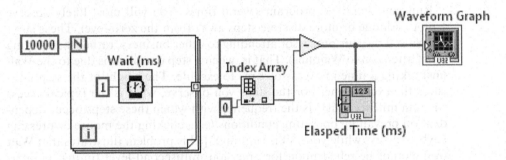

(a) Ideally, one would expect that this program produces a straight-line plot with slope equal to 1 and a final array element value equal to 9999 (the index of this element is 9999). Run your program a few times and demonstrate that this ideal expectation is close to, but not exactly, what is observed.

(b) To explain your observations, modify your block diagram as shown next. Here, an array of integers, which begins at 0 and then increments by 1, is generated and subtracted from the array of part (a). If the ideal expectation of part (a) is met, this new program should produce an N-element array of zero values.

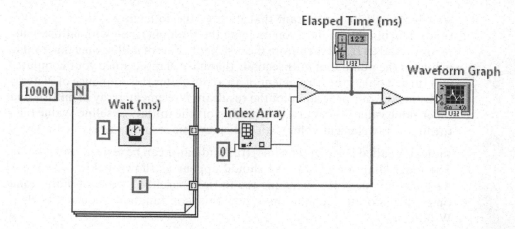

Run your modified program several times. You will most likely observe a plot with one or more discrete steps away from the zero-level. These steps are caused by your processor attending to other business, rather than paying strict attention to **Wait (ms)**. That is, when a step occurs, it is due to the **Wait (ms)** taking a time in excess of 1 ms to execute. The height of the step measures the excess time. For the steps you observe, what is the typical excess time (in milliseconds)? Is the frequency with which these steps occur dependent on prevailing operating conditions (e.g., clicking the mouse or pressing keyboard keys while your VI is running)? [This problem illustrates that **Wait (ms)** cannot be relied upon for precision millisecond-level timing. If interested, try carrying out a similar investigation of the performance of **Wait (ms)** when **milliseconds to wait** equals *10* or *100*.]

5. **Dragging Dresser.vi** It's moving day and you are dragging a heavy dresser at constant velocity across the horizontal floor of your new apartment as shown in Figure 3.1. The force you are applying to the dresser has magnitude *F*, and it is directed at angle θ above the horizontal. The weight of the dresser is *W*, and the coefficient of kinetic friction between it and the floor is μ.

FIG. 3.1 Problem 5.

Newton's laws can be used to show that

$$F = \frac{\mu W}{\cos \theta + \mu \sin \theta}$$

or, expressing the force as a fraction of the dresser's weight (i.e., $F' = F/W$),

$$F' = \frac{\mu}{\cos \theta + \mu \sin \theta}$$

(a) Write a For Loop-based program that, given μ from a **Numeric Control** labeled **Coefficient of Kinetic Friction**, calculates F' from $\theta = 0$ to $\theta = 90°$ in 1° increments and then displays the resulting array of F' values on a Waveform Graph, as well as in an Array Indicator labeled **Fractional Force**. Suggested icons: **Sine, Cosine, Degrees to Radians** (this icon is only available in newer versions of LabVIEW).

(b) Run your VI with $\mu = 0.1$, $\mu = 0.4$, and $\mu = 1.0$ and determine the angle θ_{min} at which F' is smallest in each case (i.e., θ_{min} is the angle at which the minimum magnitude force is required to drag the dresser at constant velocity for the given μ).

6. **Distribution of Random Number Icon.vi** Each time it is called, the **Random Number (0–1)** icon generates a double-precision floating-point number in the range from 0 to (but not including) 1, where the number is equally likely to be anywhere within this range (i.e., the probability distribution is uniform).

(a) Demonstrate that the distribution of values produced by **Random Number (0–1)** is uniform as follows: Write a VI that employs the icon to generate an array of N random numbers and then uses **Histogram.vi** to plot the histogram distribution of these values. The front panel of the VI is shown below, where it has been run with $N - 100,000$ and the random-number range from 0 to 1 has been divided into 100 bins for the histogram plot. The **Histrogram Graph**, which has a calibrated x-axis, can be produced on the block diagram by popping up on Histogram.vi's **Histogram Graph** terminal and selecting **Create>>Indicator**.

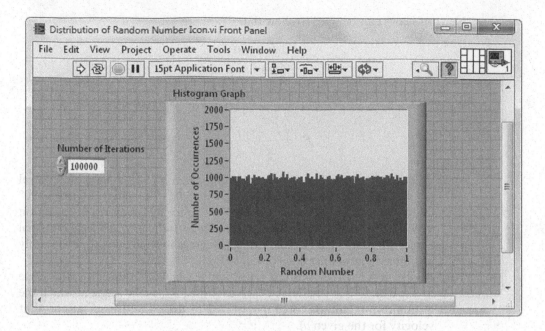

(b) During a run in which **Random Number (0–1)** produces N random numbers in the range $0 \le x < 1$, the number of occurrences of a random number in the range from x to $x + dx$ is given by $dN = \rho\, dx$, where ρ is a constant. Given that $\int_{x=0}^{x=1} dN = N$, show that $\rho = N$. Then, from the definition of the distribution's mean $\bar{x} = \dfrac{\int_{x=0}^{x=1} x\, dN}{N}$ and standard deviation

$$\sigma = \sqrt{\frac{\int_{x=0}^{x=1} (x - \bar{x})^2\, dN}{N - 1}},$$ show that $\bar{x} = 0.5$ and $\sigma \approx 1/\sqrt{12}$ (assuming $N > 1$). Finally, add the **Std Deviation and Variance.vi** icon to your block diagram so that the distribution's mean and standard deviation are calculated and displayed on the front panel. Run your VI, and show that $\bar{x}$ and σ have the expected values.

7. **Iterations Until Integer Equals Five (For Loop).vi** Problem 8 in Chapter 2 explains how to code the following task: Randomly generate an integer in the range from 1 to 10, and iterate this random process until the integer equals 5.

The number of iterations required until the randomly generated integer equals 5 is output as a quantity called **Required Iterations**.

Place the code that carries out the above-described task as a subdiagram within a For Loop so that, when executed, the For Loop will repeat the task N times and store the resulting N values of **Required Iterations** in an array. Then, use the **Mean.vi** icon to calculate the average of the N elements in this array and display this average in a front-panel **Numeric Indicator** labeled **Average Number of Required Iterations**. Make $N = 1,000,000$ on your block diagram.

Run your VI. Is the value you obtain for **Average Number of Required Iterations** equal to the value that you expect?

8. **Sine Wave Graph (Waveform).vi** The *waveform* data type provides an alternate method for producing a Waveform Graph with its x-axis calibrated properly, where it is assumed that the x-axis quantity is *time*. To investigate the use of this data type, write a program that plots the function $y = \sin(2\pi t)$ from time $t = 0$ to $t = 0.99$ in increments of $\Delta t = 0.01$. The block diagram for this VI is shown next.

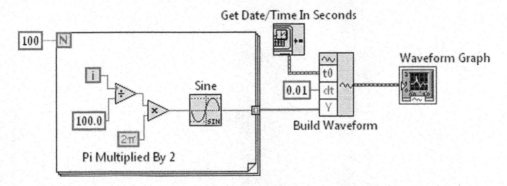

Here, the **Build Waveform** icon combines an initial time **t0** (called a *time stamp*), the sampling time increment **dt**, and the array of y-values. **Build Waveform** must be resized to expose three input terminals, and then each of these terminals is defined using the 🖑. The time stamp is created using the **Get Date/Time In Seconds** icon. Run your completed VI and verify that it executes as expected. Suggested icons: **Sine, Get Date/Time In Seconds, Pi Multiplied by 2**.

9. **Numerical Integration.vi** It is easy to show analytically that the integral $\int_0^1 3x^2\,dx$ equals 1 exactly. Write a program that calculates the integral $\int_0^1 3x^2\,dx$

numerically as follows: Using a For Loop, evaluate the quantity $3x^2$ at N values of x, starting at $x = 0$ and ending at $x = 1.000$ in steps of $\Delta x = 0.001$, and (using the auto-indexing feature) store this sequence of values in an N-element array (think carefully about the proper choice for N). Then use the **Numerical Integration.vi** icon to evaluate $\int_{0}^{1} 3x^2 \, dx$, and display the result in a front-panel **Numeric Indicator**.

Determine the deviation of your numerical integration result from the integral's known value (i.e., exactly $1.0000\ldots$). Express the deviation you observe in terms of a percentage. If you need to observe more decimal places in the numerical integration result, pop up on the **Numeric Indicator** and use the **Display Format...** option to increase **Digits of precision**. (You may find it interesting to explore this problem further by changing the value of Δx to, say, 0.01 or 0.0001.)

CHAPTER 4

The MathScript Node and XY Graph

In this chapter, you will gain experience with a LabVIEW feature called the **MathScript Node**. If the MathScript Node does not appear in your **Functions>>Programming>> Structures** subpalette, rerun your LabVIEW installation package, and when the window appears in which you choose the programs to be installed, select *MathScript RT Module*. If you find that your installation package does not include the MathScript RT Module, or if you are using the LabVIEW Base Development System (which does not support the MathScript RT Module), the chapter exercises can be carried out using LabVIEW's **Formula Node**. Instructions for this alternate approach are given throughout the chapter and in Appendix A.

4.1 MATHSCRIPT NODE BASICS

In previous chapters, we have seen that the graphical nature of the LabVIEW programming language greatly simplifies the coding of desired computer operations. However, in the case of encoding mathematical relations, graphical programming is much more cumbersome than text-based languages. For this reason, LabVIEW offers the *MathScript Node*, a resizable box within which you can enter text-based mathematical formulas directly on the block diagram. To demonstrate the utility of the MathScript Node, consider the relatively simple equation $y = 3x^2 + 2x + 1$. If you code this equation using LabVIEW arithmetic icons from **Functions>> Programming>>Numeric**, the subdiagram will appear as follows.

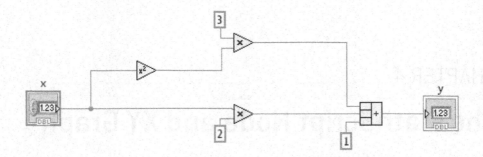

Alternatively, you can program the same equation using a MathScript Node, which is found in **Functions>>Programming>>Structures**. The resulting subdiagram is shown in the following illustration. As you can see, the MathScript Node is much easier to write and to read.

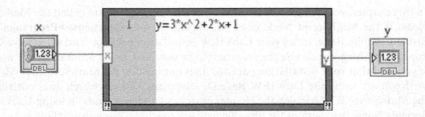

An alternate (but less powerful) text-based structure called the *Formula Node* is also available in **Functions>>Programming>>Structures**. The MathScript Node and Formula Node are similar in that both can be used to carry out common mathematical operations, including evaluating trigonometric and logarithmic functions as well as implementing Boolean logic, comparisons, and conditional branching. Here, the parity between the two structures ends, however. Where the Formula Node is capable of performing only these relatively simple tasks, we will see that the MathScript Node functions as a high-level computing language (called *MathScript*) with wide-ranging functionality in digital signal processing and data analysis.

To illustrate the power of MathScript-based programming, consider coding the problem we solved in the previous chapter: Evaluate sin (x) at 100 evenly spaced x-values in the range from $x = 0$ to $x = 99/5 = 19.8$. Using graphical programming solely, we found that the solution to this problem is as shown next.

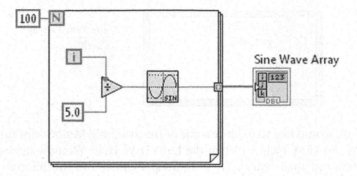

The MathScript language contains the text-based sine function sin (x), so we can simply transliterate the above code using the MathScript Node as follows.

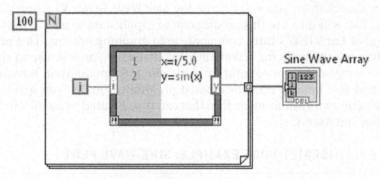

The above program will indeed execute as desired, but it fails to take advantage of the full power of the MathScript language. Namely, in MathScript, the command $x=start:step:stop$ generates a 1D array of x-values, beginning at $x=start$ and ending at $x=stop$, with the x-values evenly spaced by intervals $\Delta x=step$. Also, given x as an array of N elements, the command $y=\sin(x)$ will evaluate the sine function at each of the N x-values and return the results in an N-element array called y. Thus, when using a MathScript Node to solve our stated problem, a graphical For Loop is unneeded. The following diagram is sufficient.

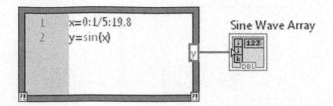

If you would like to explore some of the available MathScript functions, select **Help>>LabVIEW Help...** When the LabVIEW Help Window opens, click on the **Index** tab and then search for *MathScript*. Under the MathScript heading, you will find many topics of possible interest, including *Classes of Functions*, which provides an organized listing of the (over 800) available functions, along with the proper syntax for implementing them.

In this chapter, you will rewrite the **Sine Wave Graph** VI using a MathScript Node. You will then use this subdiagram to supply data to an *XY Graph*, the most general of LabVIEW's three commonly used graphing options. This new program will become the basis for **Waveform Simulator**, a VI you will use to simulate discretely sampled data in your future work with file input/output, waveform generation, and fast Fourier transform-based programs. Finally, you will learn how to create your own custom-made icon that can then be used as a *subVI* in higher-level "calling" programs.

4.2   QUICK MATHSCRIPT NODE EXAMPLE: SINE-WAVE PLOT

> **Formula Node Users:** Please review this section, then refer to Appendix A for instructions on writing the Formula Node-based version of **Sine Wave Graph**.

To introduce ourselves to programming the MathScript Node, let's first write the program just discussed (i.e., a VI that evaluates sin (x) at 100 evenly spaced x-values in the range from $x = 0$ to $x = 19.8$). Open a new VI and then place a **Waveform Graph** on its front panel. Using the [A] (or, alternatively, Waveform Graph's **Scale Legend**), label the x- and y-axes as **x (radians)** and **sin(x)**, respectively. By default, autoscaling will be activated on both axes.

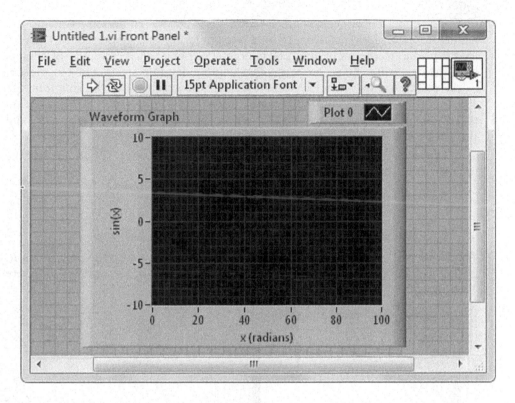

Switch to the block diagram and place a **MathScript Node** on it (found in **Functions>>Programming>>Structures**).

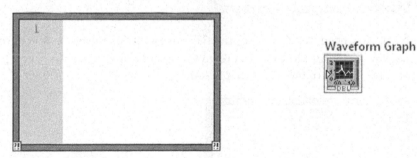

Waveform Graph

We'll now program our MathScript Node with text-based commands and equations, which will generate the desired sine-wave data. As you write coded commands and equations inside the MathScript Node, you can be as creative as you like in naming necessary variables within the limitation of the following rules: A valid variable name must begin with a letter and, after that, be composed of only letters, digits, or underscores. Names are case sensitive (i.e., the capitalized version of a particular letter is distinct from its lowercase rendering) and, if desired, can be very long (up to a maximum of about 60 characters). Long variable names, of course, have the disadvantage of occupying considerable diagram space. For our present work, we require two variables: one to denote the argument of the sine function and the other to represent the sine-wave value. Let's conserve valuable diagram real estate by giving these two variables uncreative short names—how about x and y, respectively.

Using the 🖑 or Ⓐ, click on the interior of the MathScript Node and enter the following two lines of text-based code:

$$x = 0 : 1/5 : 19.8$$

$$y = \sin(x)$$

This code will create the two desired 1D arrays x and $y=\sin(x)$. After typing it, place the cursor outside the MathScript Node and then click the mouse to secure your entry.

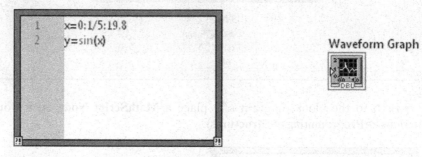

Next, we need to output y, the 1D array of sine-wave values, from the MathScript Node for plotting. To create this output, first place the ⬉ at a location on the right border of the MathScript Node.

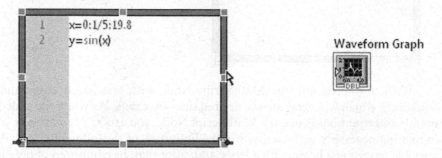

Then pop up on this border and select **Add Output>>y** from the MathScript Node's pop-up menu.

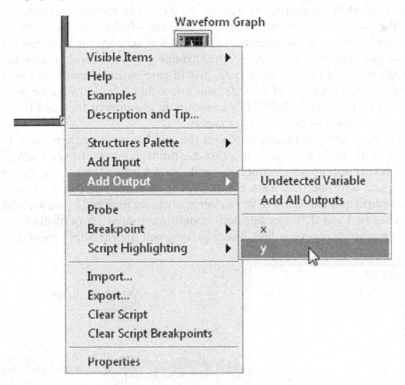

An automatically labeled indicator for this *y* output will appear in the form of a small rectangular box perched on the MathScript Node's border (for earlier versions of LabVIEW, the *y* label must be added manually using the).

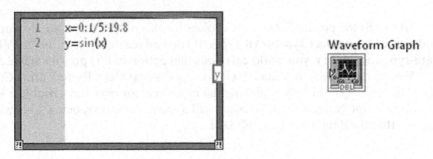

A variety of data types are possible for output terminals at the border of a MathScript Node (e.g., real and complex scalars, real and complex arrays). By default, LabVIEW automatically assigns the data type for such terminals based on how the associated variables are used inside the MathScript Node. In making its automated data-type assignment, LabVIEW will account for all possible values of the output variable. For example, the variable $z = \sqrt{x}$ would be assigned a complex (as opposed to real) data type, just in case its argument x is ever negative. However, if in your use of $z = \sqrt{x}$ you know that x will always be positive, you might wish to override LabVIEW's automatic assignment to avoid the complication of working with complex data types.

On your block diagram, note that the y indicator is orange; thus, LabVIEW has correctly surmised that y is a floating-point numeric. To see LabVIEW's full assignment, pop up on the output y and select **Choose Data Type**. In the pop-up menu, you will find that **Auto Select Type** is selected by default (denoted by the check mark) and that LabVIEW's automated choice (marked by an asterisk) is **DBL 1D**. That is, LabVIEW has selected a one-dimensional array of double-precision floating-point numerics, a fine choice to represent the variable $y=\sin(x)$.

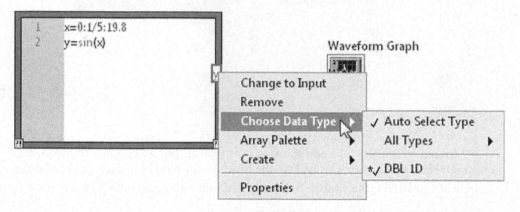

To see all the possible data-type choices for this output terminal, pop up on it and select **Choose Data Type>>All Types**. If you had reason to override LabVIEW's data-type choice for y, you could carry out that action in this pop-up menu.

You can also check the data type of y using Context Help. By activating Context Help via *<Ctrl+H>* and then placing the mouse cursor over the variable y within the MathScript Node, a Help Window will appear. As shown below, this window reports that the data type of y is **1D DBL**.

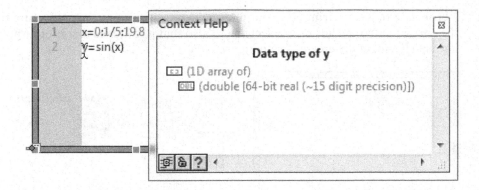

In addition, Context Help can be used to obtain useful information about other items within the MathScript Node. For example, by placing the mouse cursor over *sin(x)*, the Help Window describing this MathScript function will be activated as shown next.

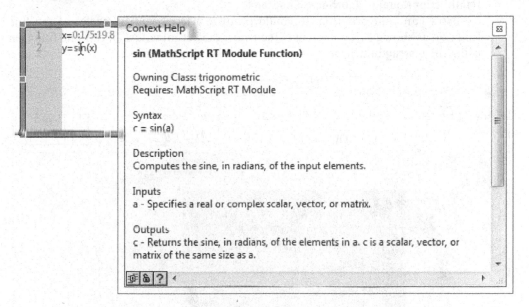

Finally, complete the block diagram as shown below. **Bundle** is found in **Functions>>Programming>>Cluster, Class, & Variant**. The y array indicator can be created with **Create>>Indicator** after popping up on the thick array wire. If you get a broken wire when wiring the *y* output, pop up on this object and make sure that the correct data type has been selected and also that it is indeed an output (that is, it wasn't accidentally created as an input).

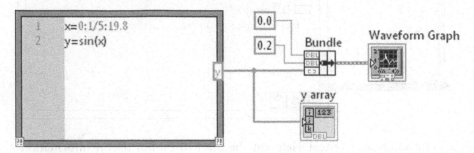

Return to the front panel. Using **File>>Save**, create a folder named **Chapter 4** within the **YourName** folder, and then save this VI under the name **Sine Wave Graph (MathScript Node)** in **YourName\Chapter 4**.

Now run your program. It should produce a few beautiful cycles of the sine function as shown below. You may then view the actual values of the sine function using the **y array** indicator.

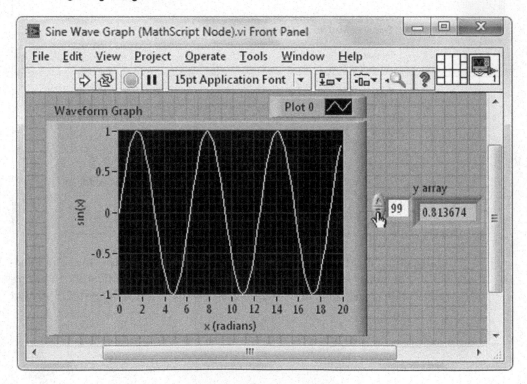

If interested, you can view the arrays produced by this VI using the MathScript Probe. To activate this tool, go to the block diagram, pop up anywhere within the MathScript Node, and select **Probe**.

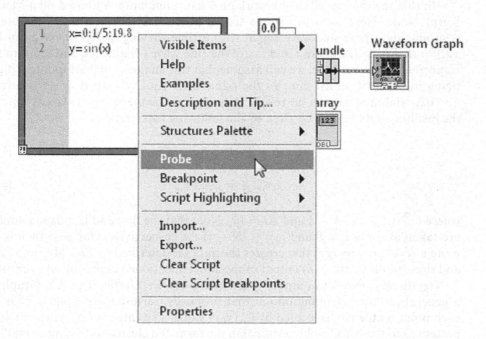

After the MathScript Probe Window appears, run your VI. Then explore all of the array information and plotting features this tool makes available.

4.3 WAVEFORM SIMULATOR USING A MATHSCRIPT NODE AND XY GRAPH

> **Formula Node Users:** Build the front panel described in this section, then refer to Appendix A for instructions on writing the Formula Node-based version of the block diagram constructed in Sections 4.3–4.4. Run your VI as described in Section 4.5.

For our studies ahead, we will need a data simulator VI, that is, a program that outputs digitized data of the type produced by a data acquisition device in a real experiment. In a typical laboratory experiment, a time-varying analog (continuous) signal $x(t)$ is input to an instrument, which is programmed to measure ("sample") this signal at N equally spaced times t_i. If Δt is the time interval between one sample and the next, then $t_i = i\,\Delta t$, where $i = 0, 1, 2, \ldots, N-1$. Since there will be one data sample taken per time interval Δt, the *sampling frequency* is defined to be

$$f_s = \frac{1}{\Delta t} \tag{4.1}$$

In this section, we will begin work on a data simulation VI based on a Math-Script Node. Here, we will assume that a physical system under investigation is producing a pure sine-wave signal with a time-varying *displacement* given by $x(t) = A \sin(2\pi f t)$, where f and A are the sine wave's frequency and amplitude, respectively. Furthermore, we will assume that this analog signal is input to a digitizing instrument, which samples the signal at N equally spaced times at a rate $f_s = 1/\Delta t$ and then outputs the result. Defining x_i to be the sample taken at time t_i, the resulting data set will be given by the following two relations:

$$t_i = i\,\Delta t \tag{4.2}$$

$$x_i = A \sin\left(2\pi f t_i\right) \tag{4.3}$$

where $i = 0, 1, 2, \ldots, N - 1$ and $\Delta t = 1/f_s$. Note that the first and last data samples are taken at times $t_0 = 0$ and $t_{N-1} = (N - 1)\,\Delta t$, respectively. Our goal then is to write a software program that creates the data set described by Eqs. [4.2] and [4.3] and thus simulates the (t, x) output of the above-mentioned digitizing equipment.

To display your VI's output, you will use an XY Graph. The XY Graph is a general-purpose graphing option that produces Cartesian-style plots. That is, each point on the plot is located by its (x, y) value. The entire set of (x, y) values is presented to the XY Graph's terminal in the form of a cluster consisting of two 1D arrays of numeric values. The first and second arrays in the cluster correspond to the sequence of x- and y-values, respectively.

Open a new VI and place an **XY Graph** (from **Controls>>Modern>>Graph**) on its front panel. Label the x- and y-axes as **Time** and **Displacement**, respectively. Autoscaling will be activated on each axis by default. Next, place four **Numeric Controls** on the front panel labeled **Number of Samples**, **Sampling Frequency**, **Frequency**, and **Amplitude**, respectively. By default, the representation for these Numeric Controls is the double-precision floating-point data type **DBL**. Pop up on **Number of Samples** and change its data type to long signed integer by selecting **Representation>>I32**. Pop up on each of the other Numeric Controls and make sure that (its default) **Representation>>DBL** is selected. Finally, save this VI under the name **Waveform Simulator** in **YourName\Chapter 4**.

As an aid in arranging the front-panel objects into a pleasing pattern, you might enjoy experimenting with the **Alignment Tool** 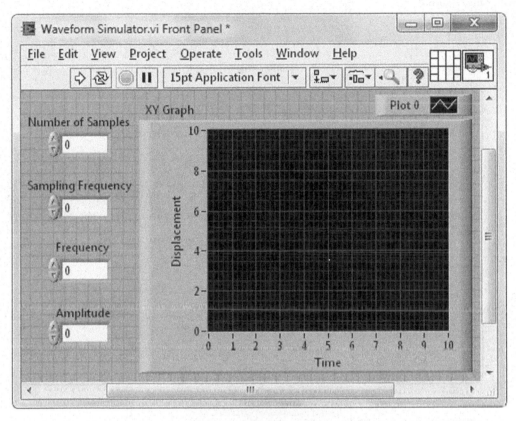 and **Distribution Tool** in the toolbar. To use these tools, you must first use the ⬎ to highlight the group of objects on which you wish to operate.

Switch to the block diagram and write MathScript-based code that evaluates Eqs. [4.1] through [4.3]. To produce the two arrays describing a time-varying sine wave's *x*- and *y*-coordinates, which we call *t* and *x*, respectively, the required code within the MathScript Node is as follows:

$$delta\_t = 1/f\_s$$
$$start = 0$$
$$step = delta\_t$$
$$stop = (N - 1)*delta\_t$$
$$t = start : step : stop$$
$$x = A * \sin(2 * pi * f * t)$$

After inputting this code, your block diagram should appear as shown next.

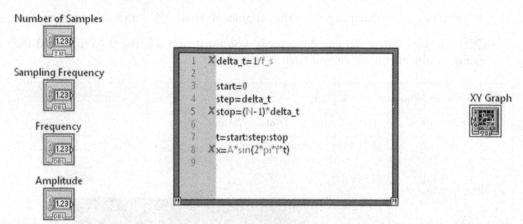

A ✖ appears at each line that involves an input variable that we have yet to assign. First, create an input for the variable *N*, which represents the value of **Number of Samples** within the MathScript Node, by popping up on the MathScript Node's left-hand border and selecting **Add Input**.

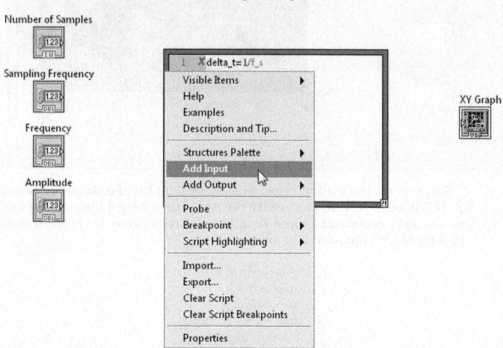

An indicator for this input will appear with its interior highlighted, awaiting your character-based name for this equation variable. If this highlighting disappears (because of an inadvertent click of the mouse), use the 🖑 to restore access to the input's interior. Label the input N and then wire it to the **Number of Samples** terminal. The input's indicator will turn blue, indicating that it has been formatted to receive the **(I32)** integer data from **Number of Samples**. Also, at the line containing N, the ✖ disappears.

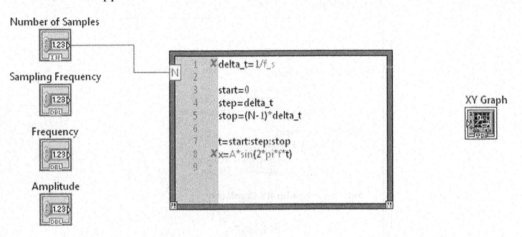

Similarly, wire the three other inputs that carry the values of **Sampling Frequency**, **Frequency**, and **Amplitude**, labeling them f_s, f, and A, respectively. Each of these (orange) inputs is a double-precision, floating-point numeric. Finally, create two outputs for the variables t and x at the right-hand border of the MathScript Node, as shown.

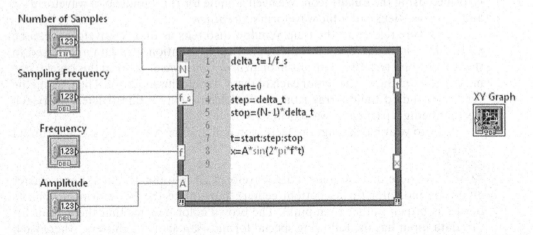

4.4 CREATING AN *XY* CLUSTER

To determine what type of input the XY Graph requires, activate its Help Window

by placing the mouse cursor over the XY Graph's icon terminal and pressing *<Ctrl+H>*. The Help Window will appear as shown next.

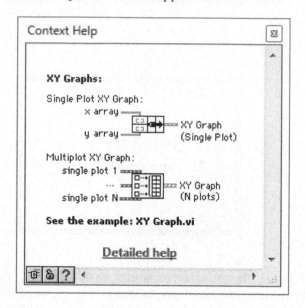

This Help Window indicates that to plot a single waveform, the XY Graph expects a cluster as input. This cluster is to contain the waveform's x and y arrays and is formed using the **Bundle** icon. We need a name for this bundle of a waveform's x and y arrays—let's call it the waveform's *xy cluster*.

For future reference, the Help Window also tells us that N waveforms can be plotted simultaneously on an XY Graph. In this situation, the data are passed to the XY Graph's terminal in the form of an *array of clusters*. To form this object, one must first form the *xy* cluster for each of the N waveforms as above. Then, using the yet-to-be-studied **Build Array** icon, an N-element array is constructed with each element being a particular waveform's *xy* cluster.

Refer to your block diagram. Here, you will find the XY Graph's icon terminal . We note that, similar to the Waveform Graph, this terminal has a cluster data-type indicator (in its bottom center), but unlike the Waveform Graph, its border is brown (rather than pink). The brown color denotes that the default for this data input has the following special format—an array of clusters, where each

142

cluster is composed of only scalar numerics. In our use of the XY Graph, we will follow the recommendation of its Help Window by inputting a cluster of numeric arrays, a format that is more general than the default.

Complete the block diagram wiring as shown below. Use the **Bundle** icon (found in **Functions>>Programming>>Cluster, Class, & Variant**) to form the sine wave's *xy* cluster. Note that when this cluster is wired to the XY Graph's icon terminal, the icon's border changes from (the specialized default cluster) brown to (the general cluster) pink. The Time and Displacement array indicators can be created by popping up on the appropriate wires and selecting **Create>>Indicator**. If an error occurs while coding a MathScript Node, click the broken **Run** button to display the **Error List** window.

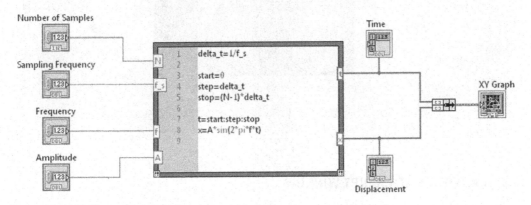

4.5 RUNNING THE VI

Return to the front panel, position the Time and Displacement indicators nicely, and then save your work. Set Number of Samples and Sampling Frequency equal to *100* and *1000*, and Frequency and Amplitude to *50* and *5*, respectively. These choices for the parameters simulate sampling a 50 Hz sine wave for 0.1 s (100 samples at a rate of 1000 samples per second) and so should result in outputting five cycles of the sine wave. Run your program, and if all goes well, you will see the five cycles. Marvel at the waveform you have created and reflect on how easy it would be to create other shapes by simply changing the equations within the block diagram's MathScript Node (which is exactly what you will do in a few moments).

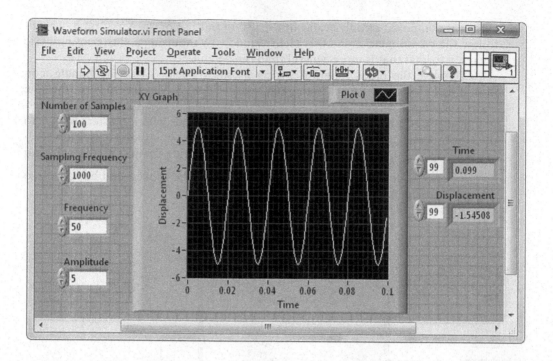

4.6 LABVIEW MATHSCRIPT WINDOW

> **Formula Node Users:** This section relates exclusively to MathScript features.

An invaluable tool while developing MathScript-based code is the *LabVIEW MathScript Window*. Let's take a moment to learn how to use this tool. With a VI open or else in the Getting Started Window, open the LabVIEW MathScript Window by selecting **Tools>>MathScript Window...** When the window appears, concentrate your attention on its left side. In the lower left, you will find the *Command Window* box where MathScript commands can be entered using the keyboard. Once a command is loaded into the Command Window, it can be executed by pressing *<Enter>* on your keyboard, and the result of this executed command is displayed directly above in the *Output Window*.

Let's test-drive the LabVIEW MathScript Window by executing commands similar to the ones just used in constructing **Waveform Simulator**. In the Command

Window, type *t=0:0.01:0.99* and then press *<Enter>*. In the Output Window, you will first see a copy of the command that you typed, followed by the result of its execution, which in this case is an array of 100 equally spaced *t*-values ranging from *t*=0 to *t*=0.99.

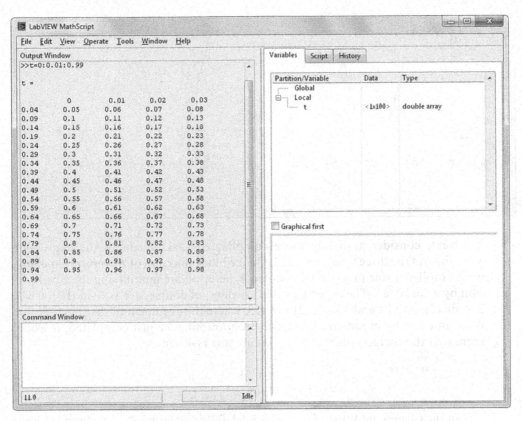

Next, in the Command Window, enter (i.e., type the command and then press *<Enter>*) *x=5*sin(2*pi*4*t)*. In the Output Window you will find the sine wave with frequency 4 and amplitude 5 sampled at the 100 equally spaced *t*-values created by the previous command.

Finally, to observe this sine wave, enter *plot(t,x)* in the Command Window. A window with a plot of *x* vs. *t* will then appear as shown in the following illustration, displaying four cycles of the given sine wave.

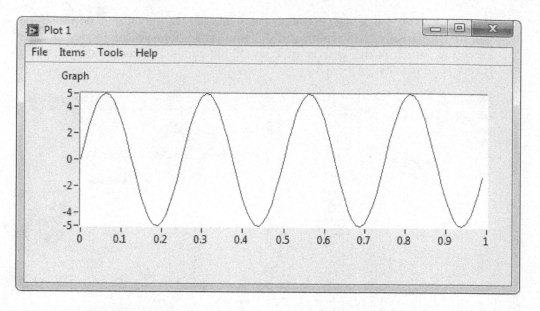

Next, consider a slightly more complicated task: Evaluate the expression $x = t\sin(t)$ at the three times $t = 1, 2, 3$. On the block diagram of **Waveform Simulator**, we multiplied a sine function by a scalar A; here, we are multiplying the sine function by a variable t. The expected three x-values, of course, are $1 \times \sin(1) = 0.841$, $2 \times \sin(2) = 1.819$, and $3 \times \sin(3) = 0.423$, where the argument of the sine function is assumed to be in radians. From the programming we just completed, it would seem that the correct commands to execute this task are

$$t = 1 : 1 : 3$$
$$x = t * \sin(t)$$

In the Command Window, enter *t=1:1:3*, followed by *x=t*sin(t)*. In the Output Window, somewhat surprisingly, you will receive an error message related to the incompatibility of the matrices involved in the requested operation. To understand the problem here, we must delve a bit deeper into the syntax of the MathScript language. In executing the commands you just input, MathScript produces the following two row vectors:

$$t = [1 \quad 2 \quad 3]$$
$$\sin(t) = [\sin(1) \;\; \sin(2) \;\; \sin(3)] = [0.841 \quad 0.909 \quad 0.141]$$

Then, in the syntax of MathScript, the command $t * \sin(t)$ requests that these two row vectors be multiplied together as matrices. That is, the command is interpreted as the following matrix multiplication operation:

$$[1 \quad 2 \quad 3] \times [0.841 \quad 0.909 \quad 0.141]$$

As you know, multiplication of two 1×3 matrices is an invalid operation; hence, the error message in the Output Window. To demonstrate a valid matrix operation (which is not the task we are trying to accomplish), try entering the command $x=t*transpose(\sin(t))$. The operation executed will then be the following valid matrix multiplication of a 1×3 row vector times a 3×1 column vector:

$$x = [1 \quad 2 \quad 3] \times \begin{bmatrix} 0.841 \\ 0.909 \\ 0.141 \end{bmatrix} = (1)(0.841) + (2)(0.909) + (3)(0.141) = 3.08$$

Instead of the above matrix multiplication operation, we want MathScript to take the three-element row vectors t and sin (t) and form a three-element row vector, whose ith element is the product of the ith elements of original row vectors. This operation is called *element-wise multiplication* and its MathScript operator symbol is a period followed by an asterisk (.*). In the Command Window, enter $t=1:1:3$, followed by $x=t.*\sin(t)$. Remember to include the all-important period before the asterisk. You will find that the commands accomplish our originally stated task of evaluating $x=t \sin(t)$. Similarly, element-wise division and power operations can be executed by the operators ./ and .^, respectively.

The LabVIEW MathScript Window is also a handy source of help regarding use of MathScript functions. These (800 or so) functions are cataloged into over 45 "classes" (e.g., basic mathematical functions, statistics, digital signal processing). To see a list of all of the classes, enter *help classes* in the Command Window. To then see a list of all of the functions in, say, the *dsp (digital signal processing)* class, enter *help dsp* in the Command Window. There, you will find the available dsp functions, which include many that perform complex filtering, waveform generation, and spectral analysis operations.

To get some experience with a MathScript function, let's try using **Periodic Signal Generator**, which is called *gensignal* for short. Enter *help gensignal* in the Command Window. The help information that appears states that the *gensignal* command generates two arrays called x and t, where t is an array of time values starting at $t=0$ and stopping at $t = duration$ with the time step $\Delta t = interval$ and x is an array of associated periodic signal values that has the functional form and period given by the parameters *type* and *period*, respectively. The possible values for *type* are *'sin'*, *'square'*, and *'pulse'*. The syntax of the command is

$$[x, t] = gensignal(type, period, duration, interval)$$

If you only need the x array of associated periodic signal values, the command becomes

$$x = gensignal(\,type, period, duration, interval)$$

Okay, let's try using *gensignal* to generate the same signal that we produced a few moments ago, which was four cycles of a sine wave with a frequency and amplitude of 4 and 5, respectively. For a sine-wave frequency of 4, the *period* is $1/4 = 0.25$. For the time variable, we want the *duration* and *interval* to be 0.99 and 0.01, respectively. So, in the Command Window, enter *[x, t] = gensignal('sin', 0.25, 0.99, 0.01)*, and then you will find that both the x and the t arrays will be listed in the Output Window. To see a plot of these variables, enter *plot(t,x)* in the Command Window. In the plot window that appears, you should see the expected signal, except the amplitude is 1, rather than the desired 5. To remedy this problem, enter *x=5*x* in the Command Window, followed by *plot(t,x)*.

As a demonstration of the power of the *gensignal* command, produce a square wave by simply changing the command entered in the Command Window to *[x, t] = gensignal('square', 0.25, 0.99, 0.01)*. Plot the result. Next, plot a periodic pulse using the command *[x, t] = gensignal('pulse', 0.25, 0.99, 0.01)*. Finally, produce a single x-array of square-wave values with amplitude 5 using the command *x=5*gensignal('square', 0.25, 0.99, 0.01)*.

Another useful MathScript function is *linramp(a,b,N)*, which generates N equally spaced samples between the lower value a and upper value b. Show that the command *t=linramp(0, 0.99, 100)* generates the same array of values as the command *t=0:0.01:0.99*. Also, show that a 100-element array, whose elements all equal 5, can be generated by *t=linramp(5, 5, 100)*. Such an array could describe, for example, the DC voltage output of a battery at 100 equally spaced times. In the program you will soon write, we will call such a constant-valued array a *DC Level* waveform.

4.7 ADDING SHAPE OPTIONS USING AN ENUMERATED TYPE CONTROL

> **Formula Node Users:** Through the end of the chapter, follow all of the instructions related to building the front panel of Waveform Simulator and the parts of the block diagram outside of the MathScript Node. Refer to Appendix A for the Formula Node-based version of the block diagram constructed in Section 4.8.

We are now positioned to turn **Waveform Simulator** into a versatile data simulator VI that can output a variety of digitized waveforms. In particular, we will program this VI to create five possible waveform shapes: *Sine, Cosine, DC Level, Square,* and *User-Defined.* So that a user can select the desired waveform, let's add a fifth control called **Shape** to the front panel of **Waveform Simulator**.

For **Shape**, we will use an *Enumerated Type Control* (nicknamed an *Enum*). Go to **Controls>>Modern>>Ring & Enum**, select an **Enum**, and then place it on the front panel and label it **Shape**. Use the ⬉ to perform any necessary resizing and/or positioning.

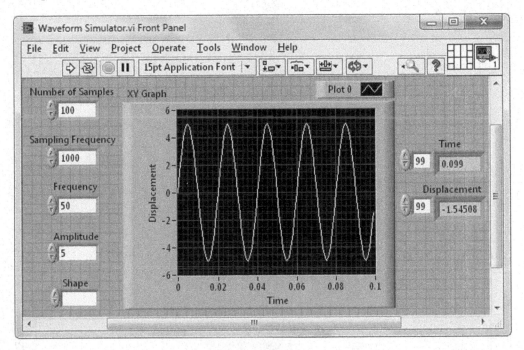

In **Controls>>Modern>>Ring & Emun**, you will find several types of *"ring"* style controls, each of which is useful when a programmer needs to present the user with a catalog of options. The **Enum** is a special type of ring control that links a sequence of text messages with a series of associated integers. By default, the data type of these integers is unsigned 16-bit (**U16**). When the user selects a particular text message on the front-panel control, its associated integer is passed to a terminal on the block diagram. In our present circumstance, the **Enum** control will provide a handy method for selecting a desired waveform shape from among the available options.

The Enum is programmed with its available options as follows. Pop up on the **Enum** and select **Edit Items....** In the dialog window that appears, type *Sine <Enter> Cosine <Enter> DC Level <Enter> Square <Enter> User-Defined*, and then click the **OK** button. If you press *<Enter>* after typing *User-Defined*, you will accidentally create a sixth option, which can be erased by clicking on the **Delete** button.

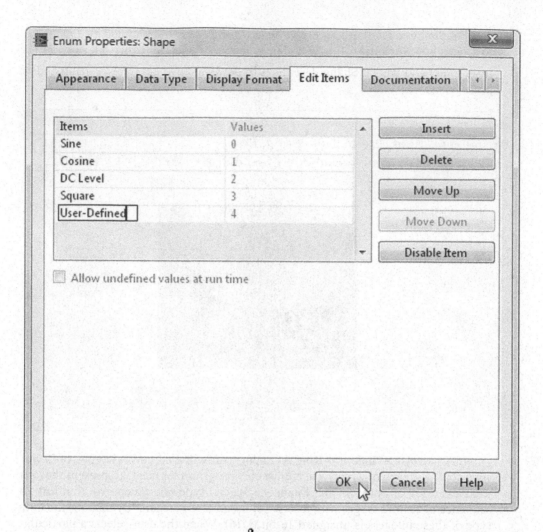

On the front panel, use the 🖑 to view the text message (Sine, Cosine, DC Level, Square, User-Defined) for each of the Enum's five programmed options. You can resize the Enum using the ⬉. To make visible the integer associated with each of these text messages, pop up on the **Enum** and select **Visible Items>>Digital Display**. After viewing the Enum's Digital Display, hide it again (by deselecting **Visible Items>>Digital Display**) and then save your work.

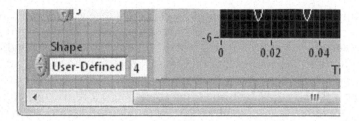

4.8 FINISHING THE BLOCK DIAGRAM

> **Formula Node Users:** Refer to Appendix A for the Formula Node-based version of the block diagram constructed in this section.

Switch to the block diagram and there you will find the **Shape** icon terminal, which is blue, denoting the (**U16**) integer data type emanating from this Enum control. Pop up on the left-hand border of the MathScript Node, select **Add Input** and label it *s*, and then complete the code as shown below.

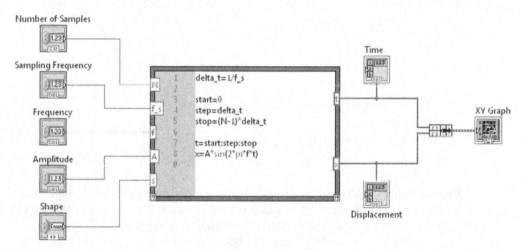

Finally, program the MathScript Node to produce the desired waveform alternatives. To accommodate the five possible options, we will implement MathScript's *if–elseif statement* syntax. For this type of statement, a logical condition is tested (in our situation, the equivalence of s to, say, zero via the command $s == 0$; note that two equal signs denotes a test of logical equivalence whose outcome is either a Boolean TRUE or FALSE), and if the condition is TRUE, a given command (or list of commands) is executed. The code we require is as given below. For the User-Defined option, we have chosen the sum of two particular sinusoids as its default. A user can then modify this choice to produce any desired waveform shape.

$$delta\_t = 1/f\_s$$
$$start = 0$$
$$step = delta\_t$$
$$stop = (N - 1)*delta\_t$$
$$t = start:step:stop$$
$$if\ s == 0$$
$$\qquad x = A*\sin(2*pi*f*t)$$
$$elseif\ s == 1$$
$$\qquad x = A*\cos(2*pi*f*t)$$
$$elseif\ s == 2$$
$$\qquad x = linramp(A,A,N)$$
$$elseif\ s == 3$$
$$\qquad x = A*gensignal(`square', 1/f, stop, step)$$
$$elseif\ s == 4$$
$$\qquad x = 4.0*\sin(2*pi*100*t) + 6.0*\cos(2*pi*200*t)$$
$$end$$

Program the MathScript Node with the above code and then click outside the Node to secure it. Note that the t and x outputs are each assumed to supply 1D DBL arrays to the block-diagram objects to which they are wired.

In my version of LabVIEW, this MathScript Node coding resulted in a bad wire at the x output. Popping up on this output, we see that the source of this problem is that LabVIEW's automatic assignment for the variable x is **DBL 2D**, rather than the expected **DBL 1D**.

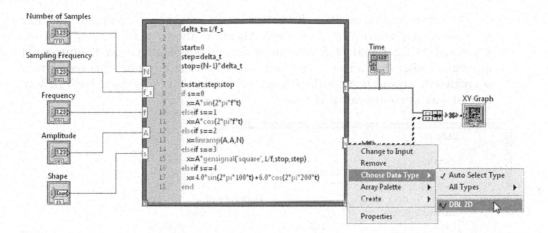

At this point, because we know that the data type for *x* should rightfully be a one-dimensional double-precision floating-point array, we may choose to override LabVIEW's assignment by popping up on the *x* output and selecting **Choose Data Type>>All Types>>1D-Array>>DBL 1D**. A red coercion dot will result, indicating that we have chosen a different data type than LabVIEW would auto-select. This coercion dot is there to strike some fear in our hearts, reminding us that LabVIEW's choice might be smarter than ours.

To discover what has motivated LabVIEW's choice this time, activate Context Help via *<Ctrl+H>*, and then place the mouse cursor over the *x* variable in the several locations it appears within the MathScript Node's *if–elseif* statement. For example, when placed over the line where *x=A*sin(2*pi*f*t)*, we are informed that the data type of *x* is **1D array** and **double**, the designation for a row vector of double-precision floating-point numbers.

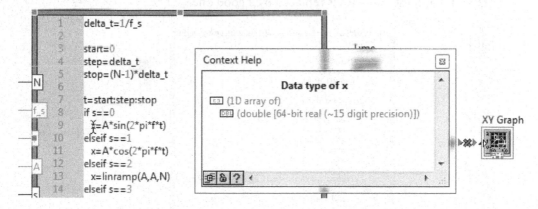

All other instances of x have this same data type, except at the line $x=A*gensignal('square', 1/f, stop, step)$. There, as shown next, we find that the x created by gensignal is a column vector of double-precision floating-point numbers, rather than a row vector. It is this data-type mismatch between how x is created at various lines of the *if–elseif* statement that has caused LabVIEW to format x as a two-dimensional array.

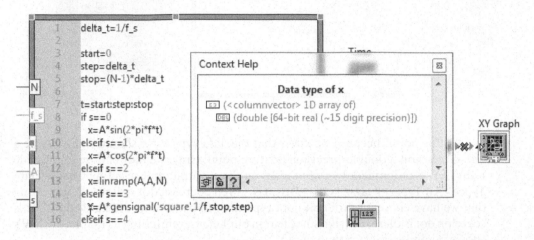

Now that we understand the source of our problem, we can correct it by transposing the column-vector array generated by gensignal so that it is transformed into a row-vector array like all of the other x-related definitions. That is, simply replace $x=A*gensignal('square', 1/f, stop, step)$ with $x=transpose(A*gensignal ('square', 1/f, stop, step))$, and then click somewhere outside the MathScript Node to secure this change. As planned, we find that LabVIEW now auto-assigns x to be **DBL 1D** and the bad wire is replaced by a good one.

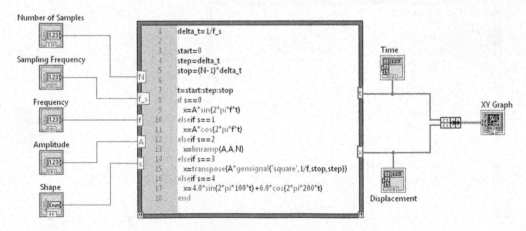

4.9 RUNNING THE VI

Save your work. Return to the front panel; set **Number of Samples** and **Sampling Frequency** to be *100* and *1000*, and **Frequency**, **Amplitude**, and **Shape** as *100*, *5*, and *Sine*, respectively; and then run your program. Assuming that the **Time-axis** is calibrated in seconds, with this choice of parameters, neighboring data samples are separated by a time interval of 0.001 s, and we are simulating an experiment that acquires 100 data samples over the time span from $t=0.000$ to $t=0.099$ s. Thus, for a 100 Hz sine-wave signal whose period is 0.01 s, we expect to observe ten of its cycles on the XY Graph. Each cycle is described by a succession of 10 samples, as can be seen by popping up on the **Plot Legend** and selecting **Point Style>>Solid Dot** (there are two closely spaced samples near the crest as well as near the trough of the sine wave).

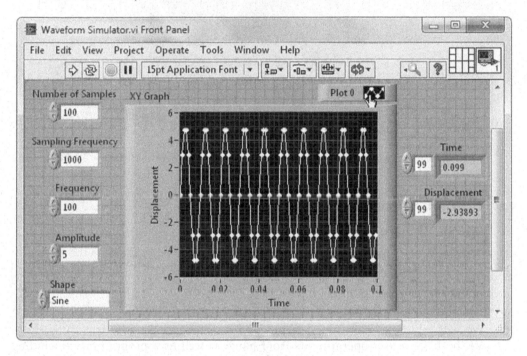

Next, without changing **Frequency**, **Amplitude**, and **Shape**, set **Number of Samples** and **Sampling Frequency** to *100* and *10000*, respectively, and then run your program. With this choice of parameters, neighboring data samples are separated by a time interval of 0.0001 s, so now the 100 samples are acquired over the time span from $t=0.000$ to $t=0.099$ s. Thus, for the 100 Hz sine-wave signal with a period of 0.01 s, we expect to observe only one of its cycles on the XY Graph, where this single cycle is well resolved because it is described by a succession of 100 samples.

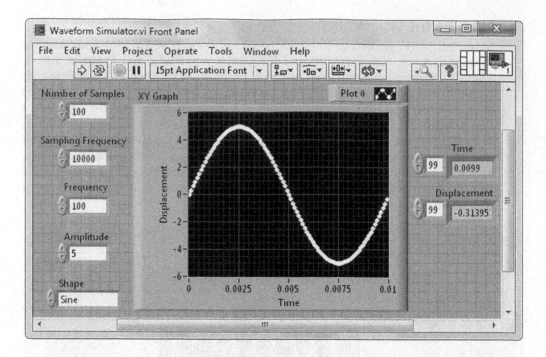

Finally, run your VI with **Shape** set to its other available options and enjoy the results.

4.10 CONTROL AND INDICATOR CLUSTERS

> **Formula Node Users:** After following the clustering methods described in this section, the completed Formula Node-based block diagram is shown in Appendix A.

Let's take this opportunity to add to your bag of LabVIEW programming tricks. We will explore a "clustering" technique that improves the user-friendliness of a VI, both for a user operating its front panel as well as a programmer (including yourself) trying to decipher its block diagram.

We first note that the Numeric Controls on our VI's front panel can be organized into two categories: *Digitizing Parameters* (**Number of Samples, Sampling Frequency**) and *Waveform Parameters* (**Frequency, Amplitude, Shape**). Likewise, both **Time** and **Displacement** Array Indicators are related to the category of *Waveform Output*.

This organizational structure within your program can be made explicit by grouping these controls and indicators on the front panel and block diagram. This feat will be accomplished by creating a cluster associated with each of the three categories through the following steps. In **Controls>>Modern>>Array, Matrix & Cluster**, obtain a **Cluster** shell and place it in an empty region of the front panel near the Numeric Controls. Label this **Cluster** shell **Digitizing Parameters**. Similarly, place another **Cluster** shell near the Numeric Controls and label it **Waveform Parameters**. Finally, place a third **Cluster** shell near the Array Indicators and label it **Waveform Output**.

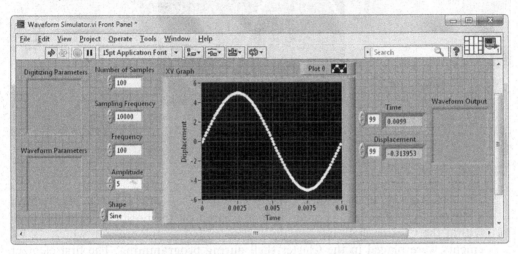

Now, using the ↖, drag the **Number of Samples** and **Sampling Frequency** controls (in that order) into the interior of the **Digitizing Parameter** cluster shell. If needed, the Cluster shell can be resized at its edges using the ↖ or (more simply) by popping up on its border and selecting **AutoSizing>>Size to Fit** in the pop-up menu (if you pop up within the interior of the Cluster shell, a Controls Palette will appear, rather than the desired pop-up menu).

Likewise, drag the **Frequency** control, then the **Amplitude** control, and then the **Shape** control into the interior of the **Waveform Parameters** Cluster shell and the **Time** and **Displacement** indicators (in that order) into the **Waveform Output** Cluster shell.

Finally, using the ↖, arrange the front-panel objects in some pleasing pattern.

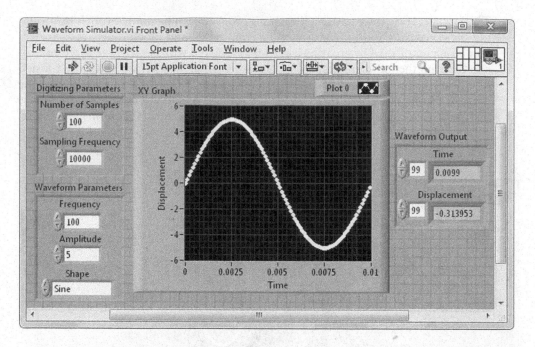

The controls or indicators (which we will generically call *elements*) within a cluster are ordered. This ordering is not based on the position of the elements within the Cluster shell, but initially is determined by the order in which the elements were placed in the Cluster shell during programming. The first element placed in the Cluster shell is indexed 0, the second element is indexed 1, and so on. Thus, for example, if you followed the directions above, the **Number of Samples** and **Sampling Frequency** controls within the **Digitizing Parameters** control cluster are indexed 0 and 1, respectively. To verify this ordering, pop up on the border of the **Digitizing Parameters** control cluster and select **Reorder Controls In Cluster…**

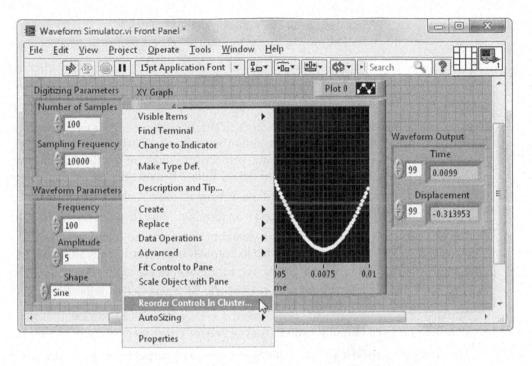

The front panel will change its appearance as shown next.

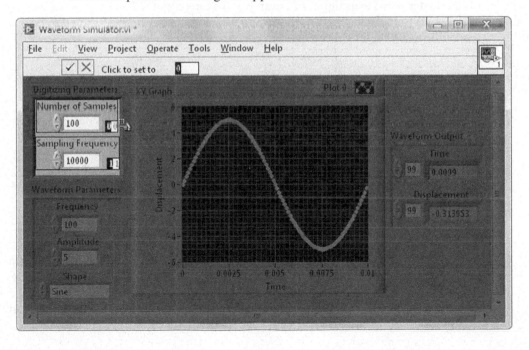

At the bottom right corner of each cluster element, its current *cluster index* is given in the white box. The mouse cursor has morphed into the *cluster order cursor* , which can be used to change the indexing of the cluster elements by clicking on their associated black boxes. Once the new index assignments have been made, they can be secured by clicking on the **Confirm** button or reverted to the original settings with the **Cancel** button . You might wish to experiment with this tool for a few moments. Then, after assigning **Number of Samples** and **Sampling Frequency** as the index 0 and 1 elements, respectively, click on the **Confirm** button.

Switch to the block diagram and select **Edit>>Remove Broken Wires**. From **Functions>>Programming>>Cluster, Class, & Variant**, select **Unbundle By Name** and place it near the **Digitizing Parameters** icon terminal. Wire this icon terminal to the left-hand input terminal of the and when the cluster–wire connection is made, the **Unbundle By Name** icon will morph so that its output terminal (on its right side) contains the name of, and gives access to, the index-zero element in the cluster . Using the , resize (at its bottom center) **Unbundle By Name** so that it has two output terminals, thus giving access to both elements . The particular element appearing in a given output terminal can be changed either by clicking on it with the or by popping up on it and choosing **Select Item**. The output terminal ordering you select for **Unbundle By Name** is not required to have the same ordering as the indexing of the cluster wired to its input: For example, the top output terminal does not have to correspond to the input cluster's index-zero element; it can be programmed to be any one of the input cluster's elements. Complete the wiring as shown below.

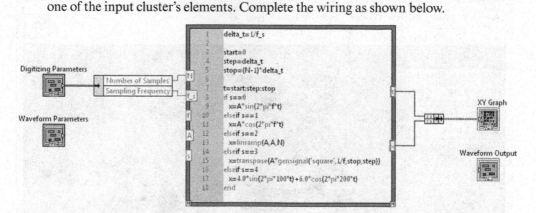

In a similar fashion, use an **Unbundle By Name** icon to wire the elements within the **Waveform Parameters** cluster to the f, A, and s inputs of the MathScript Node. Finally, since the XY Graph requires a cluster to be formed containing the t and x arrays, wire the output of the already present **Bundle** icon to the **Waveform Output** cluster's icon terminal.

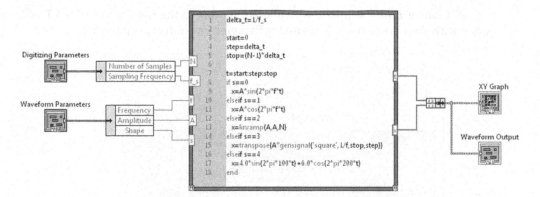

As you can see, clustering enhances the readability of a block diagram, especially because of the labeling that appears within the **Unbundle By Name** icons.

Return to the front panel. Run your VI to check that it functions as expected, and then save your work.

Finally, save the customized front-panel objects that you have created so that you can reuse them in future programs. To save the **Digitizing Parameters** control cluster, pop up on the border of this cluster and select **Advanced>>Customize...** in the pop-up menu.

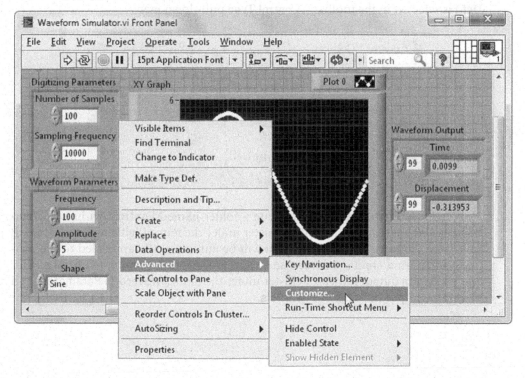

A *Control Editor* window, which looks similar to the front panel of a VI, will open with your custom-made **Digitizing Parameters** control cluster on it.

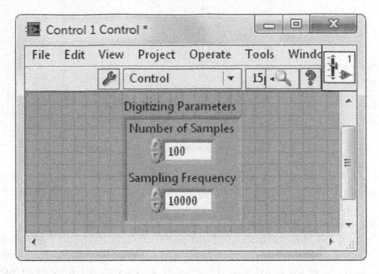

Make sure that the **Control** option (as opposed to **Type Def.** and **Strict Type Def.**) is selected in the toolbar's **Control Type** pull-down menu.

Then use **File>>Save** to first create a folder named **Controls** within the **Your-Name** folder, and save your control cluster under the name **Digitizing Parameters** in **YourName\Controls**. The extension **.ctl** will be automatically appended to the control's name. In a similar manner, save your **Waveform Parameters** control cluster and **Waveform Output** indicator cluster under the names **Waveform Parameters** and **Waveform Output** in **YourName\Controls**.

4.11 CREATING AN ICON USING THE ICON EDITOR

A well-written LabVIEW program is hierarchical in nature. It consists of a top-level VI whose front panel accepts inputs and displays outputs, while its block diagram is constructed from lower-level *subVIs*. These subVIs, which are analogous to a subroutine in a text-based language such as C, may call even lower-level subVIs, which in turn may call still lower-level subVIs. Just as there is no limit to the level of subroutine layering in a C program, there is no limit to the layers of subVIs used in a LabVIEW program. This modular approach to programming makes programs easy to read and debug.

Your program's subVIs may either be taken from LabVIEW's extensive libraries of built-in icons (found in the Functions Palette) or be custom-written by you. It is this latter point on which we now wish to focus our attention. The importance of what you are about to learn is this: Any VI that you write can then be used as a subVI in the block diagram of a higher-level VI.

To use a program as a subVI, it must have an icon to represent it in the block diagram of the higher-level ("calling") VI. There are two steps in creating this icon—designing its appearance and assigning its terminals. Let's learn these skills by creating an icon for our **Waveform Simulator** program.

4.12 ICON DESIGN

First, here is the procedure for designing the icon's appearance. Position the mouse cursor (it doesn't matter which tool it is manifesting at the moment) over the *icon pane* in the upper-right-hand corner of the front panel.

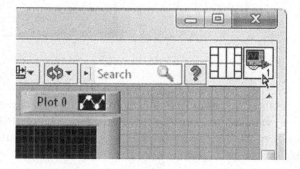

Pop up on the icon pane and select **Edit Icon...** from the pop-up menu.

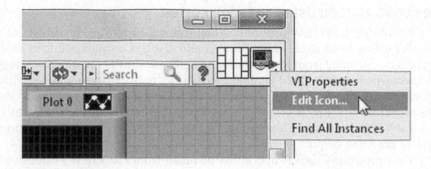

The *Icon Editor* window, shown in the following illustration, will appear. Note the **Icon Text** tab is selected.

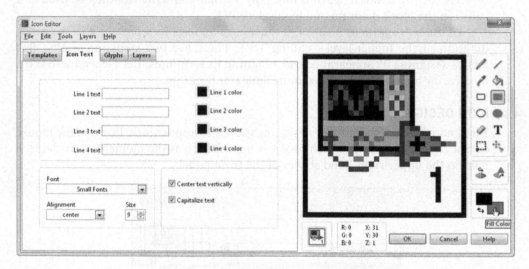

Within this window, you will find a default picture of the icon and a palette of tools that may be used to redesign its appearance. Some of the useful tools in the palette, many of which may be familiar to you from computer-based drawing programs, have the following functions:

Pencil	Draws individual pixels in the **Line Color**.
Line	Draws straight lines in the **Line Color**.
Dropper	Sets the **Line Color** to the color of a pixel you left-click or the **Fill Color** to the color of a pixel you right-click.
Fill	Fills an outlined area with the **Line Color**.
Rectangle	Draws a rectangular border in the **Line Color**. Double-click to add a one-pixel border to the entire icon in the **Line Color**.
Filled Rectangle	Draws a rectangle with a border in the **Line Color** and fills in the **Fill Color**. Double-click to add a one-pixel border to the entire icon in the **Line Color** and to fill the icon in the **Fill Color**.
Eraser	Draws individual pixels as transparent.
Select	Selects an area of the icon to cut, copy, or move. Double-click to select the entire icon.
Text	Enters text into the icon design. Double-click to access font selection dialog window.
Move	Moves all pixels you have selected.
Color	Displays the current **Line Color** and **Fill Color**. Click on each to get a palette from which to choose new colors.

You may wish to explore the use of these tools and create a sophisticated design for your icon (within the constraint of its 32 × 32 pixel size). The following step-by-step description will result in an icon with only rudimentary features.

To begin, click on the **Color** tool and set the **Line Color** and **Fill Color** to black and white, respectively.

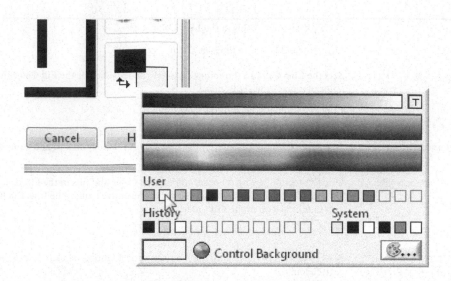

Then place the mouse cursor over the **Filled Rectangle** tool and double-click.

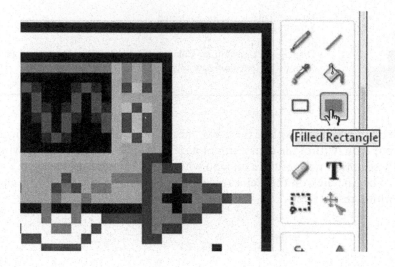

This action will frame the icon in the **Line Color** (black) and fill it with the **Fill Color** (white). You now have a framed blank canvas on which to create your icon design.

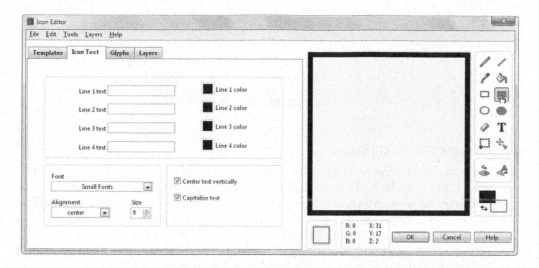

We will now enter the vertically centered text *Wave Sim* in the icon's interior as follows. In the **Line 1 text** box, enter *Wave*; in the **Line 2 text** box, enter *Sim*. Keep the **Center text vertically** box checked, but uncheck **Capitalize text**. Set the font **Size** to *11*.

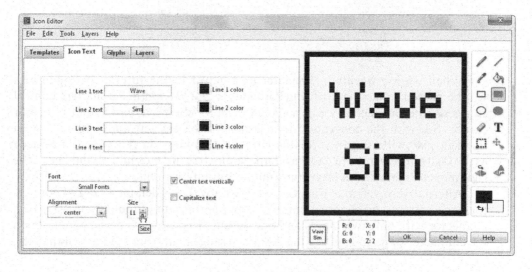

Click on the **OK** button to save the icon design. The Icon Editor window will close, returning you to **Waveform Simulator**'s front panel, with your new icon design now in the icon pane.

4.13 CONNECTOR ASSIGNMENT

Second, here's how to assign the icon's *connector pane* terminals. In the latest versions of LabVIEW, the connector pane, which consists of a pattern of rectangular terminals, is permanently visible just to the left of the icon pane; in earlier LabVIEW versions, the connector pane is accessed by popping up on the icon pane and selecting **Show Connector**. By default, LabVIEW chooses the 4×2×2×4 pattern of terminals shown in the connector pane below. We now wish to associate each of our program's inputs and outputs to a particular terminal in this pattern.

When we use **Waveform Simulator** as a subVI in the future, two quantities will be passed to it—the **Digitizing Parameters** and **Waveform Parameters** control clusters—and one quantity will be taken from it—the **Waveform Output** indicator cluster. So, with the convention that inputs are on the left and outputs are on the right, we will assign two of the leftmost terminals on the 4×2×2×4 pattern as the **Digitizing Parameters** and **Waveform Parameters** input terminals and one of the rightmost terminals as the **Waveform Output** output terminal.

Alternatively, you may wish to override LabVIEW's default 4×2×2×4 connector pane pattern and choose a pattern more appropriate to the present VI, say, a connector with only two inputs and one output. To change its pattern, pop up on the connector pane and choose the desired arrangement from the **Patterns** palette.

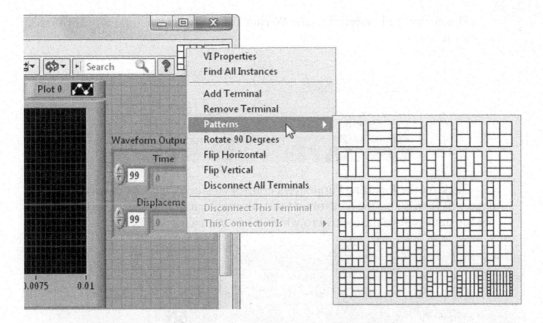

If the exact input/output pattern you desire isn't directly on the palette, you can create it by first selecting a related pattern and then applying some operation from the pop-up menu such as **Flip Horizontal**. Or, you can create your own pattern of inputs and outputs by selecting **Add Terminal** and/or **Remove Terminal** in the pop-up menu (you might take a few moments to explore the operation of these two commands).

In this book, I will always use LabVIEW's default 4×2×2×4 connector pane pattern. Many LabVIEW experts advocate this approach because it promotes uniformity between subVI icons and, as in our present case, leaves some unused terminals that may be handy if the VI is subsequently expanded. Feel free to deviate from this policy if some pattern from the **Patterns** palette better pleases you.

Now assign the **Digitizing Parameters** control cluster to one of the pattern's left-hand terminals through the following steps. Place the mouse cursor (it doesn't matter which tool it is manifesting) over one of the left-hand terminals and then click.

The cursor will morph into the Wiring Tool and the terminal will turn dark, as shown below.

Click on the **Digitizing Parameters** control cluster. A moving dashed-line marquee will then frame the control and the selected terminal will be colored, indicating the data type of the control to which it has been assigned.

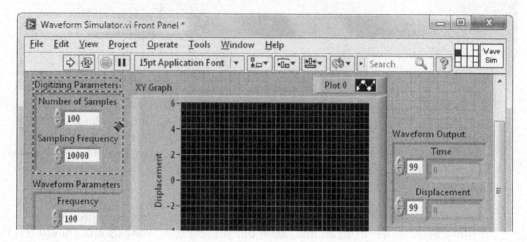

In a similar manner, assign another left-hand terminal to the **Waveform Parameters** control cluster and one of the right-hand terminals to the **Waveform Output** indicator cluster, respectively. The connector pane will then appear as below.

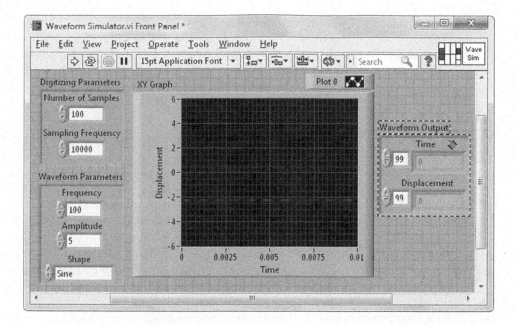

End the assignment procedure by clicking on a blank region of the front panel and then saving your work. If you have an earlier version of LabVIEW, you can return to the icon pane by popping up on the connector pane and selecting **Show Icon**.

To demonstrate the success of your icon creation, place the mouse cursor over the icon pane and activate the Help Window via *<Ctrl+H>*.

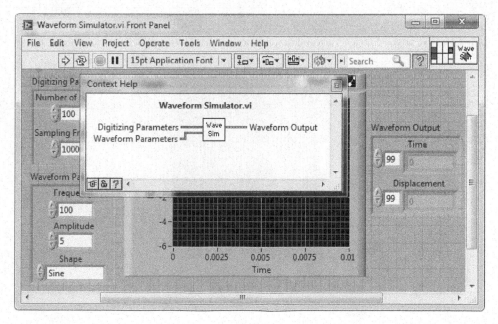

Congratulations—you've now completed your first custom-made VI that, if desired, can be used as a subVI in another program. Using **File>>Save**, store this final version of **Waveform Simulator** in **YourName\Chapter 4**. In later chapters (e.g., see Section 6.6), we'll find out how to place a custom-made VI as a subVI on the block diagram of a higher-level program.

DO IT YOURSELF

AM Wave.vi The displacement x of an amplitude-modulated (AM) wave obeys the following relation:

$$x = A\left[1 + \sin\left(2\pi f_{mod}t\right)\right] \sin\left(2\pi f_{sig}t\right)$$

where A is the amplitude, and f_{mod} and f_{sig} are the modulation and signal frequencies, respectively. Write a MathScript (or Formula) Node-based program that produces and plots an N-element array of displacement values for this AM wave. If using a MathScript Node, remember to employ element-wise operators when appropriate; Formula Node users, take care to end each equation with a semicolon. The front panel and connector assignment for your VI should appear as shown below.

Run your program with $A=1$ and some sensible choice of values for f_{mod} and f_{sig}. For a typical AM wave, $f_{sig} \gg f_{mod}$, so you might try $f_{sig} = 10 f_{mod}$. (This program will be used as a subVI in Problem 5 of Chapter 12.)

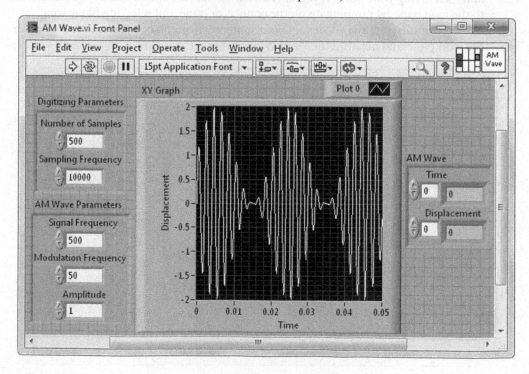

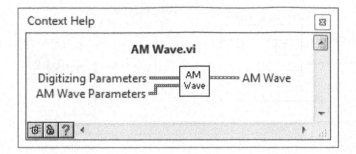

USE IT!

Built-In Icons for Waveform Simulation Within the Functions Palette, there are many icons for use in creating waveforms of various shapes. The following program implements **Basic Function Generator.vi** and, through the **Signal Type** Enum, this VI allows a user to produce four differently shaped waveforms (sine, square, triangle, sawtooth).

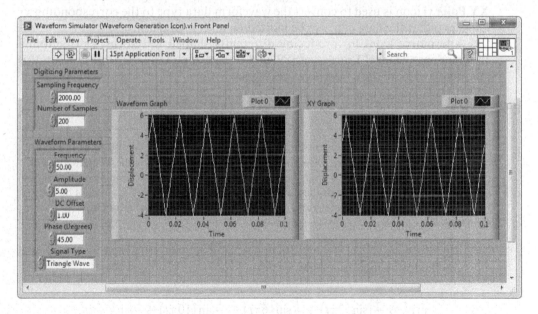

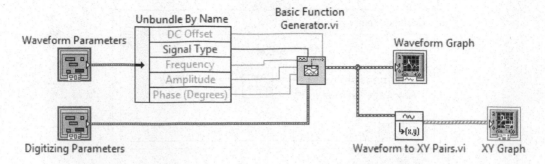

Waveform Parameters

Unbundle By Name

DC Offset

Signal Type

Frequency

Amplitude

Phase (Degrees)

Basic Function
Generator.vi

Waveform Graph

Digitizing Parameters

Waveform to XY Pairs.vi   XY Graph

BasicFunctionGenerator.viisfoundinFunctions>>SignalProcessing>>Waveform
Generation and, like all of the icons found in that palette, its output is formatted as the *waveform* data type (see Chapter 3, Problem 8, and Section 13.3). Besides the *y*-axis values that determine the waveform's shape, this data type also
includes timing information. Thus, as shown, a waveform wire connected directly
to a **Waveform Graph** produces a plot with the *x*-axis calibrated properly, where it
is assumed that the *x*-axis quantity is *time*. In the above program, the **Waveform to
XY Pairs.vi** icon is used to convert the waveform data type to the corresponding *xy*
cluster so that the data can also be plotted on an **XY Graph**.

Alternative icons for producing differently shaped waveforms can be found
in **Functions>>Signal Processing>>Signal Generation**. Also, see Problem 7 in this
chapter to explore the use of **Simulate Signal**.

PROBLEMS

> For each problem that involves writing a new program, the problem statement begins
> with a suggested descriptive name (including the **.vi** extension) for the VI that you will
> write; icons needed for the VI may be found with the aid of **Quick Drop**. Problems that
> require a MathScript Node or the LabVIEW MathScript Window (i.e., cannot be com-
> pleted using a Formula Node) are tagged with the notation **MathScript Only**.

1. **Square Wave.vi** A unity-amplitude 1 Hz square wave $y(t)$ can be approximated
    by the following sum of sine waves:

$$y(t) \approx \frac{4}{\pi}\left[\sin(2\pi t) + \frac{1}{3}\sin(6\pi t) + \frac{1}{5}\sin(10\pi t) + \frac{1}{7}\sin(14\pi t)\right]$$

    Write a program that uses a MathScript (or Formula) Node to evaluate this
    expression for *y* in the range $0 \leq t \leq 3$ s, and then plot *y* vs. *t* on an XY Graph.

2. **Spiral.vi** In polar coordinates (r, θ), a particular spiral is defined by the relation $r = \theta$, where $0 \le \theta \le 6\pi$. Build a VI that uses a MathScript (or Formula) Node to describe this spiral in terms of Cartesian coordinates (x, y), and then displays the spiral on an XY Graph. If using a MathScript Node, the function *polar_to_cart* may be useful.

3. **MathScript Root Finder** (MathScript Only) Open a LabVIEW MathScript Window, and then type *help roots* in the Command Window. After learning the syntax of the MathScript command *roots*, use it in the LabVIEW MathScript Window to find the two roots of the second-order polynomial $x^2 + x - 1 = 0$. To check that you used *roots* correctly, you can compare your result with that found using the quadratic formula.

4. **Noisy Sine (MathScript Node).vi** (MathScript Only) Write a MathScript-based VI that generates three cycles of a 100 Hz sine wave with added random "Gaussian" noise and then plots this waveform on an XY Graph. To accomplish this task, calculate N samples of the sine wave at equally spaced times from $t=0$ to $t=0.03$ s, then add a unique "Gaussian" random number to each of these sine-wave samples to create the noisy waveform. A collection of Gaussian random numbers is distributed about zero with probabilities given by a Gaussian ("bell") curve with a standard deviation of σ. This distribution accurately models the noise in many experimental situations. In your VI, use the MathScript command *randnormal(1,N)*, which produces a single row vector of N Gaussian random numbers with $\sigma = 1$.

5. **MathScript Numerical Integral** (MathScript Only) It is easy to show analytically that the integral $\int_0^1 5x^4 \, dx$ equals exactly 1. Open a LabVIEW MathScript Window and then type *help quadn_trap* in the Command Window. After learning the syntax of the MathScript command *quadn_trap*, write a VI that uses *quadn_trap* within a MathScript Node to calculate the integral $\int_0^1 5x^4 \, dx$ numerically and then display the result in a front-panel indicator with **Digits of precision** equal to 5. For your numerical result to be correct to five decimal places (i.e., the first five decimal places are all zero), how many x-values must be supplied to *quadn_trap* within the integration range $x=0$ to $x=1$?

6. **Projectile.vi** Quite commonly, a projectile lands at a level that is different than that from which it is launched. As shown below, consider a projectile launched with speed v (m/s) at angle θ above the horizontal at a height H above the ground and define range R to be the horizontal distance it travels during its flight (Figure 4.1).

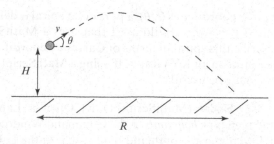

FIG. 4.1

For this situation, it can be shown that

$$R = \frac{v^2}{2g}\left[\sin\left(2\theta\right) + \sqrt{\sin^2\left(2\theta\right) + \frac{8gH}{v^2}\cos^2\theta}\,\right]$$

where θ is in the range from 0° to 90°.

(a) Write a MathScript (or Formula) Node-based program that, given v and H, calculates R for a sequence of 91 equally spaced angles in the range from $\theta=0°$ to $\theta=90°$ and then displays R vs. θ on an XY Graph as well as in an indicator cluster. The front panel of the VI is shown next.

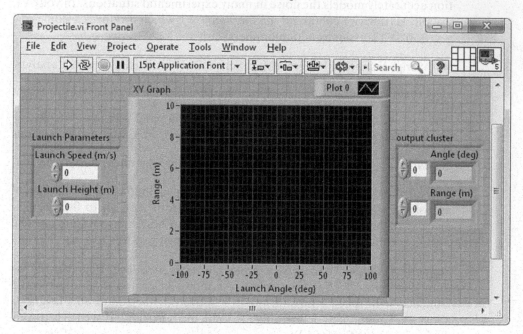

(b) At a track-and-field meet, a shot putter launches a shot (heavy metal ball) with speed 13.5 m/s. Run your VI with this launch speed, assuming the shot lands at the same level from which it was launched. Use the resulting XY Graph and output cluster to demonstrate the well-known fact that the maximum range results if the launch angle equals 45°.

(c) In reality, a shot putter launches the shot from a height close to (but above) their head. With $v = 13.5$ m/s and $H = 2.1$ m, run your VI and then determine the launch angle that produces the maximum range.

(d) Finally, if the shot putter launched the shot with $v = 13.5$ m/s from a cliff of height $H = 100$ m, run your VI and then determine the launch angle that produces the maximum range.

7. **Waveform Simulator (Express).vi** Write a VI that implements one of LabVIEW's easy-to-use Express VIs to generate a waveform. Place a **Waveform Graph** on the front panel, and then switch to the block diagram and place **Simulate Signal** there, which is found in **Functions>>Express>>Input**. Immediately after placing this Express VI on the block diagram, its dialog window will open. In this window, choose **Signal type**, **Samples per second (Hz)**, and **Number of samples** to be *Sine*, *1000*, and *100*, respectively, and then press the **OK** button near the bottom of the window. When you are returned to the block diagram, complete the code shown below. You will find that a new type of wire called *dynamic data type* emanates from the Express VI's **Sine** output and that the **Waveform Graph** automatically adapts to this format.

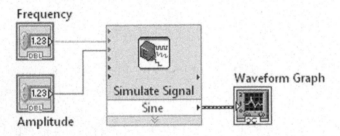

(a) Return to the front panel and run your VI with **Frequency** and **Amplitude** equal to *50* and *5*, respectively. Note that the Waveform Graph's x-axis is automatically scaled correctly, a benefit of the dynamic data type wire.

(b) On the block diagram, pop up on the **Simulate Signal** icon and select **Properties**. The Express VI's dialog window will reopen. Select **Signal type>>Sawtooth** and then click the **OK** button. Rerun your VI to view the new waveform shape.

8. **Polynomial Fit (MathScript).vi** (MathScript Only) In this problem, you will use the MathScript command *[p, s]=polyfit(x, y, n)* to fit a given set of data to a second-order polynomial function of the form $y = a_0 + a_1 x + a_2 x^2$, where a_0, a_1, and a_2 are constants. A description of this MathScript command can be found by typing *help polyfit* in the LabVIEW MathScript Window. Assume that a quantity y is measured as a function of another quantity x, resulting in the following data set:

x	1.0	2.0	3.0	4.0	5.0	6.0
y	11.0	15.1	19.8	25.1	31.0	37.5

Construct a VI as follows: Enter the given y vs. x data as row vectors in a MathScript Node. The y row vector is created with the command $y = [11.0\ 15.1\ 19.8\ 25.1\ 31.0\ 37.5]$. Then use the command *[p, s] = polyfit(x, y, n)* to fit these given data to a second-order polynomial. The values of a_2, a_1, and a_0 will be given as the first, second, and third elements of the row vector p, which can be found by the commands $a_2 = p\ (1)$, $a_1 = p\ (2)$, and $a_0 = p\ (3)$ (the integer in parentheses is the index of the desired element of p; unlike LabVIEW arrays, MathScript-array indexing begins with *1*, rather than *0*). Display the resulting values of a_0, a_1, and a_2 on the front panel.

CHAPTER 5

Introduction to Data Acquisition Devices Using MAX

In this chapter, you will be introduced to the capabilities of a National Instruments data acquisition (DAQ) device connected to your computer. Here, it is assumed that you have such a DAQ device available, along with a DC voltage source (e.g., battery, power supply, function generator), function generator, voltmeter, and required cabling. An oscilloscope would be useful, but is optional. Prior to starting this chapter, you may want to download the latest version of the device driver **NI-DAQmx** from *ni.com/support*. If you do not have access to a DAQ device, use a simulated device to carry out the exercises in this chapter (see Problem 4 at the end of this chapter).

5.1 DATA ACQUISITION HARDWARE

In this chapter, you will be introduced to some of LabVIEW's most powerful capabilities, computer-controlled data acquisition and generation. These capabilities equip an engineer or scientific investigator to reach out through a PC to monitor and to create events in the outside world. For this introduction, you will control the basic operations of a multifunction data acquisition (DAQ) device using the handy interactive utility *Measurement & Automation Explorer*, which is nicknamed *MAX*, while learning about important issues involved in acquiring and generating digitized data.

National Instruments, based in Austin, Texas, offers a wide variety of computer-controlled data acquisition devices. In LabVIEWspeak, the operations these devices perform include *Analog Input (AI)*, *Analog Output (AO)*, *Digital Input/Output (DIO)*, and *Counter Input/Output*. Respectively, these terms are shorthand for analog-to-digital conversion (acquiring a digitized representation of an incoming analog voltage); digital-to-analog conversion (generating an outgoing analog signal from a sequence of digital numbers); digital port control (setting and reading the port's HIGH/LOW state); and digital waveform operations (event timing, pulse counting, pulse train generation). Some of the available DAQ devices are quite specialized (for instance, only designed to process digital signals), while others are multipurpose, performing some or all of the above-mentioned operations. There are high-speed devices, which digitize incoming signals at rates over 10^9 samples per second (S/s), and more garden-variety systems with maximum sampling frequencies on the order of 10^6 S/s. Digital resolution varies from 8 to 24 bits, and each device connects to a PC through one of about ten possible ways (including the PCI Express and USB interfaces).

In this chapter, I will use four popular multifunction DAQ devices—PCIe-6351, USB-6002, myDAQ, and ELVIS II—to illustrate the generic features of the particular NI data acquisition product that is present in your system. The exercises in this chapter are designed so that they can be completed using a DAQ device with specifications similar to that of any of these four units. A brief description of each of these representative devices follows.

PCIe-6351: This data acquisition board is one of NI's professional-grade multifunction *X Series* DAQ devices. It resides within a desktop computer, plugged into a PCI Express expansion slot. The 6351 has 16 analog input channels that can perform 16-bit analog-to-digital conversion operations at rates up to 1.25×10^6 S/s. Also, there are two 16-bit analog output channels, which can update output voltages at a maximum rate of 2.8×10^6 S/s. Additionally, this DAQ device has 24 digital I/O ports and four 32-bit counters/timers. The PCIe-6351 is the successor to the popular PCIe-6251 *M Series* DAQ device; the USB-interfaced model USB-6351 is also available.

USB-6002: This low-cost multifunction DAQ device provides basic data acquisition functionality for applications such as simple data logging, portable measurements, and academic lab experiments. It is a small, stand-alone unit that connects to your computer via a USB port. The 6002 has eight analog input channels that can perform 16-bit analog-to-digital conversion operations at rates up to 5×10^4 S/s. There are also two 16-bit analog output channels, which can update output voltages at a maximum rate of 5×10^3 S/s. Additionally, the device has 13 digital I/O ports and a 32-bit counter (with no hardware-timing measurement capabilities). The USB-6002 is the middle unit of the USB-6001/6002/6003 product line, which has replaced the popular (but now legacy) USB-6008 and USB-6009 DAQ devices.

myDAQ: Priced to be affordable for students, this USB-interfaced unit is designed for hands-on experimentation anywhere you can take a laptop. The myDAQ has only two analog input channels, but they can perform 16-bit analog-to-digital conversion operations at rates up to 2×10^5 S/s. Impressively, it also has two 16-bit analog output channels that can update output voltages at a maximum rate of 2×10^5 S/s, which is 40 times faster than the AO capabilities of the USB-6002. Additionally, the myDAQ has eight digital I/O ports and a 32-bit counter and can be configured as a digital multimeter (DMM).

ELVIS II: Designed specifically for instructional lab use, the ELVIS II has a built-in DAQ device with specifications comparable to a PCIe-6351 (sixteen 16-bit analog input channels that operate at rates up to 1.25×10^6 S/s, and so on). ELVIS II communicates with a computer via a USB connection and includes a powered prototyping breadboard for use in the study of electronic circuits.

A conduit carrying signals to and from the outside world must somehow be connected to a DAQ device. For example, among its various capabilities, a PCIe-6351 has 16 analog input channels labeled AI 0 through AI 15, each standing ready to digitize an incoming analog signal. So how does one get the meaning-laden analog signals from an experiment to the DAQ unit's channel inputs? Typically, an experimenter connects wires from their experiment to appropriate terminals on an *I/O Terminal Block* (various styles of this item may be purchased from National Instruments). An I/O Terminal Block has screw or BNC terminals to which the experimental wires can be attached. A cable is then connected from the I/O Terminal Block to a connector at the end of the DAQ device. This connector has (depending on the particular device) 50, 68, or 100 pins, where each pin is associated with a particular function of the DAQ device. The PCIe-6351 employs the above-described connection method with a 68-pin connector, a commonly used configuration for both X Series and M Series DAQ devices. Alternately, some DAQ devices are configured with an on-board terminal block or prototyping breadboard to which wires from an experiment are directly attached. The USB-6002, myDAQ, and ELVIS II are outfitted in this scheme. Regardless of connection method, the function of each DAQ-device pin can be determined from the device's pinout diagram. The pinout diagrams for the PCIe-6351, USB-6002, and myDAQ are reproduced in Figure 5.1.

5.2 MEASUREMENT & AUTOMATION EXPLORER (MAX)

To carry out the exercises in this chapter, you must have a National Instruments data acquisition device connected to your computer, and its driver software (called *NI-DAQmx*) must be correctly installed. Let's use MAX to verify that these conditions are met on your system.

To open MAX, either select **Tools>>Measurement & Automation Explorer...** within LabVIEW or double-click on MAX's desktop icon (if available). After MAX opens, double-click on **Devices and Interfaces** under the **My System** heading

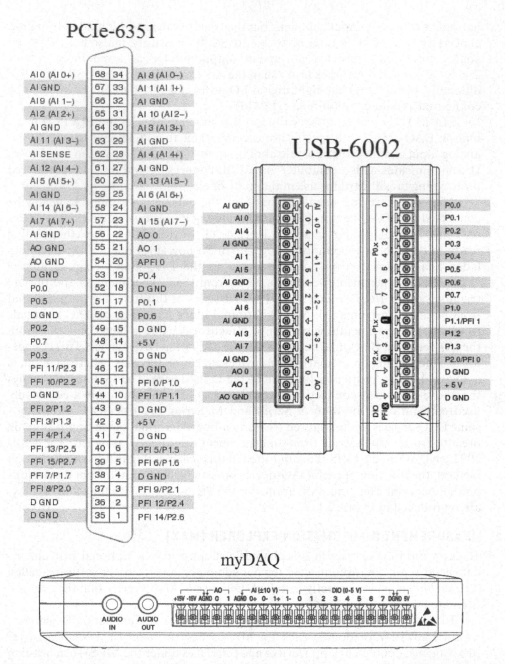

FIG. 5.1 Pinout diagrams for DAQ devices. The 68-pin configuration for the PCIe-6351 is common to all *X* Series and *M* Series DAQ devices.

at the left. This action will command MAX to determine all of the data acquisition devices present within your computing system.

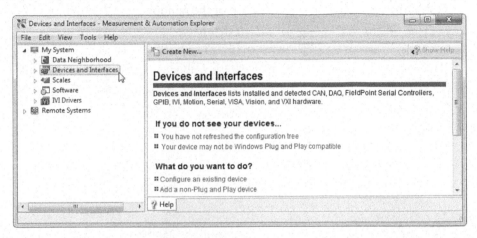

MAX will list the findings of its device survey in hierarchical tree fashion as shown next. For the system used below, we see that a *NI USB-6002* device was found to be connected to the computer (for older versions of MAX, you may have to double-click on a folder called **NI-DAQmx Devices** to view the DAQ devices that have been found). Note that the driver software has (automatically) given this device the shorthand name *Dev1*. The DAQ device on your system may have a different shorthand name. When you go through this procedure and for some reason a device that you believe is connected to your computer does not appear on the **Devices and Interfaces** list, try selecting **View>>Refresh** (or press *<F5>* on your keyboard) to command MAX to perform its device survey again. If the device in question doesn't appear then, there is something wrong with the device's connection that must be repaired.

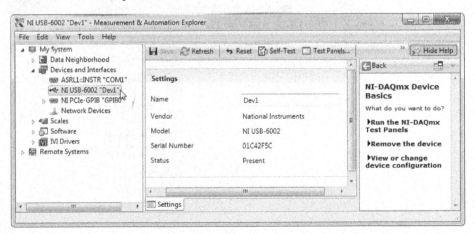

To verify that your DAQ device is properly functioning, right-click on its name and select **Self-Test** from the pop-up menu. Alternatively, you can click on the **Self-Test** button in the toolbar near the top of the window.

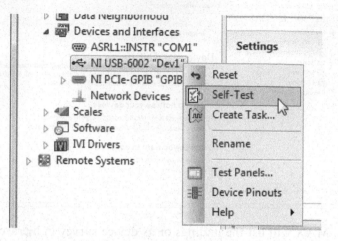

A brief test of your device's functionalities will be performed, and ideally you will receive a dialog-box message stating that all is well.

Throughout the rest of the chapter, we will use MAX to "test-drive" the various operations available on your DAQ device (e.g., digitize an analog voltage signal applied at one of its AI inputs). In order to execute these operations, we will first need to make connections to the correct pins on your DAQ device. To determine the assigned operation for each of these pins, right-click on the name of your DAQ device and select **Device Pinouts** from the pop-up menu.

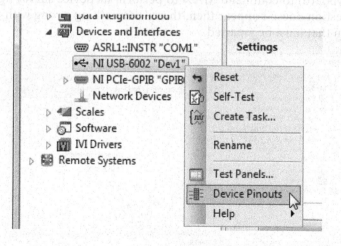

A window will appear with the pinout diagram for your particular device. Briefly examine this diagram to get a general sense of which pins are assigned to Analog Input and Analog Output operations—labels beginning with AI and AO, respectively—and which pins are dedicated to Digital and Counter I/O—labels starting with DIO or P.

Finally, right-click on the name of your DAQ device and select **Test Panels...** (or press the **Test Panels...** button in the toolbar). An interactive window named **Test Panels** will appear, which we will use to test the various functions of your DAQ device.

We'll first use **Test Panels** to explore the Analog Input operation. However, one needs to understand several important concepts to properly implement this operation and so we'll start by obtaining this needed grounding in the following three sections.

5.3 ANALOG INPUT MODES

With regard to analog input (AI) operations, the pinout diagrams in Figure 5.1 can be deciphered by noting that the PCIe-6351 and ELVIS II have 16 AI channels, while the USB-6002 has 8 (let's set aside the myDAQ for a moment). Through software settings, these inputs can be configured to operate in two distinct analog input modes—*single-ended* and *differential*.

In the single-ended mode, each available analog input pin is an AI channel, with all of these channels referenced to the same common ground. You may choose this common ground to be the building ground (at the AI GND pins), termed the *referenced singled-ended (RSE)* mode, or supply your own ground level at the AI SENSE pin, called the *nonreferenced singled-ended (NRSE)* mode. In both of these two modes, the pin assignments for the AI channels on the DAQ devices cited above are given in Table 5.1. Within the DAQ device, all AI GND pins (along with the AO GND and D GND pins) are connected to the building ground.

When measuring a voltage difference in the RSE input mode, one of the two lead wires from the voltage source is attached to the pin of a channel listed in Table 5.1;

TABLE 5.1 Channel Assignments for Single-Ended (RSE and NRSE) Analog Input Modes. (Pin labels are in uppercase, channel name in lowercase.)

Channel Name	ai0	ai1	ai2	ai3	ai4	ai5	ai6	ai7	ai8	ai9	ai10	ai11	ai12	ai13	ai14	ai15
PCIe-6351 Pin*	AI 0	AI 1	AI 2	AI 3	AI 4	AI 5	AI 6	AI 7	AI 8	AI 9	AI 10	AI 11	AI 12	AI 13	AI 14	AI 15
ELVIS II Pin	AI 0⁺	AI 1⁺	AI 2⁺	AI 3⁺	AI 4⁺	AI 5⁺	AI 6⁺	AI 7⁺	AI 0⁻	AI 1⁻	AI 2⁻	AI 3⁻	AI 4⁻	AI 5⁻	AI 6⁻	AI 7⁻
USB-6002 Pin	AI 0	AI 1	AI 2	AI 3	AI 4	AI 5	AI 6	AI 7								

*This configuration applies to all 68-pin X Series and M Series DAQ devices.

the other lead wire is attached to an AI GND pin. The DAQ device will then accurately measure the voltage difference from the source as long as the source is *floating*. Examples of floating voltage sources are transformer outputs, thermocouples, some power supplies, and battery-power devices. Voltage sources such as function generators are *ground-referenced*, that is, one of the wire leads emanating from the source is connected to the building ground. Connecting a ground-referenced source to a DAQ device in RSE mode establishes two grounding points within the measurement circuitry, introducing noise due to the creation of a "ground loop." The RSE mode is not recommended in this situation.

For the NRSE mode on the PCIe-6351 and ELVIS II, the pin assignment of each channel is the same as given in Table 5.1; however, a user-supplied ground level must be applied to the AI SENSE pin. The NRSE mode is not available on the USB-6002 or myDAQ; hence, the absence of an AI SENSE pin on the pinout diagrams for these devices.

In differential input mode, available analog input pins are paired to form eight (PCIe-6351 and ELVIS II) or four (USB-6002) independent AI channels, each of which is sensitive only to the voltage difference between paired pins. On the PCIe-6351, differential channel *ai0* is pin AI 0 paired with pin AI 8, differential channel *ai1* is AI 1 paired with AI 9, and so on (note pin labels are uppercase, channel names are lowercase). In this "differential amplifier" configuration, noise pickup by the AI channels is suppressed. Thus, if it is not a problem in your situation to halve the number of available measurement channels, the differential input mode is the preferred modus operandi. Again, as an example, the pin assignments for the differential input mode AI channels for our representative boards are given in Table 5.2.

TABLE 5.2 Channel Assignments for Differential Input Mode. (Pin labels are in uppercase, channel name in lowercase.)

Channel Name	PCIe-6351*		ELVIS II		USB-6002	
	Positive Input Pin	Negative Input Pin	Positive Input Pin	Negative Input Pin	Positive Input Pin	Negative Input Pin
ai0	AI 0	AI 8	AI 0$^+$	AI 0$^-$	AI 0	AI 4
ai1	AI 1	AI 9	AI 1$^+$	AI 1$^-$	AI 1	AI 5
ai2	AI 2	AI 10	AI 2$^+$	AI 2$^-$	AI 2	AI 6
ai 3	AI 3	AI 11	AI 3$^+$	AI 3$^-$	AI 3	AI 7
ai 4	AI 4	AI 12	AI 4$^+$	AI 4$^-$		
ai 5	AI 5	AI 13	AI 5$^+$	AI 5$^-$		
ai 6	AI 6	AI 14	AI 6$^+$	AI 6$^-$		
ai 7	AI 7	AI 15	AI 7$^+$	AI 7$^-$		

*This configuration applies to all 68-pin X Series and M Series DAQ devices.

The myDAQ has two analog input channels, which can be operated only in differential mode. The positive and negative pins for channel *ai0* are labeled AI 0⁺ and AI 0⁻, respectively; for *ai1*, the pins are labeled AI 1⁺ and AI 1 , respectively.

5.4 RANGE AND RESOLUTION

When measuring analog input signals with a DAQ device, there are several issues to keep in mind. First, a DAQ device can only measure voltages that fall within an allowed *range*, which extends from a minimum voltage V_{min} and to a maximum voltage V_{max}. If the magnitudes of V_{max} and V_{min} are the same, the range is written as $\pm V_{max}$. A NI DAQ device might have a fixed range (for the USB-6002, it is fixed at ± 10 V) or offer the choice of several software-selected ranges (the PCIe-6351 and ELVIS II have seven possible ranges starting at ± 0.1 V and ending at ± 10 V). Check the datasheet for your particular DAQ device to find the available ranges. Next, the DAQ device's analog-to-digital converter has a built-in *resolution* of n-bits, which means that this digitizer represents the analog voltage level being sampled with a binary number of n digits. This binary resolution places a limit on the resulting *voltage resolution* (i.e., the smallest detectable voltage difference ΔV). An n-bit digitizer divides the measurable voltage span $V_{span} \equiv V_{max} - V_{min}$ ($V_{span} = 20$ V when the range is ± 10 V) into 2^n divisions, and the resulting voltage resolution of the analog input signal being sampled is given by

$$\Delta V = \frac{V_{span}}{2^n} \qquad [5.1]$$

For the typical value of $V_{span} = 20$ V, the voltage resolution provided by a 14-bit and 16-bit DAQ device is 1 mV and 0.3 mV, respectively.

5.5 SAMPLING FREQUENCY AND THE ALIASING EFFECT

Another important issue peculiar to digitized data is related to the sampling frequency f_s, where f_s is the rate at which analog-to-digital conversions take place. The sampling frequency places an upper limit on the range of frequencies allowed within the original analog signal, if the digitizing process is to result in a faithful representation of the input. This upper threshold is called the *Nyquist frequency* $f_{nyquist}$, which we will show equals one-half of the sampling frequency, that is, $f_{nyquist} = f_s/2$. The physical significance of the Nyquist frequency is this: For a given sampling rate f_s, an input sine wave of frequency $f_{nyquist}$ will be sampled in the minimal manner necessary to represent the sinusoid's peaks and valleys—just twice per cycle. If the input frequency exceeds this threshold, the sampling rate becomes insufficient and the analog-to-digital conversion process becomes inaccurate in a way described next.

Fairly often, the bandwidth limitation placed on a "to-be-digitized" input analog signal does not cause you—the experimentalist—much of a problem. For

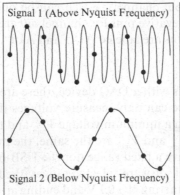
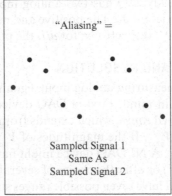

FIG. 5.2 Aliasing effect when signal with frequency greater than the Nyquist frequency is digitally sampled.

example, if dealing with audio-produced electrical signals, you might know on physical grounds that the frequency content of the analog input is bracketed by 20 and 20,000 Hz. Then, with a sampling rate of at least 40 kHz, this input can be properly acquired. Alternately, an analog signal may have passed through an amplifier that behaves as a low-pass filter because of its finite bandwidth response. In this case, the sampling rate must be twice the maximum frequency passed by the amplifier. If no natural frequency bracketing exists in your experiment, then you must impose a high-frequency cutoff by placing a low-pass filter in your data-gathering circuitry. Given the available sampling rate f_s, the filter's components then are chosen so that frequencies higher than $f_s/2$ are not passed.

What happens if a frequency higher than the Nyquist limit accidentally is input to your digitizer? Something much worse than you might expect. In a process called *aliasing*, that too-high frequency appears falsely as a lower-than-Nyquist frequency when processed by the digitizer. This phenomenon, which is peculiar to discrete sampling, is illustrated in Figure 5.2.

Quantitatively, assume that a sine wave with (high) frequency f is digitally sampled at N discrete times $t_i = i\,\Delta t$, where $i = 0, 1, 2, ..., N-1$ and Δt is the time increment between neighboring data samples. Then the sampling frequency and Nyquist frequency are $f_s = 1/\Delta t$ and $f_{nyquist} = f_s/2$, respectively. If $f > f_{nyquist}$, it can be shown that at every digitizing time t_i, the displacement of the sine wave $(2\pi f t_i)$ is equal to plus or minus the displacement of a sine wave with the "alias frequency" f_{alias}, where $0 \leq f_{alias} \leq f_{nyquist}$. By applying the condition $(2\pi f t_i) = \pm \sin(2\pi f_{alias} t_i)$ at every t_i (see Problem 1), one finds that

$$f_{alias} = |f - nf_s| \qquad \text{(Aliasing Condition)} \qquad [5.2]$$

where $n = 1, 2, 3 \ldots$.

Demonstrate the aliasing effect by opening **Waveform Simulator** (located in **YourName\Chapter 4**) and programming it to produce a 100 Hz sine wave of amplitude 5. That is, in the **Waveform Parameters** control cluster, set **Frequency**, **Amplitude**, and **Type** equal to *100*, *5*, and *Sine*, respectively. Then, in the **Digitizing Parameters** control cluster, taking **Number of Samples** and **Sampling Frequency** to be *100* and *1000*, respectively, run this VI to demonstrate what a digitized sinusoid with frequency less than the Nyquist frequency (in this case, $f_{nyquist} = 500$ Hz) looks like. Note that 10 cycles occur over the course of 0.1 second, so the frequency of the wave is 10 cycle/0.1 s = 100 Hz, as expected. Now, with the **Digitizing Parameters** unchanged, program **Waveform Simulator** to produce a 1100 Hz sine wave, and then rerun the VI. You will find that, when digitized, this "higher-than-Nyquist" frequency wrongly appears as a 100 Hz sine wave with 10 cycles over a time span of 0.1 second, consistent with the prediction of Eq. [5.2] with $n = 1$, that is, $f_{alias} = |1100 \text{ Hz} - (1)(1000 \text{ Hz})| = 100$ Hz. Predict what will happen if you program **Waveform Simulator** with some other "higher-than-Nyquist" frequencies such as 800 Hz and 2100 Hz, and then see if your expectations are realized via the use of **Waveform Simulator**.

5.6 ANALOG INPUT OPERATION USING MAX

We are now ready to program your DAQ device to digitize an analog voltage difference applied at its input. We'll start by digitizing a DC (constant) voltage that, for example, is sourced from a power supply, battery, or function generator (using its DC option). Let's perform this AI operation in differential mode using differential channel *ai0*.

To execute this analog input operation, we must first secure the connections from the DC voltage source to the correct pins on your DAQ device. As you know, you can use MAX to determine the location of these pins by right-clicking on the name of your DAQ device and selecting **Device Pinouts** from the pop-up menu. Choose a voltage source that is within the allowed range (e.g., −10 V to +10 V) of your DAQ device, then connect its two wires to channel *ai0*. For example, the high-voltage and low-voltage wire from your source, respectively, should be connected to AI 0 and AI 8 for the PCIe-6351, AI 0 and AI 4 for the USB-6002, and AI 0$^+$ and AI 0$^-$ for the myDAQ or NI ELVIS II.

Next, if you haven't done so already, right-click on the name of your DAQ device in MAX and select **Test Panels...** (or press the **Test Panels...** button in the toolbar). In the interactive **Test Panels** window that appears, select the **Analog Input** tab, and then enter the settings **Mode>>On Demand** and **Input Configuration>>Differential** as shown below. The values of **Min Input Limit** and **Max Input Limit** are appropriate for the range of +10 V. In the **Channel Name** box, the text *Dev1/ai0* selects AI differential channel *ai0* on the DAQ device with the shorthand name *Dev1* (on your **Test Panel**, use the shorthand name of your particular DAQ device, which may be different from *Dev1*).

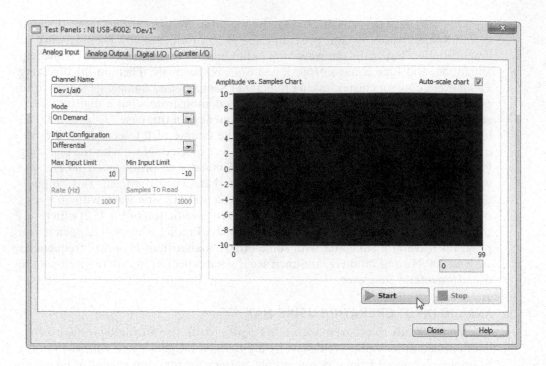

Then press the **Start** button. In the **On Demand** mode, the DAQ device will acquire one digitized reading of the applied voltage. If the connections from the voltage source to your DAQ device are properly made, this digitized value will appear in the indicator below the chart. Until you press the **Stop** button, this measurement will be repeated by **Test Panels** several times per second and the results plotted on the chart. For the plot shown below, a 2-V DC source was connected to channel *ai0*.

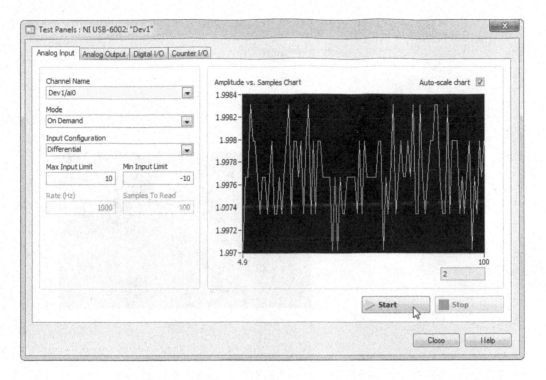

Small fluctuations in your measured voltage are expected because of experimental noise, as shown above. However, if your measured voltage undergoes very large variations as time goes on, you may be having problems resulting from the use of a "floating" voltage source. Examples of floating voltage sources are batteries, transformers, and nonreferenced power supplies. To remedy this predicament, reference your "floating" voltage source to the DAQ device's ground by connecting an extra wire (or a 100 kΩ resistor) from *ai0*'s negative input pin to the DAQ device's AI GND pin.

Let's now program the DAQ device to digitize an AC voltage input from a function generator at channel *ai0*. Adjust the settings on a function generator so that it is outputting an analog sine wave with amplitude less than 10 V and a frequency of about 100 Hz. Using the appropriate pins for your particular system, connect the positive and negative (ground) outputs from the function generator to AI differential channel *ai0* on your DAQ device. (*Note:* Most function generators are referenced to the building ground and so are not "floating" voltage sources.)

In the **Test Panels** window, make the following changes shown below: **Mode>>Finite**, **Rate (Hz)>>1000**, and **Samples To Read>>100**. In the **Finite** mode, an analog signal input at the specified AI channel is digitized at N equally spaced times at a rate f_s (i.e., the sampling frequency). The N acquired data points are then plotted. In the **Test Panels** window, **Samples to Read** and **Rate (Hz)** are used to select the values for N and f_s, respectively.

Press the **Start** button. If all goes well, you will see a plot of about 10 cycles of the 100 Hz sine wave.

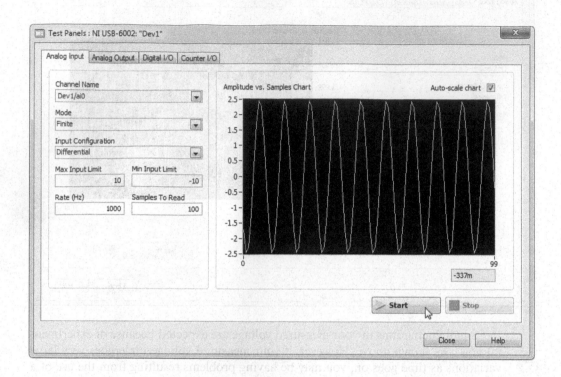

Note that the plot is formed from an array of 100 equally spaced samples and that the x-axis values are the array indices i of the samples, where i runs from 0 to 99. For the chosen sampling rate of 1000 Hz, neighboring data samples are separated by a time interval of 0.001 s, and so the x-axis covers the time span from $t = 0.000$ to $t = 0.099$ s. Thus, for the 100 Hz sine-wave signal, whose period is 0.01 s, we observe ten of its cycles.

On the function generator, set the frequency f of the input sine-wave signal to several values within the range $0 \leq f \leq 500$ Hz and press **Start** to observe the digitized waveform for each frequency setting. You should find that the frequency of each signal in this range is faithfully reproduced by the DAQ device, but that sine-wave shape becomes distorted as the signal frequency approaches 500 Hz (because of the paucity of sampled points per sine-wave cycle at these higher input frequencies).

In **Test Panels**, we set the sampling frequency equal to $f_s = 1000$ Hz; thus, the Nyquist frequency is $f_{nyquist} = f_s/2 = 500$ Hz. For input frequencies greater than the Nyquist frequency, that is, $f > 500$ Hz, we expect the aliasing effect. Demonstrate that for $f = 1100$ Hz, the digitized sine wave appears to have a frequency of 100 (rather than 1100) Hz. Use Eq. [5.2] to predict two other input frequencies f that will appear as a 100 Hz sine wave when digitized because of aliasing, and then demonstrate that your predictions are correct using **Test Panels**.

5.7 ANALOG OUTPUT

Multifunction DAQ devices manufactured by National Instruments typically have two analog output (AO) channels. Each AO channel can perform n-bit digital-to-analog conversion operations at speeds up to a *maximum update rate* given in samples per second (S/s). The possible analog voltages produced fall within a range from V_{min} to V_{max}. Within this voltage span $V_{span} \equiv V_{max} - V_{min}$, the output analog voltage can be one of 2^n possible values. The voltage resolution ΔV with which a voltage can be produced—that is, the smallest difference between one possible output voltage and the next—is given by $\Delta V = V_{span}/2^n$ (i.e., Eq. [5.1]). As a typical example, if $n = 16$ and $V_{span} = 20$ V, $\Delta V = 0.3$ mV. The specifications for the representative DAQ devices are listed in Table 5.3.

On all NI DAQ devices, the voltage difference associated with Analog Output channel *ao0* is produced between pins AO 0 and AO GND, while for channel *ao1* the output is between pins AO 1 and AO GND. Check the pinout diagram for your particular DAQ device to find the relevant pins (on the myDAQ and NI ELVIS II, AO GND is labeled AGND and GROUND, respectively).

5.8 ANALOG OUTPUT OPERATION USING MAX

Program your DAQ device to output a DC voltage difference at analog output *ao0*. To accomplish this feat, select the **Analog Output** tab in the Test Panels window, and

TABLE 5.3 Analog Output (AO) Specifications

Device	Number of AO Channels	AO Voltage Ranges	Bits	Maximum Update Rate
PCI-6351	2	±5 V, ±10 V*	16	2.86 MS/s
USB-6002	2	±10 V	16	5 kS/s
myDAQ	2	±2 V, ±10 V*	16	200 kS/s
ELVIS II	2	±5 V, ±10 V*	16	2.8 MS/s

*Ranges are software selectable.

then enter the settings as shown below with **Channel Name>>Dev1/ao0** (if different from *Dev1*, use the correct name of your particular DAQ device), **Mode>>Voltage DC**, and **Output Value (V)>>1**. Then press the **Update** button.

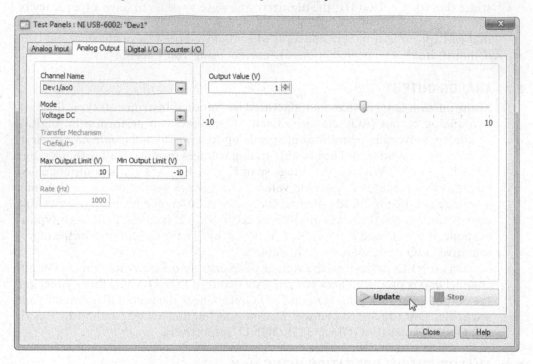

To verify that your DAQ device is outputting a 1 Volt voltage difference at AO channel *ao0*, connect the leads of a voltmeter to the AO 0 and AO GND pins. Note that this 1 V DC voltage difference will be continuously output at *ao0* until you program the DAQ device to output a different value.

For practice in making the correct connection to your DAQ device (or if you don't have a voltmeter available), connect the AO 0 and AO GND pins to the positive and negative input pins of AI channel *ai0*. Then select the **Analog Input** tab in **Test Panels**, configure the window as shown below, and press **Start**. If all connections are made properly, you will observe a real-time plot of the voltage difference being output at *ao0*.

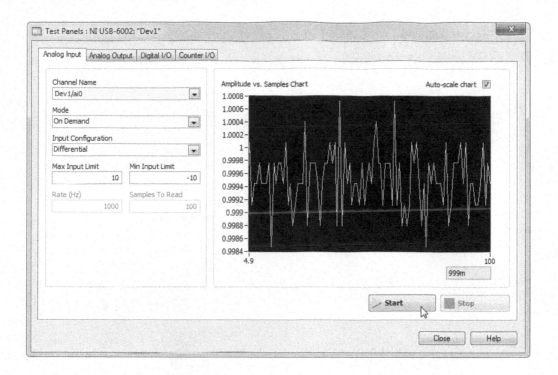

Select the **Analog Output** tab again. This time, we'll program the DAQ device to output an AC voltage at AO channel *ao0*. In the **Test Panels** window, make the following changes shown below: **Mode>>Voltage Sinewave Generation** and **Sinewave Amplitude (V)>>1**. The **Sinewave Frequency (Hz)** is fixed at 1 Hz.

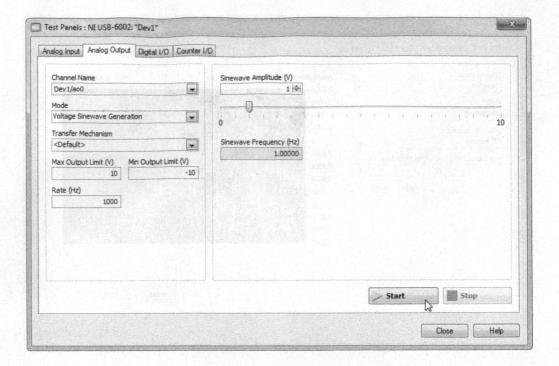

Press the **Start** button. To verify that your DAQ device is outputting a sine wave with an amplitude of 1 V and a frequency of 1 Hz at AO channel *ao0*, connect the AO 0 and AO GND pins to either an oscilloscope or your DAQ device's AI channel *ai0*. If you take the latter approach, select the **Analog Input** tab in **Test Panels**, configure the window as shown below, and press **Start**. With **Rate (Hz)>>1000** and **Samples To Read>>2000**, it will take 2 seconds to acquire the data, which will display two cycles of the 1 Hz sine wave.

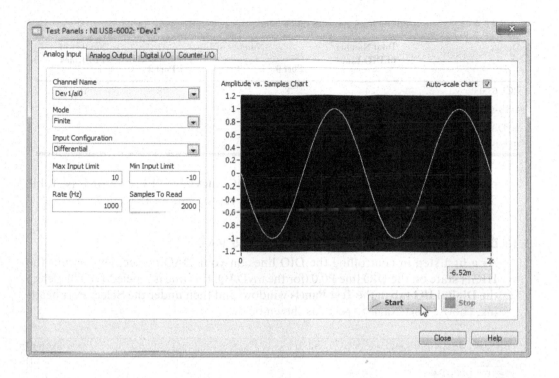

5.9 DIGITAL INPUT/OUTPUT

A digital signal can be in one of two possible states, termed HIGH and LOW. For the digital ports on NI DAQ devices, these states conform to the *transistor–transistor logic (TTL) standard*, where HIGH and LOW are defined to be within the range 2–5 V and 0 0.8 V, respectively (the exact definitions for TTL output and input states differ slightly, but are both within these ranges). When a digital signal alternates between its two states, the transition from LOW to HIGH is called the *rising edge*, while the HIGH to LOW transition is the *falling edge*.

Multifunction DAQ devices are equipped with a generous collection of *Digital Input/Output (DIO)* pins. Each of these DIO pins is called a *line*, which can be configured to either input or output a digital state, that is, ascertain ("read") whether the pin is in the HIGH or LOW state or set ("write") the pin to the HIGH or LOW state. A typical DAQ device has two or three *ports*, where each port is composed of one, four, or eight *lines*. A particular line is specified as *P port.line*, so, for example, P0.3 is line 3 of port 0 (except for the myDAQ, where each line is specified by *DIO line* with *line* running from 0 to 7). Some of the DIO lines have additional capabilities beyond that of regular DIO lines and are denoted as Programmable Function Interface (PFI) lines. PFI lines are used for triggering data acquisition, clock synchronization, counter input, and counter output. Consult the user manual for your

TABLE 5.4 Digital Input/Output (DIO) Lines Specifications

Device	Total Number of DIO Lines	Number of Lines			Number of PFI Lines
		Port 0	Port 1	Port 2	
PCI-6351	24	8	8	8	16
USB-6002	13	8	4	1	2
myDAQ	8	8	—	—	5
ELVIS II	24	8	8	8	15

particular DAQ device to determine which DIO lines have PFI functionality. Table 5.4 lists the configuration of DIO lines on our representative DAQ devices.

5.10 DIGITAL INPUT/OUTPUT OPERATION USING MAX

As a first step in controlling the DIO lines on your DAQ device, let's output the HIGH state on the DIO line P0.0 (for the myDAQ, this line is labeled DIO 0). Select the **Digital I/O** tab in the **Test Panels** window, and then under the **Select Port** heading, choose **Port Name>>port0** as shown below.

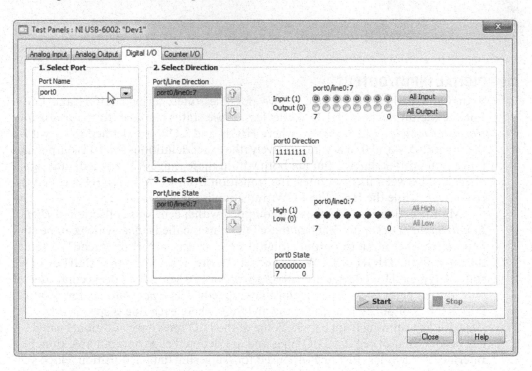

Next, under the **Select Direction** heading, make *P0.0* an output line by clicking on the output button as shown.

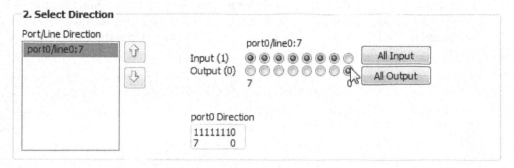

Finally, under the **Select State** heading, set the switch for P0.0 to the *HIGH* setting.

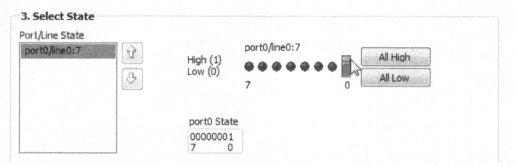

Then press the **Start** button. To verify that your DAQ device is outputting the HIGH state at the designated DIO pin, connect the leads of a voltmeter to the P0.0 and D GND pins. For TTL output signals, the HIGH state corresponds to a voltage difference in the range from 2.7 V to 5 V (common values are 3.3 V and 5 V). Use MAX to switch P0.0 to the *LOW* setting. Does the voltmeter now indicate the line is in the LOW state (for TTL output signals, the LOW state is somewhere in the range from 0 V to 0.4 V)? Alternatively, you can read the voltage using your DAQ device (rather than using a voltmeter). For this approach, connect the P0.0 and D GND pins to the positive and negative input pins of AI channel *ai0*. Then select the **Analog Input** tab in **Test Panels**, configure the window as **Mode>>On Demand**, and press **Start**.

Finally, use MAX to write the HIGH state at P0.0—a digital output operation—and read this value at P0.1—a digital input operation. First, connect these two lines on your DAQ device by attaching a wire to the P0.0 and P0.1 pins; the electrical grounds for these two lines will be automatically connected within the DAQ device, so there is no need to wire a ground connection. Second, in MAX under the **Select Direction** heading, make P0.0 an *output* and P0.1 an *input*; under the **Select State** heading, switch P0.0 to the *HIGH* position. Now, press the **Start** button. As shown below under the **Select State** heading, the LED indicator for P0.1 will light. Try toggling the P0.0 switch to the *LOW* state and repeat. Does the LED indicator for P0.1 respond as expected?

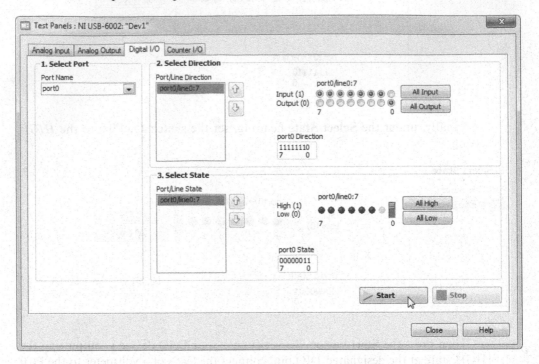

DO IT YOURSELF

Icons used in this DIY project can be found with the aid of **Quick Drop**.

Millisecond-Resolution Stopwatch Using MAX Multifunction DAQ devices come equipped with one or more counters, which can be used to count how many times an incoming digital signal makes a given transition (e.g., rising edge) between the two digital states. In this exercise, you will use MAX to transform the counter in your DAQ device into a millisecond-resolution stopwatch.

In MAX, open the **Test Panels** window for your DAQ device, then select the **Counter I/O** tab. Configure *counter 0* on your DAQ device (e.g., *Dev1/ctr0*) in

Mode>>Edge Counting as shown below. In this mode, *counter 0* will count rising digital edges at its *source* input. To find the source pin for *counter 0*, consult the datasheet for your DAQ device. For example, the counter 0 source is PFI 8 for the PCIe-6351 and ELVIS II, PFI 0 for the USB-6002, and DIO 0 for the myDAQ.

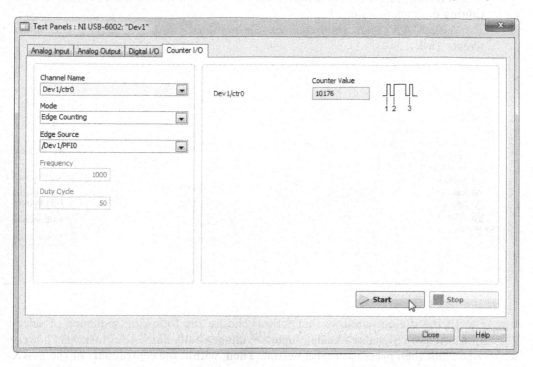

Next, configure a function generator to output a TTL digital signal with a frequency of 1000 Hz. The TTL digital signal should always be positive and toggle between (approximately) 0 V and, say, 3.5 V. Most function generators create a TTL digital signal at an output labeled *sync* (or sometimes called *TTL* or *trigger out*). If your function generator produces such an output, this is an easy option for obtaining the needed 1000 Hz digital signal.

Connect the two lead wires from your function generator, which carry the digital signal and its associated ground, to the *counter 0* source pin and the D GND pin, respectively.

Press the **Start** button. The counter will then count a rising edge once every cycle of the 1000 Hz digital waveform, that is, once every 0.001 s = 1 ms. Thus, at a later time when you press the **Stop** button, the value in the **Counter Value** indicator will equal the number of milliseconds that have elapsed between the two button presses. In the above illustration, that elapsed time was 10,176 ms = 10.176 s.

USE IT!

Creating a DAQ Task Using MAX In this exercise, you will create a DAQ *task*—that is, instructions for carrying out a particular data acquisition operation—using MAX. Once created, this task can be placed on a block diagram and used to make a functioning data acquisition VI.

With MAX open, right-click on the name of your DAQ device and select **Create Task . . . .**

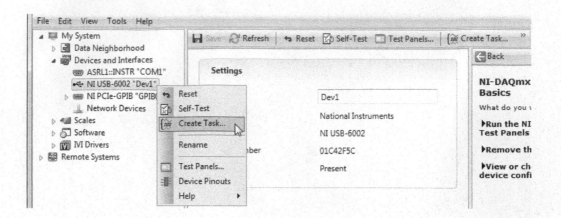

In the dialog windows that follow, choose the following sequence of selections: **Acquire Signals>>Analog Input>>Voltage>>ai0**, name your task *MyVoltage-Task*, and then press the **Finish** button. Then, within the central part of the MAX window, you will find the **Configuration** tab displayed for your task. After scrolling down, select **Acquisition Mode>>N Samples**, **Samples to Read>>100**, and **Rate (Hz)>>1k** as shown next. Finally, click on the **Save** button near the top of the window. The creation of your task is now complete.

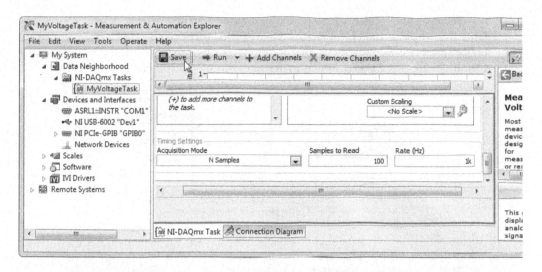

To implement your task, open a blank VI. On the block diagram, place a **DAQmx Task Name**, found in **Functions>>Measurement I/O>> DAQmx – Data Acquisition** as shown below at the left. Using the 🖑, click on its menu button and select *MyVoltageTask* as shown below at the right.

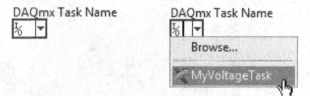

The **DAQmx Task Name** icon can automatically generate code to carry out your intended task. To implement this feature, right-click on the **DAQmx Task Name** and select **Generate Code>>Example**. Block-diagram code, which implements two DAQmx icons that we will study in Chapter 13, along with a front-panel

Waveform Graph, will be automatically created as shown next. Using a function generator, input a 100 Hz sine-wave voltage waveform to channel *ai0* of your DAQ device. Run this VI and verify that it obtains a single $N = 100$ sample trace with a sampling rate of 1 kHz, thus acquiring 10 cycles of the 100 Hz sine wave.

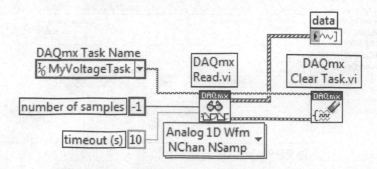

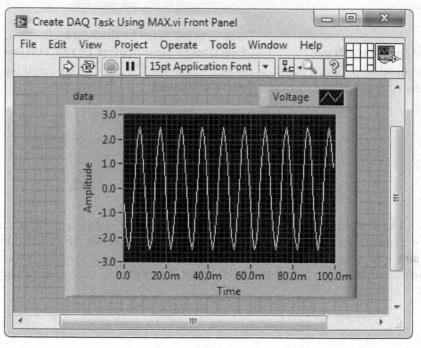

PROBLEMS

1. **Derivation of Aliasing Condition** Writing the aliasing condition as $f = \pm f_{alias} + n f_s$, show that $(2\pi f t_i) = \pm \sin (2\pi f_{alias} t_i)$ at all $t_i = i \Delta t$.

2. **Observation of Voltage Resolution** Observe the voltage resolution ΔV of an analog input operation carried out by your DAQ device and verify that your

observation is consistent with the prediction of Eq. [5.1]. Attach a DC voltage to AI channel *ai0* and then use MAX (under the **Analog Input** tab with **Mode>>On Demand**) to acquire 100 (or so) digitized samples of this input voltage. After stopping the acquisition, inspect the data samples of this "constant" input signal, which in reality will vary slightly due to electronic noise in the experimental setup and, possibly, the signal's slow time-variation. You should find that these samples are distributed in discrete voltage levels. Measure the smallest spacing ΔV between two of these adjacent levels. Given the range and resolution of your DAQ device, does your value for ΔV agree with the prediction of Eq. [5.1]? If your DAQ device can operate with a different range, repeat this procedure to verify that ΔV decreases (or increases) as expected.

3. **Pulse Train Generation Using MAX** If you have a DAQ device with a hardware digital timer (PCIe-6351 and ELVIS II have one, USB-6002 and myDAQ do not), use MAX to generate a *digital pulse train*, that is, a waveform that alternates between the HIGH and LOW digital states at a given frequency. Accomplish this feat as follows: Under the **Counter I/O** tab, for an available counter (e.g., **Channel Name>>Dev1/ctr0**) select **Mode>>Pulse Train Generation**. For **Pulse Terminal**, select the output pin for the counter being used (e.g., the output pin for *counter 0* on the PCIe-6351 is called CTR 0 OUT and is the PFI 12 pin). Connect the pin selected by **Pulse Terminal**, along with the D GND pin, to the input of an oscilloscope (or an AI channel of your DAQ device). Press **Start** and verify that a TTL-compatible digital waveform is being created with the frequency input at **Frequency** control.

4. **Creating a Simulated Device** If a DAQ device is not available, a software replica (called a *simulated device*) can be created using MAX and then used in the development of data acquisition programs.

Create a USB-6002 simulated device as follows: In MAX, right-click on **Devices and Interfaces** under the **My System** heading and select **Create New…**. In the dialog window that appears, select **Simulated NI-DAQmx Device or Modular Instrument** and press **Finish**. Then, in the **Create Simulated NI-DAQmx Device** window, choose **USB DAQ>>USB-6002** and press **OK**.

The USB-6002 simulated device will now appear listed under **Devices and Interfaces** in the same manner as a real DAQ device, except simulated devices are denoted by yellow icons.

Create a **Test Panels** window for the simulated device, then choose the **Analog Input** tab and select **Mode>>Finite**. When you press **Start**, a simulated acquisition at the specified AI channel of one cycle of a noisy sine wave will be displayed on the chart. Under the **Digital I/O** tab, the digital lines will count in binary, if **All Input** is selected.

CHAPTER 6

Data Acquisition Using DAQ Assistant

> In this chapter, you will learn how to control a National Instruments data acquisition (DAQ) device connected to your computer. Here, it is assumed that you have such a DAQ device available, along with a DC voltage source (e.g., battery, power supply, function generator), function generator, voltmeter, oscilloscope, and required cabling. Prior to starting this chapter, you may want to download the latest version of the device driver **NI-DAQmx** from *ni.com/support*. If you do not have access to a DAQ device, use a simulated device to carry out the exercises in this chapter (see Problem 4 in Chapter 5).

6.1 DATA ACQUISITION VIs

Equipped with the skills developed in the previous five chapters, you are now ready to write LabVIEW programs that automate data acquisition and generation processes. In this chapter, the programs you create will be based on *DAQ Assistant*, an easy-to-use *Express VI* that controls the operation of a multifunction data acquisition (DAQ) device.

LabVIEW's built-in data acquisition functions are designed to operate all of the features on any DAQ device manufactured by National Instruments (NI), the company that also produces LabVIEW. These features include analog-to-digital conversion, digital-to-analog conversion, digital input/output operations, event timing, digital pulse counting, and digital pulse generation. Once a NI DAQ device is connected to your computer (e.g., plugged into a PCI Express expansion slot or attached via a USB cable), LabVIEW offers you both a low-level and a high-level approach to control its operation.

In the low-level approach, one develops programs based on the *DAQmx VIs*, which are found in **Functions>>Measurement I/O>>DAQmx — Data Acquisition**. Each of the DAQmx VIs executes a key step (such as setting the sampling rate or storing acquired data in a memory buffer) necessary in performing a complete data acquisition or data generation process. By properly configuring these software building blocks on a block diagram, a LabVIEW programmer can fully control every operation involved in the data acquisition process and so write programs that provide the ultimate in flexibility and power to the user. This control, of course, comes at the price of added complexity to the block-diagram programming.

At the other extreme, LabVIEW offers the programmer a suite of high-level functions that perform common measurement tasks such as communicating with an instrument, storing acquired data in a file, and performing fast Fourier transform analysis. These functions, called *Express VIs*, are found in **Functions>>Express**. Express VIs are straightforward to use and perform many sophisticated tasks automatically, but have limited capabilities. If your programming requirements are simple, you may often find that an Express VI will do your job quite nicely.

For the high-level method of controlling the data acquisition or generation process, LabVIEW offers an Express VI called **DAQ Assistant**. In this chapter, you will explore the fundamentals of controlling a DAQ device using DAQ Assistant to perform three basic DAQ device operations—analog-to-digital conversion (acquiring a digitized representation of an incoming analog voltage), digital-to-analog conversion (generating an outgoing analog signal from a sequence of digital numbers), and digital line control (setting and reading the line's HIGH/LOW state). In the world of LabVIEW, these three operations are called *Analog Input (AI)*, *Analog Output (AO)*, and *Digital Input/Output (DIO)*, respectively. Later, in Chapter 13, we will revisit this topic and learn to control a DAQ device using the more advanced DAQmx VIs.

6.2 SIMPLE ANALOG INPUT OPERATION ON A DC VOLTAGE

Let's begin by turning your computer into a DC voltmeter by building a VI that carries out the same Analog Input operation that we ran using MAX in Section 5.6. For this exercise, apply a given DC voltage difference (I'll use 2 V) to differential channel *ai0* of your DAQ device. For example, the high-voltage and low-voltage wire from your DC source, respectively, should be connected to AI 0 and AI 8 for the PCIe-6351, AI 0 and AI 4 for the USB-6002, and AI 0^+ and AI 0^- for the myDAQ or NI ELVIS II (if you are using a "floating" voltage source, see Section 5.6).

Construct the front panel shown below in which an acquired voltage is displayed in the **Numeric Indicator** labeled **Voltage**. Through Eq. [5.1], you can determine the voltage resolution ΔV of the analog-to-digital converter in your DAQ device and use this information to select the proper number of **Digits of precision** on the **Voltage** indicator (using the pop-up menu's **Display Format...** option). For example, if $\Delta V = 0.3$ mV, **Digits of precision** should be *4*. Using **File>>Save**, first create a new folder named **Chapter 6** within the **YourName** folder, and then save this VI under the name **DC Voltmeter (Express)** in **YourName\Chapter 6**.

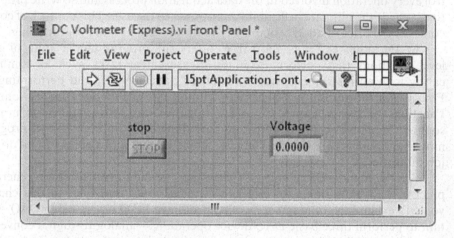

Switch to the block diagram. There, place **DAQ Assistant**, found in **Functions>>Express>>Input**.

A moment after you place DAQ Assistant on the block diagram, a dialog window called **Create New...** will open automatically. DAQ Assistant can be configured to perform any of the data acquisition operations within the capabilities of your DAQ device. In this dialog window, you will select the particular operation that you would like DAQ Assistant to perform for this VI. Since we want to digitize ("acquire") an incoming analog voltage, click on **Acquire Signals**.

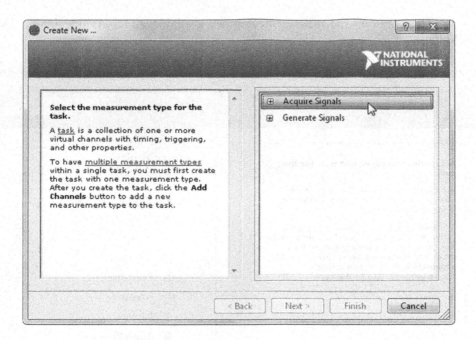

A list of options under **Acquire Signals** will appear in hierarchical tree fashion. Click on **Analog Input**.

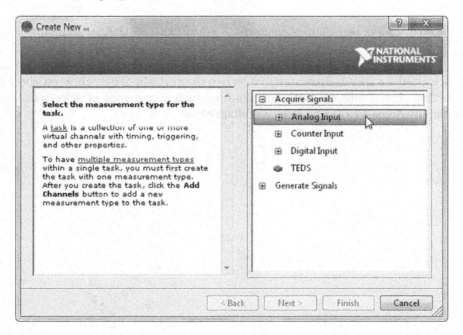

Then, among the measurement types available under **Analog Input**, select **Voltage**.

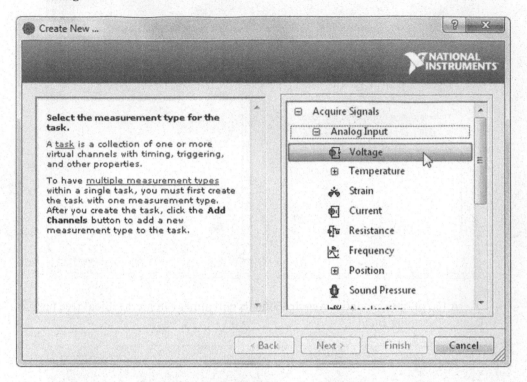

Finally, the available AI channels on your DAQ device (called *physical channels*) will be listed. From this list, select channel **ai0**. Then click on the **Finish** button. From now on, our shorthand for such a sequence of selections will be **Acquire Signals>>Analog Input>>Voltage>>ai0**.

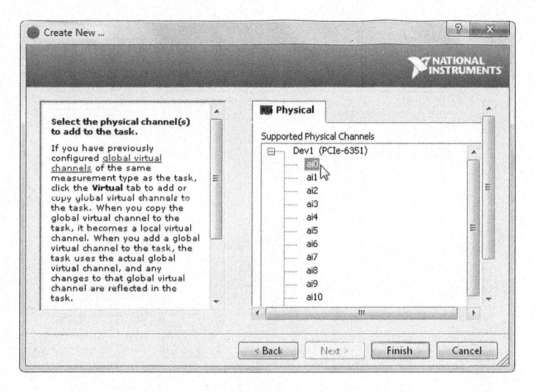

Above, we selected a particular data acquisition operation (analog input voltage measurement) to be performed on a particular physical channel (*ai0*). This combination of selections becomes part of the definition of a newly created *task*. The **DAQ Assistant** dialog window will now appear automatically. In this window, we will configure data-taking details that will complete the task's definition.

The **DAQ Assistant** dialog window opens with the **Express Task** tab selected and the (automatically given default) name of our task—**Voltage**—is highlighted. Under the **Settings** tab, enter the values shown in the following diagram. Under **Timing Settings**, select **Acquisition Mode>> 1 Sample (On Demand)**. In this mode, DAQ Assistant will acquire (only) one voltage reading every time it is executed.

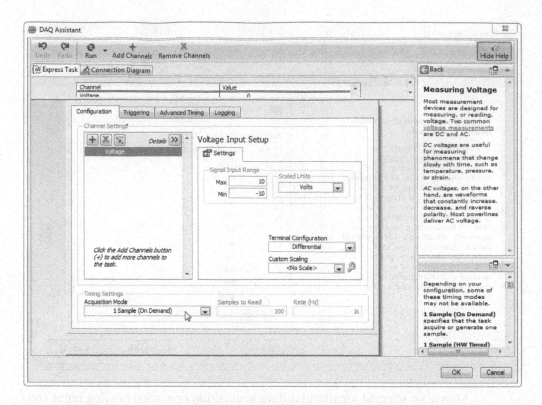

To familiarize yourself with some of the available information within this dialog window, click on the **Connection Diagram** tab. There (after possibly having to input the I/O Terminal Block used by your system), you will find a helpful list and diagram outlining the proper DAQ device pins to use when connecting to differential channel *ai0*.

Return to the original window by clicking on the **Express Task** tab. Select **Display Type>>Chart** (if the chart region is not fully visible, use the Resizing Cursor ⇕ as shown to expand the window), and then click on the **Run** button near the upper left corner of the window. DAQ Assistant will execute repeatedly, acquiring one voltage sample per execution, and the sequence of acquired voltage readings is plotted on the chart. If this plot consists of a sequence of values, each (approximately) equal to the DC voltage difference you applied at channel *ai0*, then you have successfully configured DAQ Assistant for our intended task. When satisfied, press the **Stop** button to cease the voltage acquisition process.

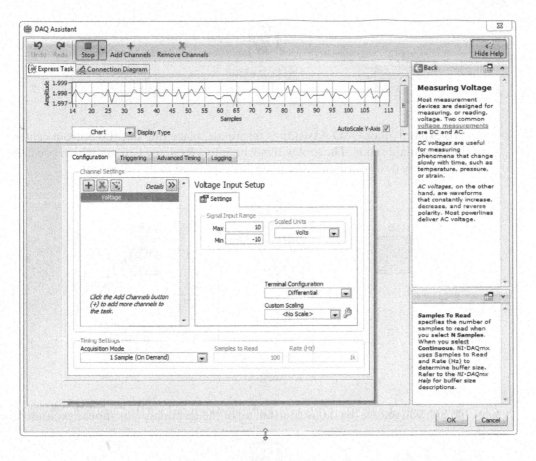

Click **OK** near the bottom right corner to finalize this configuration process. When the dialog window disappears, **DAQ Assistant** will appear on your block diagram as an *expandable node*, with inputs and outputs appropriate for the selections that you made within the DAQ Assistant dialog window. In an expandable node, some of its terminals are "unexpanded" unnamed arrows (inward direction for inputs, outward directed for outputs), while other terminals are "expanded" in their own named band near the bottom of the icon. Below, only the **data** output terminal is expanded. This terminal reports the acquired voltage reading each time DAQ Assistant executes.

To expand all of DAQ Assistant's terminals, place the ⬉ at the bottom center of the expandable node. The tool will morph into a Resizing Cursor, which you can click and drag downward until the node has expanded to its full extent.

When you release the mouse button, all of the terminals will appear expanded.

 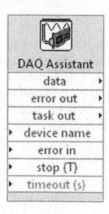

Code the block diagram as shown next. Here, the While Loop iterates once every 0.1 s until the **Stop Button** is pressed. With each iteration, DAQ Assistant acquires a single voltage reading from channel *ai0*, and this value is displayed on the front panel **Voltage** indicator.

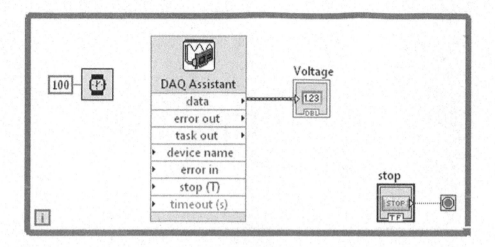

There are a few subtle features of this block diagram, which we will pause for a moment to discuss. First, the **Wait (ms)** icon is included within the While Loop so that, when executing, this program does not monopolize your computer's resources. Because DAQ Assistant alone executes very quickly, without the delay produced by **Wait (ms)**, the loop would iterate at such a high rate that your processor would have to devote most of its efforts to running this program, leaving little time for the other operations for which it is responsible (see Problem 7 in Chapter 2).

Second, look closely and you'll find that the wire connecting the **data** output of DAQ Assistant and Voltage's icon terminal is dark blue and banded, a form that we have not encountered up to this point. The format of this wire is called *dynamic data type* (abbreviated *DDT*), a data type used by Express VIs. In addition to the data associated with a signal (in our present case, a single digitized voltage value), a dynamic data type wire includes attributes of the signal such as its name and timestamp (i.e., the date and time that the data were acquired). Note that, because of the polymorphic nature of a **Numeric Indicator**, we are able to wire the dynamic data type wire directly to Voltage's icon terminal. A red coercion dot appears, however, indicating that only the data value is being passed to the icon terminal (and all additional information about this signal's attributes is being ignored). Later in this chapter, we will see that these attributes can be quite useful when plotting an acquired signal on a Waveform Graph.

Finally, all LabVIEW data acquisition icons (including DAQ Assistant) have an *error cluster* input and output that you can elect to use for runtime error reporting. Include this useful feature in your VI by popping up on the **error out** terminal and selecting **Create>>Indicator**.

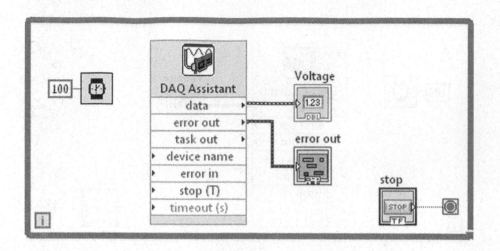

Return to the front panel, where you will now find an **error out** cluster included. The error cluster consists of three elements: **status** (Boolean TRUE if there is an error, FALSE if no error); **code** (integer identifying the error); and **source** (descriptive text identifying the function within which the error occurred). When an error does occur, its integer code can be deciphered by popping up on **code** and selecting **Explain Error**.

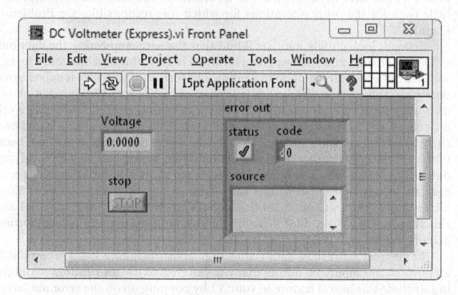

Use the ⬥ to position all the front-panel objects nicely. Then, with your chosen DC voltage difference connected to differential channel *ai0*, run **DC Voltmeter**

(Express). Does it read the input voltage difference correctly? Save your work on this VI.

If interested, add the two improvements shown below to the block diagram of DC Voltmeter (Express).

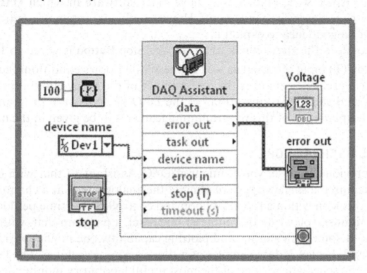

First, if your computing system has an up-to-date version of the **NI-DAQmx** device driver, **DAQ Assistant** will possess a **device name** input terminal. In the above block diagram, a **DAQmx Device Name Constant** ![Dev1] has been wired to this input terminal. To produce the ![Dev1], pop up on the **device name** terminal and select **Create>>Constant**. Then, by clicking on its *menu button* ![▼] with the ![hand] as shown below, the **DAQmx Device Name Constant** will present you with a list of available DAQ devices connected to your computer, allowing you to select the one that you wish to be controlled by **DAQ Assistant**.

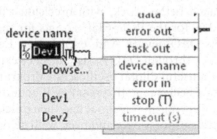

Wiring a ![Dev1] to DAQ Assistant's **device name** input has the following advantages. If more than one DAQ device is connected to your computer, your program can be easily switched from controlling one device to another. Additionally, your VI can be easily reconfigured to run on another computing system. In the subsequent VIs written in this chapter, I will not include a $\boxed{\text{Dev1} \blacktriangledown}$ on the block diagrams (since some readers may have older software in which **DAQ Assistant** does not have the **device name** input). However, if your system presents this opportunity, it's a good idea to exploit it.

Second, in the above block diagram, the **Stop Button** is wired to DAQ Assistant's **stop (T)** input terminal as well as the While Loop's conditional terminal $\boxed{\odot}$. This wiring scheme enhances the performance of the VI by eliminating unneeded start-up and shut-down operations of the DAQ device as the program is run. A detailed explanation of this programming feature will be given in the next section.

6.3 DIGITAL OSCILLOSCOPE

In the previous exercise, you configured DAQ Assistant so that with each execution it acquired a single voltage reading of the analog signal at a channel input. By putting this icon within a repetitive loop and controlling the time per loop iteration with **Wait (ms)**, you wrote **DC Voltmeter (Express)**, a program that, with the simple addition of some data storage and plotting capability, one might consider using to sample and display N data samples of a time-varying waveform. The last sentence outlines the schematic for one of the most useful laboratory monitoring systems— the *digital oscilloscope*. Perhaps you've already thought of some cool things to do with this idea and are ready to go back and start to make the needed modifications to **DC Voltmeter (Express)** . . . but unfortunately **Wait (ms)** is an Achilles' heel in this plan. **Wait (ms)** measures time by accessing a clock within your computer, which has an accuracy of 1 millisecond (ms). Thus, by using **Wait (ms)** to mark time in the data-taking process, the moment of each voltage digitization will be determined with an uncertainty on the order of 0.001 second. For very-low-frequency inputs (say, less than 10 Hz), this level of uncertainty in x-axis values might possibly be acceptable for the envisioned oscilloscope's *Voltage* vs. *Time* output, but it will be useless over the higher range of frequencies in which you would want an oscilloscope to operate.

Thankfully, a much better clock, with precision timing on the order of 10 nanoseconds (e.g., 10 ns for the PCIe-6351, 12.5 ns for the USB-6002), exists on a National Instruments DAQ device. And, even better, LabVIEW provides access and control of this clock through DAQ Assistant. Let's explore this feature of DAQ Assistant as we write a useful digital oscilloscope program. The program we wish to write, called **Digital Oscilloscope (Express)**, will acquire N equally spaced voltage samples of a time-varying analog input and then quickly plot this array of data values. By repeating this process over and over, we'll achieve a real-time display of the waveform input.

Open a blank VI and, using **File>>Save**, store it in **YourName\Chapter 6** under the name **Digital Oscilloscope (Express)**. On the front panel of **Digital Oscilloscope**

(Express), place a **Stop Button** and a **Waveform Graph** with its *x*- and *y*-axes labeled Time and Voltage, respectively. If you like, hide the Stop Button's label by toggling it off via **Visible Items>>Label**. Also, in the Controls Palette, choose **Select a Control...**, and then navigate to the YourName\Controls folder in the **Look in:** box. There, select a **Digitizing Parameters.ctl** control cluster (by either double-clicking or highlighting and then pressing the **OK** button), and then place this control cluster on the front panel. (If you haven't created and saved **Digitizing Parameters.ctl** previously, refer to Section 4.10, "Control and Indicator Clusters," and create this control cluster now.)

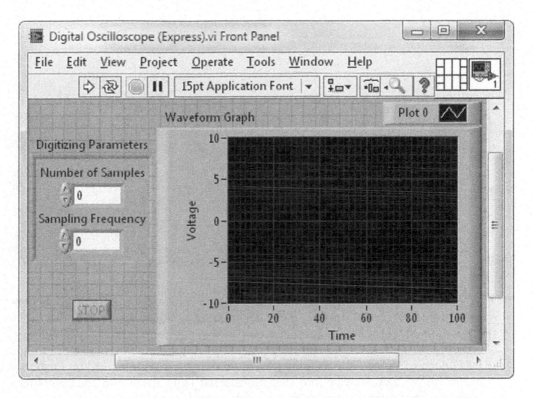

Switch to the block diagram. Place a **While Loop** there and then place a **DAQ Assistant** (from **Functions>>Express>>Input**) icon within the loop. When the **Create New...** dialog window opens automatically, select **Acquire Signals>>Analog Input>>Voltage>>ai0**. Then, when the **DAQ Assistant** dialog window appears, configure it as shown below. Note that we have set **Acquisition Mode** to **N Samples** and that in this mode, the settings for **Samples to Read** and **Rate (Hz)** are activated. These settings were disabled in the **1 Sample (On Demand)** mode we used for DC Voltmeter (Express).

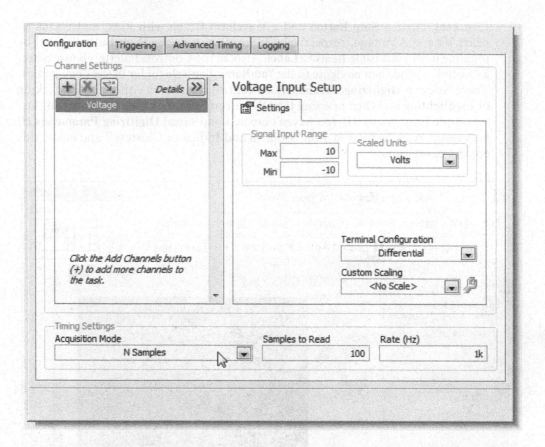

The **N Samples** acquisition mode is a *buffered hardware-timed analog input* operation. In this process, a time-dependent analog signal is digitized at N equally spaced times. The time interval between neighboring samples $\Delta t = 1/f_s$, where f_s is the sampling frequency, is very accurately controlled by a hardware clock on the DAQ device. As the data samples are acquired, they are first placed in a memory buffer on the DAQ device and then transferred from there into the computer's memory. This method assures that no data samples are lost when the computer multitasks (i.e., switches among its various chores) during the data-taking process. In the dialog window's **Timing Settings** section, **Samples to Read** and **Rate (Hz)** are used to select the values for N and f_s, respectively.

Once you have programmed the appropriate values in the DAQ Assistant dialog window, close it by pressing the **OK** button. If later you wish to reopen this dialog window, you can do so either by popping up on **DAQ Assistant** and selecting **Properties** in its pop-up menu or by simply placing the ⬚ over **DAQ Assistant** and double-clicking.

After you are returned to the block diagram, resize the DAQ Assistant icon so that all of its terminals are expanded, and observe all of the available input and output terminals for DAQ Assistant when configured in the **N Sample** mode.

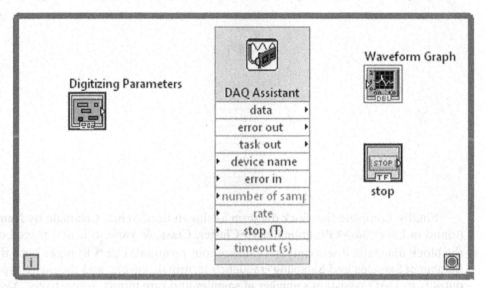

We could now proceed directly to wiring, but some convoluted wire paths would result from the present ordering of the DAQ Assistant terminals. Let's first change the terminal ordering to something more desirable. Pop up on the (top) **data** terminal and choose **Select Input/Output>>number of samples**. DAQ Assistant will then appear as shown next.

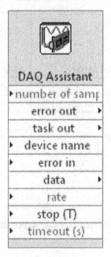

Through a similar procedure, order the terminals as shown below and then resize DAQ Assistant so that **error in**, **task out**, and **timeout (s)** are no longer expanded (if **device name** appears on your icon, you may wish to keep it expanded).

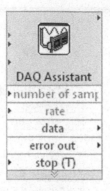

Finally, complete the block diagram as shown next. When **Unbundle By Name** (found in **Functions>>Programming>>Cluster, Class, & Variant**) is first placed on the block diagram, it will only have one output terminal. Use ⬈ to make both the **Number of Samples** and **Sampling Frequency** terminals visible, and then wire these outputs to DAQ Assistant's **number of samples** and **rate** inputs, respectively. Also, create the **error out** cluster by popping up on DAQ Assistant's **error out** terminal and selecting **Create>>Indicator**. Note that when you connect DAQ Assistant's **data** output to the Waveform Graph's terminal, the resulting wire is of the dynamic data type.

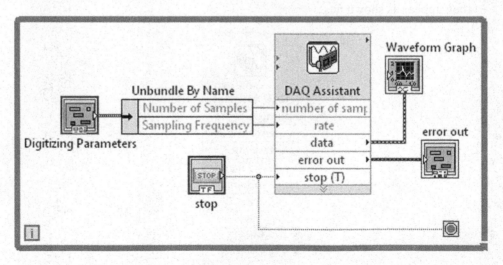

Besides being wired to the While Loop's conditional terminal 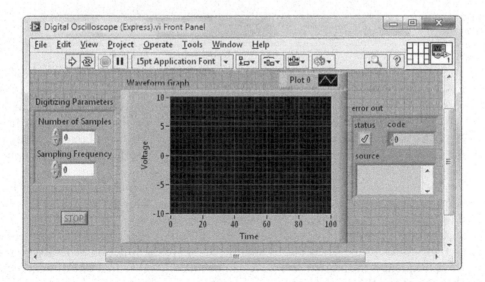, the **Stop Button** is wired to DAQ Assistant's **stop (T)** input for the following reason. Within DAQ Assistant, several operations are carried out that either start up or shut down the requested data acquisition task. Thus, these particular operations need only be called during the first or last of the multitude of While Loop iterations. However, by default, the **stop (T)** input of DAQ Assistant is TRUE, causing these start-up and shut-down (collectively termed *overhead*) operations to be executed each time DAQ Assistant executes. Since your data-taking system is not able to digitize the incoming stream of real-time data during overhead operations, needlessly including them increases the program's *dead time*, the percentage of each iteration period during which the system is not sensitive to the incoming data. With increasing dead time, more and more cycles of the incoming repetitive data will stream past your system's input undetected. In certain situations, such as when collecting a large number of data cycles with the intent of adding them together so as to average out random noise, a programming inefficiency that causes you to miss a significant percentage of the incoming data cycles can easily extend the completion time of your experiment by minutes (or sometimes even hours). By wiring **stop (T)** to the **Stop Button**, this input will be FALSE during all but the last While Loop iteration. Then DAQ Assistant will perform its start-up operations only during the first iteration and its shut-down operations solely during the last, thus enhancing the performance of this program.

Return to the front panel, neaten the arrangement of objects using the ⬆, and then save your work.

Let's test-drive **Digital Oscilloscope (Express)**. Adjust the settings on a function generator so that it is outputting an analog sine wave with amplitude less than 10 V and a frequency of about 50 Hz. Using the appropriate method for your particular system, connect the positive and negative (ground) outputs from the function generator to the positive and negative inputs of differential channel *ai0* on your DAQ device. (*Note:* Most function generators are referenced to the building ground and so are not "floating" voltage sources.) Then set **Number of Samples** and **Sampling Frequency** equal to *100* and *1000*, respectively, and run **Digital Oscilloscope (Express)**. If all goes well, you will see a plot of about five cycles of the 50 Hz sine wave.

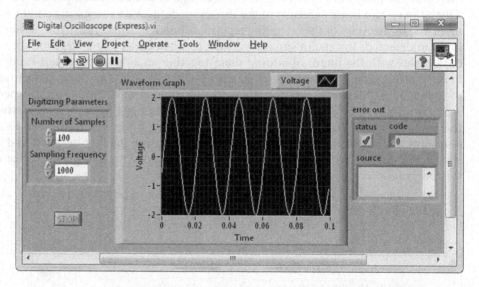

Note that, without any coding effort on our part, the **Plot Legend** is now labeled with the task name **Voltage** and the *x*-axis is properly scaled to reflect the time increment Δt between successive sample acquisitions. These automatic actions are a benefit of the dynamic data type format used to pass the acquired data from DAQ Assistant to the Waveform Graph. Attribute information about these data is included within the dynamic data type wire and is used by the Waveform Graph to accomplish these automatic actions.

Be impressed by the relative ease with which you wrote this program, which acquires and plots a succession of digitized sine waves. However, I'm sure that one flaw in this computer-based instrument is apparent to you. Most likely, the sine-wave plot you are observing does not appear stationary but instead is moving either to the right or to the left. And, by slightly changing the function generator's sine-wave frequency, the plotted sine wave can be made to move first one direction and then the other. Why is this so?

In your program, DAQ Assistant is configured so that with each execution, it outputs an array of N equally spaced data samples. Let's call this array of N samples a *trace*. As you know from coding the block diagram, the sine-wave image viewed on the Waveform Graph is not just a single trace but is actually a sequence of many traces, where a new trace is generated and plotted with each iteration of the While Loop. Since nothing in your code forces each trace to begin at an equivalent point on the repetitive incoming waveform, the successive traces are displaced from each other along the *x*-axis, resulting in a motion picture-style moving image or, worse, a jumbled mess. The issue we are broaching here is called *triggering* and, for **Digital Oscilloscope (Express)** to be a useful program, we must build in this capability.

In a commercially available oscilloscope, triggering is accomplished in the following manner. The input signal is monitored by an analog *"level-crossing"* circuit. The purpose of this circuit is to determine each time the incoming signal passes through a specified threshold voltage level and then immediately trigger the scope's data acquisition process. Using knobs on the scope's front panel, the oscilloscope user sets the circuit's threshold level and specifies whether acquisition should be initiated when the level is passed through starting from above (*negative*, or *falling, slope*) or starting from below (*positive*, or *rising, slope*).

Many National Instruments DAQ devices are capable of performing the analog level-crossing triggering procedure just described. For example, the PCIe-6351 and ELVIS II have this capability; however, the lower-cost USB-6002 and myDAQ do not. So that the broadest range of readers can include triggering in their **Digital Oscilloscope (Express)** VI, we'll implement an alternate mode—*digital edge triggering*—in our program because this mode is available on (almost) all NI DAQ devices. At the time of this writing, the exception to this rule is the myDAQ; myDAQ owners, please see Problem 6 at the end of this chapter for a software triggering alternative to the hardware digital triggering described here.

To execute a data acquisition operation under digital edge triggering, one connects a digital signal to the appropriate pins on a DAQ device. Then, by choosing the proper software settings, the desired data acquisition operation can be begun ("triggered") whenever the digital signal has a rising or, alternatively, falling edge.

Stop **Digital Oscilloscope (Express)**, if it's running, and then configure it for digital edge triggering as follows. On the block diagram, double-click on **DAQ Assistant**. When the DAQ Assistant dialog window opens, click on the **Triggering** tab, and select **Trigger Type>>Digital Edge** and **Edge>>Rising**. **Trigger Source** will identify the possible DAQ device pins at which the digital signal can be applied; choose one of the available pins. In the window below, the PFI 0 pin is chosen on a PCIe-6351. The ground for the digital signal would also need to be connected to the D GND. Once you have made your selection, click on the **OK** button.

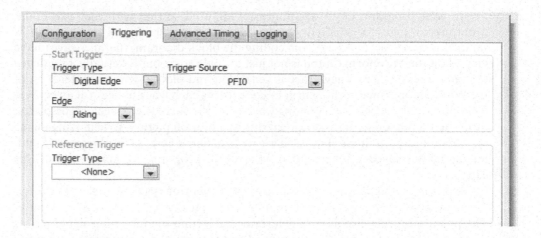

That's it. **Digital Oscilloscope (Express)** is now configured for digital edge triggering. Return to the front panel and save your work.

Besides producing a sine wave with frequency f at its *main* output, your function generator creates a TTL digital signal of the same frequency f, which is available at its *sync* (sometimes called *TTL* or *trigger out*) output. The digital edges of this TTL signal coincide in time with particular points on the sine-wave's cycle. For example, on some function generator models, the rising edge of the TTL signal occurs when the sine wave is at its peak; for other models, the rising TTL edge coincides with the positive-going zero-crossing of the sine wave. Thus, the sync output provides a convenient digital signal with which one can control the acquisition of sine-wave data (i.e., always begin at the same equivalent point on the cycle) via digital edge triggering.

Connect the sync output from your function generator to the digital triggering pins on your DAQ device. For example, sync's positive and negative (ground) terminals would connect to PFI 0 and D GND, respectively, for the software selection in the illustration given above.

On the front panel, set **Number of Samples** and **Sampling Frequency** equal to *100* and *1000*, respectively, and set the sine-wave frequency on your function generator to approximately 100 Hz. Run **Digital Oscilloscope (Express)**. If all goes well, you will see a "stationary" plot of about 10 cycles of the 100 Hz sine wave. (For USB-6001/6002/6003 devices, you may have to delete the Boolean wire connected to the **stop (T)** input of **DAQ Assistant** to make digital triggering functional.)

On the function generator, set the frequency f of the input sine-wave signal to several values within the range $0 \leq f \leq 500$ Hz. You should find that frequencies of signals in this frequency range are faithfully reproduced by **Digital Oscilloscope (Express)**, but that sine-wave shape becomes distorted as the signal frequency approaches 500 Hz (because of the paucity of sampled points per sine-wave cycle at these higher input frequencies).

We have set the sampling frequency equal to $f_s = 1000$ Hz; thus, the Nyquist frequency is $f_{nyquist} = f_s/2 = 500$ Hz. For input frequencies greater than the Nyquist frequency, that is, $f > 500$ Hz, we expect the aliasing effect. Demonstrate that for $f = 1100$ Hz, the digitized sine wave appears to have a frequency of 100 (rather than 1100) Hz. Use Eq. [5.2] to predict two other input frequencies f that will appear as a 100 Hz sine wave when digitized because of aliasing, and then demonstrate that your predictions are correct using **Digital Oscilloscope (Express)**.

Finally, this VI has one constraint, which reflects a limitation of the **DAQ Assistant** Express VI. Namely, **DAQ Assistant** sets the DAQ device's sampling frequency during its first execution within the program (i.e., during the first iteration of the While Loop), and this value cannot be changed as the program continues to run. You can verify this point as follows: Set the sine-wave frequency on your function generator to 50 Hz and start **Digital Oscilloscope (Express)** with **Number of Samples** and **Sampling Frequency** equal to *100* and *1000*, respectively. On the front panel, you will see a plot of five cycles of the 50 Hz sine wave. Now, with the VI continuing to run, keep **Number of Samples** equal to *100*, but change **Sampling Frequency** to *5000*. By increasing the sampling rate by a factor of five, one would expect to now acquire only one cycle of the 50 Hz sine wave. However, you will still observe five cycles because the DAQ device's sampling frequency did not change from its initial value of 1000 Hz. Stop the VI and then start it again with the new settings: **Number of Samples** and **Sampling Frequency** equal to *100* and *5000*, respectively. Now you will observe the expected single cycle.

Stop the VI and then reset **Number of Samples** and **Sampling Frequency** to *100* and *1000*, respectively. Start **Digital Oscilloscope (Express)** so that you again see the plot of five cycles of the 50 Hz sine wave. Now, with the VI continuing to run, keep **Sampling Frequency** as *1000*, but change **Number of Samples** to *200*. By increasing the number of samples by a factor of two, one would expect to now acquire ten cycles of the 50 Hz sine wave, but again the plot remains of just five cycles.

To summarize, we have found the following properties of **DAQ Assistant** when performing an AI operation in an iterative loop: After being set during this icon's initial execution, the values of **rate** (i.e., the sampling frequency) and **number of samples** remain fixed. To changes these values, one must stop and then restart **DAQ Assistant**. In Chapter 9, we will find a way to work around this limitation of **DAQ Assistant** (see Section 9.4).

6.4   DC VOLTAGE SOURCE

Now let's explore the Analog Output (AO) operations available on your DAQ device. To familiarize ourselves with these processes, we'll start by writing a program that simply outputs a requested voltage value. On a blank VI, place a **Numeric Control** labeled **Voltage** and a **Stop Button**. If you like, hide the Stop Button's label (by toggling it off via **Visible Items>>Label**) and make the Voltage control's **Digits of precision** appropriate for the voltage resolution of your particular DAQ device. Save this VI in the **YourName\Chapter 6** folder under the name **DC Voltage Source (Express)**.

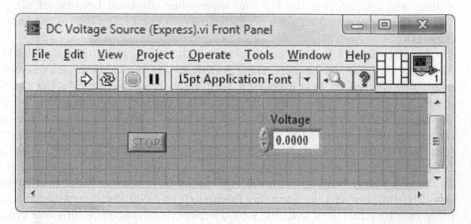

Switch to the block diagram and place **DAQ Assistant** within a While Loop there. When the **Create New...** window opens, make the following sequence of selections: **Generate Signals>>Analog Output>>Voltage>>ao0**, and then press **Finish**.

After the **DAQ Assistant** dialog window opens, input the selections shown in the following diagram, and then click on the **OK** button. Note we are programming V_{min} and V_{max} to be -10 and $+10$ V, respectively, values that are appropriate for all current DAQ devices (on the USB-6009 legacy device, the appropriate values are 0 and $+5$ V, respectively). With the **Generation Mode>>1 Sample (On Demand)** option selected, the DAQ device will update the voltage output at channel *ao0* every time that DAQ Assistant is executed in your program. This method of updating the AO channel is called *software timing*.

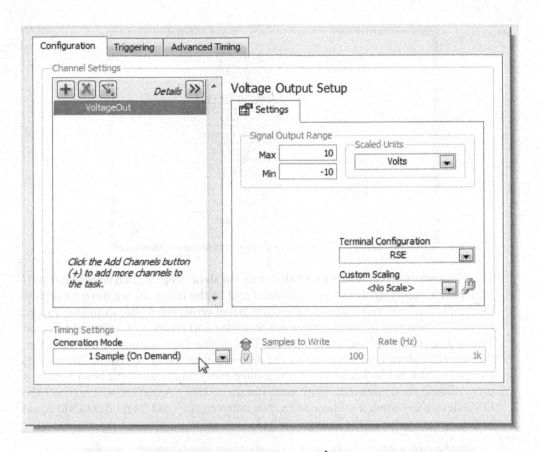

Upon return to the block diagram, use the ↖ to expand all of the terminals of DAQ Assistant when programmed for a **1 Sample (On Demand)** Analog Output operation.

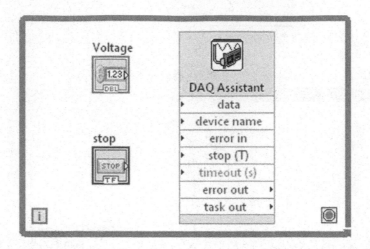

Then resize DAQ Assistant so that only the **data**, **stop**, and **error out** (and possibly **device name**) terminals are expanded and in the order shown next. Complete the block diagram, remembering the front-panel **error out** cluster indicator can be created by popping up on the **error out** terminal and selecting **Create>>Indicator**. The red coercion dot indicates that the DBL value from **Voltage** is converted to the dynamic data type required at the **data** input to DAQ Assistant. When this code runs, the While Loop will execute once every 10 ms until the **Stop Button** is pressed (and then released). With each iteration, DAQ Assistant will instruct the DAQ device to output a voltage difference between pins AO 0 and AO GND equal to the current value of the **Voltage** control.

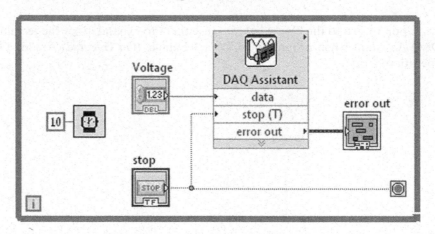

Switch to the front panel and arrange the objects as you wish.

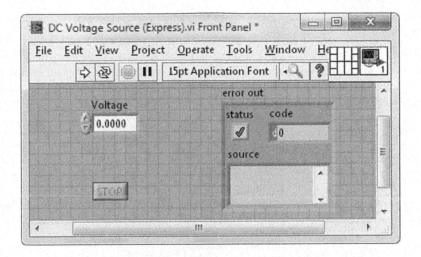

Connect the positive and negative (ground) input of a voltmeter (or oscilloscope in autotriggering mode) to the AO 0 and AO GND pins of your DAQ device, respectively.

Then set the **Voltage** control to some value within the range −10 V to +10 V and press the **Run** button. The voltmeter should inform you that the requested voltage difference is being produced by your DAQ device. With the VI still running, set **Voltage** to a few other values within the −10 V to +10 V range to convince yourself that the VI is functioning properly.

Next, let's purposely cause a runtime error in this program and see what happens. Remembering that we programmed the DAQ device to output voltages only within the range from −10 to +10 V, set **Voltage** equal to a value outside of this range, say, 12 V.

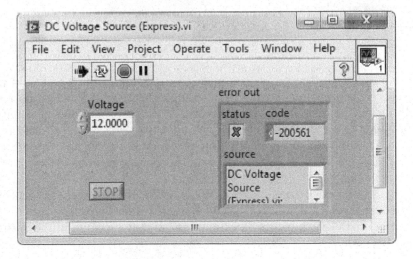

The **error out** cluster indicator informs us that an error has occurred with the ✖ in its **status** box, which corresponds to a Boolean value of TRUE. The **source** box tells us where within the program the error occurs, and **code** identifies the cause of the error via an integer code. To decipher this integer code, pop up on the **code** box and select **Explain Error**. A dialog window will appear, describing the reason for the error; in the present case, the cause is that the requested voltage is "too large or too small."

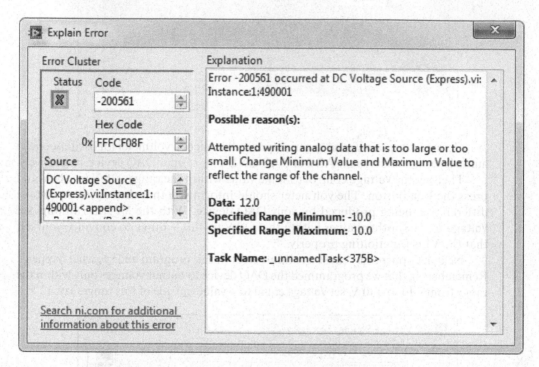

Note that your program continues to run, despite this error, and if you set **Voltage** back to within the acceptable range, the VI will function correctly as if nothing out of sorts had ever happened. In some programming situations, the "continue-to-run-as-if-nothing-is-wrong" behavior you observed here is fine, but more commonly one wants a program to respond in some appropriate way (such as to cease operation) if an error occurs. Appropriate response to an error is called *error handling*.

Let's build error handling into **DC Voltage Source (Express)** by making the VI cease operation when an error occurs. Press the **Stop Button**, if the VI is still running, and then switch to the block diagram so that we can add the desired error-handling code.

Currently, there is only one way to stop our VI. When the **Stop Button** is pressed, a TRUE Boolean value is passed to the ⏻, which causes the While Loop to stop iterating. Consider the following alternate way to stop the program. When an error occurs, the **status** element within the **error out** cluster becomes TRUE. By passing this value to the ⏻, we could stop the While Loop. You can include both of these two possible program-stopping methods in your VI with the following coding.

Place an **Unbundle By Name** (found in **Functions>>Programming>>Cluster, Class, & Variant**) on the block diagram and wire it to the **error out** terminal of DAQ Assistant. Once wired to the **error out** terminal, **Unbundle By Name** can be resized to have output terminals for all three of the elements within the **error out** cluster (status, code, source) if desired. However, since we only need **status** in our present program, keep **Unbundle By Name** sized to a single terminal. The element associated with this terminal can be selected by popping up on the right (output) side of **Unbundle By Name** and using **Select Item**.

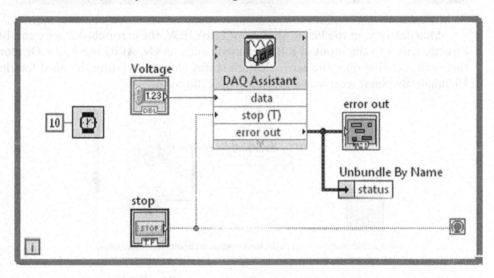

On your block diagram, allow a TRUE Boolean value from either the **Stop Button** or the error cluster's **status** element to cease the While Loop iteration by including an **Or** icon (found in **Functions>>Programming>>Boolean**) as shown next.

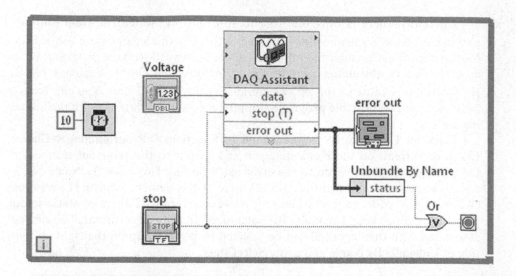

Alternatively, in the latest versions of LabVIEW, the error cluster wire can be directly wired to the input of a Boolean icon such as **Or**. At its input, the **Or** icon then will monitor only the error cluster's **status** value, obviating the need for the **Unbundle By Name** icon in the previous block diagram.

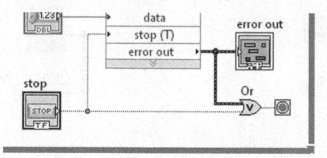

Return to the front panel and save your work. Run the VI with various values for **Voltage**. You will now find that when **Voltage** is set to a value outside of the range of −10 V to +10 V, the program ceases operation.

When you are finished, run **DC Voltage Source (Express)** one more time with **Voltage** set to *0*, so the channel *ao0* is left in the zero-voltage state.

6.5 HARDWARE-TIMED WAVEFORM GENERATOR

By continually updating an analog output channel with appropriately chosen voltage values, one can turn a DAQ device into a waveform generator with a time-dependent voltage output of any desired shape (e.g., sine wave). To obtain an output voltage whose time dependence accurately represents that desired shape, one must assure that the AO channel changes ("updates") to particular values at precisely the right moments. The most accurate method for timing these required updates involves the use of a high-frequency hardware clock. Such clocks are contained internally on all of our representative DAQ devices and can be used to create high-quality analog waveforms (a hardware clock was not included on the USB-6009 legacy device, and so only the much less accurate method of software-timed waveform generation is possible on that device; see Problem 3).

If your DAQ device supports hardware-timed analog output operations, try building the following high-quality waveform generator. On the front panel of a blank VI, place a **Stop Button** and an **XY Graph** with *x*- and *y*-axes labeled **Time** and **Voltage**, respectively. Then save this VI under the name **Waveform Generator (Express)** in **YourName\Chapter 6**.

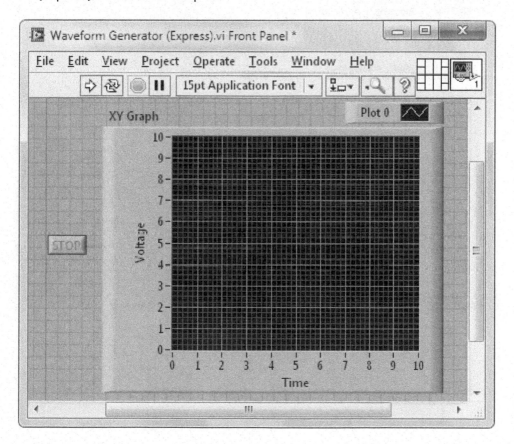

Switch to the block diagram. Place a **While Loop** there and then put a **DAQ Assistant** within it. When the **Create New...** window opens, make the following sequence of selections: **Generate Signals>>Analog Output>>Voltage>>ao0**, and then press **Finish**.

When the **DAQ Assistant** dialog window opens, make the selections shown in the following diagram. Be sure to uncheck the **Use Waveform Timing** box (located next to the ✋ in the illustration). Note we are programming V_{min} and V_{max} to be -10 and $+10$ V, respectively. In the **Generation Mode>>Continuous** mode, a 1D array consisting of a sequence of voltage values is first written into a memory buffer associated with DAQ Assistant, and then the DAQ device outputs this sequence of voltages at channel *ao0* in a *circular fashion* (i.e., after the last array value is output, the next output is the first array value). When all of the selections are properly made, click on the **OK** button.

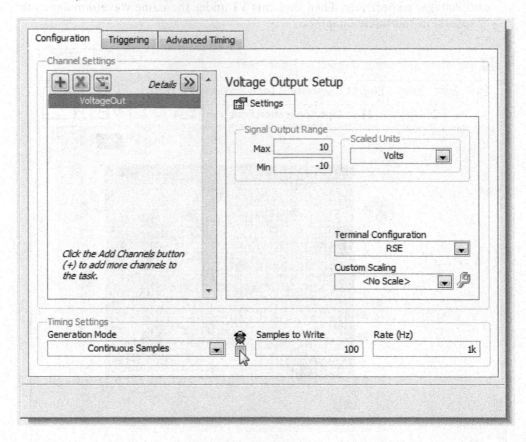

Upon return to the block diagram, expand the terminals of DAQ Assistant and order them as shown next using **Select Input/Output** in each terminal's pop-up menu.

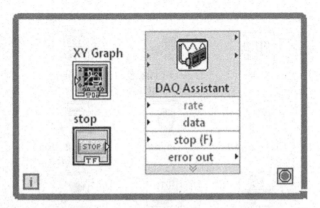

The plan for our VI is this: Use the **Waveform Simulator** program written in Chapter 4 to generate a 1D array of voltage values that describes a desired waveform. Then pass this 1D array to the **data** input of **DAQ Assistant**, which initiates the process of outputting the voltage waveform at channel *ao0*.

6.6 PLACING A CUSTOM-MADE VI ON A BLOCK DIAGRAM

First, we must place **Waveform Simulator** (as a subVI) on our block diagram. Here's how to do that. On the Functions Palette, click on the **Select a VI...** category.

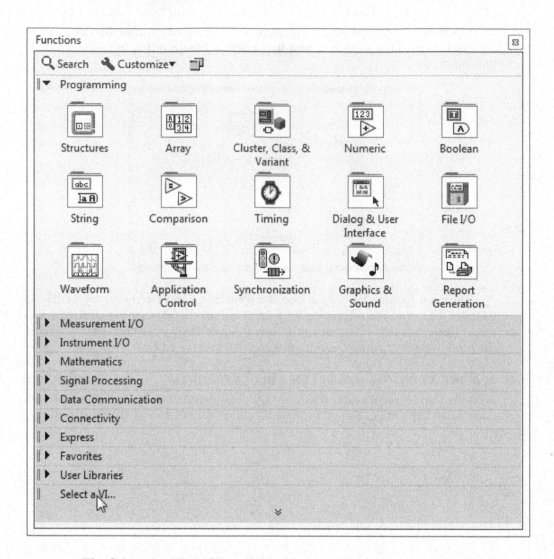

The **Select the VI to Open** dialog window will appear. In the top box, navigate to the **YourName\Chapter 4** folder. Once the file list for this folder is displayed, double-click on **Waveform Simulator** or highlight it, and then click on the **OK** button in the dialog box.

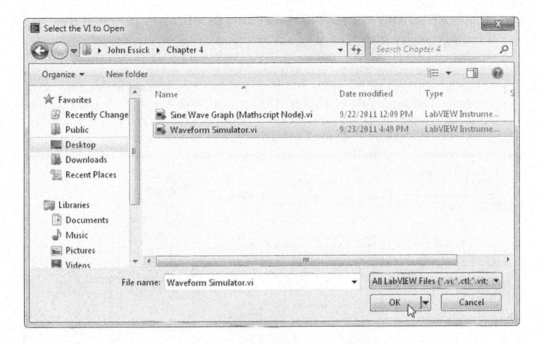

You will then be returned to the block diagram, where you can place your custom-written icon in a convenient position. Then, with the ✒, create **Digitizing Parameters** and **Waveform Parameters** control clusters by popping up on these terminals and using **Create>>Control**.

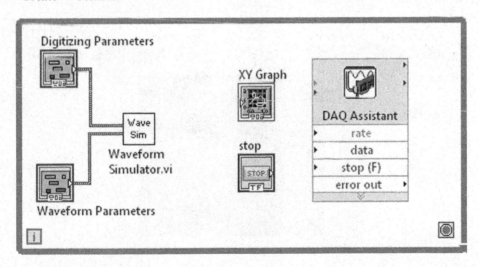

6.7   COMPLETING AND EXECUTING WAVEFORM GENERATOR (EXPRESS)

Finally, complete the block diagram as shown next. When you wire Unbundle By Name's **Displacement** output to DAQ Assistant's **data** input, the **Convert to Dynamic Data** icon will be inserted into the wire automatically. This icon converts the 1D floating-point numeric array produced by **Waveform Simulator** to the dynamic data type used by Express VIs such as DAQ Assistant. If you ever need to place **Convert to Dynamic Data** on a block diagram manually, it is found in **Functions>>Express>>Signal Manipulation**.

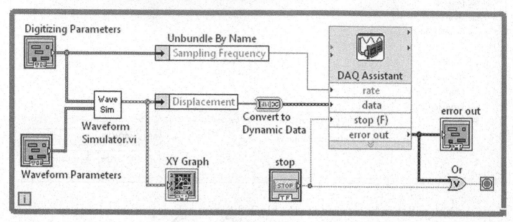

Return to the front panel, tidy it up, and then save your work.

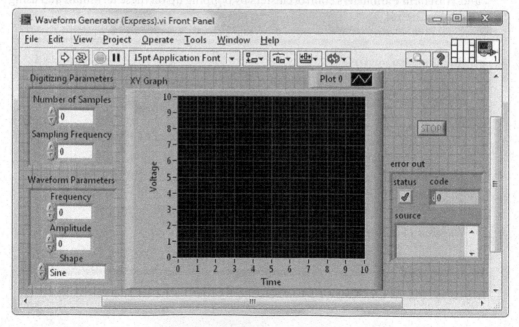

Here's how to program your VI to output a continuous waveform. Using the **Digitizing Parameters** and **Waveform Parameters** control clusters, choose values that instruct **Waveform Simulator** to produce one (or, possibly, a few) complete cycle(s) of the desired periodic waveform. This software representation of the waveform's cycle will be written into a memory buffer by **DAQ Assistant**. The array of waveform values will then be output sequentially (in a circular fashion) by your DAQ device at a rate given by the **Sampling Frequency** until a TRUE value is sent to DAQ Assistant's **stop** input.

Connect the positive and negative (ground) input of an oscilloscope to the AO 0 and AO GND pins of your DAQ device, respectively.

Test-drive your VI by using it to output a 50 Hz sine wave of amplitude 5 V. First, in the **Waveform Parameters** control cluster, program **Frequency**, **Amplitude**, and **Shape** to be *50*, *5*, and *Sine*, respectively. Next, assuming 100 points are required to well define each cycle of the sine wave, the sampling frequency should be 100 times larger than the sine wave's frequency. Thus, for **Waveform Simulator** to create one complete cycle, in the **Digitizing Parameters** control cluster let **Number of Samples** and **Sampling Frequency** be *100* and *5000*, respectively.

Run your VI. The software waveform produced by **Waveform Simulator** will be displayed on the **XY Graph** for you to view. Using your oscilloscope, observe the voltage waveform output at channel *ao0*. Is it a (quasi-) continuous sine wave with a 50 Hz frequency and 5 V amplitude?

With your VI still running and keeping all other parameters the same, change **Frequency** (in the **Waveform Parameters** control cluster) to *75*. From the XY Graph, you will find that **Waveform Simulator** produces 1 1/2 cycles of this waveform. Using your knowledge of the circular fashion in which the DAQ device sequences through the memory buffer values, can you explain the discontinuities that you observe in the voltage waveform being output at channel *ao0*?

With the VI still running, try changing **Frequency** to *100*, *200*, *250*, and *500*. With each of these choices, **Waveform Simulator** will produce an integer number of complete cycles of the waveform and so produce voltage waveform outputs with no discontinuities. Why does the waveform output at channel *ao0* become less well defined for the higher frequencies? Also, you might enjoy changing **Shape** and **Amplitude** as the VI is running.

Finally, this VI has one constraint, which reflects a limitation of the **DAQ Assistant** Express VI. Namely, **DAQ Assistant** sets the DAQ device's sampling frequency during its first execution within the program (i.e., during the first iteration of the While Loop), and this value cannot be changed as the program continues to run. You can verify this point as follows: Start the VI with **Frequency**, **Amplitude**, and **Shape** equal to *50*, *5*, and *Sine*, respectively, and **Number of Samples** and **Sampling Frequency** equal to *100* and *5000*, respectively. **Waveform Simulator** will create one cycle of a 100 Hz sine wave, and you will observe a 50 Hz sine wave at *ao0*. Now, with the VI continuing to run, change **Frequency** and **Sampling Frequency** to

25 and *2500*, respectively. **Waveform Simulator** will create 100 points that represent one cycle of a 25 Hz sine wave if output at a rate of 2500 Hz, but you will observe a 50 Hz sine-wave output at channel *ao0* because the DAQ device's sampling frequency did not change from its initial value of 5000 Hz.

To summarize, we discovered the following properties of **DAQ Assistant** when performing an AO operation in an iterative loop: After being set during this icon's initial execution, the value of **rate** input (i.e., the sampling frequency) remains fixed. However, the array read into its **data** input can be changed during any execution of the icon, allowing the parameters **Frequency, Amplitude, Shape,** and **Number of Samples** to be adjusted as an iterative program runs.

DO IT YOURSELF

Icons used in this DIY project can be found with the aid of **Quick Drop**.

Alternating LEDs.vi In this exercise, you will write a VI that alternately lights one LED and then another. The signals that toggle these LEDs on and off will be provided by Digital Input/Output (DIO) pins on a DAQ device.

First, light a single LED under computer control as follows: Choose a DIO line on your DAQ device, for example, P0.0 (on the myDAQ, this line is labeled DIO 0), and then construct the hardware circuit shown in Figure 6.1. In this circuit, when the line goes HIGH, it supplies a DAQ device-dependent voltage difference (common values are 3.3 V or 5 V) across the resistor–LED series combination. Assuming this voltage difference is 3 V, the predicted current through the LED will be 3 V/220 $\Omega \approx 14$ mA (for a more precise prediction of the current, the voltage drop across the forward-biased LED should be taken into account). If this predicted current is larger than the current-sourcing capability of your DAQ device's digital output line, the current will top out at the maximum allowed value. This maximum value can be found on your DAQ device's datasheet (for the PCI-6351 and USB-6002, this value is 24 mA and 4 mA, respectively).

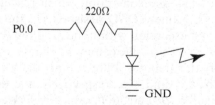

FIG. 6.1 Hardware circuit for Do It Yourself project. In this figure, the Digital I/O line P0.0 of the data acquisition device is connected to the (current-limiting) resistor in series with LED.

Write a VI that can be used to illuminate this single LED via a Boolean switch (e.g., **Push Button**) on its front panel. On the program's block diagram, place **DAQ Assistant** and configure it as **Generate Signals>>Digital Output>>Line Output>>port0/line0** and **Generation Mode>>1 Sample (On Demand)**. With this configuration, when a Boolean TRUE (FALSE) is passed to DAQ Assistant's **data** input, P0.0 will go HIGH (LOW). The complete block diagram is shown next. Since a DAQ device has N DIO lines, **data** is formatted to accept an array of Boolean values with up to N elements. Hence, the **Build Array** icon is necessary to convert the Boolean switch's single Boolean value to a Boolean array with one element.

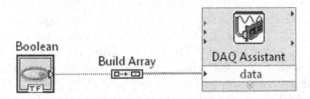

Run your VI several times and show that you can toggle the LED in your hardware circuit on and off via the setting of the Boolean switch.

Next, construct the hardware circuit shown in Figure 6.2 with two LEDs, which we will call A and B. Then write a While Loop-based VI such that during even iterations of the While Loop, A is lit and B is dark, and during odd iterations, B is lit and A is dark. The LEDs alternately light until a front-panel **Stop Button** is clicked.

Suggested icons: **Quotient & Remainder, Select, Equal To 0?.**

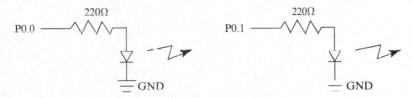

FIG. 6.2 Hardware circuits controlled by Alternating LEDs VI. In this figure, digital I/O lines P0.0 and P0.1 are used.

USE IT!

Anti-Aliasing Filter By passing an analog signal through a low-pass filter prior to inputting it to a digitizer, the aliasing effect can be suppressed. When used in this manner, the low-pass filter is called an *anti-aliasing filter*. The gain of a filter is defined as its output voltage divided by its input voltage: $G \equiv V_{out}/V_{in}$. An ideal anti-aliasing filter would transmit all signals with frequencies less than the Nyquist frequency to the digitizer at unity gain $\left(V_{out} = V_{in}\right)$, but would completely block—that is, attenuate to zero gain—all signals with frequencies greater than the Nyquist frequency. The two frequency ranges $f < f_{nyquist}$ and $f > f_{nyquist}$ are called the *pass band* and *stop band*, respectively. For real filters, the shift from the pass band to the stop band is not abrupt, but takes place over an intermediate range of frequencies called the *transition band*; for frequencies within the transition band, gains are in the range $0 < G < 1$.

Butterworth low-pass filters of various orders n are commonly used as anti-aliasing filters. This class of filter provides "maximally flat" signal transmission (i.e., close to constant unity gain) up to a given cutoff frequency f_c. In the transition region above f_c, the filter's gain changes by (approximately) a factor of $(1/2)^n$ for each doubling of signal frequency. A doubling of frequency is called an *octave*. For example, if $f_c = 100$ Hz, a 200 Hz signal is one octave, and a 400 Hz signal is two octaves, above the cut-off frequency. Thus, for a first-order $(n = 1)$ Butterworth filter with $f_c = 100$ Hz (assuming $G \approx 1$ at the cutoff frequency), the gain at 200 Hz and 400 Hz will be $(1/2)^1 = 0.5$ and $(1/2)^1 (1/2)^1 = 0.25$; for an eighth-order $(n = 8)$ Butterworth filter with $f_c = 100$ Hz, the gain at 200 Hz and 400 Hz will be $(1/2)^8 = 4 \times 10^{-3}$ and $(1/2)^8 (1/2)^8 = 2 \times 10^{-5}$, respectively.

To demonstrate the design of an anti-aliasing filter, construct the op-amp circuit shown in Figure 6.3, which is a first-order Butterworth low-pass filter. The cutoff frequency (also known as the *3-dB frequency*) for this circuit is given by $f_c = 1/2\pi RC$; for your circuit, use $R = 10$ kΩ and $C = 0.1$ μF so that $f_c \approx 150$ Hz. Attach a function generator to the filter's input and connect its output to the analog input channel pins programmed into **Digital Oscilloscope (Express)**. Also attach the function generator's sync output to the proper DAQ device pins to facilitate digital triggering.

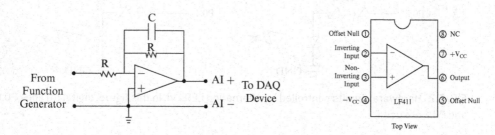

FIG. 6.3 First-order Butterworth low-pass filter. Suggested op-amp is LF411.

On the front panel of **Digital Oscilloscope (Express)**, set **Number of Samples** and **Sampling Frequency** equal to *200* and *2000*, respectively, making $f_{nyquist} = 1000$ Hz. Configure your function generator to input a (smaller than Nyquist frequency) 50 Hz sine wave to amplitude 1 V to the low-pass filter. Run **Digital Oscilloscope (Express)**. Since the sine-wave frequency is less than the filter's cutoff frequency, you will see a plot of about five cycles of the unattenuated $(G \approx 1)$ 50 Hz sine wave. Turn off autoscaling on the y-axis.

Next, input a 1200 Hz sine wave of amplitude 1 V to the filter. This frequency is greater than the Nyquist frequency as well as the filter's cutoff frequency and so, instead of a 1200 Hz waveform, your data acquisition system will produce an aliased waveform of 800 Hz, but with an amplitude attenuated from the 1 V input value. Since 1200 Hz is three octaves above $f_c \approx 150$ Hz, we expect the gain due to this first-order filter to be approximately $(1/2)^1 (1/2)^1 (1/2)^1 = 0.125$. Thus, this simple filter suppresses aliasing, but not to the level that we might hope. Since 1200 Hz exceeds $f_{nyquist}$, this input sine wave is in the stop band and so the gain would ideally be much closer to zero.

To be optimally effective, an anti-aliasing filter would receive a large-amplitude signal within the stop band and attenuate it to below the DAQ system's smallest resolvable voltage level. Thus, for a 16-bit digitizer with a range of $+10$ V and voltage resolution $\Delta V = 0.3$ mV, the optimal gain for this filter within its stop band would be on the order of $G \equiv V_{out}/V_{in} = 0.0003$ V/10 V $= 3 \times 10^{-5}$. From our above calculations, we see that an eighth-order Butterworth filter can achieve this order of attenuation for frequencies greater than two octaves above f_c. Such high-order Butterworth filters are available as low-cost integrated circuits. For example, the Maxim MAX291 chip is an eighth-order Butterworth low-pass filter, whose cutoff frequency can be chosen anywhere within the range from 0.1 Hz to 25 kHz. When placed in the circuit shown in Figure 6.4, the cutoff frequency (in Hertz) is given by $f_c = 333/C$, where C is in nF. Build this circuit with $C = 1$ nF so that $f_c \approx 300$ Hz, and demonstrate its ability to completely suppress the aliasing of a 1200 Hz input sine wave (1200 Hz is two octaves above 300 Hz).

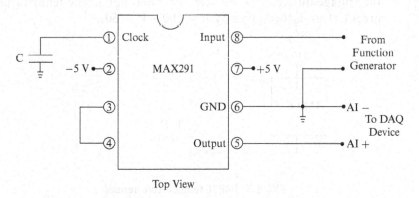

Top View

FIG. 6.4 MAX291 eighth-order Butterworth low-pass filter. Capacitor C determines cutoff frequency.

PROBLEMS

For a problem that involves writing a new program, the problem statement begins with a suggested descriptive name (including the **.vi** extension) for the VI that you will write; icons needed for the VI may be found with the aid of **Quick Drop**. For a problem that involves the use of a VI already written as part of the chapter text, the problem's topic is given in bold at the beginning of the problem statement.

1. **Dual Digital Oscilloscope (Express).vi** Build a two-channel digital oscilloscope. With **Digital Oscilloscope (Express)** open, use **File>>Save As...** to create a new VI. On the block diagram, you must reprogram **DAQ Assistant** to read two channels rather than just one. Open **DAQ Assistant** and add the new channel, and then close the dialog window. On the front panel, the **Plot Legend** can be expanded using a Resizing Cursor to accommodate two plots. Your VI is now complete.

   Attach a voltage waveform and the sync out digital signal from a function generator to the two AI channels of your DAQ device programmed on the block diagram. Additionally, attach the sync out digital signal to the PFI 0 channel of your DAQ device. Run your VI and verify that it simultaneously displays traces of the two inputs.

2. **Temperature Sensor (Express).vi** The Analog Devices TMP36 temperature sensor (available from adafruit.com and sparkfun.com) is a low-cost three-terminal semiconductor device, which can be used to measure temperatures in the range from $-40\ ^\circ$C to $125\ ^\circ$C. As shown in Figure 6.5, to operate the TMP36, simply connect a power supply voltage to pin 1 (this voltage can be anything in the range from $+2.7$ to 5.5 V) and ground to pin 3. Then measure the voltage difference V between pin 2 and pin 3. The temperature T in degrees Celsius is then given by $T = 100 \times V - 50$.

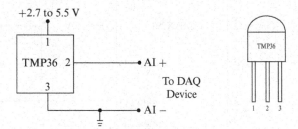

FIG. 6.5 TMP36 temperature sensor.

Connect a powered TMP36 device to the AI differential channel *ai0* pins of your DAQ device. Using **DAQ Assistant**, write a program that, every 250 ms until a **Stop Button** is pressed, reads the voltage V, converts this value to the corresponding temperature in Celsius, and then displays this temperature in a front-panel indicator. Program DAQ Assistant as follows: **Acquire Signals>>Analog Input>>Voltage>>ai0** and **Acquisition Mode>>1 Sample (On Demand)**. To extract the measured voltage from the dynamic data type wire output from DAQ Assistant, use the **Convert from Dynamic Data** icon (found in **Functions>>Express>>Signal Manipulation**). When **Convert from Dynamic Data** is first placed on the block diagram, its dialog window will open; select **Conversion>>Single scalar** and **Scalar data type>>Floating point numbers (double)**, and then press the **OK** button.

Run your program and use it to measure room temperature as well as the temperature of your skin.

3. **Sine Wave Generator (Software-Timed).vi** Build the following program, which creates a sine-wave voltage output using the software timing provided by a **Wait (ms)** icon. As the While Loop's iteration terminal increments, the code within the MathScript (or Formula) Node evaluates the function $y(x) = 2 + \sin x$ at N equally spaced arguments x in each sine-wave cycle. The spacing in radians between adjacent arguments is $2\pi/N$, where N is the number of samples per cycle. On the block diagram, pop up on the **data** output and make sure that **Choose Data Type>>All Types>>Scalar>>DBL** is selected. Program DAQ Assistant with the following selections: **Generate Signals>>Analog Output>>Voltage>>ao0** and **Generation Mode>>1 Sample (On Demand)**.

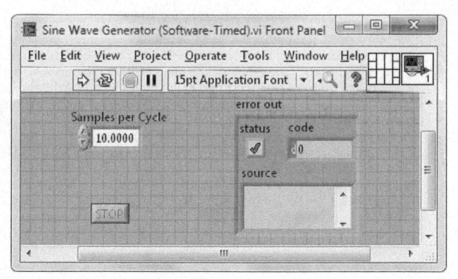

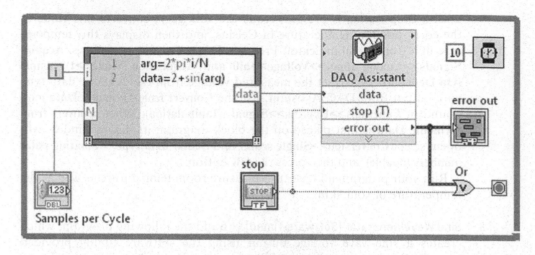

If using a Formula Node instead, substitute the following coding.

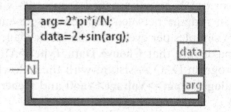

Connect the AO 0 and AO GND pins of your DAQ device to the positive and negative (ground) input of an oscilloscope, respectively. Then run your VI with **Samples per Cycle** set to *10* so that each sine cycle is represented by ten voltage levels. Given that each While Loop iteration takes 10 ms, the period of the observed sine wave should be (10 ms/update) × (10 updates) = 100 ms, making its frequency 10 Hz. You will find that, although the underlying sine pattern is apparent, the ten voltage levels used to form this waveform are too well resolved to give the appearance of a continuous function.

Try increasing **Samples per Cycle** to *20*, then *30*, and so on. Since the While Loop iteration time remains constant, as you increase **Samples per Cycle**, the frequency of the output waveform will decrease. You will need to adjust your oscilloscope's *Time/Div* control appropriately to view a complete cycle of the waveform. At what value for **Samples per Cycle** does the output waveform start to take on a continuous appearance? (This problem can be done with all DAQ devices, including the USB-6009 legacy device.)

4. **Bode Magnitude Plot.vi** If your DAQ device supports hardware-timed analog output operations, write a program that performs the following process: Apply an AC voltage with frequency f and 1 V rms amplitude to the input of a circuit, measure the resulting rms voltage amplitude V_{out} at the circuit's output, and then calculate the gain (in decibels) $G = 20 \log \left(V_{out} / V_{in} \right)$, where $V_{in} = 1 \text{ V}_{rms}$. Construct the VI such that this process is repeated for frequencies in the range $f = 10$ Hz to $f = 2010$ Hz in 25 Hz increments, and the resulting data are displayed on a plot of G vs. $\log(f)$. Build your VI according to the following guidelines.

    Block Diagram: The workhorse of this VI is formed by configuring three Express VIs as shown below. Here, **Simulate Signal** is used to create a sine wave of frequency f and 1 V rms amplitude, which is passed to a **DAQ Assistant** configured to generate this waveform at an analog output channel. Assuming this sine-wave signal is applied at the input of a circuit, the second **DAQ Assistant** is configured to read 500 samples of the response at the circuit's output at a sampling frequency of $50f$ (i.e., 10 cycles of the output will be read and made available at the **data** terminal). Finally, these 10 cycles are supplied to **AC & DC Estimator.vi**; when you wire the second DAQ Assistant to AC & DC Estimator.vi, the **Convert from Dynamic Data** icon will be inserted into the wire automatically. In the dialog window for Simulate Signal, set **Samples per second (Hz)>> 100000** (or **40000**, if using USB-6009). For **Number of samples**, select both **Automatic** and **Integer number of cycles**. For the analog output DAQ Assistant, select **Generation Mode>>Continuous** and check **Use Waveform Timing**. For the analog input DAQ Assistant, select **Acquisition Mode>>N Samples**.

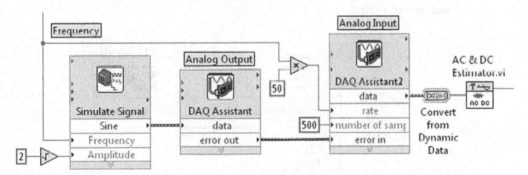

    Front Panel: Place an **XY Graph** on the front panel and supply it with bundled arrays of f and G from the block diagram. To obtain a log scale on the x-axis, pop up on the XY Graph and select **X Scale>>Mapping>>Logarithmic**.

When completed, use your VI to obtain a Bode plot on the *RC* circuit shown in Figure 6.6, where $R = 4.7$ kΩ and $C = 0.1\,\mu$F.

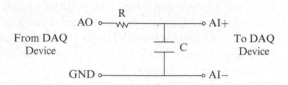

FIG. 6.6 Problem 4.

5. **Music Box.vi** If your DAQ device supports hardware-timed analog output operations, write a program that plays the scale of A major, with each note sounding for 500 ms. The eight notes in this scale have the following frequencies in Hertz: 440.0, 493.9, 554.4, 587.3, 659.3, 740.0, 830.6, and 880.0. These frequencies can be stored either in a front-panel **Array Control** or in a block-diagram **Array Constant** (created by placing a **Numeric Control** within an **Array Constant** shell). On the block diagram, input each frequency f one at a time into a For Loop that contains the subdiagram shown below (see the Use It! example in Chapter 3). Here, the **Simulate Signal** Express VI creates an array representing an integer number of cycles of the desired sine wave with frequency f, which is then output as a real voltage waveform on an AO channel by DAQ Assistant. When programming **Simulate Signal**, set **Samples per second (Hz)>>100000** (or **40000**, if using USB-6009); for **Number of samples**, select both **Automatic** and **Integer number of cycles**.

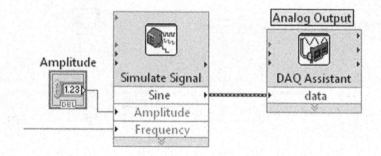

To play the scale, attach an amplifier and speaker to the AO channel such as shown in Figure 6.7, and then run your program.

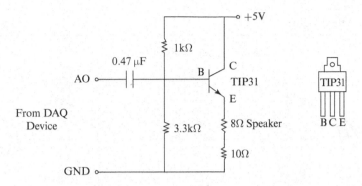

FIG. 6.7 Amplifier circuit for Problem 5.

6. **myDAQ Digital Oscilloscope (Express).vi** For myDAQ owners, include digital triggering in your digital oscilloscope program using software rather than hardware. Assume that you wish to use your oscilloscope to observe a waveform produced by a function generator and that this waveform is being digitized on AI channel 0 of your myDAQ. Further assume that the sync output from the function generator is simultaneously being digitized on AI channel 1 of the myDAQ.

First, create a VI called **Digital Trigger (Software)**, whose front panel is shown next. To create the 2D Array Control **Acquired Data**, first create a 1D Array Control (select an **Array** shell first, then place a **Numeric Control** inside), and then pop up on its index display and select **Add Dimension**. The representation of **Number of Samples** is **I32**.

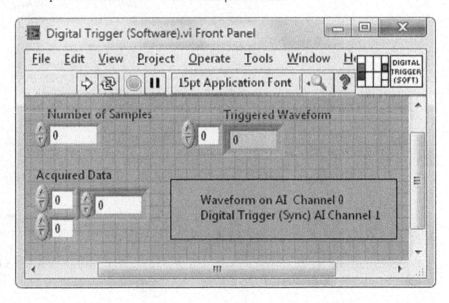

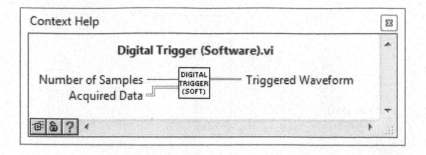

Next, code the block diagram of **Digital Trigger (Software)** as shown. The workhorse of this code is **Basic Level Trigger Detection.vi**, which searches the acquired data from the Digital Sync channel to find the location (i.e., the array index) at which the trigger condition occurs (which is taken to be about halfway up the rising edge as the digital signal transitions from its LOW to HIGH level). The **Index Array** icons extract the first (index-0) and second (index-1) rows of the 2D array, which are the Waveform and Digital Sync 1D arrays, respectively. Consult the Help Windows of **Index Array** and **Array Subset** to determine how these icons function; also see Sections 8.7 and 11.6.

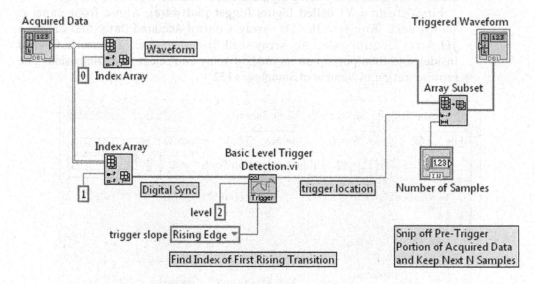

Finally, modify the block diagram of **Digital Oscilloscope (Express)** as shown below. Here, you must program **DAQ Assistant** to read both channels AI 0 and AI 1. As you configure this Express VI, it will instruct you how to select multiple AI channels as you define a task. Also, when you wire DAQ Assistant's **data** output to Digital Trigger (Software)'s **Acquired Data** input, the **Convert from Dynamic Data** icon will be inserted into the wire automatically. This icon converts the dynamic data type produced by Express VIs to the 2D floating-point numeric array used by **Digital Trigger (Software)**. If you ever need to place **Convert from Dynamic Data** on a block diagram manually, it is found in **Functions>>Express>>Signal Manipulation**.

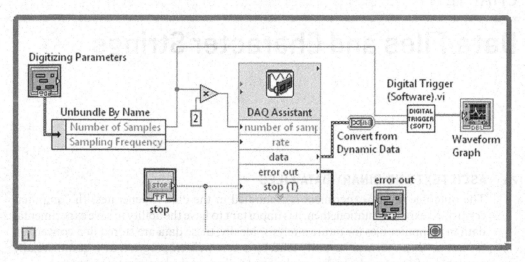

Once completed, connect a function generator's main and sync output to the AI 0 and AI 1 pins of your myDAQ. Configure the generator to produce a sine wave of frequency 50 Hz with amplitude less than 10 V, and then run with **Number of Samples** and **Sampling Frequency** set to *1000* and *10000*, respectively. Is the observed waveform on your oscilloscope properly triggered (i.e., stationary in appearance)?

CHAPTER 7

Data Files and Character Strings

7.1 ASCII TEXT AND BINARY DATA FILES

The outcome of any experiment is embodied in the data it generates. In computer-controlled experimentation, then, it is important to have the ability to save experimental data in computer files for future analysis. Ideally, these data are stored in a convenient format so that they can be easily read by a user-written program or a commercially available software package that analyzes and displays the experimental results.

Computers commonly communicate alphanumeric data using the *American Standard Code for Information Interchange* (ASCII). In this coding scheme, seven bits of a byte are used to represent a character, while the byte's eighth "parity" bit is used for error checking when the data are received by a reader. The ASCII set of $2^7 = 128$ distinct states is used to represent all of the keyboard's alphanumeric characters as well as non-displayable control characters such as carriage return (CR) and line feed (LF). In the Extended ASCII set, all eight bits of the byte are utilized for character coding, yielding 256 distinct states.

A sequence of ASCII characters is called a *string*. Character strings, of course, can be used to represent text messages. However, strings also are useful in passing commands and data to and from the computer and stand-alone instruments. Additionally, as you will find in this chapter, strings can be used to store numeric data in files on a computer drive (e.g., hard drive, solid-state drive, flash drive).

The closest approximation to a universally readable data file format is the *ASCII text file*. Storing data in this manner has the following advantages. Your files will be read accurately by computers of all manufacturers. Additionally, the files can be viewed with a word processor and, if desired, easily cut and pasted into a document for report generation. Finally, your data will be easy to import into commercially available data analysis software packages. A large number of these application programs prefer that your file is in tab-delimited ASCII text, or what is commonly called the *spreadsheet format.* In this format, tabs separate columns and end of line (EOL) characters separate rows as shown next.

0.00 → 0.4258¶

1.00 → 0.3073¶   → = Tab¶

2.00 → 0.9453¶   ¶ = EOL¶

3.00 → 0.9640¶

4.00 → 0.9517¶

Opening this file using a spreadsheet program such as Microsoft Excel yields the following:

	A	B	C
1	0.00	0.4258	
2	1.00	0.3073	
3	2.00	0.9453	
4	3.00	0.9640	
5	4.00	0.9517	
6			

Alternatively, numeric data may be stored as a *binary byte stream file.* This file format is simply a bit-for-bit image of the data that reside in your computer's memory. Thus, when data are exchanged between a binary file and computer memory, little or no data conversion is required, so you get maximum performance. Also, binary data files provide the most memory-efficient storage method for numeric data. To demonstrate this fact, consider the number of bytes necessary to store the integer *54321.* Since this number is less than $2^{16} - 65636$, it will require only two bytes of memory in a binary-format file. However, in an ASCII text file, this number will occupy five bytes of memory, one byte for each of the characters *5, 4, 3, 2,* and *1.* The disadvantage of binary files is their lack of portability. They cannot be viewed by a word processor, and they cannot be read by any program without a detailed knowledge of the file's format.

LabVIEW contains built-in VIs that facilitate data storage and retrieval using either the ASCII text or the binary file format. If ease of use and compatibility with a spreadsheet application program are of most concern, use the ASCII-based file storage. If, on the other hand, efficient memory use and high speed are desired, a binary file is the best choice.

In this chapter, you will learn how to store data in an ASCII-based spreadsheet file. You will write a program that generates data using your custom-made **Waveform Simulator** as a subVI and then store these data by implementing icons from **Functions>>Programming>>File I/O**.

7.2   STORING DATA IN A SPREADSHEET-FORMATTED FILE

Here's the plan for our data-storage VI: Use the **Waveform Simulator** program created in Chapter 4 to generate some data, and then store these data in a spreadsheet-formatted data file, which is saved on your computer's drive.

Start by opening a new VI. Using **File>>Save**, create a new folder called **Chapter 7** within the **YourName** folder, and then save this VI under the name **Spreadsheet Storage** in **YourName\Chapter 7**.

Now switch to the block diagram. First, you must place **Waveform Simulator** (as a subVI) on your block diagram. In summary, here's how to do that (you may wish to review Section 6.6, "Placing a Custom-Made VI on a Block Diagram," which illustrates the steps in the process). On the Functions Palette, click on the **Select a VI...** category. The **Select the VI to Open** dialog window will appear. In the top box, navigate to the **YourName\Chapter 4** folder. Once the file list for this folder is displayed, double-click on **Waveform Simulator**, or highlight it, and then click on the **OK** button in the dialog window. You will be returned to the block diagram, where you can place the custom-written icon in a convenient position.

Next, with the ✒, create **Digitizing Parameters** and **Waveform Parameters** control clusters on the front panel by popping up on these terminals of the **Waveform Simulator** icon and using **Create>>Control**.

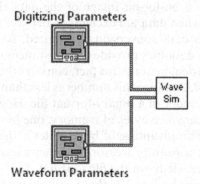

Then place a **Write Delimited Spreadsheet.vi** icon, found in **Functions>> Programming>>File I/O**, on the diagram (earlier versions of LabVIEW have an equivalent icon named **Write To Spreadsheet File.vi**).

Digitizing Parameters

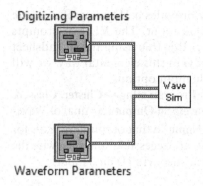

Write Delimited Spreadsheet.vi

Waveform Parameters

7.3 STORING A ONE-DIMENSIONAL DATA ARRAY

To understand how the **Write Delimited Spreadsheet.vi** icon functions, view its Help Window, as shown below. Here, you will find that this VI has the potential to function in numerous ways. As is generically true for Help Windows, default values for this icon's inputs are shown in parentheses. Also, inputs are labeled in boldface, plain, or dimmed text to denote whether each input is *required, recommended*, or *optional*, respectively. Required inputs (**Write Delimited Spreadsheet.vi** has none of these) must be wired, or else the icon will not execute. Recommended and optional inputs control features are available for your use, if desired. Optional inputs are less commonly used and so their labeling only appears when you click (as I have done) on the **Show/Hide Optional Terminals and Full Path** button 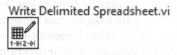 in the Help Window's bottom left-hand corner.

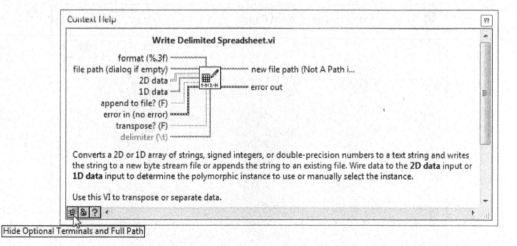

257

In its most basic operating mode, one simply provides a 1D array of numeric values to Write Delimited Spreadsheet.vi's **1D data** input. The VI then prompts the user for a file name and stores the values under this title using the spreadsheet format. By operating **Write Delimited Spreadsheet.vi** in this elemental way, we will gain insight into the necessity of its other available input options.

Place **Unbundle By Name** (found in **Functions>>Programming>>Cluster, Class, & Variant**) on your diagram, and then wire it to the **Waveform Output** terminal of Waveform Simulator. Using the ⬉, expand **Unbundle By Name** so that output terminals for both elements of the cluster (**Time** and **Displacement**) are accessible, and then wire the **Displacement** terminal to the Write Delimited Spreadsheet.vi's **1D data** input.

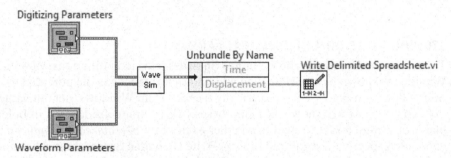

Now return to the front panel, tidy up the object arrangement, and then save your work. Program the inputs to create a small number (such as *10*) of displacement values for a 100 Hz sine wave of amplitude 4.0 with a sampling frequency of 1000 Hz. Then run your program.

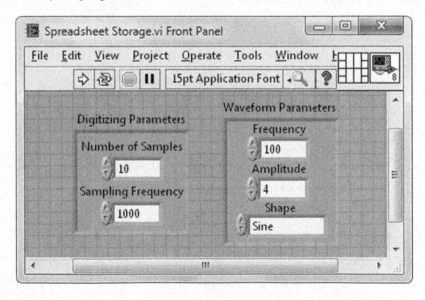

A dialog window called **Choose file to write** will appear. Navigate to the **YourName\ Chapter 7** folder in the top box, name the file **Sine Wave Data.txt** in the **File name:** box, and then click on the **OK** button. The program will then complete execution by creating the requested spreadsheet file in **YourName\Chapter 7.**

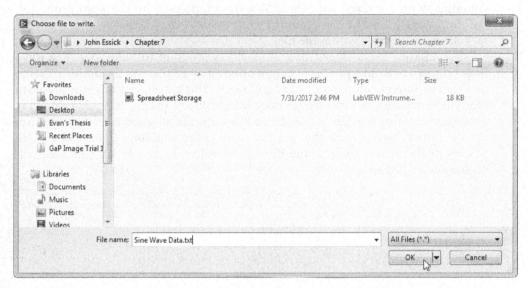

Use a word-processing program to open the ASCII-based **Sine Wave Data.txt** spreadsheet file (in the Open dialog window, you may have to choose the file type as **All Files** to see **Sine Wave Data.txt** listed). Shown next is what I found by viewing this data file in Microsoft Word (the appearance of your file may vary slightly because of the particular tab settings of your word processor). Here, I have activated Word's **Show ¶** command, which allows one to see the file's non-displayable characters such as Tab (→) and EOL (¶). If your word processor has a similar option, use it to view these usually hidden characters.

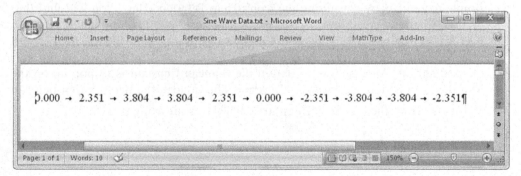

We see that the 10 sine-wave values are delimited by tabs and that the sequence concludes with an EOL. From our knowledge of the spreadsheet formatting convention, we conclude that a spreadsheet application program will interpret this string as a row of data, placing these 10 numeric values in a sequence of columns contained within a single row. Close **Sine Wave Data.txt** with the word processor, but don't save it (or else the program will possibly embed its own characteristic formatting statements within the file).

If available, use a spreadsheet application program to read your **Sine Wave Data.txt** file. To do this, your application program may give you the option of using an **Open** or **Import** command. Either will work. You may have to tell the program that you're reading in a text file (as opposed to binary, Excel, etc.). Also, you may encounter a dialog window in which you must tell the program that your data are tab-delimited. Next, I show the result of reading the **Sine Wave Data.txt** file using Microsoft Excel. As expected, the 10 numeric values appear in 10 sequential columns of a single row.

	A	B	C	D	E	F	G	H	I	J	K
1	0	2.351	3.804	3.804	2.351	0	-2.351	-3.804	-3.804	-2.351	
2											
3											

7.4 TRANSPOSE OPTION

If you have a bit of experience with spreadsheet applications, the above illustration will cause some concern. In using such programs to generate plots, perform curve fitting, and other useful data analysis operations, the array of values for a particular quantity (e.g., a sine-wave displacement) is expected to reside in a single column, with the array's elements indexed by the row numbers. This organizational scheme for data is often called *column-major order*. We have seen that, in its default setting, the **Write Delimited Spreadsheet.vi** icon does not function in a manner consistent with this convention. However, once this problem has been identified, it is simple to remedy by consulting the icon's Help Window. Here, you will find that **Write Delimited Spreadsheet.vi** possesses a **transpose?** input, whose default value is FALSE. By wiring a TRUE **Boolean Constant** to this input, the file will be recorded in a "column-like" manner, rather than "row-like." Perform this wiring on your block diagram. An easy way to obtain the **Boolean Constant** is to pop up on the **transpose?** input and use **Create>>Constant**. The value of the **Boolean Constant** can be toggled from the FALSE (default) to the TRUE state using the 🖑.

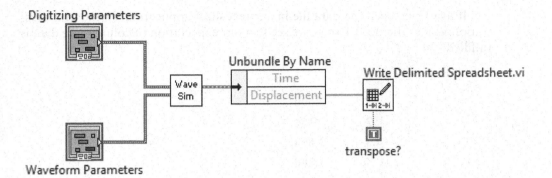

Digitizing Parameters

Unbundle By Name

Write Delimited Spreadsheet.vi

Wave Sim

Time
Displacement

transpose?

Waveform Parameters

Run your program to produce a spreadsheet file of 10 sine-wave values. You can either reuse the file name **Sine Wave Data.txt** (by opening this name in the **Choose file to Write** dialog window, assuming the file is initially closed) or invent a new name. View this file using a word-processing program. As shown next (using Microsoft Word), we see that each data value is separated from its neighbor by an EOL character. Thus, when read by a spreadsheet application, this array of values will be placed in a single column.

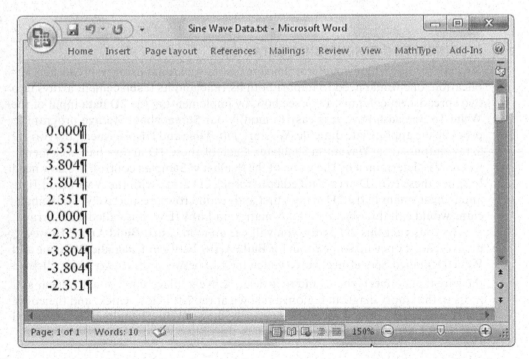

If available, open the data file in a spreadsheet application. In the next illustration (using Microsoft Excel), we see that our expectation of column-like data is fulfilled.

	A	B	C
1	0.000		
2	2.351		
3	3.804		
4	3.804		
5	2.351		
6	0.000		
7	-2.351		
8	-3.804		
9	-3.804		
10	-2.351		
11			

7.5   STORING A TWO-DIMENSIONAL DATA ARRAY

To analyze and/or plot **Waveform Simulator**'s sine-wave data using a spreadsheet application, one would need to import both its **Time** and its **Displacement** arrays into two spreadsheet columns. Let's see how, by implementing the **2D data** input of the **Write To Spreadsheet.vi**, it is easy to modify our **Spreadsheet Storage** program to produce the appropriate data file. We start with **Time** and **Displacement**, the two 1D array outputs from **Waveform Simulator**. Each of these 1D arrays has N elements, where N is determined by the value of the **Number of Samples** control. We now need to splice these two 1D arrays together to form a 2D array with the N values of **Time** and **Displacement** in the 2D array's first and second rows, respectively. Mathematicians would call this object a 2-by-N matrix; in LabVIEW, it is called a 2D array.

To construct this 2D array, you will use an icon called **Build Array**. However, first you must open up some room for **Build Array** between **Unbundle By Name** and **Write Delimited Spreadsheet.vi**. To widen the gap between these two icons, click on the wire that connects them, and then delete it. Next, place the ⬚ (it will appear as a cross in the empty diagram region) as shown at the left below. Click, and then drag the cursor until a dotted rectangle frames both the **Write Delimited Spreadsheet.vi** and the **Boolean Constant** icon as shown in the right.

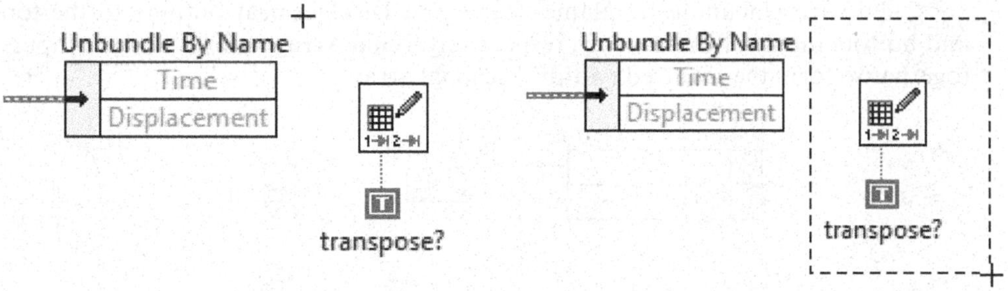

When you release the mouse button, both icons will be highlighted with a marquee. You can then move them to the right as one object by either using the keyboard's <*Right Arrow*> key, dragging with the Positioning Tool while depressing the <*Shift*> key (allowing only horizontal motion), or simply dragging with the ↖.

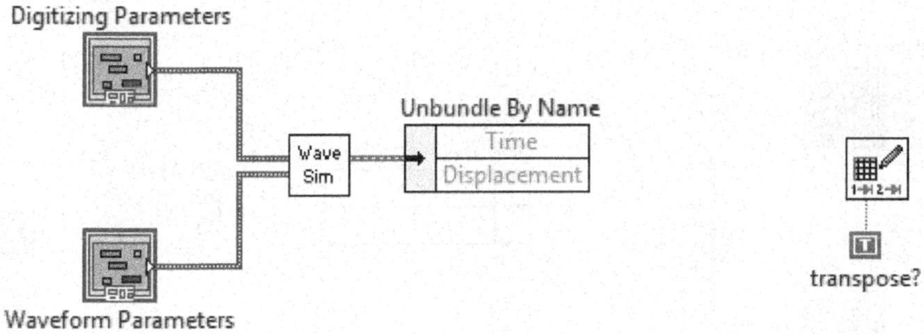

Now select the **Build Array** icon from **Functions>>Programming>>Array**. When you put this icon on the block diagram, it will initially appear with only one input.

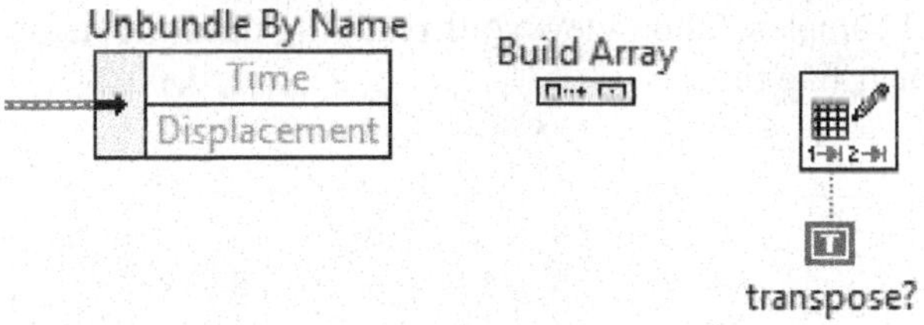

Place the ↖ at the bottom center of the icon until it morphs into a Resizing Cursor. Resize the icon so that it can accommodate two inputs.

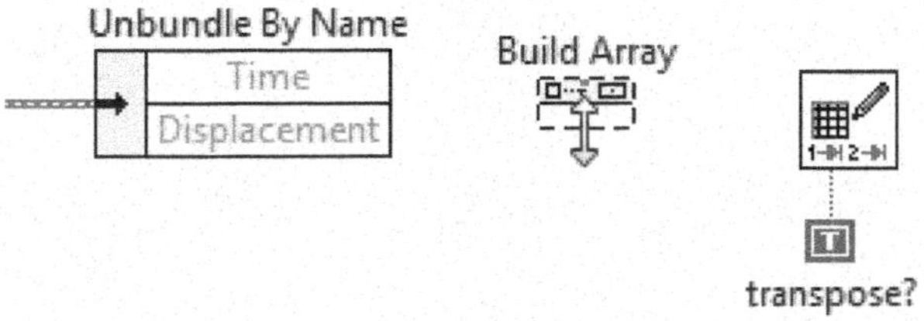

Then wire Unbundle By Name's **Time** and **Displacement** outputs to the top and bottom inputs of **Build Array**, respectively. **Build Array** will splice these inputs together to form the required two-dimensional array.

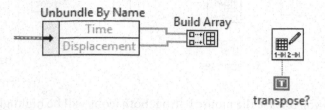

Wire the output of **Build Array** to the **2D data** input of Write Delimited Spreadsheet.vi. Note that this connection appears as a double wire, the LabVIEW designation for a 2D array.

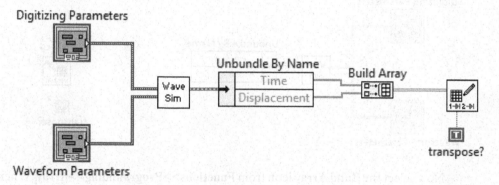

Return to the front panel and run your program with a small value such as *10* for **Number of Samples**. Then view your resulting data file using a word processor. It should appear like this.

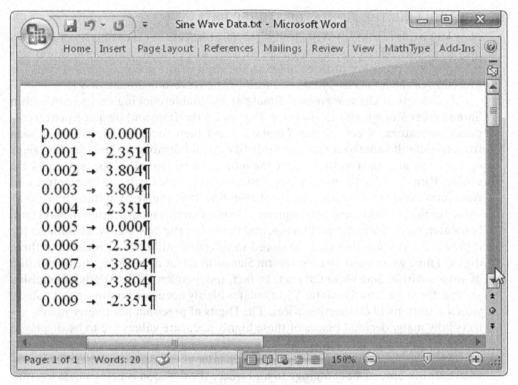

This formatting appears correct for importing the **Time** and **Displacement** data into two adjacent columns of a spreadsheet sheet. To verify this expectation, open your file with a spreadsheet application program. It should appear as below. If you're familiar with the operation of the spreadsheet application, you may enjoy plotting column *B* vs. column *A* or curve fitting the column *B* data to a sine function.

	A	B	C
1	0.000	0.000	
2	0.001	2.351	
3	0.002	3.804	
4	0.003	3.804	
5	0.004	2.351	
6	0.005	0.000	
7	0.006	-2.351	
8	0.007	-3.804	
9	0.008	-3.804	
10	0.009	-2.351	
11			

7.6   CONTROLLING THE FORMAT OF STORED DATA

Now let's explore some of **Write Delimited Spreadsheet.vi**'s other options. First, you may have noticed that, when viewing your data file, each of its numeric values was stored with a precision of three digits to the right of the decimal point. Is this the precision of the numeric values produced by the **Waveform Simulator VI**?

Try this test: Open **Waveform Simulator** by double-clicking on its icon within **Spreadsheet Storage**'s block diagram. Pop up on the **Time** and **Displacement** front-panel indicators, select **Display Format...**, and then increase **Digits of precision** from its default value to something large like *10* and disable the **Hide trailing zeros** option. You may then want to resize the indicators so that all of the new digits are visible. Run the VI with some appropriate choices for the **Digitizing Parameters** and **Waveform Parameters** inputs. You will then find that the array indicators display values of the precision that you requested. Now, with this modification in **Waveform Simulator**, rerun **Spreadsheet Storage**, and then view the new **Sine Wave Data.txt** file it produces. You will find that the stored values there still have a precision of three digits. Thus, we surmise that **Waveform Simulator** is not influencing the formatting of values within **Sine Wave Data.txt**. In fact, independent of its **Digits of precision** setting, the **Waveform Simulator** VI calculates highly accurate floating-point values, which it outputs to connecting wires. The **Digits of precision** parameter merely selects how many decimal places of these highly accurate values are to be displayed on the VI's front-panel indicator.

Since we have eliminated **Waveform Simulator** as a suspect, we must look elsewhere within **Spreadsheet Storage** to find where the decision is being made to truncate highly precise **Waveform Simulator**-produced values in the thousandths place upon storage in the file. A quick glance at Write To Spreadsheet.vi's Help Window identifies that the decision is made at this icon's **format** input. By leaving this input unwired, we have instructed the icon to store numeric values in the default format of *%.3f*. To understand this instruction, we need to know that the **format** string obeys the following syntax:

$$\% \; [\text{WidthString}] \; [\text{.PrecisionString}] \; \text{ConversionCharacter}$$

Here, optional features are enclosed in square brackets. Each syntax element, and how it affects the form of the stored number within the data file, is explained in Table 7.1.

Thus, we see that *%.3f* instructs **Write Delimited Spreadsheet.vi** to store numbers in the fractional floating-point format with each number's total width auto-adjusted under the restriction of maintaining three digits to the right of the decimal point.

Place a **String Constant** on the block diagram, define it to be *%8.5f*, and then wire it to Write Delimited Spreadsheet.vi's **format** input. The easiest way to create the **String Constant**, of course, is to pop up on the **format** input and then select **Create>>Constant**. You can also find this icon in **Functions>>Programming>>String**.

TABLE 7.1 Syntax Elements of **format** String

%	Character that denotes the beginning of a format specification
WidthString (Optional)	Integer specifying the total number of ASCII characters to be used to represent the stored number. If your specified value of **WidthString** exceeds the number of characters actually required to represent a particular number, the excess portion of the string will be padded with spaces. If **WidthString** is not specified (or the stored number requires more characters than the specified **WidthString**), the numeric string will expand to as long as necessary to represent the stored number.
.PrecisionString (Optional)	Period (.) followed by an integer specifying the number of digits to the right of the decimal point in the stored number. If **.PrecisionString** is not specified, six digits to the right of the decimal point are stored. If only the period (.) appears and **PrecisionString** is missing or zero, no digits to the right of the decimal point are stored.
ConversionCharacter	Single character that specifies in what manner the number is to be stored with the following code: d decimal integer x hex integer o octal integer f floating point with fractional format e floating point with scientific notation

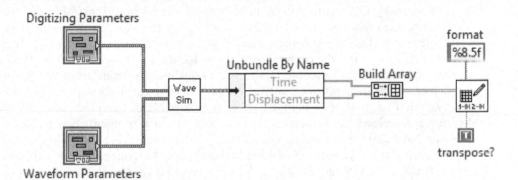

Run your program and store the data in a file. View the file using a word processor. Do the file's numeric values appear in the format that you specified? You might try exploring the effect of changing the **ConversionCharacter** to *d* (decimal integer) or *e* (floating point with scientific notation).

7.7 THE PATH CONSTANT AND PLATFORM PORTABILITY

Next, let's look more closely at the act of naming the data file. We note that, in some cases, it would be desirable to hardwire the name of a data file into our program code. Glancing back at the Write Delimited Spreadsheet.vi's Help Window, we find that this option is available via the **file path (dialog if empty)** input. Previously, we have left this input unwired; thus, the default **dialog if empty** (i.e., a dialog window prompting us for a file name) has been operational. Now we will take the alternate approach of programming in a specific file name.

As you probably know, the *path* to a particular file in your computer system is specified by a hierarchical directory structure. The path is typically denoted by a **drivename**, followed by **directory names** (commonly called **folders**), followed by the **filename** itself. In Windows, these various levels of the hierarchy are separated by backslashes (\); on the Macintosh, forward slashes (/) are used. Thus, on a Windows machine with a drive named C:, the file **Sine Wave Data.txt** within the directory **Users\ Account\Desktop\YourName\Chapter 7** has the path C:\Users\Account \Desktop\Your Name\Chapter 7\Sine Wave Data.txt. On a Mac with a drive named **Macintosh HD**, a similarly named file on the **Desktop** has the path **Macintosh HD/Users/Account/Desk top/YourName/Chapter 7/Sine Wave Data.txt**. To hardwire a specific file name into a program, one might anticipate that the path appropriate to your computing platform (Windows or Macintosh) is simply inserted into a **String Constant**, and then this object is wired to the **file path** input of the Write Delimited Spreadsheet.vi icon. However, in deference to the issue of *platform portability*, this expectation is slightly modified.

Although it has a notable limitation, platform portability is a feature of LabVIEW programming. Here is the ideal in this regard: You write a LabVIEW program called **Widget** on a Windows-based system and then send it (e.g., electronically, on a flash drive, or on a disk) to a colleague for use in a Macintosh-based laboratory. Once your colleague has copied the VI onto their Mac system, **Widget** is opened from within LabVIEW running on the Macintosh system and, while opening it, LabVIEW detects that **Widget** was written on another platform and recompiles it to use the correct instructions for the local processor. One small part of this complicated recompilation task might include translating any hard-wired paths that are contained within **Widget** into the form appropriate for the new system. The good news is that LabVIEW is capable of carrying out much of the above-described ideal, including (as we will see in a moment) being able to translate hardwired file paths across platforms. However, there is also some bad news. Because the device drivers for the Windows and Macintosh systems are different (**DAQmx** vs. **DAQmx Base**), LabVIEW icons that control the functioning of data acquisition devices are not platform portable. So, in our above example, if the Windows-based version of **Widget**, say, contained **DAQ Assistant**, that icon could not be automatically translated for the Mac-based **Widget**. Your colleague in the Macintosh-based laboratory would have to manually replace such DAQmx-related icons with equivalent DAQmx Base coding. As you can imagine, this limitation can make the porting of certain programs nontrivial.

LabVIEW facilitates the cross-platform translation of hardwired path names through the use of a special object called the **Path Constant** (found in **Functions>>Programming>>File I/O>>File Constants**). As a consequence of dedicating the **Path Constant** to the sole use of enclosing paths, these special strings are distinguished from other character strings when the program is being ported.

By popping up on the **file path** input of Write Delimited Spreadsheet.vi, create a **Path Constant** on your block diagram, and then enter a path appropriate to your computing platform. Here, I show the resulting diagram on my Windows system. If you encounter an error when your VI tries to open the data file, carefully check that the file's path within the **Path Constant** is perfectly correct (to find the exact path for an existing Windows file, open the folder containing the file, and then right-click on the file's icon or name and select **Properties**).

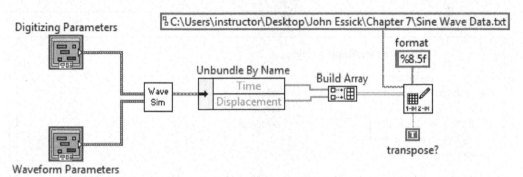

Save your work and then run this program. Verify that it performs as expected by viewing the resulting data file with a word processor or spreadsheet application program.

7.8 FUNDAMENTAL FILE I/O VIS

Write Delimited Spreadsheet.vi is actually a high-level VI constructed from more fundamental File I/O functions. To gain more control of the file input/output process, let's first use the fundamental File I/O functions to write a program with the same capabilities as **Write Delimited Spreadsheet.vi** and then use our deepened understanding to enhance its functionality.

All data storage ("file output") operations must perform the following three steps: *open a file, write data to the open file, close the file.* Retrieving previously stored data ("file input") is performed in the same three steps, except that data are read from the open file (rather than written to it). In **Functions>>Programming>>File I/O**, the following three icons are available to perform the three required steps of data storage: **Open/Create/Replace File**, **Write To Text File**, and **Close File**. To understand how to wire these three icons together to accomplish data storage, we will first briefly describe the function of each individual icon.

The Help Window for **Open/Create/Replace File** is shown next. Given a file defined at its **file path** input, the job of this icon is either to open that existing file or to create it (if it doesn't already exist). If nothing is wired to **file path**, a dialog window will appear when the icon executes to allow one to input the file's path. Using the given file path as well as other configurational information such as the location within the file of the last I/O operation and degree of user access (e.g., read only), this icon also produces a "reference number" (called *refnum* for short). To pass this needed information to other File I/O icons, the refnum is available at the **refnum out** output terminal. Note that all of the inputs for this icon are (plain text) recommended or (dimmed text) optional inputs; if there were required inputs, they would be shown in boldface text. Thus, when **Open/Create/Replace File** executes, each input left unwired will take on its default value, which is shown in parentheses.

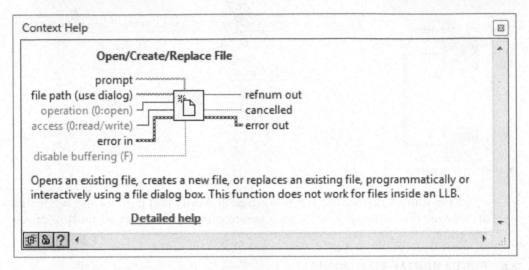

Write To Text File, whose Help Window follows here, performs the actual data storage. Once presented with the open file's *refnum* configurational information at its **file** input, this icon writes the ASCII string at its (required) **text** input into the open file. Additionally, the refnum is available at the **refnum out** output terminal.

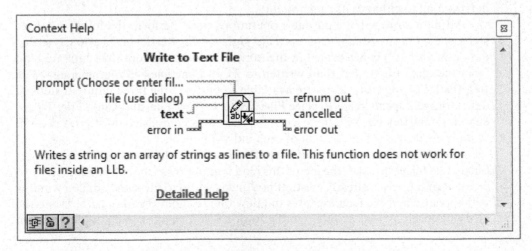

Finally, **Close File**'s Help Window is given below. This icon closes the file specified at its **refnum** input. The file path, which is contained within the input refnum, is made available at the **path** output terminal.

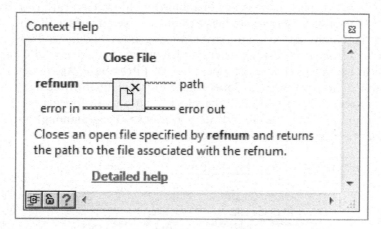

Note that all three of the above icons include error reporting via the error cluster, which appears at the **error in** and **error out** terminals.

The three-step data storage process is accomplished by wiring these three icons together as follows.

This wiring scheme takes advantage of the principle of LabVIEW programming called *data dependency*. Simply stated, data dependency means that an icon cannot execute until data are available at *all* of its inputs. In this diagram, all of the inputs to **Open/Create/Replace File** are wired (the **Path Constant** is explicitly wired to **file path**, and the other optional inputs are implicitly wired to their default values). So, when this diagram is run, **Open/Create/Replace File** executes immediately. Upon completion, **Open/Create/Replace File** outputs a refnum at its **refnum out** terminal, which is passed through the wiring to Write to Text File's **file** input. Because of data dependency, **Write to Text File** cannot execute until it receives the refnum from **Open/Create/Replace File**. Then, after **Write to Text File** completes its execution, it passes the refnum from its **refnum out** terminal to Close File's **refnum** input. Only then can **Close File** execute. Thus, through this programming scheme, we are assured that the icons will execute in the desired sequence: **Open/Create/Replace File** followed by **Write to Text File** followed by **Close File**.

Also, the correct manner of chaining together File I/O icons for error reporting is shown above. If an error does occur at one point in the chain, subsequent icons will not execute and the error message will be passed to the **General Error Handler .vi**, which will display the message in a dialog window. **General Error Handler.vi** is found in **Functions>>Programming>>Dialog & User Interface**. Rather than using this dialog-window approach to error reporting, one could instead pass error information to a front-panel error cluster for viewing (as you did in your Chapter 6 VIs).

With **Spreadsheet Storage** open, select **File>>Save As...** and then use it to create a new VI named **Spreadsheet Storage (OpenWriteClose)** in the **Your Name\Chapter 7** folder. Modify the block diagram to appear as shown below. The required icons are found in **Functions>>Programming>>File I/O** and **Functions>>Programming>>Dialog & User Interface**.

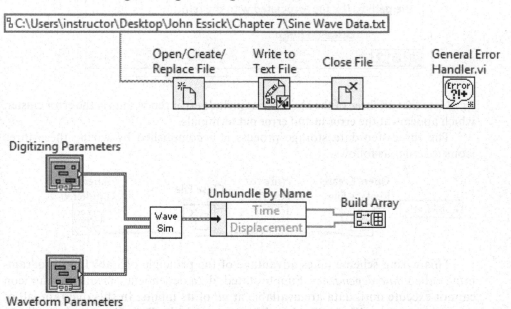

Pop up on the **operation** input of Open/Create/Replace File and select **Create>>Constant**. A block-diagram **Enumerated Constant** ("Enum") will be created. Place the 🖑 atop the **Enum**, click the mouse, and then select the **replace or create** option. This option will be secured when you release the mouse button. With the path input at **file path**, this icon will then either create and leave open a new file (if a file with that path does not previously exist) or open the existing file with that path and instruct the subsequent icons to overwrite ("replace") its existing data with new data. This mode of operation is the same as that programmed into **Write Delimited Spreadsheet.vi**. Also, note that by leaving the other inputs (**prompt, access, error in**) to **Open/Create/Replace File** unwired, these inputs will be automatically programmed to their default values (**no prompt, read/write, no error**).

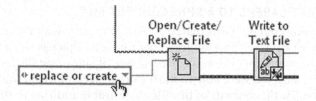

Finally, we need to supply **Write to Text File** with a string containing the 2D array of waveform values in spreadsheet formatting. This string can be created through the use of **Transpose 2D Array** (found in **Functions>>Programming>>Array**) and **Array To Spreadsheet String** (found in **Functions>>Programming>>String**). Complete the block diagram shown next, with the spreadsheet values formatted as *%8.5f*.

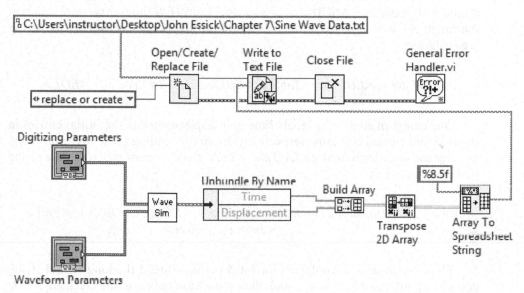

Save your work. Run **Spreadsheet Storage (OpenWriteClose)** with the same choice of values for its front-panel controls as when you ran **Spreadsheet Storage** a few minutes ago. Then use a word processor to verify that the file **Sine Wave Data.txt** appears the same as when it was created using **Spreadsheet Storage** (which implements **Write Delimited Spreadsheet.vi**).

If interested, place **Write Delimited Spreadsheet.vi** on a block diagram, and then open this VI by double-clicking on it. Look at its block diagram and open (by double-clicking) its subVI **Write Spreadsheet String.vi**. You will find that **Open/Create/Replace File**, **Write to Text File**, and **Close File** are the basic building blocks from which **Write Delimited Spreadsheet.vi** is constructed.

7.9   ADDING TEXT LABELS TO A SPREADSHEET FILE

In the previous exercise, you created an ASCII text file with two tab-delimited columns of (x, y) data, the first column containing a sequence of **Time** values and the next column containing the associated values of sine-wave **Displacement**. Because of the file format, you were able to use a word processor or spreadsheet application program to view the contents of this file. A desirable addition to this data file would be the ability to provide descriptive text at the top of each column, labeling each experimental quantity. More generally, one might wish to include all sorts of text statements within the file that would provide a complete record of the experimental conditions and choice of parameters under which the data were taken. Let's see how to provide labels at the top of each data column. Once this skill is in place, it's a small step to include any other desired text in the file.

In our text-based method of data storage, the entire set of (x_i, y_i) values is contained within one long ASCII character string. LabVIEW indexes these N values in the range of $i = 0$ to $N - 1$, so the structure of this spreadsheet-formatted string is as follows:

$$x_0 <Tab> y_0 <EOL> x_1 <Tab> y_1 <EOL> \ldots . x_{N-1} <Tab> y_{N-1} <EOL>$$

One can then attach the labels **Time** and **Displacement** as the initial entries in the first and second columns, respectively, by simply adding these text characters as a spreadsheet-formatted prefix *Time <Tab> Displacement <EOL>*. The string will then appear as

$$Time <Tab> Displacement <EOL> x_0 <Tab> y_0 <EOL> x_1 <Tab> y_1 <EOL>$$
$$\ldots . x_{N-1} <Tab> y_{N-1} <EOL>$$

Thus, our task is to construct the label prefix, append the long ASCII string containing all the data values, and then store this entire string (termed a *byte stream*) in a file. Let's see how to do this.

The required byte stream is constructed using the **Concatenate Strings** icon, which is found in **Functions>>Programming>>String**. This icon splices together all its input strings into a single output string.

Delete the pink wire connecting **Array To Spreadsheet String** to **Write to Text File**, and then place **Concatenate Strings** on your block diagram as shown next. You will find that at first this icon has only two inputs. Position the ⬎ at the icon's bottom center, where ⬎ will morph into a Resizing Cursor, and then expand the icon until it has five inputs.

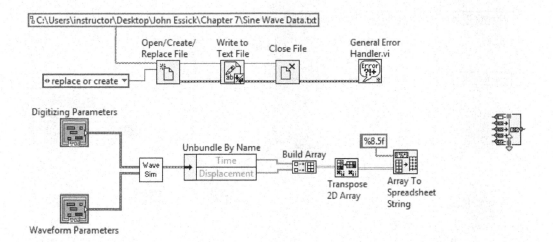

You will now construct the desired label prefix using the top four inputs of **Concatenate Strings**. Wire a **String Constant** (found in **Functions>>Programming>>String** or by popping up on the input) containing the text *Time* and *Displacement* to the first and third inputs of **Concatenate Strings**, respectively. Obtain a **Tab Constant** and an **End of Line Constant** icon from **Functions>>Programming>>String** and wire them to the second and fourth **Concatenate Strings** inputs, respectively.

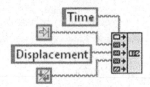

Complete construction of the desired byte stream by wiring Array To Spreadsheet Strings' **spreadsheet string** output to the fifth input of **Concatenate Strings**. Then wire Concatenate Strings' **concatenated string** output to Write to Text File's **text** input.

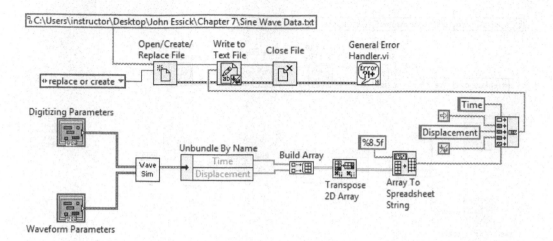

Save your work. Run **Spreadsheet Storage (OpenWriteClose)** with some choice of values for its front-panel controls. Then view the resulting file using a word processor. As shown next, with Microsoft Word, I found the desired labels atop each column just as we had planned. You may have to play with your word processor's tab settings to get the labels to nicely align with the numeric data.

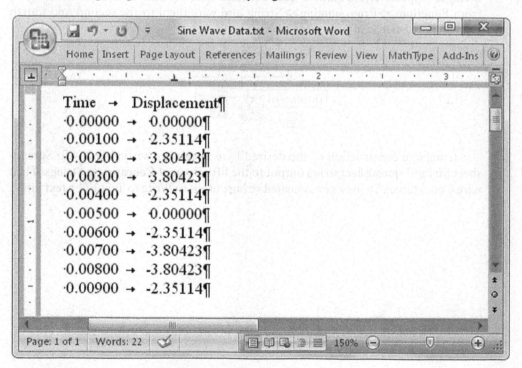

If available, try opening this text-based data file into a spreadsheet application program. In Excel, after a sequence of dialog windows within which I selected the option to begin data import at the file's first line, the data appeared as shown below. If the column labels are not needed, begin the data import at the file's second line. With the data input in this way, an experimentalist can then analyze and graph their experimental results in Excel.

	A	B	C
1	Time	Displacement	
2	0.000	0.00000	
3	0.001	2.35114	
4	0.002	3.80423	
5	0.003	3.80423	
6	0.004	2.35114	
7	0.005	0.00000	
8	0.006	-2.35114	
9	0.007	-3.80423	
10	0.008	-3.80423	
11	0.009	-2.35114	
12			

7.10 BACKSLASH CODES

To conclude, let's explore a shortcut method for including non-displayable ASCII characters within a **String Constant**. The trick is to pop up on the **String Constant** icon and then select **'\' Codes Display** (as opposed to **Normal Display**) from the menu that appears. When this option is activated, LabVIEW will interpret characters immediately following a backslash (\) as a code for non-displayable characters in a manner listed in Table 7.2. You must use uppercase letters for the hexadecimal characters and lowercase letters for the special characters such as Tab and CR.

Let's demonstrate the convenience of backslash codes by employing them in the construction of the labeling string *Time <Tab> Displacement <EOL>*. To build this string, I must tell you that the EOL marker is platform-dependent. The definitions for EOL peculiar to each computing systems are as follows:

Windows: Carriage Return, then Line Feed (\r\n)
Macintosh: Line Feed (\n)
Linux: Line Feed (\n)

TABLE 7.2 Backslash Codes

Code	LabVIEW Interpretation
\00–\FF	Hexadecimal value of an eight-bit character
\b	Backspace (ASCII BS, equivalent to \08)
\f	Formfeed (ASCII FF, equivalent to \0C)
\n	Line Feed (ASCII LF, equivalent to \0A)
\r	Carriage Return (ASCII CR, equivalent to \0D)
\t	Tab (ASCII HT, equivalent to \09)
\s	Space (equivalent to \20)
\\	Backslash (ASCII \, equivalent to \5C)

Thus, by enabling the '\' **Codes Display** option, the entire column-labeling text can be constructed by inserting *Time\tDisplacement\r\n* within a single **String Constant**, assuming we are working on a Windows system. However, if in the future an attempt is made to port this program to another platform, the explicit expression of the EOL character will cause problems. Because of this portability issue, it is good programming practice to take the following slightly modified approach to the string construction: Enclose *Time\tDisplacement* within a single **String Constant**, and then concatenate this object with the **End of Line** icon found in **Functions>>Programming>>String**. The **End of Line** icon is a "smart" EOL character. At compile time, it checks which platform is currently being implemented and automatically generates the appropriate EOL character(s), thus infusing the program with portability.

Here is a step-by-step procedure for modifying **Spreadsheet Storage (Open WriteClose)** to accomplish the above. Start by replacing all of the labeling-string construction code by a **Concatenate Strings** icon with three inputs. Wire an empty **String Constant** to the top input (by popping up and selecting **Create>>Constant**), an **End of Line** icon (from **Functions>>Programming>>String**) to the middle input, and Array To Spreadsheet String's **spreadsheet string** output to the bottom input.

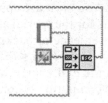

Next, pop up on the empty **String Constant** and enable backslash coding by selecting **'\' Codes Display** from the pop-up menu. Using the 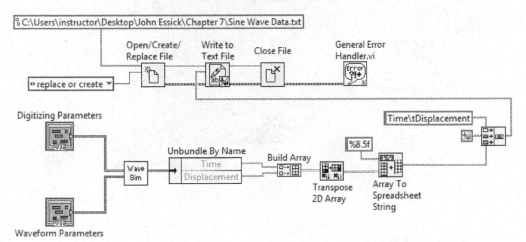 or , enter *Time \tDisplacement* into the **String Constant**, and then press the numeric keypad's *<Enter>* or click on a blank region of the block diagram. When you are finished, you can resize and reposition this object using the , if needed. Then wire the **Concatenate Strings** output to the Write Character To File.vi's **character string** input.

Save your work and then run the program. Use a word processor to verify that backslash coding produces a data file with the desired column labeling.

DO IT YOURSELF

Read Spreadsheet.vi Write a VI that reads and displays data from a two-column spreadsheet-formatted computer file. Name the first and second columns of data within the spreadsheet file *X* and *Y*, respectively. Construct your program to do the following: Open the desired two-column spreadsheet file, read its contents, and then plot *Y* vs. *X* on an **XY Graph** as well as display these arrays in an indicator cluster labeled **XY Cluster**. Save this VI in **YourName\Chapter 7**.

Build the front panel and block diagram of your program as shown below. Refer to Sections 11.5 through 11.7 for help in understanding the relevant icons and how the program functions.

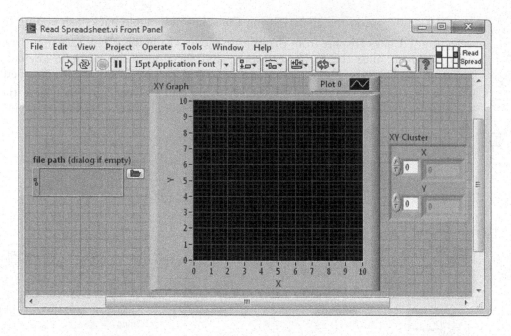

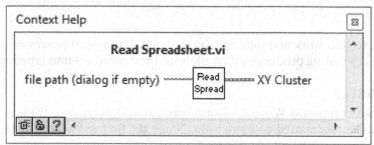

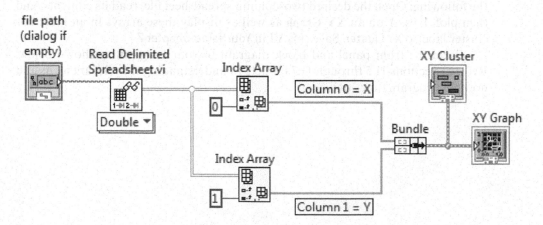

When your VI is completed, run **Spreadsheet Storage** with its front-panel controls set to the values shown below. After **Spreadsheet Storage** executes, there will be a two-column spreadsheet data file named **Sine Wave Data.txt** in the **YourName\Chapter 7** folder with its first and second column containing Time and Displacement values for five cycles of a 50 Hz sine wave of amplitude 5.

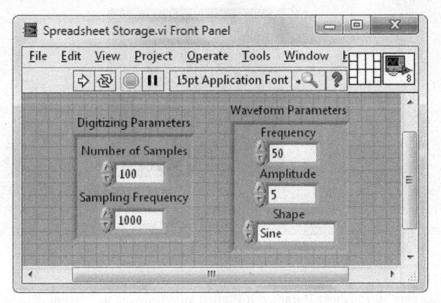

Next, use your newly created VI to read **Sine Wave Data.txt**. After successfully running this VI, its front panel should appear as below.

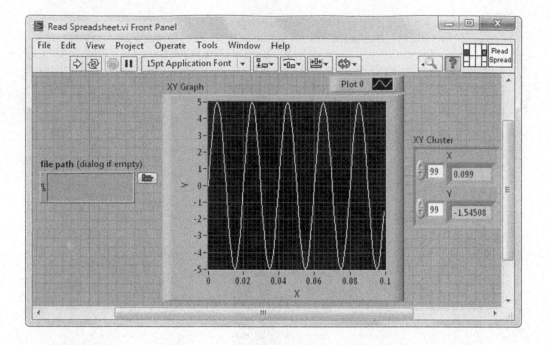

USE IT!

Runtime Spreadsheet Storage.vi It is often an advantage to "stream data to disk" during a data acquisition program's runtime (i.e., add each data sample to an open computer file at the instant it is acquired), rather than store the entire collection of data samples at the close of the VI's execution. With this approach, if your computing system crashes before the data acquisition program has completed its full execution, the data up to the time of the crash have been saved, a feature that is especially valuable for long data runs.

Using the block diagram shown below, create a VI that streams sine-wave data to a file. **The Set File Position** icon (with its **from** input wired to **end**) configures the file storage process so that each new data sample is appended to the end of the open file.

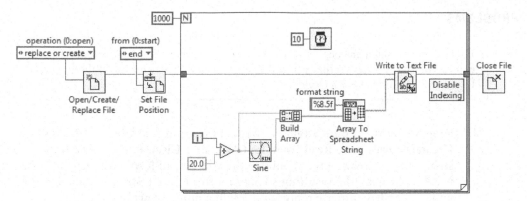

Once completed, run your VI. As the program starts, a dialog box will appear, asking for the desired file path of the to-be-created spreadsheet file. After you enter the path, the VI will execute its 10-second runtime. When the program completes execution, use **Read Spreadsheet** to view the created spreadsheet file. If all goes well, you should see about eight cycles of the sine wave as shown here.

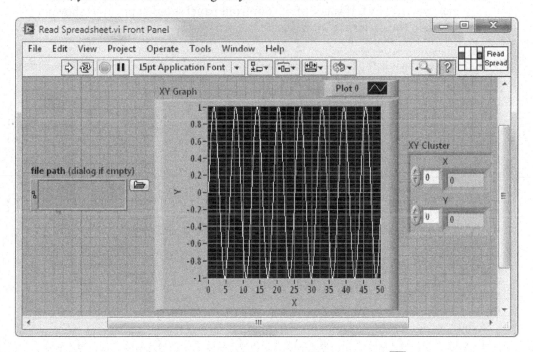

Finally, run your VI and abort its execution (using the ⬤ button in the toolbar) before it has fully completed its 10-second run. Then use **Read Spreadsheet** to verify that the spreadsheet file produced contains all of the data samples up to the time that the VI was halted.

PROBLEMS

Each problem statement begins with a suggested descriptive name (including the **.vi** extension) for the program that you will write. Suggested icons for use in the VI can be found with the aid of **Quick Drop**.

1. **Read Spreadsheet Byte Stream.vi** Write a program that uses the icons **Open/Create/Replace File**, **Read from Text File**, and **Close File** to read the entire string ("byte stream") that contains the contents of a spreadsheet-formatted text file. Display the resulting string in a front-panel **String Indicator**, along with a decoded numeric representation in a numeric array indicator.

    Once your VI is complete, create a spreadsheet-formatted text file by running **Spreadsheet Storage** with its front-panel controls set to create 10 samples of a 50 Hz sine wave. After **Spreadsheet Storage** executes, there will be a two-column spreadsheet text file with 10 rows named **Sine Wave Data.txt** in the **YourName\Chapter 7** folder. Then read **Sine Wave Data.txt** by running your newly written VI. If successful, your front panel should appear as shown next. Explain the pattern of characters that you now see.

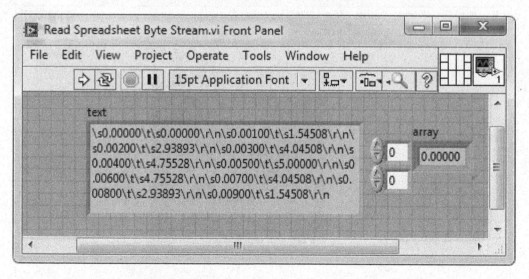

A few tips for coding your VI are as follows:

*   Pop up on the **text** output terminal of **Read from Text File** and uncheck **Convert EOL** (or else the icon will abbreviate all platform-dependent end-of-line characters in the **text** output as simply line feed characters).

- Pop up on the front-panel **String Indicator** and select **'\' Codes Display** so that non-displayable ASCII characters will now be visible in the indicator.
- On the block diagram, use the **Spreadsheet String To Array** icon to convert the text string to a 2D double-precision floating-point array, with a format consistent with that used in creating **Sine Wave Data.txt**.

2. **Spreadsheet Storage with Labels.vi** A LabVIEW programmer modifies **Spreadsheet Storage** as shown next in an attempt to add column labels to the stored spreadsheet data.

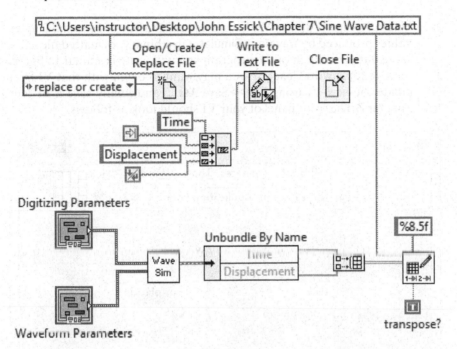

With **Spreadsheet Storage** open, use **File>>Save As...** to create a copy of this VI and store it under a new name in **YourName\Chapter 7**. Modify the block diagram of this new VI so that it replicates the code shown above. Run this program and demonstrate (with the help of a word processor) that it does not work as hoped.

(a) Explain why this code does not accomplish its stated goal.
(b) Fix the block diagram code so that it functions properly. You should just have to change one wire.
(c) Once the code functions properly, explain why this approach to achieving column labeling is less efficient than that used in **Spreadsheet Storage (OpenWriteClose)**.

3. **Four Text Files.vi** Write a For Loop-based program that (with no runtime input from a user) creates four separate files named **Text0.txt**, **Text1.txt**, **Text2.txt**, and **Text3.txt**, each containing a unique text message, and stores these files in the **YourName\Chapter 7** folder. Create one file during each iteration of the For Loop. The file's path can be constructed using the **Build Path** and **Format Into String** icons. The **Format Into String** icon can be expanded so it has multiple inputs, for example, one for an **I32** integer and another for the string **.txt**. Make the text messages in **Text0**, **Text1**, **Text2**, and **Text3** be *This is file 0*, *This is file 1*, *This is file 2*, and *This is file 3*, respectively.

4. **Binary Storage.vi** Build a VI that stores the 2D array of **Time** and **Displacement** values produced by **Waveform Simulator** in a binary-formatted file named **Sine Wave Data.txt**. Much of the program you will write is similar to **Spreadsheet Storage (OpenWriteClose)**, so you may want to begin with that VI open, then create the new VI using **File>>Save As...**, and finally save it in **YourName\Chapter 7**. The front panel of your VI should look as follows.

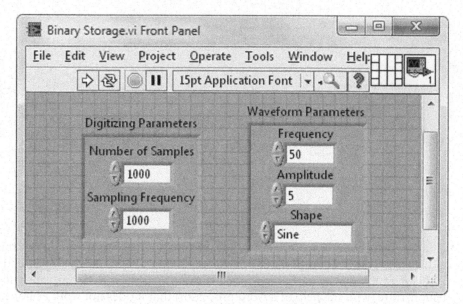

On the block diagram, implement the required three-step data storage process using the **Open/Create/Replace File**, **Write to Binary File**, and **Close File** icons. Wire a **True Constant** to Write to Binary File's **prepend array or string size?** input. This selection instructs LabVIEW to begin the binary file with an eight-byte header, which records the number of rows and columns into which the data are arranged.

(a) Run your VI with its front-panel controls programmed as shown above. Find the icon (or name) for the binary-formatted **Sine Wave Data.txt** file in the **YourName\Chapter 7** folder and determine its size (in kilobytes). The file's exact size is found by right-clicking on the icon and selecting **Properties**. Given the front-panel input values and binary formatting, explain why **Sine Wave Data.txt** has the size it does.

(b) Open **Sine Wave Data.txt** with a word processor. Offer a qualitative explanation for what you observe. Close **Sine Wave Data.txt**.

(c) Run **Spreadsheet Storage** with its front-panel controls programmed as shown above. Determine the size (in kilobytes) of the spreadsheet-formatted **Sine Wave Data.txt**. You will find that this file is now slightly larger than its previous binary-formatted size. Given the front-panel input values, explain why **Sine Wave Data.txt** has the size it does under spreadsheet formatting.

(d) By changing a particular parameter on the block diagram of **Spreadsheet Storage**, you can cause spreadsheet-formatted **Sine Wave Data.txt** to be approximately twice as large as binary-formatted **Sine Wave Data .txt**. Predict the block-diagram change needed to result in this file-size doubling. Make this change and verify your prediction.

5. **Read Binary.vi** Create a binary-formatted file called **Sine Wave Data.txt** by running the binary-storage VI written in Problem 4 with the choice of front-panel control values shown next. In your binary-storage program, make sure that the Write to Binary File's **prepend array or string size?** input is wired TRUE so that an eight-byte informational header is inserted at the beginning of the **Sine Wave Data.txt** file.

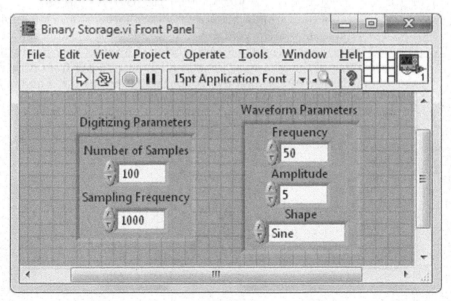

Your assignment is to write a VI that reads the binary-encoded **Time** and **Displacement** data within **Sine Wave Data.txt** and then plots these data on an **XY Graph** as well as displays them in a front-panel XY Cluster as shown below. For this program, the file path must be entered in the **File Path Control** prior to running the VI; this task can be easily accomplished using the **Browse Button** 📁. When the completed VI is run, its front panel should look as follows.

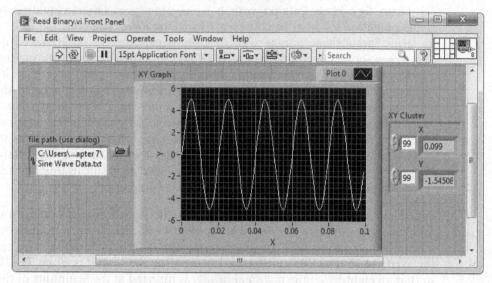

Here are some guidelines to follow in writing your program:

- Retrieval of data from a file is a three-step (*open*, *read*, *close*) process. These three steps are accomplished by configuring the appropriate File I/O icons as shown in the diagram below. The eight-byte header at the beginning of the **Sine Wave Data.txt** file informs **Read from Binary File** about the size of the data file to be read as well as the required form (i.e., the number of rows and columns) of the data array output.

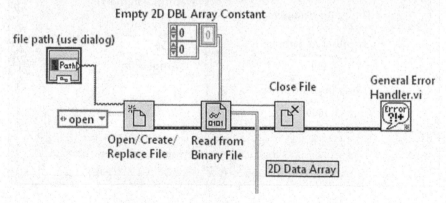

- The fact that the data consist of a 2D array of double-precision floating-point numerics is communicated to **Read from Binary File** by wiring an **Empty 2D DBL Array Constant** to its **data type** input as shown in the above diagram. This constant inputs no data to **Read from Binary File**; rather, it simply specifies the desired data type for the output data. To form this Array Constant, first place an **Array Constant** shell on the block diagram. Then place a **Numeric Constant** within the Array Constant shell, pop up on the **Numeric Constant**, and select **Representation>>DBL**. Finally, pop up on the Array Constant's **index display** and select **Add Dimension**. The icon is now an **Empty 2D DBL Array Constant**, where the top and bottom index displays show the row and column index, respectively.

- The 2D data array emanating from Read from Binary File's **data** output has two rows, where the first and second rows contain the **Time** and **Displacement** values, respectively (note that this 2D array has never been transposed, in contrast to the 2D array involved in the spreadsheet VIs that we wrote). Each of these rows must be sliced off in preparation for plotting on an **XY Graph**. To accomplish this slicing operation, use **Index Array** with a **Numeric Constant** of appropriate value wired to its top (row) **index** input. Leave the bottom (column) index input unwired (see the DIY Project in this chapter and Section 11.6).

6. **Palindrome Detector.vi** Write a program that determines whether an input word, phrase, or sentence is a palindrome, that is, reads the same backward as forward. As shown below, a **Round LED** indicator is lit on the front panel when the input string is determined to be a palindrome.

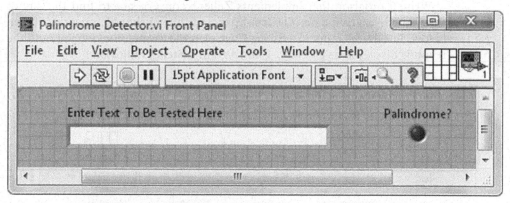

Construct your block diagram as follows: Before comparing the equality of the input string with its reverse, first change the input to all lowercase characters and also remove all spaces. The spaces can be removed by replacing each space character with an **Empty String Constant**. Suggested icons: Use icons from **Functions>>Programming>>String** and its subpalette **Additional String Functions**.

When completed, test your VI with the following inputs: *radar, Able was I ere I saw Elba.*

7. **Day of the Week.vi** The day of the week for a given (Gregorian calendar) date can be found using the following formula called Zeller's congruence:

$$day = \text{mod}\,(I, 7)$$

where

$$I = D + floor\left[(M + 1) \times 2.6\right] + Y + floor\left[\frac{Y}{4}\right] + floor\left[\frac{C}{4}\right] + 5C$$

Here, the *floor*[x] function truncates its argument *x* to the closest lower integer and the modulo function mod(*I, 7*) equals the remainder when the integer *I* is divided by 7. The quantities appearing in Zeller's convergence are defined as follows:

day = day of the week (0 = Saturday, 1 = Sunday,..., 6 = Friday)
  D = day of the month
  M = month (3 = March, 4 = April, 5 = May,..., 12 = December)
  C = century given by *floor(year/*100*)*
  Y = year within century given by mod(*year*, 100)

For the months of January and February, $M = 13$ and $M = 14$ for the previous year, that is, *year → year −1*.

Write a program that implements Zeller's convergence to find the day of the week for a given date. The front panel of this VI should appear as shown below. The date is input in a **String Control** in the format *mm/dd/yyyy*, and the result is displayed as text (e.g., *Thursday*) in a **String Indicator**.

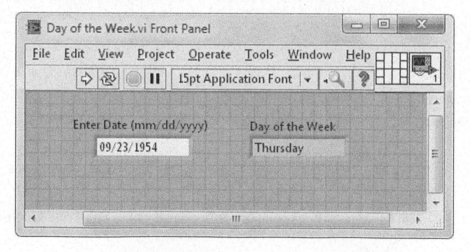

On the block diagram, parse the input string and convert it into numeric representations of *M*, *D*, and *year*. Also, create a seven-element array of strings (e.g., using **Build Array** or a block-diagram **Array** shell containing a **String Constant**), where the index-zero, index-one,..., index-six elements are the strings *Saturday, Sunday,..., Friday*, respectively. Then use **Index Array** to select the correct element from this array to output to the front panel. Suggested icons: Use icons from **Functions>>Programming>>String** and its subpalettes (e.g. **Match Pattern, Decimal String To Number**), **Select, Quotient & Remainder, Round Toward -Infinity**.

Verify that your VI works correctly by finding the day of the week for July 4, 1776 (Thursday) and January 1, 2000 (Saturday).

8. **Parse and Convert Multimeter String.vi** Computer-controlled scientific instruments often report their measurement results in the form of an ASCII string. As an example, assume a multimeter sends the string *VOLTS DC +1.345E+02* in which *VOLTS DC* identifies the type of measurement being reported and *+1.345E+02* is the actual measured value. For this measured value to be useful within a program (e.g., plotted on a chart, input to a calculation), the string received from the multimeter must be "parsed" (i.e., the identifier portion separated for the measured value portion) and the measured value portion converted from its ASCII string representation to a numeric format.

Write a program in which *VOLTS DC +1.345E+02* is input in a front-panel **String Control**. On the block diagram, this string is parsed appropriately and the measured value portion is converted to the double-precision floating-point format. This DBL-formatted number is then displayed on the front panel in a **Numeric Indicator** labeled **Measured Value**. Suggested icons: Use icons from **Functions>>Programming>>String** and its subpalette, **String/ Number Conversion**.

When properly functioning, the front panel of your VI should appear as shown below. In the String Control, **'\' Codes Display** is selected so that non-displayable characters are made visible.

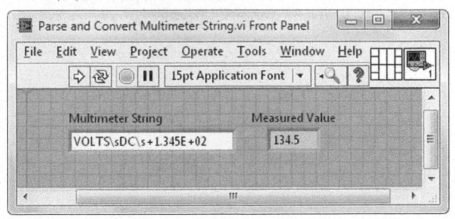

CHAPTER 8

Shift Registers

8.1 SHIFT REGISTER BASICS

A looping structure allows one to implement a common requirement in computer programming—repeating the same operation many times. In earlier chapters, we found that LabVIEW provides two such looping structures: the For Loop and the While Loop. We discovered that, through the creation of a tunnel at its border, each of these loop structures possesses a form of memory. By activating the tunnel's auto-indexing feature, the loop (once it has completed its full execution) will output an array in which is stored the sequence of values created by the succession of loop iterations. Often, however, another form of memory is needed—that which interconnects successive loop iterations. In this guise, a value created by the previous iteration is transferred for use within the calculations of the present iteration. This form of memory is called a *local variable* and can be accomplished in Lab-VIEW's looping structures via the creation of a *shift register*.

Shift registers are created through the use of a pop-up menu at a loop boundary. For example, you may create a shift register by popping up on a For Loop's vertical border and then selecting **Add Shift Register**.

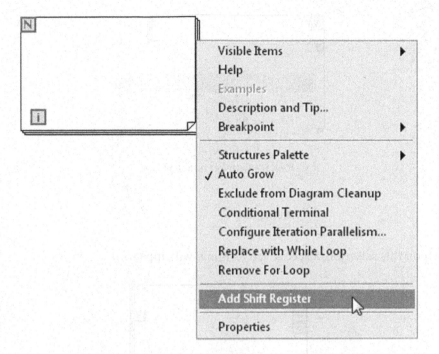

When you release the mouse button, the shift register will appear. It consists of a pair of terminals directly opposite each other on the vertical sides of the loop border.

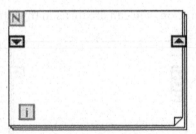

The right terminal stores a value upon completion of an iteration. This stored value is then shifted to the left terminal for use in calculations during the next iteration. Such a feature is useful, for example, in summing the components of a quantity calculated over the span of the entire set of loop iterations.

You can configure the shift register to remember values from more than one previous iteration. To accomplish this feat, pop up on the shift register's left terminal and select **Add Element** from the menu.

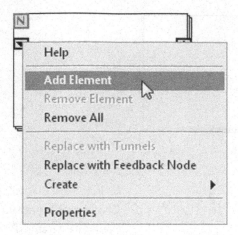

Upon this selection, a second left terminal will appear.

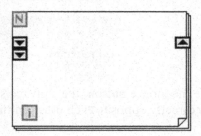

By repeating this operation, you can create as many left terminals as desired.

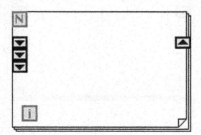

In the previous illustration, we have created three left terminals that will function as follows. When a subdiagram within the For Loop (not shown) is calculating some quantity during the ith loop iteration, the top left terminal contains the value of this quantity calculated during the $(i-1)$th iteration, the middle left terminal the value from the $(i-2)$th iteration, and the bottom left terminal the value from the

($i-3$)th iteration. Over the course of the complete For Loop execution, this set of left terminals will behave analogously to a First In, First Out (FIFO) digital shift register (if that is a familiar concept to you).

Here is a shortcut method for creating multiple shift-register terminals: First, pop up on the border of a For Loop (or While Loop) and select **Add Shift Register** to create a shift register. Then place the ⬉ over the shift register's left terminal until it morphs into a Resizing Cursor ⇕. Next, depress and hold down the mouse button so that the terminal is outlined by a marquee and drag the ⇕ downward until you have created the number of terminals you desire. When you release the mouse button, the multiple terminals will appear. This process is illustrated from left to right in the following diagram.

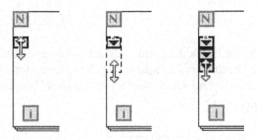

In the next section, we will get introductory experience with the shift register by writing a program that sums a sequence of integers. Then, after a short tutorial on experimental noise, we will write a VI that simulates the noisy data one might acquire in a real experiment and go on to use this program as a subVI in two sophisticated programs that implement shift registers to carry out a commonly used signal-processing technique—the *moving average*.

8.2 QUICK SHIFT REGISTER EXAMPLE: INTEGER SUM

We'll first explore the use of local variables in LabVIEW programming with the following quick example. A famous story begins like this: In the late 1700s, a German elementary schoolteacher wanted to keep the kids in his unruly classroom occupied for 30 minutes so he gave them the following "busy work" assignment: Find the sum of all the integers from 1 to 100, that is, determine $S = 1 + 2 + 3 + \cdots + 98 + 99 + 100$.

Let's show that this assignment would have taken the kids much less than 30 minutes if LabVIEW had been available to them. To write this summing program, first open a blank VI, use **File>>Save** to create a folder named **Chapter 8** within the **YourName** folder, and then save this new program under the name **Sum to 100** in **YourName\Chapter 8**. As shown next, on the front panel place a single **Numeric Indicator** labeled **Value of Sum**, whose representation is **I32**.

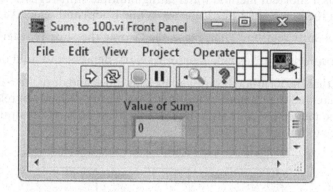

Now switch to the block diagram. Add a **For Loop** to your diagram and place a shift register on its boundary. To add the shift register, pop up on either of the For Loop's vertical borders and select **Add Shift Register**. You can then reposition the terminal pair using the 🖑, if desired.

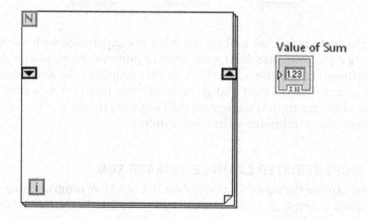

We are going to use the For Loop to compute the sum *S* of the integers *I* where $I = 1, 2, 3, \ldots, 100$. In a text-based language, we would write something along the lines of the following:

$$S = 0.0$$
$$\text{For } i = 0, 99$$
$$I = i + 1$$
$$S = I + S$$

To implement this code in LabVIEW, *initialize* the shift register to zero by wiring a **Numeric Constant** (defined as *0* with representation **I32**) to its left terminal. Place an **Add** icon within the loop and wire one of its inputs and its output to the shift register's left and right terminals, respectively.

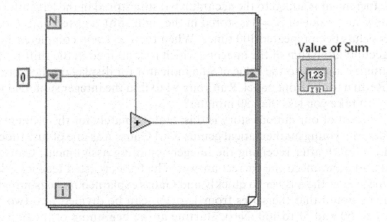

Complete the block diagram as shown next. The convenient **Increment** icon, which takes the input *i* and then outputs $i + 1$, is found in **Functions>>Programming>> Numeric**. Save your work.

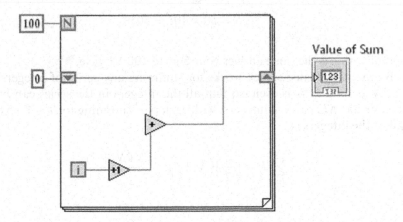

The above diagram works like this. The For Loop steps through the set of integers from 1 to 100, accumulating the sum of these numeric values as it goes. During a particular iteration, the shift register's left terminal provides the sum S accumulated up through the last iteration. The new integer I is added to this sum and the result is stored in the shift register's right terminal for use in the next loop iteration.

That is, on the first loop iteration when ⬛ equals 0, the integer $I = 1$ is constructed by **Increment** and added to the accumulated sum S, which comes from the left shift register and has the initialized value of zero. The new value of $S = 1$ is stored in the right shift register. On the second iteration when ⬛ equals 1, the new integer $I = 2$ from **Increment** is added to the accumulated sum stored in the left shift register and the new sum value of $S = 3$ is stored in the right shift register, and so on until this process has been repeated 100 times. When the For Loop completes its operation, the accumulated sum of the integers, which is contained in the shift register's right terminal, is output to the **Value of Sum** indicator for display on the front panel.

Return to the front panel. Run your VI to find the integer sum. Did writing this program take you less than 30 minutes?

The end of our famous story is this: Unfortunately for the elementary schoolteacher, the young mathematical genius Karl Gauss was one of his students. In less than a minute after receiving the integer-summing assignment, Gauss raised his hand and announced the correct answer. The flabbergasted teacher asked how he had arrived at the answer so quickly and Gauss explained his reasoning as follows. First, he noted that the series from 1 to 100 can be divided into two 50-element sets: 1 to 50 and 51 to 100. Next, starting at the beginning of the first set and the ending of the second, 50 pairs of integers can be formed, where the sum of each pair equals 101. That is, $1 + 100 = 101$, $2 + 99 = 101$, $3 + 98 = 101$, and so on. Thus, given 50 pairs of integers, where each pair sums to 101, the sum of all of the pairs is

$$S = 50 \times 101 = 5050$$

Hopefully, this is the same answer your **Sum to 100** VI gave.

By the way, Gauss's trick works for summing any series of integers $I = 1, 2, 3,\ldots, N$ as long as N is even (so that all the integers in the series can be paired). Then one has $N/2$ pairs of integers, with each pair summing to $N + 1$, so the sum S of all of the integers is

$$S = \sum_{I=1}^{N} I = \left(\frac{N}{2}\right)(N + 1) = \frac{N(N + 1)}{2} \qquad [8.1]$$

Even better than that, because of a little bit of mathematical luck, $S = \sum_{I=1}^{N} I = N(N+1)/2$ is valid even if N is odd. You might have fun proving that fact.

8.3 NOISE AND SIGNAL AVERAGING

For the rest of this chapter, we will demonstrate the use of local variables while writing two of our most ambitious programs to date, each of which is designed to carry out the widely used signal-processing technique called *signal averaging*. In signal averaging, a measurement is repeated under the same experimental conditions and the resulting set of values is averaged. As we will find, such averaging suppresses random noise that obscures the quantity one wishes to determine. In our first program, multiple shift registers will be implemented to recall past data values needed for the averaging process, while, in the second program, a single shift register will be used to store a two-dimensional array of past data. Each of these programs will be carried out through the coordination of several smaller component programs. We will use this opportunity to illustrate one of the important "best practices" of LabVIEW programming—modularity.

When determining the value of a physical quantity experimentally, the quantity's "true" value will always be masked as a result of errors introduced by the measurement process. Some of these errors are *systematic*, biasing the measured value only in one direction (i.e., either too high or too low). Sources of systematic errors are, for example, miscalibrated voltmeters, slow-running clocks, and bent rulers. Once identified, these flaws in the measurement process can be corrected, eliminating (or greatly diminishing) these errors in subsequent measurement runs. Other sorts of errors are *random*, causing repetition of the same measurement to produce a set of values that fluctuate both above and below the measured quantity's true value. Sources of random error include statistical fluctuations of current in an electrical circuit, unpredictable mechanical vibrations in a lab room, and irregular heat loss to the environment. Random errors are always present in an experimental measurement and cannot be eliminated. However, if the errors are due to truly random sources, a set of repeated measured values is distributed about a mean value according to the well-known Gaussian ("bell-curve") distribution. When governed by this distribution, if a quantity Q is measured M times, roughly two-thirds of the measured values will fall within the range $\bar{Q} - \sigma$ to $\bar{Q} + \sigma$, where $\bar{Q}$ and σ are the mean and standard deviation of the set of M values.

Signal averaging allows one to reduce the influence of random error by averaging a measurement over a number of observations. To explain how this method works, assume that the measured value of a quantity Q consists of two components: the unchanging true value of Q (termed *signal S*) and the time-varying random error (called *noise n*). Then, for a given measurement, $Q = S + n$, where n is the value of the random error at the instant that the measurement is made. For that single measurement, we define the *signal-to-noise ratio* (SNR_1) to be

$$SNR_1 \equiv \frac{S}{\sigma} \qquad [8.2]$$

where σ is the standard deviation of the random noise (which, for example might be deduced from measurements taken when the signal is suppressed).

Now, assume that one performs M different measurements of Q, generating the set of values $Q_0, Q_1, Q_2, \ldots, Q_{M-1}$, each of which contains a component of the signal and a component of noise. If one then adds these M values together, the signal components will add coherently, yielding the total signal component

$$S_{tot} = S_0 + S_1 + S_2 + \cdots + S_{M-1} = MS \qquad [8.3]$$

where we assume that the signal is unvarying in time so that each term in the sum has the same value S. However, due to the random nature of the noise components, all of the values of n are uncorrelated with each other. Then, rather than simply being summed, statistical theory dictates that these values will add "in quadrature" (in analogy with how the squares of two sides of a right triangle add via the Pythagorean Theorem to produce the square of the hypotenuse). Thus, the resulting standard deviation of the "summed" noise is given by

$$\sigma_{tot} = \sqrt{\sigma_0^2 + \sigma_1^2 + \sigma_2^2 + \cdots + \sigma_{M-1}^2} = \sqrt{M\sigma^2} = \sqrt{M}\sigma \qquad [8.4]$$

where we assume that the standard deviation associated with each noise value is the same value σ. Finally, defining the average signal and standard deviation as $\bar{S} \equiv S_{tot}/M$ and $\bar{\sigma} \equiv \sigma_{tot}/M$, respectively, the signal-to-noise ratio for the average of M measurements is

$$SNR_M \equiv \frac{\bar{S}}{\bar{\sigma}} = \frac{S_{tot}/M}{\sigma_{tot}/M} = \frac{MS}{\sqrt{M}\sigma} = \sqrt{M}\,\frac{S}{\sigma} \qquad [8.5]$$

or, using Eq. [8.2],

$$SNR_M = \sqrt{M}\,SNR_1 \qquad [8.6]$$

Thus, we find from Eq. [8.6] that by averaging the values of, say, four measurements, the signal-to-noise ratio can be improved by a factor of 2 (i.e., the influence of noise has been suppressed by this factor). Likewise, averaging 9, 16, or 25 measurements will increase the SNR by a factor of 3, 4, or 5. From this analysis, we see that signal averaging exploits the fact that if one adds many measurements of the same quantity, the signal components within the measurements will accumulate while the noise components (which are equally likely to be positive or negative) will partially cancel each other out.

8.4 NOISY SINE VI

To demonstrate the signal averaging technique, we will need a source of noisy data that accurately simulates the array of discrete samples obtained in an actual computer-controlled experiment. In a typical such experiment, upon the receipt of a trigger (e.g., the falling edge of a TTL pulse), a quantity x is sampled at N equally spaced times $t_0, t_1, t_2, \ldots, t_{N-1}$, yielding an array of values $x_0, x_1, x_2, \ldots, x_{N-1}$, where each of these x-values is a composite of the desired signal and unwanted experimental noise. Let's call this array of samples a *trace*.

In this section, we will write a program called **Noisy Sine (Unity SNR)**, which will be used as a subVI in our subsequent programs to provide the needed simulated data. The trace produced by this VI is (arbitrarily) chosen to be two cycles of a sine wave to which Gaussian-distributed noise is added. The amplitude of the sine wave and standard deviation of the Gaussian noise are both (again, arbitrarily) chosen to be unity so that the signal-to-noise ratio of these data is $SNR = 1$.

Open a blank VI and use **File>>Save** to store this new program under the name **Noisy Sine (Unity SNR)** in YourName\Chapter 8. Place a **Waveform Graph** on the front panel with its x- and y-axes labeled **Time** and **Displacement**, respectively.

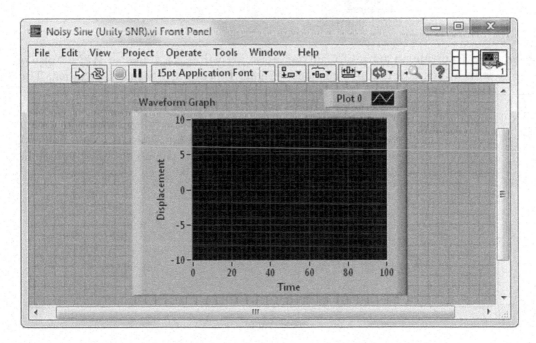

Switch to the block diagram. Here, we will write code that adds an array of unity-amplitude sine-wave values to an equal-sized array of Gaussian noise values (with standard deviation $\sigma = 1$) and outputs the result to the front panel. To produce the sine-wave array, use **Functions>>Select a VI...** to place **Waveform Simulator** (found in **YourName/Chapter 4**) on the block diagram. In the pop-up menus of each of its two inputs, select **Create>>Constant** to create a block-diagram **Digitizing**

Parameters and **Waveform Parameters** cluster constant. Then, using the ✋, choose **Number of Samples** and **Sampling Frequency** to be *200* and *10000*, respectively, in the **Digitizing Parameters** constant. For the **Waveform Parameters** constant, make **Frequency**, **Amplitude**, and **Shape** to be *100*, *1*, and *Sine*, respectively. With these inputs, **Waveform Simulator** will produce two cycles of a unity-amplitude sine wave, which is sampled at 200 equally spaced times.

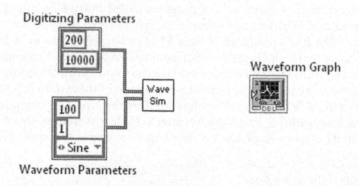

To create an array of Gaussian noise, we will use a **Gaussian White Noise.vi** icon, which is found in **Functions>>Signal Processing>>Signal Generation**. The Help Window for this icon is shown next. Here, from its thick orange wire, we see that the **Gaussian noise pattern** output is formatted to produce a 1D array of double-precision floating-point numerics, which is the same format as the arrays output from **Waveform Simulator**. (*Note:* **Gaussian White Noise.vi** is not included as part of the LabVIEW Base Development System. If you do not find this icon in your Functions Palette, an alternative approach will be given below.)

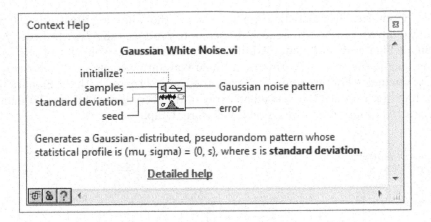

Place **Gaussian White Noise.vi** on your block diagram and wire it to produce an array of equal size to the sine-wave array (i.e., **samples** equals *200*) and a Gaussian distribution with a **standard deviation** of *1*. Then use **Add** to sum the sine-wave array from **Waveform Simulator** and Gaussian noise array from **Gaussian White Noise.vi** as shown next. Note that, because of its polymorphic property, when two *N*-element arrays are input to **Add**, the icon outputs an *N*-element array whose *i*th element is the sum of the *i*th elements of the two input arrays. That is, if *y* and *y'* are the two 1D arrays input to **Add**, the *i*th element of the output array is equal to $y_i + y_i'$.

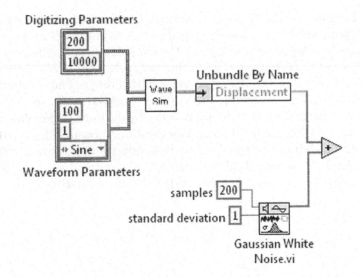

Complete the block diagram as shown in the following illustration. Pop up on the array wires and use **Create>>Indicator** to produce the Array Indicators labeled **Signal Sine** and **Noisy Sine**. Additionally, for a comparison plot of the sine-wave signal and the noisy sine wave on a single **Waveform Graph**, use **Build Array** (found in **Functions>>Programming>>Array**) to form a 2D array with the **Signal Sine** array in the first row and the **Noisy Sine** array in the second row. Such a 2D array is the required input for multiplots on a **Waveform Graph**.

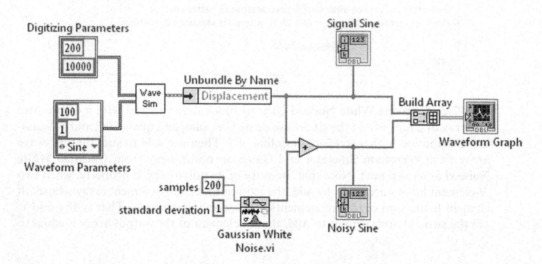

If **Gaussian White Noise.vi** is not available in your Functions Palette, use the following substitute code on your diagram to construct an approximation to Gaussian-distributed noise with $\sigma = 1$. Here, **Random Number (0-1)** creates a random number x in the range from 0 up to (but not including) 1, where there is an equal probability of producing any given value in this range. Then, within the **Formula Node**, $(2*x - 1)$ creates a value in the range from -1 to $+1$ and the factor of $\sqrt{3}$ assures that the 200-element array produced by the For Loop has a standard deviation of 1 (see Problem 6 in Chapter 3 for details on how to calculate the $\sqrt{3}$ factor). Although the 200 values produced by this code are uniformly distributed (rather than Gaussian distributed) about their mean, the resulting array will serve well for our purposes. **Random Number (0-1)** is found in **Functions>>Programming>> Numeric**.

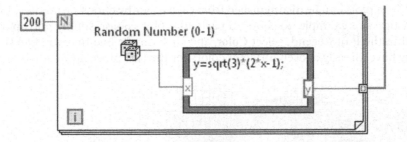

Return to the front panel. Using the , neaten the arrangement of objects. Then resize the **Plot Legend** downward to accommodate information for two plots. The first and second row of **Build Array**'s 2D array output are plotted with the default label **Plot 0** and **Plot 1**, respectively.

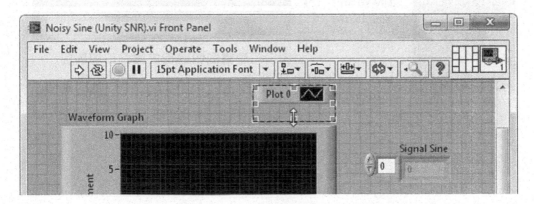

Using the , highlight the text **Plot 0** in the Plot Legend, and then enter the text *Signal Sine*. In a similar way, replace **Plot 1** by *Noisy Sine*.

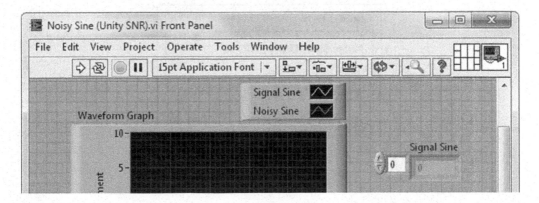

The plots can be distinguished from each other by choosing unique plot characteristics. For example, to choose a particular plot's color, simply pop up on its icon within the Plot Legend, select **Color**, and click on the hue that you like (LabVIEW may have already chosen different colors for the two plots automatically).

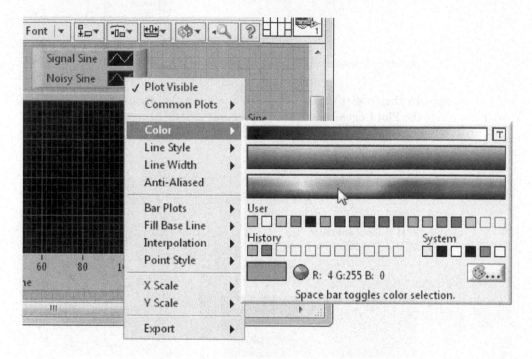

Finally, use the **Icon Editor** to create an icon and assign the two output terminals consistent with the Help Window shown next (for a reminder of this process, see Sections 4.12 and 4.13). Save your work.

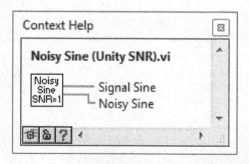

Now run your program and verify that it produces a dual plot of the noise-free sine-wave signal as well as a noisy sine wave with a signal-to-noise ratio of 1.

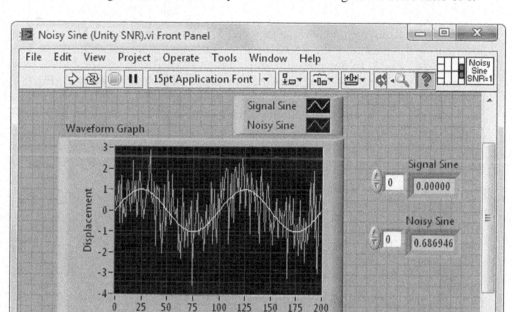

8.5   MOVING AVERAGE OF FOUR TRACES

We're equipped now to demonstrate the benefit of signal averaging. This is our plan: Place **Noisy Sine (Unity SNR)** within a While Loop so that this subVI generates one trace each loop iteration. Then, during a given loop iteration, average each data sample in the current trace with its equivalent counterparts in each of the three most recent traces. Finally, show that, in comparison to the unity-SNR trace generated by **Noisy Sine (Unity SNR)**, the level of noise is suppressed in the averaged trace. In more quantitative terms, our scheme is this: In each trace generated by **Noisy Sine (Unity SNR)**, the noisy sine-wave displacement x is sampled at N equally spaced times generating the discretely sampled data $x_0, x_1, x_2, \ldots, x_{N-1}$. Let's concentrate our attention on one of these data samples—call it x_k. Furthermore, let's define the value of this data sample x_k during the While Loop's ith iteration and during the three previous iterations to be $x_{ki}, x_{k(i-1)}, x_{k(i-2)}$ and $x_{k(i-3)}$, respectively. The four-trace average of this particular data sample is then determined via

$$\bar{x}_k = \frac{x_{ki} + x_{k(i-1)} + x_{k(i-2)} + x_{k(i-3)}}{4}$$ [8.7]

Since the SNR for the single trace generated by **Noisy Sine (Unity SNR)** is 1, from Eq. [8.6] we expect that the SNR for this four-trace averaged value $\bar{x}_k$ will be SNR $= \sqrt{4} \cdot 1 = 2$. In our program, during each While Loop iteration, data from the current trace and the three most recent traces will be used to evaluate Eq. [8.7] for each of the N data samples. Since the iteration number i continues to increase as the program executes, this method of signal averaging is called a *moving average*.

Open a blank VI and use **File>>Save** to store this new program under the name **Moving Average (Four Traces)** in **YourName\Chapter 8**. Place a **Waveform Graph** on the front panel with its x- and y-axes labeled **Time** and **Displacement**, respectively.

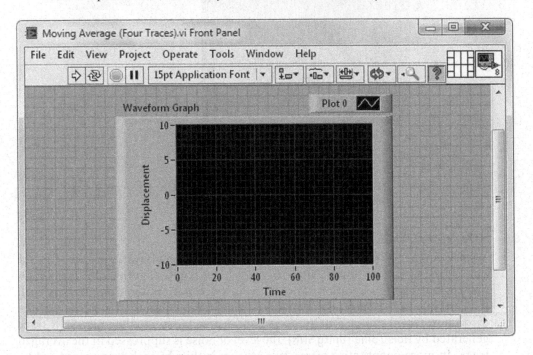

Switch to the block diagram and start building your code as shown below. Within a While Loop, use **Functions>>Select a VI...** to place **Noisy Sine (Unity SNR)**. Our goal is to write code that adds the trace produced by this subVI in the current iteration with the traces produced during the previous three iterations. To recall these three previous traces, pop up on the While Loop border and select **Add Shift Register**, then create three left terminals on the left-hand border of the loop. During the ith loop iteration, the top, middle, and bottom left terminals will contain the trace from the $(i-1)$th, $(i-2)$th, and $(i-3)$th iteration, respectively. Finally, include **Compound Arithmetic** (found in **Functions>>Programming>> Numeric**), which will be used to add the four traces. Noisy Sine (Unity SNR)'s **Noisy Sine** output is the trace for the current (ith) loop iteration; connect this wire to **Compound Arithmetic**'s top terminal.

Next, we will initialize the shift register's three left terminals so that, when the VI begins execution, each starts in a known state and is formatted correctly as a 1D array of double-precision floating-point numerics. For this operation, perhaps the most obvious approach is the one we took in our **Sum to 100** VI—initialize these terminals with zeros. If we adopt this approach in this present situation, since each terminal carries a trace of 200 data samples, our initialization object should be a 200-element array with all of its elements equal to zero. This object is easily created using the **Initialize Array** icon (found in **Functions>>Programming>>Array**) and the initialization process can be accomplished as follows.

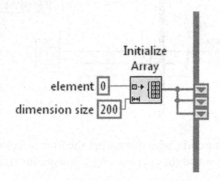

However, for this VI, we will opt for an alternate initialization method using an *Empty Array*. An Empty Array carries the formatting of a chosen data type (in our case, a 1D array of elements with **Representation>>DBL**) but has zero elements. Thus, this object can be used to format the data type of the shift register without being required to specify the size of the array it will carry. We chose this approach because it nicely generalizes to the VI we will build in the next section.

Create an **Empty 1D DBL Array Constant** as follows: First, place an **Array Constant** (found in **Functions>>Programming>>Array**) on the block diagram. Then select a **DBL Numeric Constant** (in older LabVIEW version, use a **Numeric Constant** and change its representation to **DBL**) from **Functions>>Programming>>Numeric** and place it within the Array Constant's element display. This procedure, which is illustrated from left to right below, creates the desired **Empty 1D DBL Array Constant**.

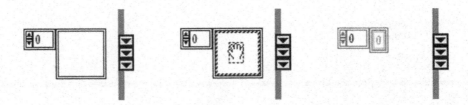

Use the created **Empty 1D DBL Array Constant** to initialize the shift register's three left terminals. Note that when these connections are made, the shift register's terminals turn orange, indicating that they are now formatted to hold floating-point numerics. Then wire the left terminals to **Compound Arithmetic** as shown next.

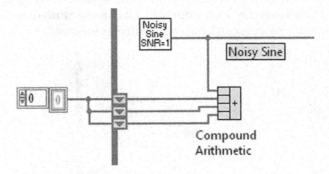

Add the following code, which averages the four traces and then plots both the resulting averaged trace and the current single **Noisy Sine** trace on the same Waveform

Graph for comparison. When wiring the **Divide** icon, if you use **Create>>Constant** to produce the required **Numeric Constant** programmed with *4.0*, create this constant first (i.e., before connecting the array wire from **Compound Arithmetic**), or else an **Array Constant** will be created rather than a scalar **Numeric Constant**. This wiring procedure results in the code shown below, where each element of the array input to the **Divide** icon is divided by the same value of 4.0.

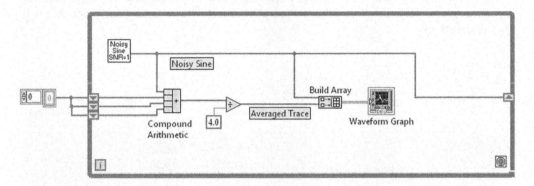

Complete the block diagram as shown next. Pop up on the 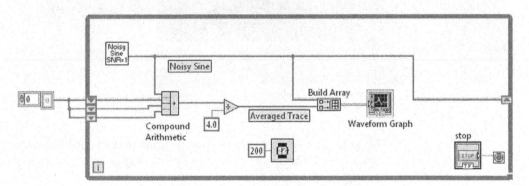 and use **Create>>Control** to make the **Stop Button**. The **Wait (ms)** icon slows the While Loop iteration rate to 5 times per second, allowing the front-panel viewer to discern each update of the **Noisy Sine** trace as well as reducing the load this program places on the computer's processor when it is run.

311

Before leaving the block diagram, I have one final comment regarding the Empty Array method of initializing the shift registers. Although **Compound Arithmetic** outputs an Empty Array when given an Empty Array at one of its inputs, by the fourth While Loop iteration, all of the shift registers will be populated with 200-element arrays and the VI will produce valid results from that point onward.

Return to the front panel and resize the **Plot Legend** to accommodate two plots. Relabel Plot 0 and Plot 1 as Single Trace and Four-Trace Average, respectively.

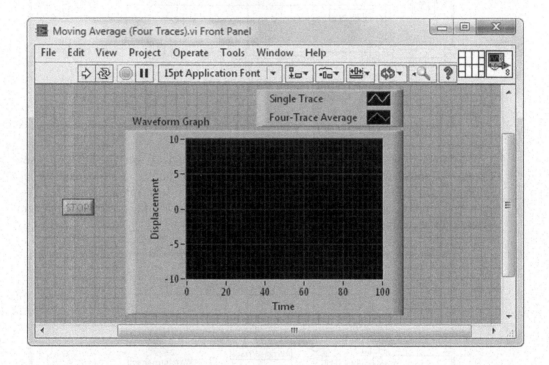

Run your VI. Do you observe smaller noise fluctuations in the **Four-Trace Average** data when compared with the **Single Trace**? Your plots may be easier to view if you turn off autoscaling on the *y*-axis and choose the scaling manually.

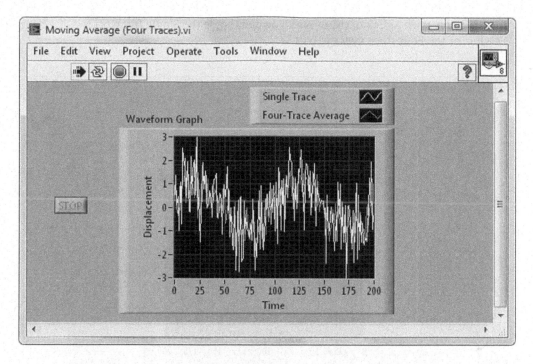

If you have one of the latest versions of LabVIEW, a handy feature of the Waveform Graph may be helpful for comparing the two plots. Pop up on the **Plot Legend** and select **Visible Items>>Plot Visibility Checkbox**.

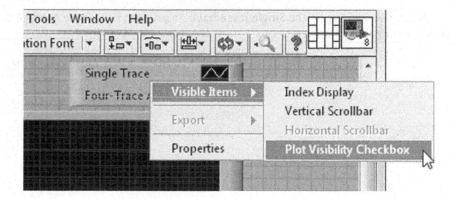

Then, with a click of the mouse, you can toggle the **Single Trace** off to get a clear view of the **Four-Trace Average** plot.

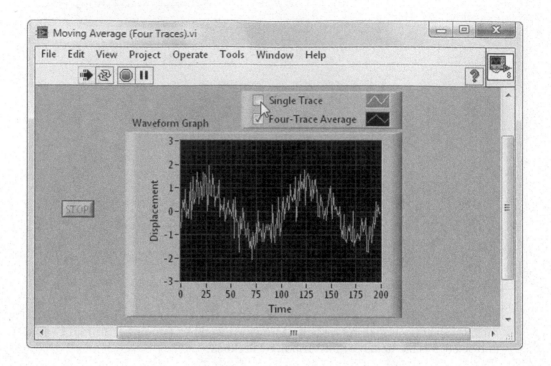

Next, you can toggle the **Single Trace** back on and the **Four-Trace Average** off to view the (hopefully) noisier plot.

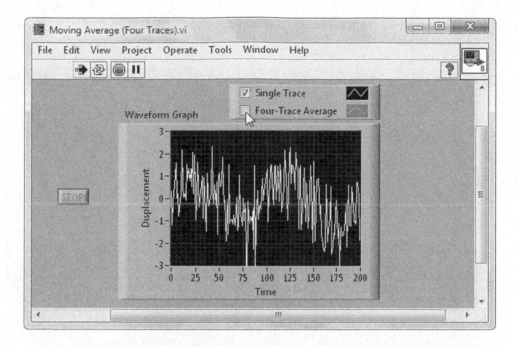

For a quantitative determination of the noise in the **Four-Trace Average** array, add the following code to your block diagram. Here, the sine-wave signal is subtracted from the four-trace average array, leaving an array consisting of only the noise. Then the **Std Deviation and Variance.vi** icon (found in **Functions>>Mathematics>> Probability and Statistics**) is used to determine the standard deviation σ of this sampling of the noise distribution. Finally, since the signal amplitude used in **Noisy Sine (Unity SNR)** is 1, the **Signal-to-Noise Ratio** is found via $SNR = 1/\sigma$.

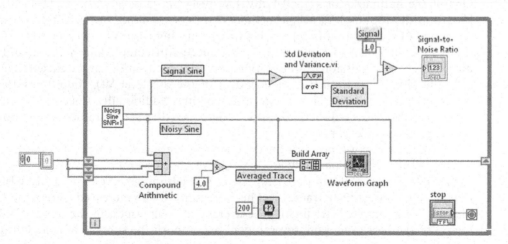

Run your VI from the front panel. Do you obtain the expected value for **Signal-to-Noise Ratio?**

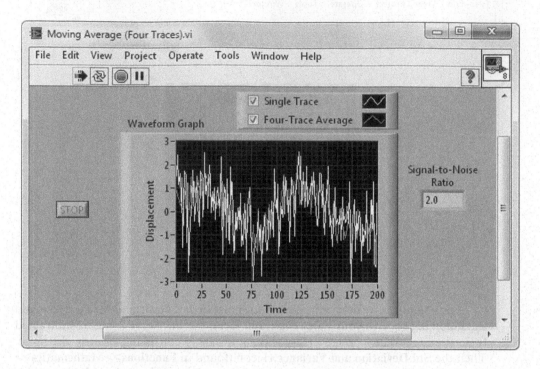

8.6 MODULARITY AND AUTOMATIC SUBVI CREATION

One of the hallmarks of a good LabVIEW program is its *modularity*. In such a program, the top-level VI acts as a supervisor, coordinating the execution of its collection of subVIs. In this approach to programming, a subVI is written to carry out each component task required for the successful completion of the greater program's overall mission. Modularity means that each subVI can be executed as a "stand-alone" program (i.e., independent of the top-level VI), simplifying the programmer's inevitable job of program debugging. Additionally, since subVIs are designed to perform well-defined tasks, many times they can be reused as components in other top-level programs.

In the exercise we have just finished, **Moving Average (Four Traces)** is the top-level VI, while **Noisy Sine (Unity SNR)** is a subVI and **Waveform Simulator** is a sub-subVI. When viewed from this perspective, a programmer might suddenly wish that **Moving Average (Four Traces)** had been written with an increased level of modularity. For example, after finding the average of four traces, in the upper right region of the block diagram, several icons accomplish the subtask of calculating the signal-to-noise ratio of this averaged trace. Thus, this group of icons performs a well-defined task and is an excellent candidate for becoming a subVI.

When such a realization is made, here's an editing trick that is most helpful.

First, use the ⬉ to select the group of block-diagram objects that you wish to include within a subVI.

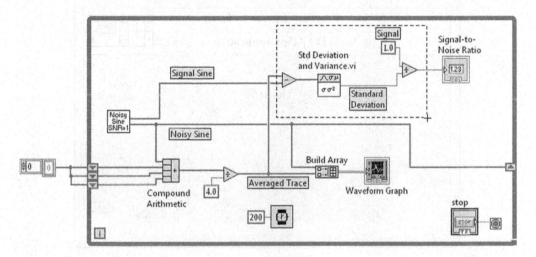

When you release the mouse button, this group will be highlighted with a marquee. Then, in the **Edit** pull-down window, select **Create SubVI**. The highlighted group will be automatically converted into a subVI with a default icon as shown next.

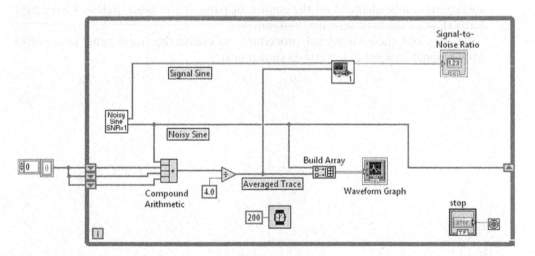

To view this subVI, double-click on its icon. The front panel will open and you will find the new subVI's automatically created controls and indicators with default labels. The subVI itself will also have a default name.

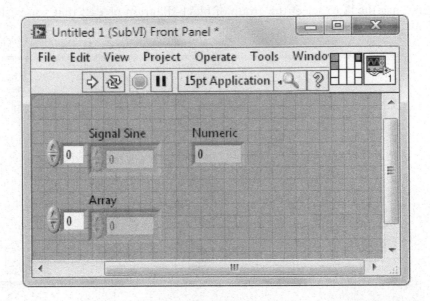

You can then use the [A] to create more accurate labels. Also, a name can be given to the subVI using **File>>Save**, and the subVI's icon can be designed by popping up on the icon pane and selecting **Edit Icon . . .** . You will find that the terminals have been automatically associated with the controls and indicators, but these associations can be changed on the connector pane, if you wish. Below, I have kept LabVIEW's automatic terminal assignments.

Carry out these suggested procedures to create the front panel of a subVI called **Signal-to-Noise Calculator** as shown next.

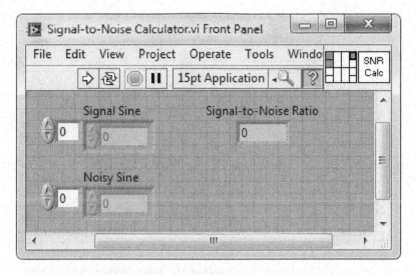

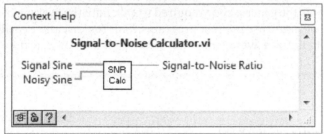

When you switch to the block diagram of **Signal-to-Noise Calculator**, you will find code that you selected when using **Edit>>Create SubVI**. The ⟶ can be used to tidy up the diagram as shown below.

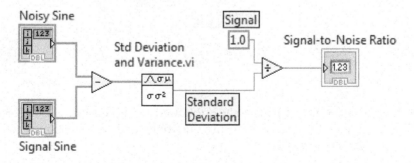

Upon closing this subVI, the block diagram of **Moving Average (Four Traces)** will appear as shown next.

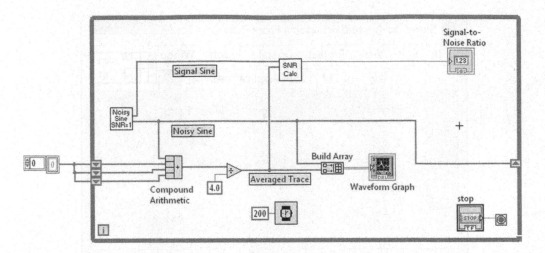

One then might suddenly realize that the lower left region of **Moving Average (Four Traces)** is the generic code required to calculate the average of four traces and so might be of future use in other programs. This region can be easily morphed into a VI called **Four Trace Averager** by first selecting it as shown next.

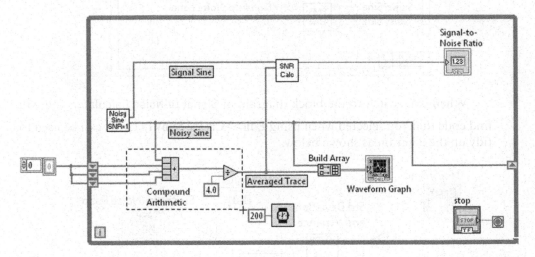

Then, using **Edit>>Create SubVI**, followed by some work with Tools Palette tools and the Icon Editor, you will have the front panel and block diagram for **Four Trace Averager** as shown below.

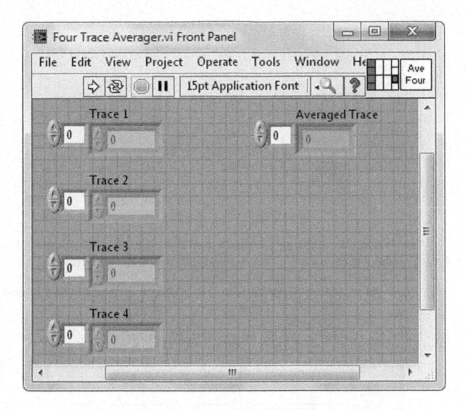

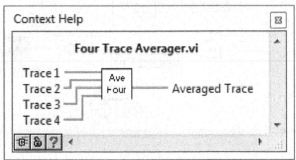

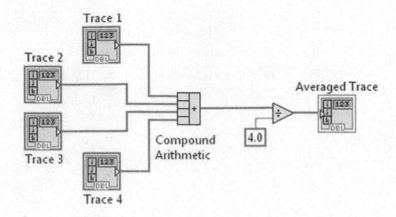

After saving and closing **Four Trace Averager**, the block diagram of **Moving Average (Four Traces)** then is a model of modularity, as shown here.

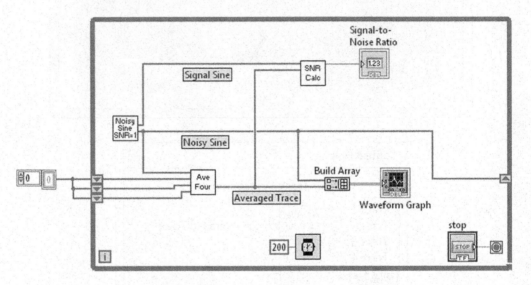

Return to the front panel of this top-level VI and run it to verify that it performs as expected. Then save your work.

8.7 MOVING AVERAGE OF ARBITRARY NUMBER OF TRACES

Hopefully, you are impressed with the performance of **Moving Average (Four Traces)**, but I'm sure you will agree it has this limitation: The approach we have used would be very cumbersome to generalize if we wished to increase the number of averaged traces from 4 to, say, 25. Let's then revise our scheme for implementing signal averaging, this time using just a single shift register to remember any arbitrarily chosen number of previous traces. We will accomplish this trick by storing the traces as the rows in a two-dimensional array.

Open a blank VI and use **File>>Save** to store a new program under the name **Moving Average (Multiple Traces)** in YourName\Chapter 8. Place a **Waveform Graph** on the front panel with its *x*- and *y*-axes labeled **Time** and **Displacement**, respectively. Resize the **Plot Legend** to accommodate two plots, renaming **Plot 0** and **Plot 1** as **Single Trace** and **Multiple-Trace Average**, respectively. Finally, include a **Numeric Control** labeled **Number of Traces** with representation selected to be **I32** and a **Stop Button** as shown.

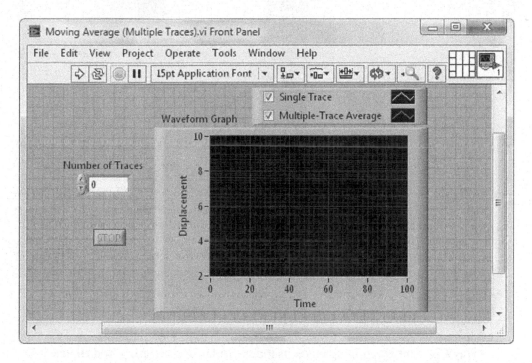

Switch to the block diagram, where we will carry out the following strategy: A $M \times N$ array will be used to store M traces, each of which has N elements. The trace produced in the current While Loop iteration will be in the first row of this 2D array, while the $M-1$ most recent traces will occupy remaining rows of the array in time-ordered fashion—in other words, the trace from the previous itera-tion in the second row, the trace from two iterations ago in the third row, and so on.

Start coding the block diagram as shown next. Here, **Noisy Sine (Unity SNR)** is placed within a While Loop, which is outfitted with a single shift register. **Noisy Sine (Unity SNR)** produces the current **Noisy Sine** trace and it is wired to the top input of **Build Array**. This trace will then be the first row in the 2D array output from **Build Array**.

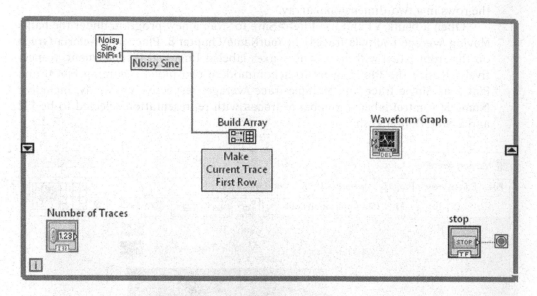

The shift register will be used to remember the previous traces, stored in the form of a 2D array of double-precision floating-point numerics. Let's initialize this shift register so that it begins in a known state when the program commences ex-ecution and is formatted for the 2D DBL array data type. For this initialization, we will use the Empty Array approach because this method does not require us to specify the array size. This feature of the method allows us to place the **Number of Traces** control within the While Loop so that we will be able to change this param-eter interactively during the VI's runtime.

Create the needed **Empty 2D DBL Array Constant** as follows: First, make an **Empty 1D DBL Array Constant** as described in Section 8.5 (place **DBL Numeric Constant** within an **Array Constant**). Next, change the icon into an **Empty 2D DBL Array Constant** by popping up on its index display and selecting **Add Dimension**.

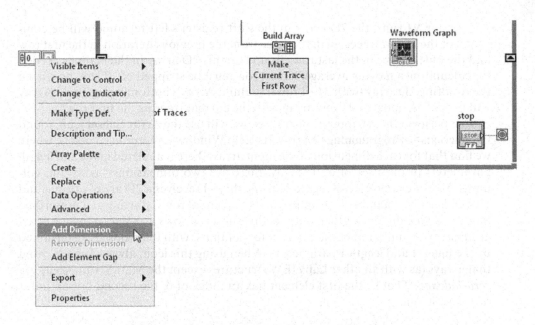

Then wire the resulting icon to the left terminal as shown. Note that this connection appears as a double wire, LabVIEW's designation for a 2D array. Also, note that this double wire and the shift-register terminals are orange, indicating that they are formatted for the floating-point numeric data type.

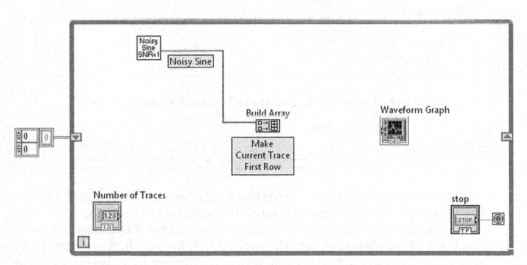

As our VI runs, the 2D array in the shift register's left terminal will be composed of the last M traces, with the trace from the previous iteration in the first row and the oldest trace in the last row. To update this 2D array in the manner needed for calculating a moving average, its last row must be stripped off. Then, by wiring the resulting 2D array (with $M-1$ rows) to Build Array's bottom input, Build Array will output an updated M-row array with the current trace as the first row.

To perform the "stripping" operation, we will use the **Array Subset** icon, found in **Functions>>Programming>>Array**. Its Help Window is reproduced below, where we find that for each dimension of the input array, there is a related **index** and **length** parameter. In our case of a 2D array, there are two dimensions—rows and columns. So, for example, with regard to rows, this VI receives a 2D array as input and sets aside a given number (= **length 0**) of its sequential rows, starting at a prescribed index (= **index 0**). The VI then outputs this subset of rows as a part of a new array at **subarray**. A similar operation occurs for columns, with the cropping determined by the **index 1** and **length 1** parameters. When using this icon, always keep in mind that arrays (as with all other LabVIEW structures except the MathScript Node) are *zero-indexed*. That is, the first element has an index of 0, the second has an index of 1, and so on.

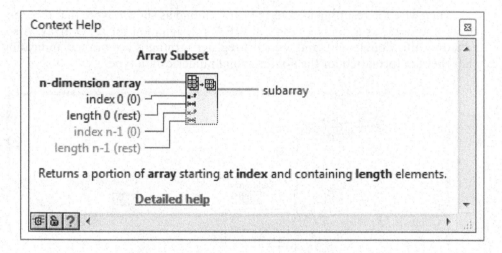

Place an **Array Subset** icon on your block diagram and wire it as shown below. The **Number of Traces** front-panel control allows the user to select the value of M (i.e., the number of traces to be averaged), and so the output of **Decrement** (found in **Functions>>Programming>>Numeric**) equals $M-1$. By default, **Array Subset** will be configured for one dimension (and thus only have one set of **index** and **length** inputs when first placed on the block diagram). However, when the 2D array wire is attached to its **n-dimension array** input, this icon will automatically create a second set of these parameters. For the row-related (top) inputs, wire **length 0** to the output of **Decrement** and keep **index 0** unwired so that it takes on its default value of zero. For the column-related (bottom) inputs, keep both **index 0** and **length 0** unwired so that they take on their default values of *zero* and *rest*, respectively. With these selections, **Array Subset** will simply delete the last row of the input array, creating the $(M-1) \times N$ array we desire.

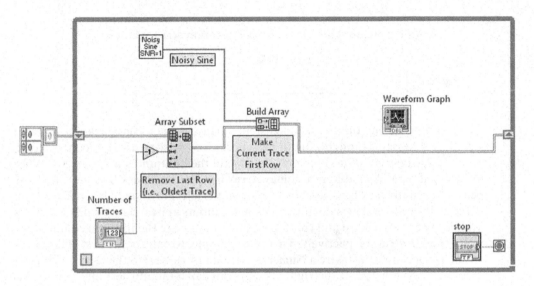

Using the 2D array output from **Build Array**, we are now ready to perform the moving average. In each trace (i.e., row of the 2D array), the data sample x_k taken at time t_k is located in column k. Thus, to carry out the moving average for a particular x_k, we simply need to sum the values in the 2D array's kth column and then divide this sum by the number of times that x_k was measured (i.e., divide by M, the number of traces). To carry out this procedure, we will use the **Index Array** and **Add Array Elements** icons.

The Help Window for **Index Array** (found in **Functions>>Programming>>Array**) is shown below. With a one-dimensional array input, **Index Array** is exceedingly easy to use. One simply wires the 1D array and an integer-containing **Numeric Constant** to the **n-dimension array** and **index 0** inputs, respectively. The integer denotes the index of the array element to be extracted. Once extracted, the numeric value of this element appears at the icon's **element** output terminal.

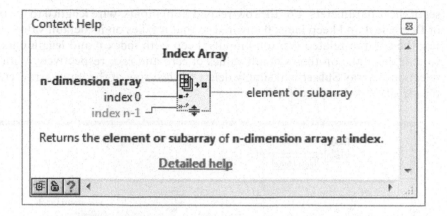

When operating on a two-dimensional array, proper configuration of **Index Array** requires some thought. In our present case, we wish to use this icon to extract a particular column of a 2D array (i.e., isolate the column as a 1D array). When first placed on a block diagram, **Index Array** has just one **index 0** input (which appears as a small black box), as shown below at the left. However, when a 2D array is wired to its **n-dimension array** input, two **index** inputs appear as shown below in the middle. The (top) **index 0** and (bottom) **index 1** inputs are the row and column indices of the 2D array, respectively. Then, for example, to extract the first column (indexed as column 0), just wire a **Numeric Constant** of value *0* to the (column) **index 1** input, as shown below at the right. The (row) **index 0** input is left unwired so that row indexing is disabled (i.e., no particular row is selected). The **subarray** output of Index Array will then be a 1D array containing all rows in the selected column 0.

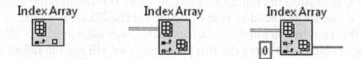

Once a 1D array has been produced, all of its elements can be easily summed using **Add Array Elements**, which is found in **Functions>>Programming>>Numeric**. The Help Window of this icon is shown next.

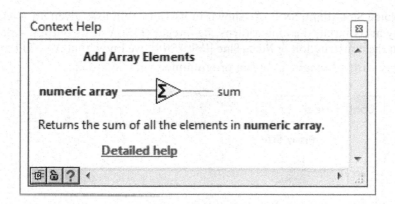

Add the following code to your block diagram, which performs a moving average on each of the N data samples contained in a trace. Here, by popping up on the tunnel and selecting **Disable Indexing**, the entire 2D array is passed into the For Loop (see the Use It! example in Chapter 3). During each iteration, the 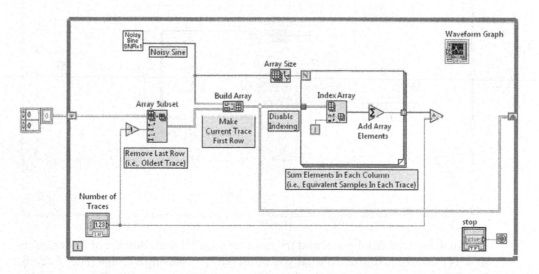 selects a particular column using **Index Array** and the values for the associated data sample x_k from all of the M traces are summed. Over the course of the For Loop's N iterations, the x_k-sums are accumulated in an array at the For Loop border (using auto-indexing). Upon completion of the For Loop, each element in this array is divided by M (i.e., the value of the **Number of Traces** control) to complete the averaging calculation.

In the above diagram, **Array Size** (found in **Functions>>Programming>>Array**) is used to determine N, the number of data samples contained in a trace produced by **Noisy Sine (Unity SNR)**. As shown in its Help Window, given an N-element 1D array at its input, the icon returns the integer (**I32**) N at its output. Alternatively, from the construction of **Noisy Sine (Unity SNR)**, we know that $N = 200$ so we could simply wire a **Numeric Constant** programmed with *200* to 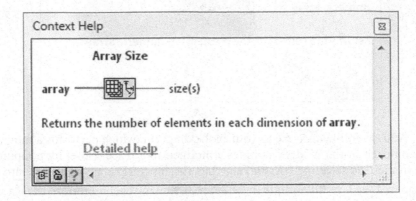.

Complete the block diagram so that the **Waveform Graph** plots the moving-average trace, along with a single noisy trace for comparison. The **Wait (ms)** icon causes the While Loop to execute at a rate of 10 iterations per second.

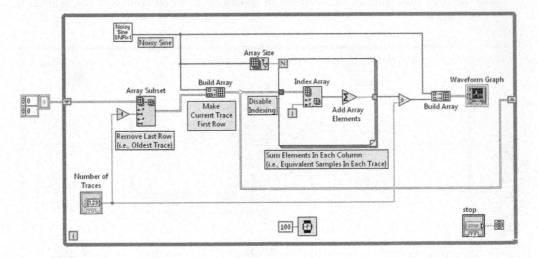

Switch to the front panel and try running your VI with **Number of Traces** equal to *4*. Is the output similar to what you observed with **Moving Average (Four Traces)**? Try increasing the value of **Number of Traces** to *25* or even *100*. Remember that traces are produced at a rate of only 10 per second. So, for example, if $M = 100$, you will have to wait 10 seconds for the moving-average trace to be valid.

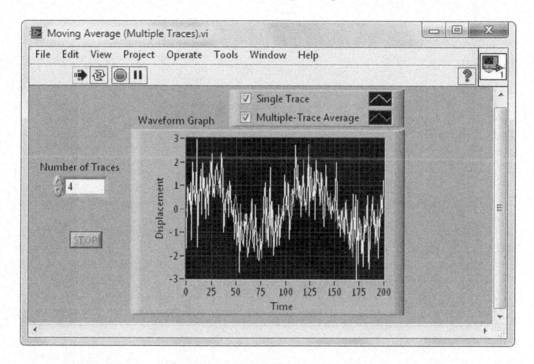

Okay, let's return to the block diagram and see if we can improve its readability. After staring at this somewhat complicated code for a moment, it becomes clear that there are two calculational steps being carried out upon receipt of a new trace from **Noisy Sine (Unity SNR)**: First, the 2D array is updated and, second, the moving average is calculated. Thus, each of these steps is a good candidate for conversion to a subVI.

To make these two conversions, first use the 🔺 to select the array-updating group of icons as shown below. Then, using **Edit>>Create SubVI**, convert these objects to a subVI.

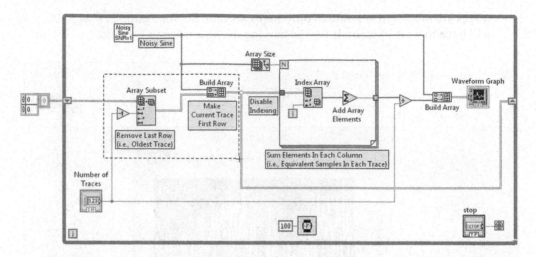

Double-click on the resulting subVI to open its front panel and organize it in a pleasing pattern. Create an icon and assign the terminals consistent with the Help Window shown below. Save this subVI under the name **Update Moving Average 2D Array** in **YourName/Chapter 8**.

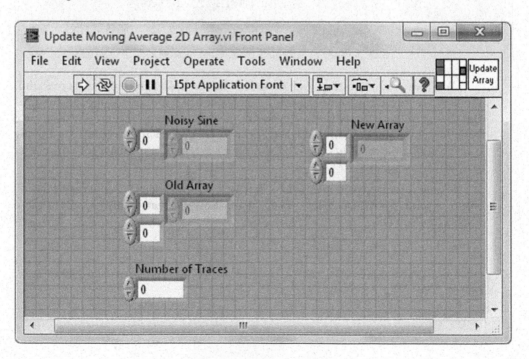

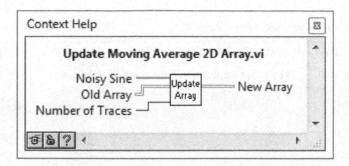

After you close **Update Moving Average 2D Array** and are returned to the block diagram of **Moving Average (Multiple Traces)**, do any needed neatening and then, with the 🅺, select the average-calculating group of icons as shown below. Once selected, use **Edit>>Create SubVI** to convert these objects to a subVI.

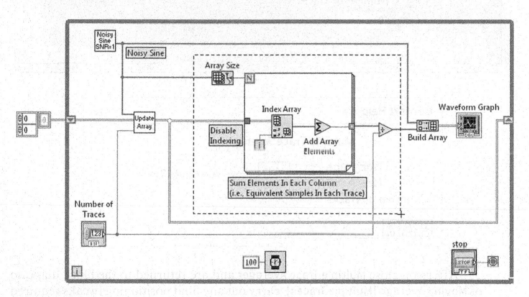

Double-click on the resulting subVI to open its front panel and organize it nicely. Create an icon and assign the terminals consistent with the Help Window shown below. Save this subVI under the name **Multiple Trace Averager** in **YourName/Chapter 8**.

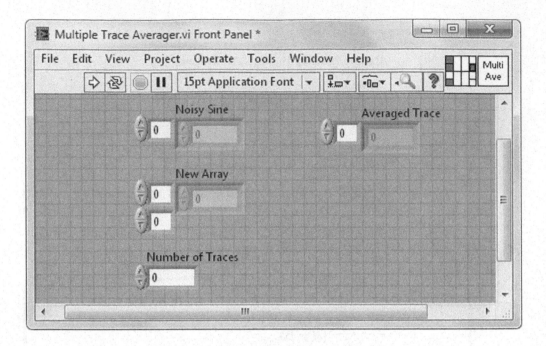

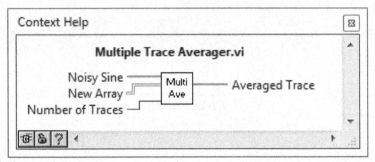

After you close **Multiple Trace Averager** and are returned to the block diagram of **Moving Average (Multiple Traces)**, carry out any final positioning tweaks required for a well-organized block diagram.

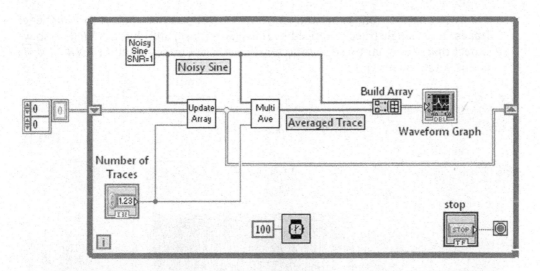

Recalling that we previously created a subVI to compute the signal-to-noise ratio of an averaged trace, let's reuse this code in our present program. Employing **Functions>>Select a VI...**, place **Signal-to-Noise Calculator** on your block diagram and wire it as shown. Use **Create>>Indicator** to make the front-panel **Signal-to-Noise Ratio** indicator. Save your work.

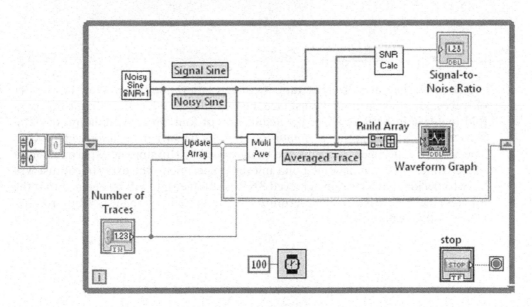

Switch to the front panel and run your VI with various choices for **Number of Traces**. Each single trace produced by **Noisy Sine (Unity SNR)** has a $SNR = 1$, so we expect that the signal-to-noise ratio for the averaged trace should be $SNR = \sqrt{M}$. Is this what you observe?

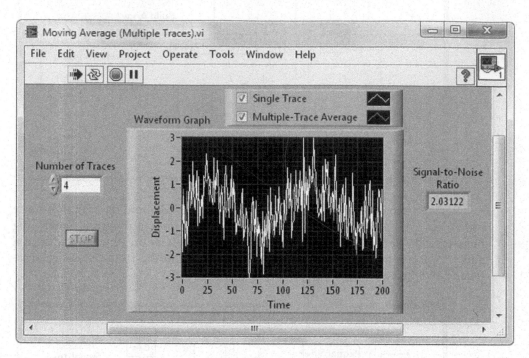

Because the signal-to-noise ratio increases with the square root of the number of traces, an experimentalist must face the following harsh reality. To glean a two-fold increase in SNR requires the acquisition of four times the amount of data. This fact leads to the notion of *diminishing returns*, which can be described like this: The SNR in experimental results can be increased by running the experiment for a longer time. However, at some point in every experiment, the experimentalist will have to decide that a certain achieved SNR is sufficient. It will be deemed that the fourfold increase in experimental runtime required to increase the SNR by another factor of two is not worth it.

DO IT YOURSELF

Icons used in this DIY project can be found with the aid of **Quick Drop**.

(a) **Factorial.vi** The factorial of integer N is defined to be $N! \equiv 1 \times 2 \times 3 \times \cdots \times (N-1) \times N$ with $0! \equiv 1$. Write a program that, given N, uses a shift register-equipped For Loop to calculate $N!$ To check that your VI works, show that $10! = 3,628,800$ and $15! = 1,307,674,368,000$. If your program does not give the correct answer for both of these cases, figure out the problem within your code and fix it. Your program should be able to produce correct values up to about 170!

(b) **Stirling Test.vi** Stirling's approximation states that, for large N,

$$\ln(N!) \approx N \ln(N) - N \qquad [8.8]$$

Using your factorial-calculating program as a subVI, write a program that plots the percentage deviation of Stirling's approximation from the actual value of ln $(N!)$ over a range of N-values. Use your program to determine the minimum value of N so that the approximate ln $(N!)$ deviates from the actual value by less than 1%. Suggested icon: **Natural Logarithm**.

USE IT!

Real-Time XY Graph.vi In some cases, it is desirable to view an **XY Graph** plot while data are being acquired. The following program uses a shift register to build an array of sine-wave data within a For Loop, and this ever-growing array is plotted on a XY Graph during each iteration. [*Caution:* This VI is a poor performer when run for a large number (e.g., 10,000) of iterations.]

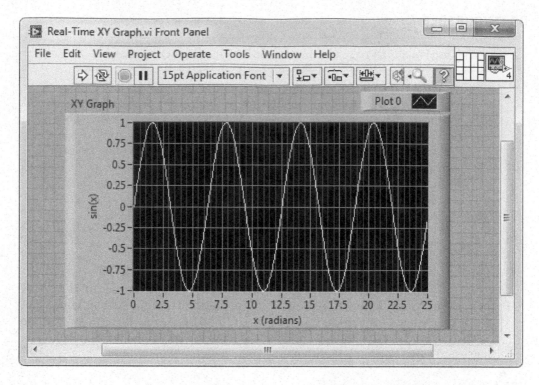

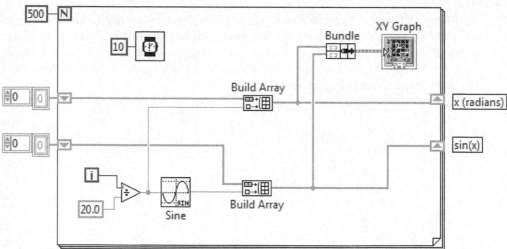

For scaling the *x*-axis, there are a few options. By popping up on the XY Graph and selecting **X Scale>>AutoScale X**, the plot will automatically rescale the axis with each iteration. Alternatively, since in the above example the range of the *x*-variable during the data run is predetermined, on the front-panel XY Graph you can turn off *x*-axis autoscaling and use the 🖑 to manually fix the *x*-axis scale to run over the appropriate range (in the above case, from *0* to *25*). Finally, in the case of a predetermined range, a *Property Node* (see Section 9.2) can be used to fix the scale of the *x*-axis. For this method, pop up on the block-diagram XY Graph terminal and select **Create>>Property Node>>X Scale>>Range>>Maximum**. This action will create a Property Node for the maximum value of the *x*-axis scale as shown below at the left. Next, resize this Property Node to include the minimum value of the *x*-axis scale as shown in the middle illustration. Then pop up on the Property Node and select **Change All To Write** so that the Property Node is in its *write mode* as shown at the right.

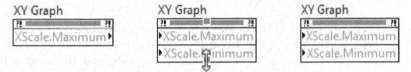

Finally, place this Property Node outside the For Loop and wire it to **Numeric Constant**s with the desired minimum and maximum *x*-axis values as shown next.

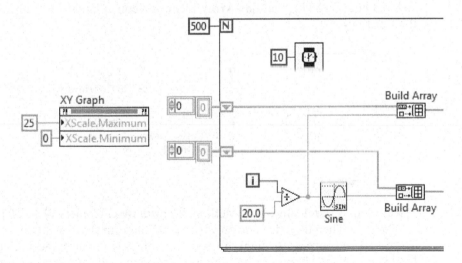

PROBLEMS

Each problem statement begins with a suggested descriptive name (including the **.vi** extension) for the program that you will write. Suggested icons for use in the VI can be found with the aid of **Quick Drop**.

1. **Month's Total Pay (Shift Register).vi** I have agreed to hire you for a 30-day month. I pay you 1 penny on the first day, 2 pennies on the second day, and continue to double your daily pay on each subsequent day up to (and including) Day 30. How many total dollars do you earn by the end of the month? Write a For Loop-based program with a shift register that determines the answer to this question. Suggested icons: **Power Of 2**.

2. **Thirtieth Power of Three.vi** Write a VI that calculates 3^{30} by two methods: (Method A) Using a For Loop to build a 30-element array of 3s and then forming a product of these elements, and (Method B) using a For Loop equipped with a shift register to multiply 3 times itself 30 times. Begin by coding Method A with the following block diagram, which employs the auto-indexing feature of the For Loop to create a 30-element array, where each element equals 3. The **Multiply Array Elements** icon calculates 3^{30} outside of the For Loop.

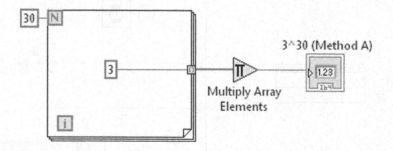

Run your VI and show that Method A correctly calculates $3^{30} = 205{,}891{,}132{,}094{,}649$. Then include code on your block diagram that calculates 3^{30} via Method B and outputs the final shift-register value on a front-panel indicator. Run your VI and show that both methods produce the same answer for 3^{30}. Think carefully about the proper value to use to initialize the shift register and the proper representation of numeric objects.

3. **Alphabet String.vi** Create a VI that displays a 26-character string, where the string consists of the letters of the alphabet in proper order. When run, the VI's front panel should appear as shown below.

Construct the alphabet string on the block diagram using a For Loop equipped with a shift register and a subdiagram that includes the icons **To Unsigned Byte Integer**, **Type Cast**, and **Concatenate Strings**.

The **Type Cast** icon is used to create the symbol for each letter. In ASCII code, the unsigned integers 65 through 90 represent the uppercase letters A through Z. Thus, for example, the following code shows how **Type Cast** is used to decode the integer 65 into the symbol for the letter A.

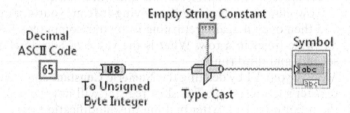

Here, the **Empty String Constant** wired to Type Cast's **type** input alerts the icon to convert the integer into its associated ASCII symbol.

4. **Sum To Five.vi** In this problem, you will explore how LabVIEW handles an uninitialized shift register.

(a) Write the following VI.

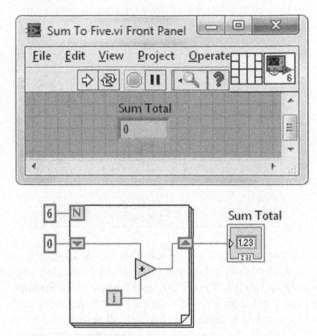

With your program open, run it three times in a row. What is the value of **Sum Total** after the first, second, and third runs?

Now close your program (removing it from your computer's RAM), and then open it again (returning it to your computer's RAM). Run the VI three times in a row. What is the value of **Sum Total** after the first, second, and third runs?

(b) Modify your VI by deleting the **Numeric Constant** connected to the shift register's left terminal. The shift register will now be *uninitialized* when the program is run. After making this modification, save your VI, close it, and then reopen it.

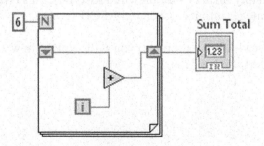

With your VI open, run it three times in a row. What is the value of **Sum Total** after the first, second, and third runs? Now close your program, and then open it again. Run the VI three times in a row. What is the value of **Sum Total** after the first, second, and third runs?

(c) Based on your observations, what is the default value of an uninitialized shift register when the program is first loaded into RAM (i.e., before the program has been run the first time)?

(d) Based on your observations, what is the value for an uninitialized shift register at the beginning of a program's second run? How about at the beginning of the program's third run? (Uninitialized shift registers are occasionally used in programs as a form of memory.)

5. **Newton's Square Root.vi** Let x be a given positive number. Newton's iterative method for determining the square root $y = \sqrt{x}$ is as follows: As an initial guess for y, take $y_{init} = x/2$. Then, for the first iteration ($i = 0$), calculate the value for y to be $y_0 = \dfrac{1}{2}\left[y_{init} + \dfrac{x}{y_{init}}\right]$. Iterate this process, where on the ith iteration, $y_i = \dfrac{1}{2}\left[y_{i-1} + \dfrac{x}{y_{i-1}}\right]$, until the desired accuracy for y is obtained. Write a VI that implements this method. Noting that $\sqrt{2} = 1.414213562\ldots$, how many iterations are required to obtain a value $y = \sqrt{2}$ that is accurate to the fourth decimal place (i.e., $\sqrt{2} = 1.4142$)?

6. **Real-Time Waveform Graph.vi** Employing the Use It! example in this chapter as a guide, write a program that uses a shift register to build an array of sine-wave data within a For Loop, and plots this ever-growing array on a Waveform Graph (rather than an XY Graph) during each iteration. Use a Property Node to properly scale the x-axis.

7. **Reverse Array Elements.vi** The icon **Reverse 1D Array**, found in **Functions>> Programming>>Array**, produces an output array whose elements are in reverse order to that of the array supplied at the icon's input. That is, if the array (0, 1, 2, 3) is input, then the array (3, 2, 1, 0) is output. Using any of the array-related icons available in **Functions>>Programming>>Array**, except for **Reverse 1D Array**, write a program that accomplishes the same task as **Reverse 1D Array**. The front panel of **Reverse Array Elements** should contain an **Array Control** and an **Array Indicator** labeled **Input Array** and **Output Array**, respectively. The elements of an array can be input one at a time into a For Loop using the loop's auto-indexing feature (see the Use It! example in Chapter 3). Program **Input Array** with the array (10, 20, 30, 40), and then run

your VI to demonstrate that it performs correctly. Suggested icons: **Initialize Array**, **Array Size**, **Replace Array Subset**.

8. **Sum of Sines.vi** Fourier analysis tells us that a square wave $y(x)$ with a period and amplitude of 1 is given by the following sum of sine waves:

$$y(x) = \frac{4}{\pi} \sum_n \frac{\sin(2\pi nx)}{n} \qquad n = 1, 3, 5, 7, \ldots, \infty \qquad [8.9]$$

Write a VI that approximates the above sum by evaluating only its first N terms (i.e., if $N = 3$, then only the $n = 1$, $n = 3$, and $n = 5$ terms are included in the sum). To observe three high-resolution cycles of the square wave, evaluate the sum over the range from $x = 0$ to $x = 3$ with $\Delta x = 0.001$.

The front panel of your VI should appear as shown next with a **Numeric Control** called **Number of Terms** to input the desired value of N and an **XY Graph** to plot y vs. x.

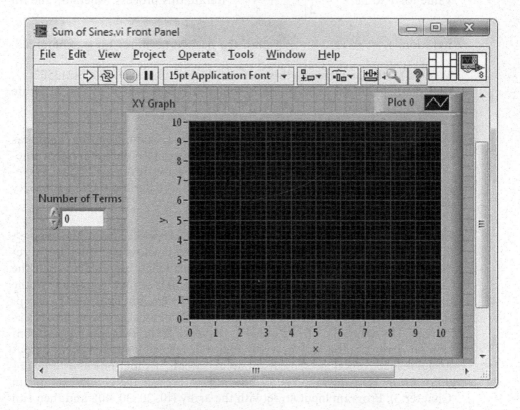

In writing your block diagram, consider the following:

- All positive odd integers can be determined by $n = 2i + 1$, where $i = 0, 1, 2, 3, \ldots$.
- If individual terms in a sum can be calculated one at a time, these terms can be summed using a For Loop with a shift register. For the term associated with the odd integer n, form a 3001-element 1D array of values consisting of the quantity $\dfrac{4}{\pi} \sin(2\pi nx)$ evaluated over the range from $x = 0$ to $x = 3$ with $\Delta x = 0.001$. This 1D array can be summed with another similar-sized 1D array using the **Add** icon and the result stored in a shift register. Suggested icons: **Sine**, **Compound Arithmetic**, **Pi**, **Pi Multiplied By 2**.
- The shift register will have to be initialized as a 3001-element array of zero values (the **Add** icon outputs an Empty Array if one of its inputs is an Empty Array, so initializing the shift register with an Empty Array will not work for this situation). To initialize an N-element array with each element equal to zero, use **Initialize Array** with **element** and **dimension size** wired to *0* and *N*, respectively.

Run your program with a fairly small value of **Number of Terms**. As shown in the next illustration, near the transitions between $y = +1$ and $y = -1$, you will observe "overshoot" peaks (also known as ringing artifacts), an effect called the *Gibbs phenomenon*. As **Number of Terms** is increased, you should be able to minimize this effect. For what value of **Number of Terms** does the Gibbs phenomenon become negligible (e.g., overshoot peak is only 5% of square wave's amplitude)? (With our choice of Δx, the maximum allowed value for **Number of Terms** is 500. Why?)

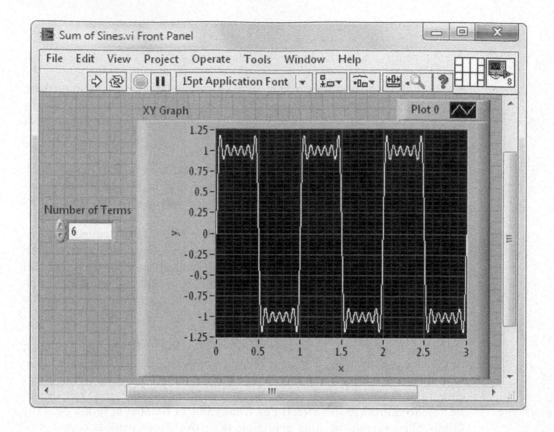

9. **Falling In Air.vi** In the presence of air, a falling object's acceleration a ($=$ change in velocity per unit time) can be described by the following relation:

$$a = g - \alpha v^2$$

[8.10]

where $g = 9.8$ m/s^2 is the acceleration caused by gravity in a vacuum, v is the object's instantaneous speed in meters per second, and α is a constant with units of m^{-1}. The velocity-squared term describes the effect of air resistance, and it becomes more and more significant as the object attains greater speeds. When the object achieves terminal velocity v_T, the object no longer accelerates because, at that speed, $g = \alpha v_T^2$.

Assume an elevated object is released from rest at time $t = 0$ and that the subsequent time is divided into N small intervals, each of duration Δt. Then the object's fall can be sampled at N times given by $t_n = n\Delta t$, where $n = 0, 1, 2, \ldots, N-1$. Let v_n be the object's instantaneous speed at time t_n. Then, if the object's speed v_{n-1} at time t_{n-1} is known and Δt is small, Eq. [8.10] can be used to determine v_n as follows:

$$v_n = v_{n-1} + a\,\Delta t = v_{n-1} + \left(g - \alpha v_{n-1}^2\right)\Delta t \qquad \text{[8.11]}$$

(a) Write a program that implements Eq. [8.11] to determine a falling object's speed v_n at the N times t_n and then plots *Speed* vs. *Time* on a Waveform Graph, whose x-axis is calibrated. Take $\Delta t = 0.001$ s. The front panel of your VI should appear as follows.

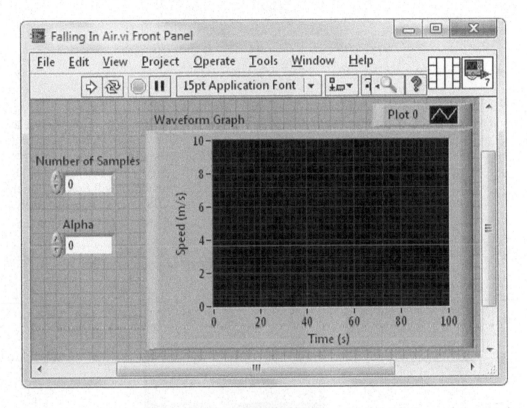

(b) For a 1.0 kg steel ball, a human skydiver, and a raindrop, the numeric value of α is about 0.00092, 0.0035, and 0.11, respectively. Run your program with each value of α. Determine the resulting terminal velocity v_T for each object. After being released, about how long does it take each object to attain v_T?

CHAPTER 9

The Case Structure

9.1 CASE STRUCTURE BASICS

In a LabVIEW program, conditional branching is accomplished through the use of the *Case Structure*. This structure is analogous to an "if–else" statement in a text-based programming language. The Case Structure is found in **Functions>> Programming>>Structures** and is Boolean by default. That is, by wiring a TRUE or FALSE Boolean value to its *selector terminal* ⊡, the structure will execute either the code within its TRUE window or that within its FALSE window, respectively. Using the ⬧, the selector terminal can be placed anywhere along the Case Structure's left border.

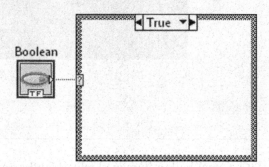

You may view only one of the Case Structure windows at a time. In the above illustration, the TRUE window is visible. To view the FALSE window, simply click the mouse cursor on the decrement (left) or increment (right) button in the *case selector label* ◀ True ▼▶ at the top of the structure or click on the label's central region to access a selection menu. The FALSE window will then appear as shown here.

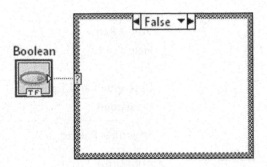

The Case Structure will automatically change its character from Boolean to numeric when you wire a numeric quantity to its selector terminal, such as the **I32** integer control shown next.

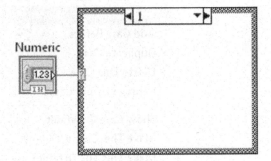

Initially, only two Case windows (**0** and **1**) are available. Above, the case selector label indicates that the **Case 1** window is currently visible. You can easily add another case by popping up anywhere on the Case Structure's border and then selecting **Add Case After** in the pop-up menu.

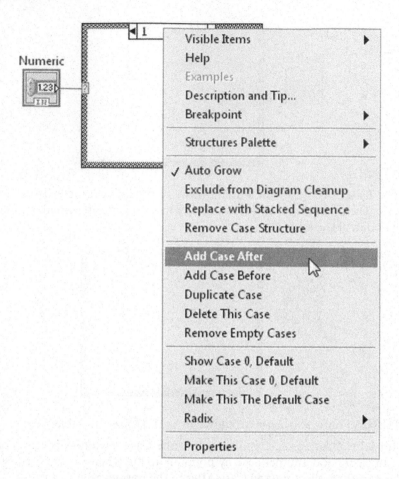

You will then find that the structure now has three (**0, 1,** and **2**) cases.

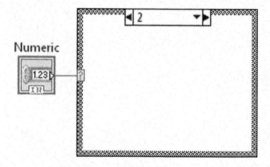

By repeating the above procedure, you may add as many (positive-integer) cases as you desire, unless you desire more than the maximally allowed

$2,147,483,648 \left(= 2^{31}\right)$ cases. If you wire a floating-point number to the selector terminal, LabVIEW will round that number to the nearest integer value. If the number supplied to the selector terminal is negative or if the number is larger than the highest-numbered case, the case designated (using the pop-up menu) as *Default* will be selected.

Finally, when wired to a selector terminal, an *Enumerated Type Control* can be used to self-document each of a Case Structure's cases, often making it the best choice for the case-selecting control. An Enumerated Type Control (called an *Enum*, for short) is a ring-style control, which associates a unique integer value with each item on a list of text descriptors. When an Enum is wired to the selector terminal of a Case Structure, its text descriptors (rather than the associated integers) appear in the case selector label. If you define these text descriptors wisely, they will intuitively describe the purpose of each Case Structure diagram, obviating the need to document every case with, say, free labels. Such documentation is invaluable to other programmers (and your future self!) attempting to decipher your code.

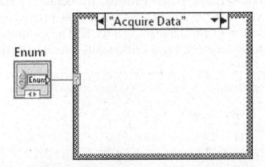

In the next section, we will explore the use of both a Boolean- and Enum-controlled Case Structure while adding runtime options to a VI. This exercise will also introduce an important LabVIEW programming tool called the *Property Node*. Then, for the remainder of the chapter, another powerful LabVIEW programming tool—the *state machine* program architecture—will be studied. Through two examples, we will learn how the state machine, which consists of a Case Structure nested within a While Loop, is used to properly sequence the steps of an algorithm designed to carry out a computer-based task.

9.2 QUICK CASE STRUCTURE EXAMPLE: RUNTIME OPTIONS USING PROPERTY NODES

As an initial example of the use of the Case Structure, let's add two runtime options to the sine-wave generating program **Sine Wave Chart (While Loop)** that you wrote in Chapter 2. Here is our plan: The features of a front-panel object such as the Waveform Chart can be controlled on the block diagram by means of *Property Nodes*. First, with a **Boolean Case Structure** and **History Data** Property Node, we will enable a user to clear **Sine Wave Chart (While Loop)**'s Waveform Chart while the VI is running. Second, using an **Enum Case Structure** and **Plot Area** Property Node, we'll allow an operator to change the Waveform Chart's background color during runtime.

Start by creating a clone of **Sine Wave Chart (While Loop)** through the following procedure. Open the **Sine Wave Chart (While Loop)** in the **YourName\Chapter 2** folder and then select **File>>Save As….** In the dialog window that appears, select **Copy>>Substitute copy for original**, and then press the **Continue...** button. In the next window, first create a folder called **Chapter 9** within the **YourName** folder, and then save this copied VI under the name **Sine Wave Chart (While Loop with Runtime Options)** in **YourName\Chapter 9**. The original file **Sine Wave Chart (While Loop)** will be closed (and safely saved in **YourName\Chapter 2**), while the newly created file **Sine Wave Chart (While Loop with Runtime Options)** is open and ready for you to modify.

On the open VI's front panel, add a **Push Button** (found in **Controls>> Modern>>Boolean**) and label it **Clear Chart?**. Pop up on this control and select **Mechanical Action>>Latch When Pressed**. By default, **Push Button** has a Boolean value of FALSE. Its mechanical action is now programmed as follows: When pressed, Push Button's value changes to TRUE. Then, immediately after this value is read on the block diagram, Push Button will return to its default FALSE value.

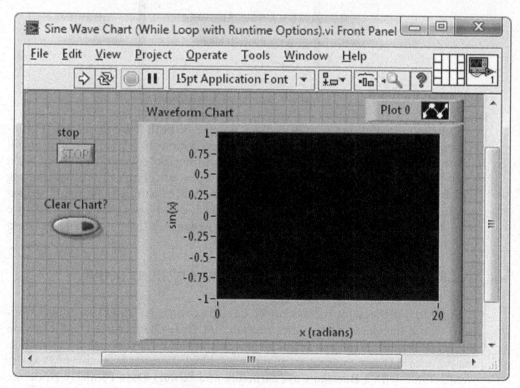

Switch to the block diagram. There, add a **Case Structure** (from **Functions>> Programming>>Structures**) and wire the Push Button's Boolean terminal to its selector terminal as shown. The Case Structure is now Boolean in nature; to toggle between its TRUE and FALSE windows, click the mouse cursor on the ◄ True ▼► at the top of the structure.

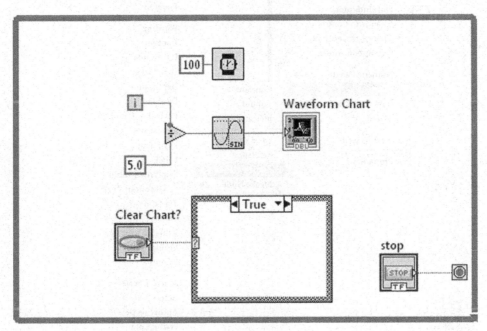

Next, we will place the appropriate Property Node that clears the Waveform Chart within the Case Structure's TRUE window. Then, when the Push Button is pressed on the front panel, this chart-clearing Property Node will be executed. The appropriate Property Node is called **History Data**, which is created by popping up on the Waveform Chart's icon terminal and selecting **Create>>Property Node>>History Data** as shown next. While making this selection, you might take a moment to peruse the pop-up menu's list of Waveform Chart properties that can potentially be controlled using a Property Node. It's an impressive list.

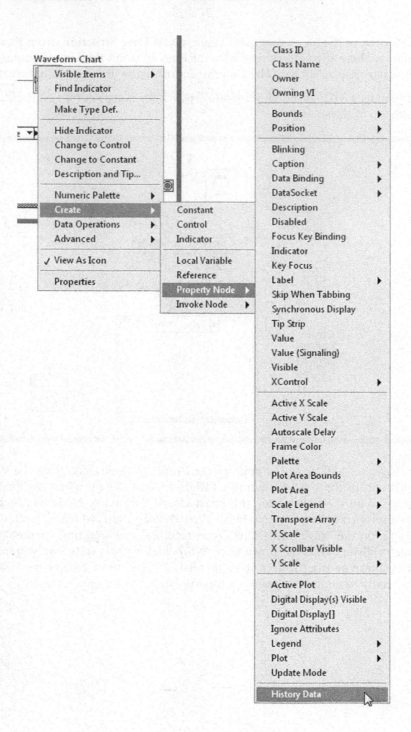

After selecting **History Data**, place this Property Node within the Case Structure's TRUE window. As with most Property Nodes, **History Data** is created in its *read mode* (i.e., it is enabled to "get" the current value of **History Data**). The small outward-directed black arrow on its right side indicates that, in this current configuration, the Property Node outputs a value.

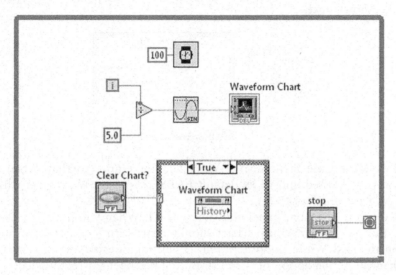

Change this icon to its *write mode* (i.e., enable it to "set" the value of **History Data**) by popping up on its lower section and choosing **Change To Write**.

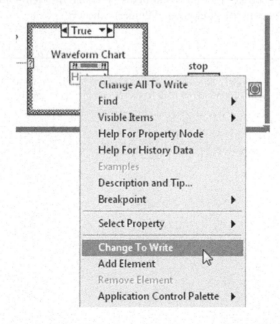

The black arrow will become inward-directed on its left side, indicating that **History Data** is now configured to accept an input value. The needed input to clear the Waveform Chart is an **Empty 1D Array** with double-precision floating-point numeric formatting. Pop up on the region containing the black arrow and then, using **Create>>Constant**, produce the required **Empty 1D DBL Array Constant** input as shown next.

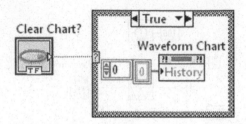

Leave the Case Structure's FALSE window empty so that, when the Push Button is unpressed and in its default FALSE state, the Waveform Chart is not cleared. Save your work.

Return to the front panel and run your VI. When you press the **Clear Chart?** Push Button, the Waveform Chart should clear, with the sine-wave plot resuming on the next While Loop iteration. If plotting doesn't resume, check that the **Mechanical Action** option of the Push Button has been correctly selected as **Latch When Pressed**.

Next, we will control the plot's background color programmatically. Start by placing an **Enum** control (found in **Controls>>Modern>>Ring & Enum**) on the front panel. Label this control **Background Color** as shown.

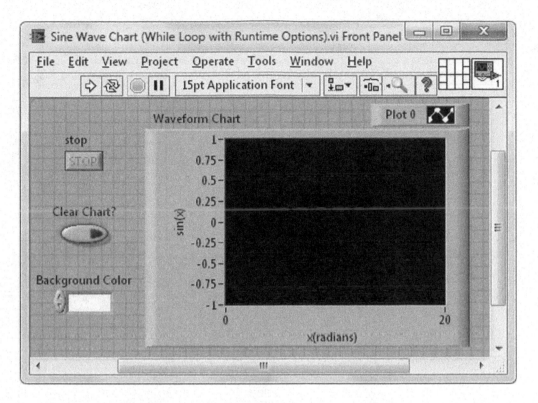

Pop up on the **Enum** and select **Edit Items...**. In the dialog window that appears, program this control with the following four items—*Black*, *Red*, *Green*, *Blue*— and then click the **OK** button. For a review of this programming procedure, see Section 4.7, "Adding Shape Options Using An Enumerated Type Control."

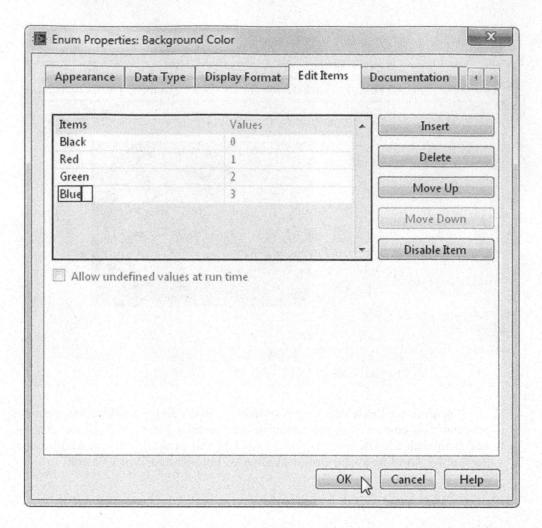

Switch to the block diagram. Add a **Case Structure** and wire **Background Color**'s Enum terminal to its selector terminal as shown.

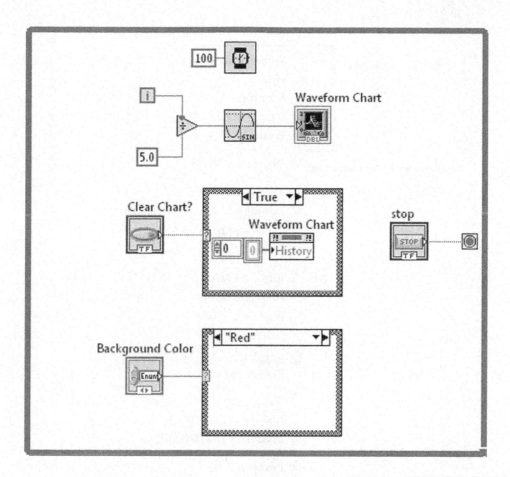

When the four-item Enum is initially wired to the selector terminal, the Case Structure will only contain two windows corresponding to the first two items in the Enum's list of items. To create the other required windows, pop up anywhere on the case Structure's border and choose **Add Case for Every Value**.

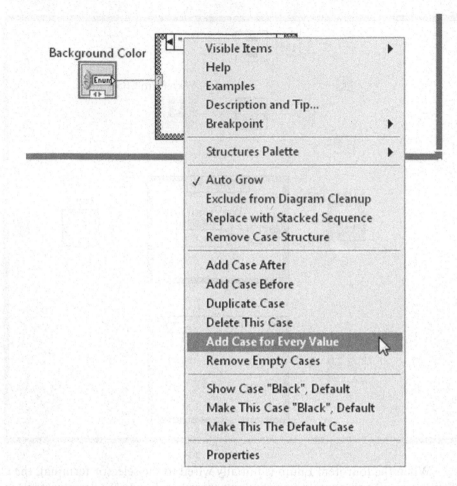

By clicking on the case selector label, check to see that the Case Structure now contains four case windows labeled **Black**, **Red**, **Green**, and **Blue**.

The Waveform Chart's background color is generated by the RGB color method. In this approach, varying intensities of the three additive colors red, green, and blue are combined to produce a broad array of colors. For the LabVIEW implementation of RGB, eight bits are used to denote the desired intensity of each additive color, which corresponds to a decimal number in the range from 0 to 255. Each of these three decimal numbers is converted to a two-digit hexadecimal equivalent in the range from 00 to FF. The red, green, and blue hexadecimals—called *RR*, *GG*, and *BB*—are then packaged as a single six-digit hexadecimal number *RRGGBB*. As an example, in this scheme, *FF0000* is pure red, and *0000FF* is pure blue. Thankfully, LabVIEW supplies an icon called **RGB to Color.vi**, which carries out the details of the RGB color method. **RGB to Color.vi** is found in **Functions>> Programming>>Numeric>>Conversion** and its Help Window is shown next. At the icon's **R**, **G**, and **B** inputs, you supply three decimal numbers, each in the range from 0 to 255 with unsigned eight-bit integer representation **U8**. These numbers denote the relative intensities of the three additive colors needed to generate the desired color. The icon then does the mathematical manipulations required to produce the *RRGGBB* hexadecimal number, which is supplied at the **Color** output.

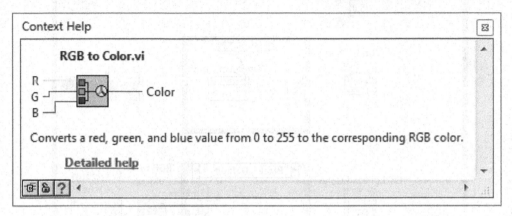

The Property Node that controls the background color of a Waveform Chart is called **BG Color**. Pop up on the Waveform Chart's icon terminal and create this Property Node using **Create>>Property Node>>Plot Area>>Colors>>BG Color**. Place **BG Color** on the block diagram, and then change it from its default read mode to the write mode by selecting **Change To Write** in its pop-up menu. Finally, place the appropriate three **U8** integers within each of the four Case Structure windows and wire them as shown next. For black, red, green, and blue, (**R**, **G**, **B**) are (0, 0, 0), (255, 0, 0), (0, 255, 0), and (0, 0, 255), respectively. The next block diagram shows the **Black** Enum case.

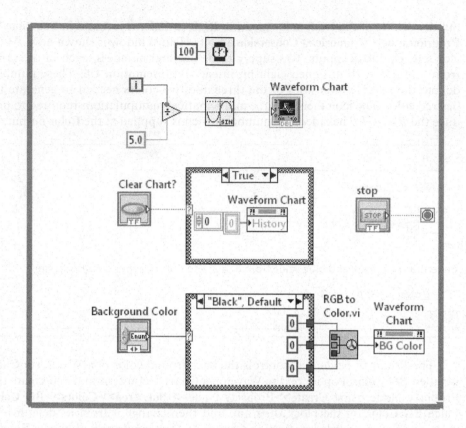

The three other Case Structure windows should be programmed as follows.

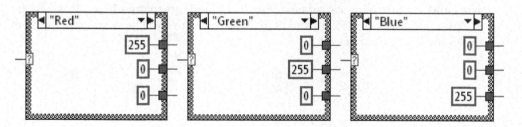

Save your work and then return to the front panel. Run **Sine Wave Chart (While Loop with Runtime Options)** and try out each of the plot area's four possible background colors. My favorite is blue—how about you?

9.3 STATE MACHINE ARCHITECTURE: GUESSING GAME

For the rest of this chapter, we will explore a commonly used design in LabVIEW programming—the *state machine*. With this design, each step of the algorithm being programmed is called a *state*, and the order in which the set of states executes is determined by built-in decision making based on external inputs and/or actions in previously executed states. A LabVIEW-based state machine consists of a Case Structure nested within a While Loop, where each of the Case Structure's cases contains the code associated with one of the state machine's states.

The While Loop executes continuously until its ⊙ is set to TRUE, and with each loop iteration, one of the Case Structure cases (i.e., a state) executes. The state that executes during a particular iteration performs some operation (e.g., reads data) and, in addition, selects which state will be executed during the next iteration. The block-diagram template for a state machine is shown in the next illustration. The state selection is made using an **Enum Constant**, which is stored in a shift register. Information that needs to be passed from one state to another is similarly stored in other shift registers.

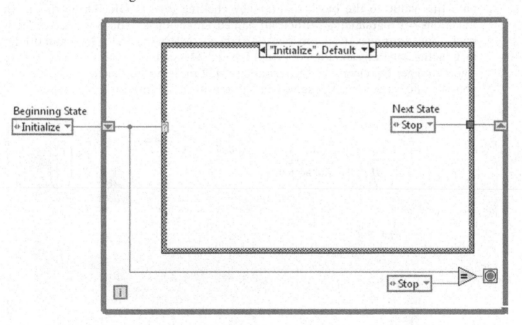

To familiarize ourselves with the state machine architecture, let's write a simple example program called **Guessing Game (State Machine)**. This VI will randomly generate an integer in the range from 1 to 10 and then guide the user in correctly guessing the integer. A state diagram representation of the algorithm for **Guessing Game (State Machine)** is shown next. Here, we see that the algorithm is carried out through five distinct states, which can be labeled *Generate Integer*, *Enter Guess*, *Check Guess*, *Give Hint*, and *Report Success*.

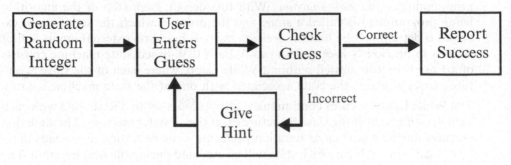

Open a new VI and construct the front panel as shown next. Here, when the VI is running, you will type your guessed integer into the **Numeric Control** (found in **Functions>>Programming>>Numeric**) labeled **Guess** and then pass this value to the block diagram by clicking on the **OK Button** (found in **Functions>>Programming>>Boolean**) named **Enter Guess**. You will receive feedback about your guess (e.g., too low, too high, correct) in the **String Indicator** (found in **Functions>>Programming>>String**) labeled **Message**. Format **Guess** as a long signed integer by selecting **Representation>>I32** in its pop-up menu and then save your VI under the name **Guessing Game (State Machine)** in **YourName\Chapter 9**.

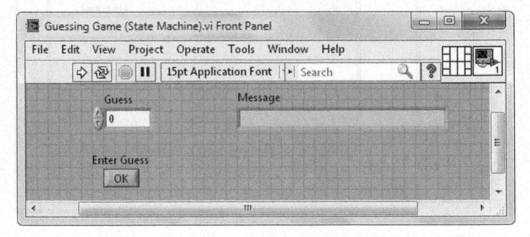

Switch to the block diagram, where we will write code based on the state machine architecture. For our guessing-game state machine, we require five states: *Generate Integer*, *Enter Guess*, *Check Guess*, *Give Hint*, and *Report Success*. Thus, we need to create an **Enum Constant** on our block diagram programmed with these five items.

Similar to the method we used to create a front-panel **Enum** control in the previous section, one can simply obtain an **Enum Constant** icon in **Functions>> Programming>>Numeric**, place it on the block diagram, select **Edit Items...** in its pop-up menu, and then program the five state names in the dialog window that appears. After closing the dialog window, the completed **Enum Constant** will appear on the block diagram. Multiple copies of this icon can then be obtained simply by cloning it (pressing *<Ctrl>* while clicking on the object with the).

I'll outline an alternate method for creating the required **Enum Constant** that involves the **Control Editor** and includes an option that is very handy. Open the Control Editor as follows: First, select **File>>New...** at the top of an open VI window or the Getting Started Window. The **New** dialog window will open. In the **Create New** box, find **Custom Control** in the **Other Files** folder and double-click on it. The Control Editor window will open, which appears similar to the front panel of a VI.

With the Control Editor open, activate the Controls Palette, and then place an **Enum** (found in **Controls>>Modern>Ring & Enum**) on it.

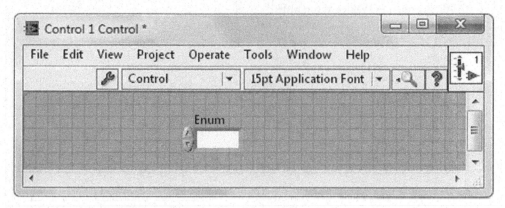

Pop up on the **Enum** and select **Edit Items...**, and then program the **Items** box with the first three state names—*Generate Integer, Enter Guess, Check Guess*. In constructing a state machine, it is not uncommon to realize part of the way through the process that an unanticipated state needs to be added to the code. To illustrate how to easily solve this frequently occurring problem, let's neglect to include the last two states—*Give Hint* and *Report Success*—in our Enum for now, and then show how to add them later.

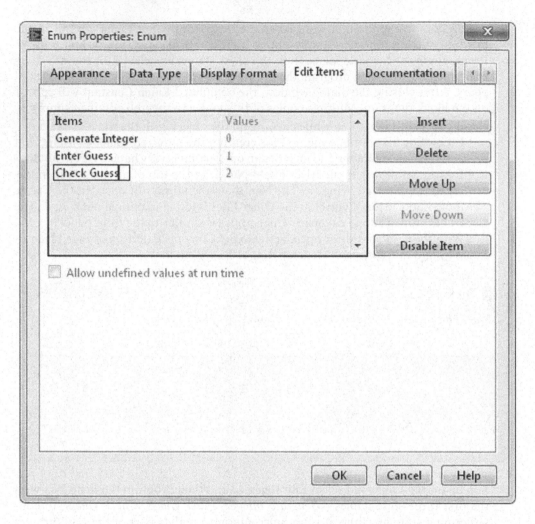

After programming the three states, press the **OK** button. You will be returned to the Control Editor window, which now contains the completed Enum. You may wish to resize the Enum to make all of the text within it visible.

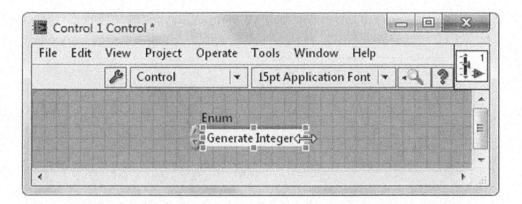

One crucial selection is needed before we save our customized object. In the Control Editor's toolbar, note that (by default) the **Control** option is selected for our Enum in the **Control Type** pull-down menu. In this mode, if we place two copies of this customized Enum on the block diagram, each copy will be independent of the other; in other words, we may use **Edit Items…** to add states to one without changing the other. For our work, however, it will be more beneficial to save our Enum in the **Type Definition** mode by selecting **Type Def.** as shown below before saving the customized control.

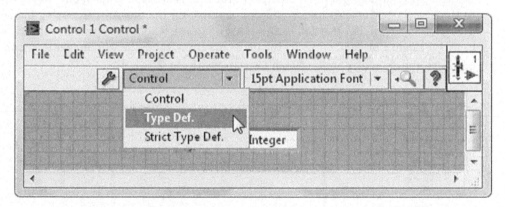

When saved in the **Type Definition** mode, this Enum becomes a master file that is linked to every related Enum placed on the block diagram. Then any change made to the master file will be transmitted to every related Enum Constant. As we will see, this is the feature that is quite labor-saving when writing a state machine and finding well into the project that an additional state must be added.

Select **Type Def.** in the **Control Type** pull-down menu and then, using **File>>Save**, save this customized control under the name **Guessing Game States** in

YourName\Controls (if this folder doesn't already exist, create it). The extension .ctl will be added automatically. Then close the Control Editor window.

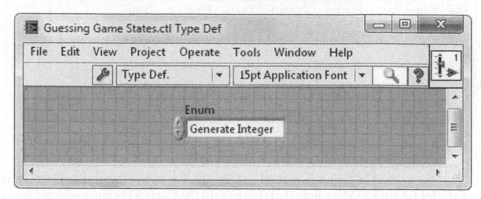

Begin coding the block diagram of **Guessing Game (State Machine)** by constructing the **Generate Integer** state as shown in the next diagram. The shift register, which is created by popping up on the While Loop border and selecting **Add Shift Register**, will be used to store the state-selecting **Enum Constant** (the free label **State** is created using the [A]). To obtain a copy of the customized Enum, click on **Select a VI...** in the Functions Palette. In the dialog window that opens, navigate to the YourName\Controls folder and double-click on the filename **Guessing Game States.ctl**. You will be returned to the block diagram, where you can place the customized **Enum Constant**. Once on the block diagram, the **Generate Integer** item can be selected by clicking on the **Enum Constant** with the 🖑.

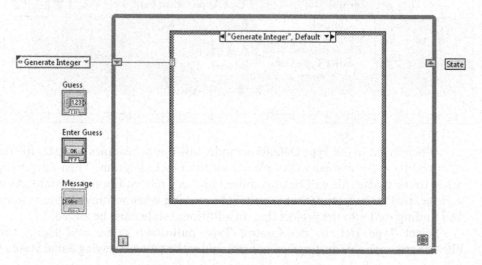

When the **Enum Constant** is connected to the Case Structure's selector terminal, two cases will be activated that are associated with the first two items programmed on the **Enum Constant** (*Generate Integer* and *Enter Guess*). Pop up on the Case Structure's border and choose **Add Case for Every Value**. The Case Structure will then create a case for each of the three states.

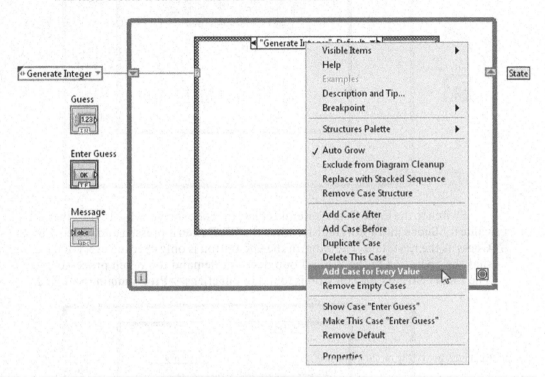

Complete the rest of the code for the **Generate Integer** state as shown in the next illustration, which randomly generates an integer in the range from 1 to 10, stores this number in a shift register, and selects **Enter Guess** as the next state to be executed. The **Random Number (0-1)** and **Round Toward + Infinity** icons are found in **Functions>>Programming>>Numeric**, while **To Long Integer** is in **Functions>> Programming>>Numeric>>Conversion**. The **String Constant** icon is found in **Functions>>Programming>>String**. The second Enum Constant (with **Enter Guess** selected) can be obtained most easily by cloning the other Enum Constant (that has **Generate Integer** selected). Use Help Windows to understand the functioning of any unfamiliar icons.

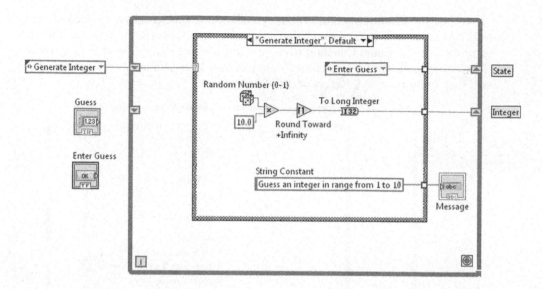

Switch to the **Enter Guess** state and create the code shown, which enters the user's value for Guess into a shift register when the **OK Button** is pressed and selects **Check Guess** as the next state. The value of the OK Button is only checked every 100 ms so that the operation of the While Loop does not demand too much processor time. The **Empty String Constant** icon is found in **Functions>>Programming>>String**.

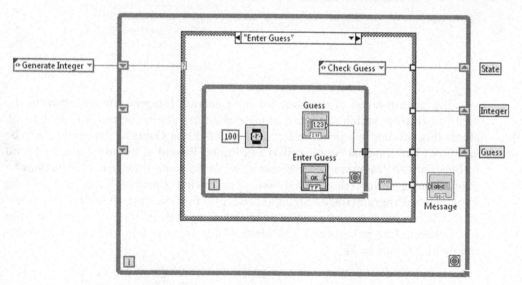

Switch to the **Check Guess** state, where we will write code that compares the randomly generated integer with the user's current guess. Begin by building the code shown. The **Equal?** and **Select** icons are found in **Functions>>Programming>> Comparison**.

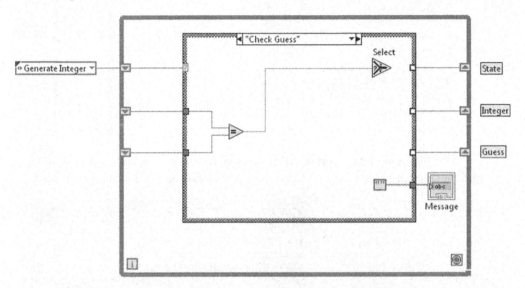

Okay, here is where we confront the frequently occurring problem when well into the construction of a state machine—we need to add more states. In the above diagram, when the output of the **Equal?** icon is TRUE (i.e., the user's guess is the same as the randomly generated integer), the **Select** icon should direct the program execution to a state that announces the user's success. On the other hand, if **Equal?**'s output is FALSE, **Select** should choose the next executed state to be one that assists the user in making their new guess. Unfortunately, neither of these two needed states currently exists and we now realize they need to be added to the collection of Case Structure cases as well as appended to the items listed in each Enum Constant already on the block diagram. These required modifications could involve a lot of work, but fortunately we have anticipated this possible problem and we will be able to make the needed changes easily. Here's how: Pop up on one of the Enum Constants, for example, the Enum Constant outside of the While Loop as shown in the next diagram, and select **Open Type Def.**

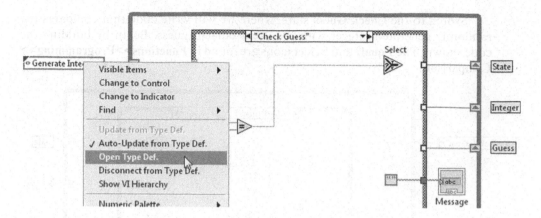

The **Guessing Game States.ctl** file in the **YourName/Controls** folder will open and be standing ready for editing in the Control Editor window as shown next.

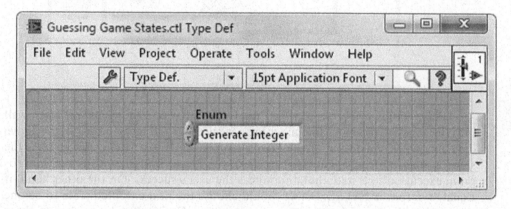

Pop up on the **Enum** and select **Edit Items…** in the pop-up menu. Enter the two new items needed—*Give Hint* and *Record Success*—and press the **OK** button.

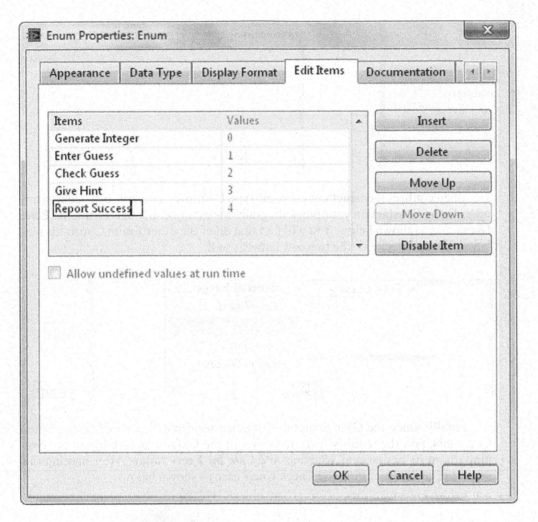

When returned to the Control Editor window, make sure that **Type Def.** is still selected in the **Control Type** pull-down menu, and then save your work as you close this window.

With the ☝, verify that the Enum Constant that you used to open the Control Editor (by selecting **Open Type Def.** in its pop-up menu) now includes the two newly entered items.

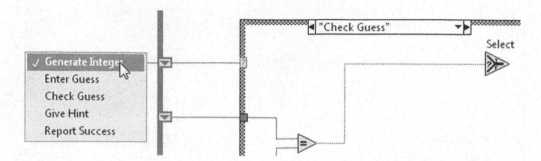

Then confirm the benefit of creating Type Definition controls by checking any other **Enum Constant** on your block diagram, for example, the one within the **Enter Guess** case as shown below. You will find that all of the other Enum Constants now have been updated with the two new items as well.

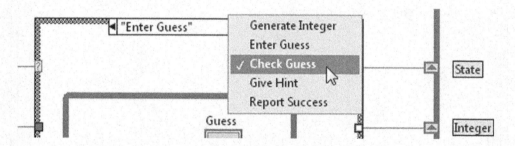

Finally, since the Case Structure's selector terminal ⬚ is wired to the Enum Constants, add the required two new cases to the Case Structure by simply popping up on its border and selecting **Add Case for Every Value**. After making this selection, complete coding the **Check Guess** case as shown below.

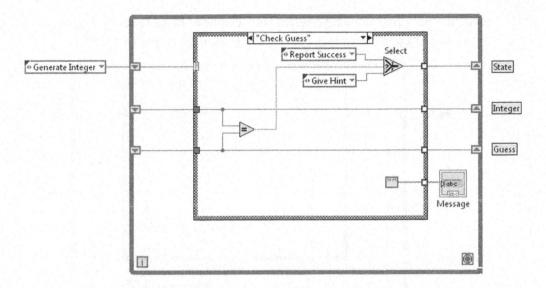

Next, code the **Give Hint** state, which alerts the user if their guess is too high or too low in comparison to the randomly generated integer and selects **Enter Guess** as the next state. The **Greater?** icon is found in **Functions>>Programming>> Comparison**.

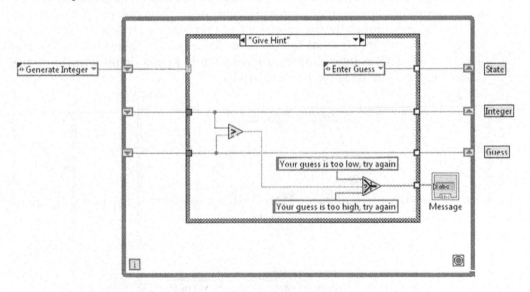

Finally, switch to the **Report Success** state and complete the diagram as shown. With this code, after the **Report Success** state executes, the While Loop will cease execution because an Enum Constant with a value of **Report Success** will be passed to the **Equal?**'s upper terminal, causing a TRUE to be output to the ▣.

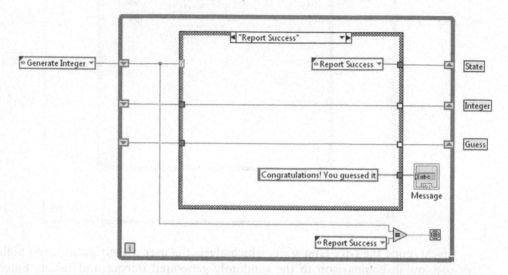

Note that two of the Case Structure's output tunnels are not solidly colored, indicating that these terminals are not wired in all five cases. These unfilled tunnels each constitute a programming error that must be rectified before the VI can run.

In clicking through the five cases, we find that the *Guess*-related output tunnel is unwired in only the **Generate Integer** state. Since the user's guess has not yet been logged at this early point in the program's execution, we can simply wire this tunnel as shown next, with no adverse repercussions.

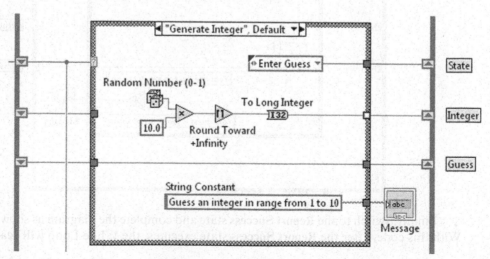

For the *Integer*-related output tunnel, the **Enter Guess** state is the culprit. Resolve this problem by including the wire shown (since the value of Integer remains unchanged after the **Generate Integer** case).

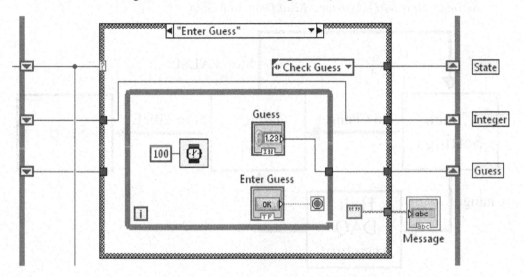

The program is finished now. Return to the front panel and save your work. Run the VI and verify that it works as advertised.

9.4 STATE MACHINE ARCHITECTURE: EXPRESS VI-BASED DIGITAL OSCILLOSCOPE

Let's see how the state machine architecture can help us in the laboratory. In Chapter 6, we found that an intrinsic feature of **DAQ Assistant** imposed a limitation on the functioning of programs that contained this Express VI. In particular, we discovered that the choice of **Sampling Frequency** is set when **DAQ Assistant** first executes within a program and this important digitizing parameter cannot be changed from that point on as the program continues to run. Thus, in Chapter 6, our only recourse for changing **Sampling Frequency** was to stop our program (in order to halt the execution of DAQ Assistant), reload a new value for this parameter on the front panel, and then restart the program. The good news is that, with our new knowledge of the state machine design, we now have a powerful way to work around this limitation. To demonstrate this fact, we will build a state machine-based VI called **Digital Oscilloscope (Express State Machine)**, which uses DAQ Assistant to acquire analog input data, while continually checking for changes to the front-panel **Digitizing Parameters** cluster. If changes to the cluster are detected, then the program execution is directed to a sequence of steps that first halt DAQ Assistant, reload it with the new **Digitizing Parameters** values, and then restart the

Express VI for a new round of data-taking. A state diagram of the algorithm for **Digital Oscilloscope (Express State Machine)** is shown next. Here, we see that the algorithm is carried out through four distinct states, which can be labeled *Check Settings*, *Halt DAQ Assistant*, *Read Data*, and *Stop*.

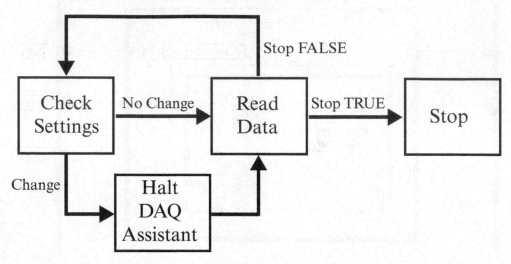

Ready to get started? Create a new VI named **Digital Oscilloscope (Express State Machine)** and store it in **YourName\Chapter 9**. On the front panel, place a **Waveform Graph**, whose *x*- and *y*-axes are labeled **Time** and **Voltage**, respectively. Also, using **Controls>>Select a Control…**, locate a **Digitizing Parameter** cluster in **YourName\Controls** and place it on the front panel (if you haven't stored a **Digitizing Parameter** cluster in **YourName\Controls** previously, construct the cluster now as described in Section 4.10, "Control and Indicator Clusters"). Save your work.

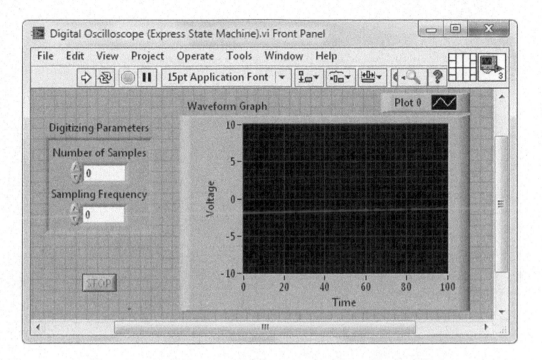

For our digital oscilloscope state machine, we require four states: *Check Settings*, *Halt DAQ Assistant*, *Read Data*, and *Stop*. Create an Enum programmed with these four items as follows: Open the Control Editor (by selecting **File>>New...** and then, in the **Create New** box, **Other Files>>Custom Control**) and place an **Enum** on it. Pop up on the **Enum**, select **Edit Items...**, and program it as shown below.

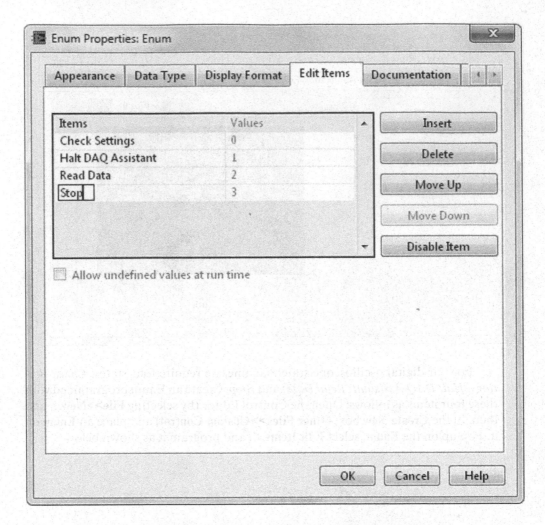

Press the **OK** button to return to the Control Editor window. You may wish to resize the Enum to make all of the text within it visible. Select **Control Type>>Type Def.**, then save this control under the name Oscilloscope Express States in YourName/Controls as you close it.

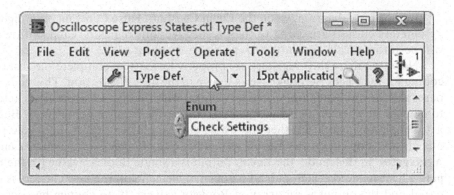

Begin coding the block diagram of **Digital Oscilloscope (Express State Machine)** by constructing the **Check Settings** state as shown in the next diagram. The **False Constant** is found in **Functions>>Programming>>Boolean**. To obtain a copy of your customized Enum, choose **Functions>> Select a VI...** and then navigate to the **YourName\Controls** folder and double-click on the filename **Oscilloscope Express States.ctl**. When the **Enum Constant** is connected to the Case Structure's selector terminal, two cases will be activated that are associated with the first two items programmed on the **Enum Constant** (*Check Settings* and *Halt DAQ Assistant*). Pop up on the Case Structure's border and choose **Add Case for Every Value**. The Case Structure will then create a case for each of the four states.

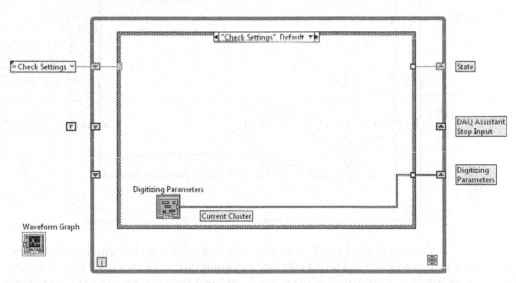

In the above diagram, when the **Enum Constant**, **False Constant**, and Digitizing Parameters clusters are wired to their respective shift registers, each shift register will be automatically formatted to the appropriate data type.

Now that these shift registers are properly formatted, complete the rest of the coding for the **Check Settings** case as shown in the next illustration. This code compares the current Digitizing Parameters cluster with that from the previous iteration to determine whether a change has occurred in the front-panel settings. Because it is simplest to check whether the whole current cluster equals the whole previous cluster, pop up on the **Equal?** icon and select **Comparison Mode>>Compare Aggregates**. In this mode, the **Equal?** output is a single Boolean value (in the **Compare Elements** mode, each pair of associated elements in the two clusters is compared and the output is an array of Boolean values). If no change has occurred (**Equal?** is TRUE), **Read Data** is chosen as the next state to execute. Alternatively, if a change has occurred (**Equal?** is FALSE), **Halt DAQ Assistant** is chosen as the next state. To create the **Cluster Constant**, which initializes the **Digitizing Parameters** shift register, (since it is already formatted) simply pop up on this shift register and select **Create>>Constant**. For the next state, **Halt DAQ Assistant** is selected if a change in the Digitizing Parameters cluster is detected; otherwise, **Read Data** is selected as the next state.

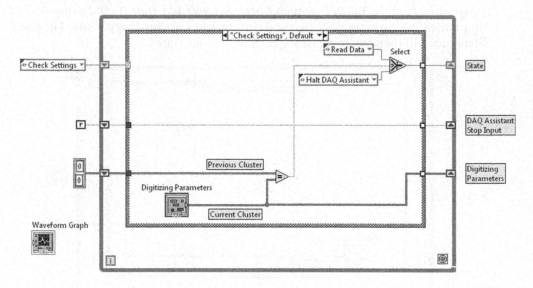

Switch to the **Halt DAQ Assistant** state and code it as shown. The purpose of this state is to load the **DAQ Assistant Stop Input** shift register with a **True Constant**. This Boolean will be used in the subsequent **Read Data** state to stop **DAQ Assistant** so that this Express VI can then be programmed to operate at a different sampling frequency.

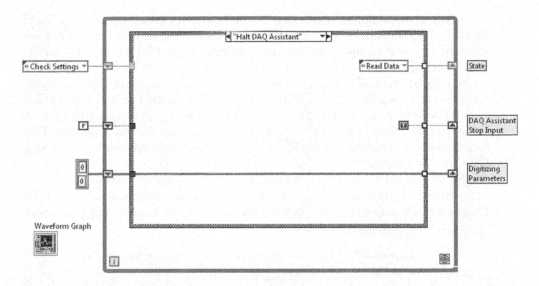

Next, code the **Read Data** state. The **Or** icon is found in **Functions>>Program ming>>Boolean**. When programming **DAQ Assistant**, make the selections **Acquire Signals>>Analog Input>>Voltage>>ai0**, and **Acquisition Mode>>N Samples**. Under the **Triggering** tab, in the **Start Trigger** box, select **Trigger Type>>Digital Edge**, **Trigger Source>>PFI0**, and **Edge>>Rising**.

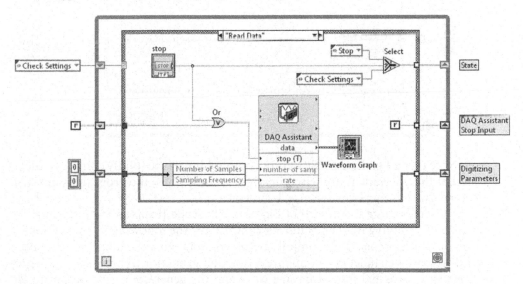

In the above diagram, if the **Stop Button** has been pressed (so that its value is TRUE), **DAQ Assistant** is stopped and **Stop** is chosen as the next state to execute. Alternatively, if the **Stop Button** has not been pressed (so that its value is FALSE), **Check Settings** is chosen as the next state (which commences another cycle of data taking). In addition, if a change in settings has been detected in the most recent execution of **Check Settings**, a TRUE Boolean will be passed to the **stop (T)** input of **DAQ Assistant**, halting the Express VI's execution. Upon exiting this case, the **DAQ Assistant Stop Input** shift register is reloaded with a **False Constant**; DAQ Assistant will be restarted with a new value for **Sampling Frequency** when the **Read Data** state is executed next.

Finally, switch to the **Stop** state and complete the diagram as shown. With this code, after the **Stop** state executes, the While Loop will cease execution because an **Enum Constant** with a value of **Stop** will be passed to the **Equal?**'s upper terminal, causing a TRUE to the ⬛.

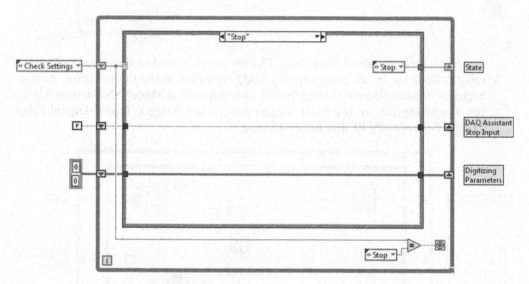

Note that all of the Case Structure's output tunnels are solidly colored, indicating that these terminals are wired in all four cases. Thus, the program is finished. Return to the front panel and save your work.

To demonstrate the ability of **Digital Oscilloscope (Express State Machine)** to change its sampling frequency during runtime, try the following: Since we configured our oscilloscope to be digitally triggered and read an analog signal input at AI differential channel *ai0*, configure a function generator to produce a (close to) 50 Hz sine wave and input this sine wave and the generator's sync output to *ai0* and PFI 0 (and D GND) pins of your DAQ device, respectively. On the VI's front panel, select **Number of Samples** and **Sampling Frequency** equal to *100* and *1000*, respectively, and then run **Digital Oscilloscope (Express State Machine)**. You will see (about) five cycles of the digitized sine wave. [For USB-6001/6002/6003 devices, you may have to delete the Boolean wire connecting the **Or** icon to DAQ Assistant's **stop (T)** input to make digital triggering functional.]

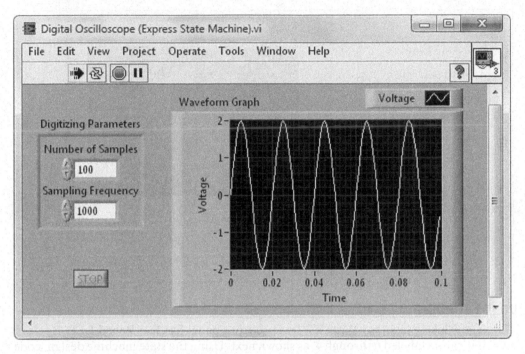

Now, show that during runtime you can "zoom in" on one cycle of the sine wave by simply increasing the sampling frequency by a factor of five, while keeping the number of samples the same. With **Digital Oscilloscope (Express State Machine)** still running, change **Sampling Frequency** to *5000* (to secure this choice, you will have click on the ✓ or press <*Enter*>) and enjoy your VI's capacity to maneuver around DAQ Assistant's limitations.

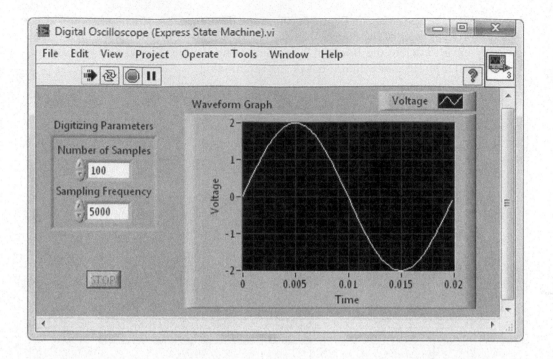

DO IT YOURSELF

Five Blinking Lights.vi Write a VI whose front panel has five **Round LED** Boolean indicators labeled *0* through *4*, as shown next. Using the state machine design, code the block diagram so that these LEDs light one at a time in the order 0-1-2-3-4 and keep repeating this sequence until the **Stop Button** is pressed (when the **Stop Button** is pressed, the current 0-1-2-3-4 sequence should be completed before the program halts execution). Make each LED stay lit for 0.2 second, so that an entire 0-1-2-3-4 sequence occurs once every second. Note that the value of a Boolean Constant can be changed using the ⌖. *Hint:* Place the LED icon terminals outside the Case Structure.

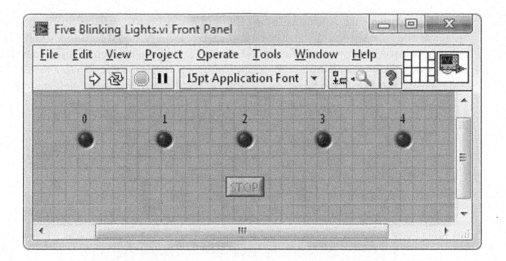

USE IT!

Selective Array Building In laboratory work, there is often the need to sort the output of a measuring device or a calculation into different categories. Consider one such example: You have designed an experimental setup to monitor the decay of a radioactive material. The output of this experiment is presented as an oscilloscope trace—when a decay occurs, the trace contains a large pulse; when no decay occurs, the trace has no pulse. The LabVIEW program needed to run this experiment might then do the following: (1) acquire an oscilloscope trace; (2) determine if the trace contains a pulse; (3) for later analysis, store the trace in a dedicated "decay array" if it contains a pulse, or store it in dedicated "no decay array" (or delete it) if the trace has no pulse; and (4) repeat this process.

The following simplified LabVIEW program demonstrates how to use a Case Structure to sort a repeatedly "acquired" quantity. In this VI, a sequence of integers is created and sorted into two arrays representing two different categories—even and odd. Here, we use the 🅸 to generate the sequence of integers from 1 to 30. For each integer, **Quotient & Remainder** is used to determine if it is even or odd. Then, for example, if the integer is determined to be even, the Case Structure's **True** case is selected, where the integer is appended to the **Even Array** via **Build Array** and the **Odd Array** is passed through the Case Structure unchanged. Likewise, the Case

Structure's **False** case is used to build the **Odd Array**. Shift registers store the two arrays as they are being constructed over the span of For Loop iterations. Each shift register is initialized with the **Empty 1D Array Constant** with **Representation>>I32** as shown, which can be created by popping up on the left shift register (after the code within the For Loop has been completed) and selecting **Create>>Constant**.

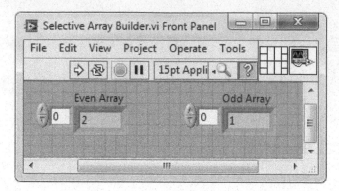

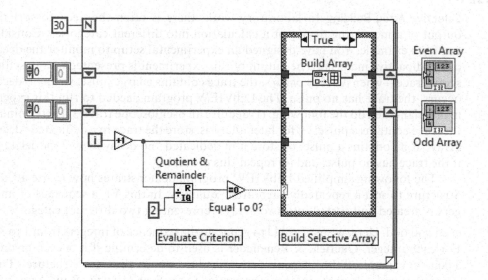

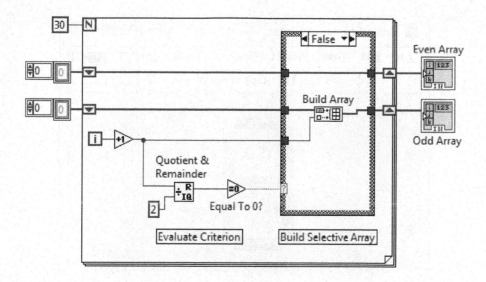

PROBLEMS

Each problem statement begins with a suggested descriptive name (including the **.vi** extension) for the program that you will write. Suggested icons for use in the VI can be found with the aid of **Quick Drop**.

1. **Seven-Segment Counter.vi** As shown below, use seven **Square LED** Boolean indicators, which have been appropriately resized and labeled A through G, to form a seven-segment display on the front panel of a VI. Then code the block diagram so that, when the Run button is pressed, this front-panel display counts from 0 to 9 with each digit illuminated for 1.0 second.

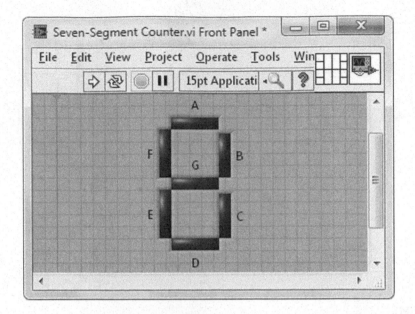

2. **Multiples of 3, 4, or 5.vi** Employing the Use It! example in this chapter as a guide, write a program that forms a single array containing all of the integers up to 30 that are evenly divisible by either 3, 4, or 5 and then displays this array in a front-panel indicator. (When your VI is run, you should find that the array contains 18 integers.) Suggested icon: **Quotient & Remainder**, **Compound Arithmetic**.

3. **Digital Oscilloscope with Spreadsheet Storage (Express State Machine).vi** Add the option of storing the final trace in your digital oscilloscope state program as follows: With **Digital Oscilloscope (Express State Machine)** open, use **File>>Save As...** to create a new VI. In the state machine code of this program, add a new state called **Store Data** between the **Read Data** and **Stop** states. When a user presses the Stop Button, make **Store Data** the next executed state after **Read Data**, where the code shown below is carried out. Here, the **Two Button Dialog** icon asks the user if the last trace acquired by DAQ Assistant is wished to be saved and, if so, the dynamic data type trace is converted to a 2D (Time-Voltage) array and stored as a spreadsheet file. For the conversion process, the initial DDT wire must first be converted to a waveform data type via the **Convert from Dynamic Data** icon with the **Conversion>>Resulting data type>>Single waveform** option selected in its dialog window. Then, using the time increment **dt**, the For Loop constructs the Time array.

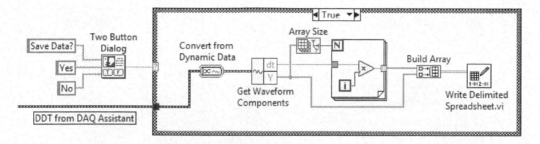

4. **Simple Calculator.vi** Write a VI that, given two floating-point numbers X and Y as input, provides a user with the choice (via an Enum) of either adding $(X+Y)$, subtracting $(X-Y)$, multiplying $(X*Y)$, or dividing (X/Y) these inputs and then displays the result. The front panel of this VI is shown next.

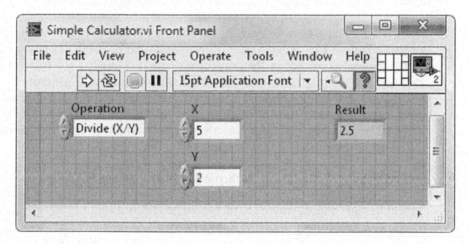

As shown below, include block-diagram code that lights a Boolean indicator labeled **Divided By 0?** when a division operation with the divisor equal to zero is requested. Use a Property Node to make this Boolean indicator visible only when the illegal "divide-by-zero" request is made. To create the required Property Node, pop up on the indicator's block-diagram terminal and select **Create>>Property Node>>Visible** (a second Property Node will be required to make this indicator not visible when the VI commences execution).

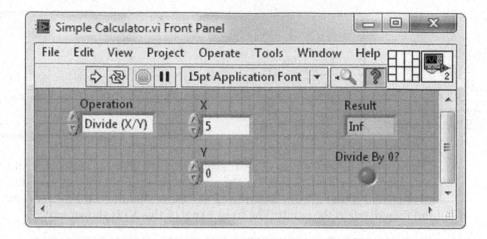

5. **Two-Try Guessing Game.vi** Using the state machine architecture, write a guessing-game VI in which a user is given only two chances to guess a randomly generated integer in the range from 1 to 10. Here are the rules of the game: When the program commences, the random integer is generated, and the user makes an initial guess. If the guess is correct, success is reported; if the guess is incorrect, the user is given a hint and asked to guess again. After being given this one hint, if the user's guess is correct, success is reported; otherwise, the user is told that they lost the game.

6. **Prime Detector.vi** Write a Case Structure-based program that determines whether an input integer is a prime number. The front panel of this VI should appear as shown next, where the integer to be tested is input at **Integer** and, if it is determined to be prime, the **Round LED** labeled **Prime?** is lit.

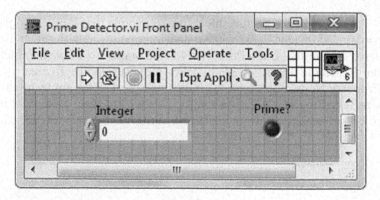

To construct your VI, consider the following properties of prime numbers. By definition, a prime number is evenly divisible by 1 and itself solely (i.e., only these two divisors result in a remainder of zero). Since even integers are evenly divisible by 2, even integers (except 2) are not prime. An odd integer N is not prime if it can be written as the product of two integers, where one integer is less than or equal to $\sqrt{N}$ and the other is greater than or equal to $\sqrt{N}$. Thus, to test whether an odd integer is prime, it is sufficient to show that it is not evenly divisible by all of the odd integers less than or equal to $\sqrt{N}$, with the exception of 1. Note that a sequence of odd integers, starting at 3, can be constructed from $N = 2i + 3$, where $i = 0, 1, 2, \ldots$. Suggested icon: **Quotient & Remainder**, **Square Root**, **Or**.

To test your VI, try inputting 104717 (prime) and 99763 (not prime). Which of the following integers are prime: 101467, 102703, 97861?

7. **Temperature Scale Converter (with Runtime Captioning).vi** Write a program that converts given temperatures between the Celsius and Fahrenheit scales continuously, until a **Stop Button** is pressed, and properly adapts the captioning of front-panel objects in response to the conversion mode selected by a user.

(a) Place an **Enum** control on the front panel and program it with the following two items: *C to F* and *F to C*. On the block diagram, wire the Enum's terminal to the selector terminal of a Case Structure, and then program the **C to F** and **F to C** case to execute the calculations $F = 1.8C + 32$ and $C = (F - 32)/1.8$, respectively. The completed front panel should appear as shown next. Run your VI with several inputs and verify that it functions correctly, for example, 68°F $\leftrightarrow$ 20°C, 100°C $\leftrightarrow$ 212°F.

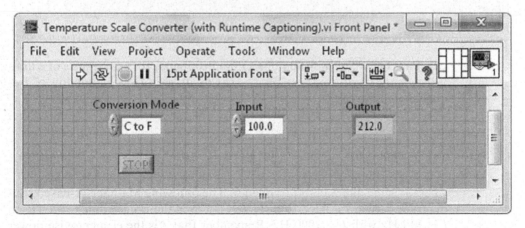

(b) Add code to your program so that the texts above the **Input** control and **Output** indicator appropriately adapt to the **Conversion Mode** selection as the program runs. To accomplish this feat, note that the owned label of a control or an indicator cannot be changed during runtime, but its caption can be changed. Thus, in the front-panel pop-up menus of **Input** and **Output**, deactivate **Label** and activate **Caption**. Then, on the block diagram, pop up on the **Input** terminal and select **Create>>Property Node>>Caption>>Text**. Place the resulting **Property Node** within the **C to F** Case, pop up on it, and select **Change To Write** to switch it from its read to write mode. Pop up on the Property Node again and select **Create>>Constant**, and then enter the text *Input deg C* into the resulting **String Constant**. Repeat this procedure to make the caption for **Output** be *Output deg F*. Then program the **F to C** Case so that the captions for **Input** and **Output** are *Input deg F* and *Output deg C*, respectively. Run your VI and verify that the captioning performs properly, for example, as shown below.

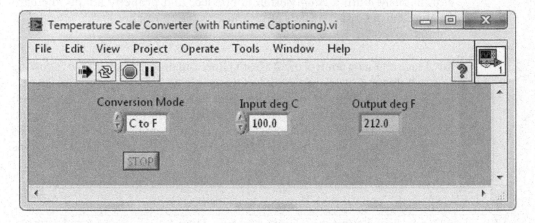

8. **Low-Pass Filter Plot (with Runtime XY Graph Options).vi** The response of a first-order Butterworth low-pass filter—that is, its gain G as a function of frequency f—is given by

$$G = \frac{1}{\sqrt{1 + \left(f/f_c\right)^2}}$$

where f_c is the filter's cutoff frequency. The following MathScript-based subdiagram executes a linear plot of G vs. f over the range from $f = 1$ Hz to $f = 10$ kHz with $f_c = 2000$ Hz. Remember that .^is the element-wise power operator. Manually choose the data type for the G output to be **DBL 1D**.

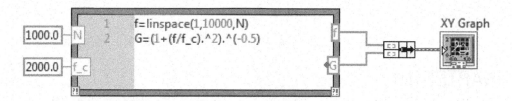

The Formula Node version of this subdiagram is as follows. Remember semicolons are required at the end of each line.

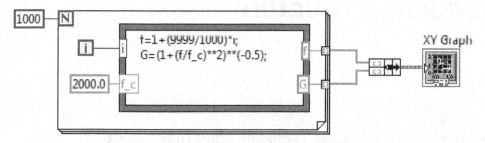

The gain in decibels is defined as $dB = 20 \log_{10} (G)$. A *Bode magnitude plot* graphs *Gain (in dB)* vs. *f*, where the frequency axis is logarithmic. By modifying the subdiagram given above, write a program that allows a user to graph the low-pass filter response on an XY Graph either as a linear plot or as a Bode magnitude plot. Let the user choose the type of plot via a Boolean switch labeled **Bode?**, which selects between two cases in a block-diagram **Case Structure**. For the logarithmic frequencies needed for the Bode magnitude plot, you can either "manually" take the base-10 logarithm of all the frequencies or (preferably) create a Property Node by popping up on the XY Graph terminal, selecting **Create>>Property Node>>X Scale>>Mapping Mode**, popping up on the Property Node and selecting **Change To Write**, and then wiring a constant *1* to it. For the linear case, you will then need a similar Property Node with a *0* wired to it. For extra flair, program the *y*-axis label as *Gain* or *Gain (dB)* for the Linear or Bode plot, respectively, using the Property Node **Y Scale>>Name Label>>Text**.

Verify that the gain at cutoff frequency is 3 dB smaller than the (constant) gain at low frequencies. Because of this fact, the cutoff frequency is often called the *3 dB point* or *3 dB frequency*.

CHAPTER 10

Data Dependency and the Sequence Structure

10.1 DATA DEPENDENCY AND SEQUENCE STRUCTURE BASICS

Sequencing—that is, controlling the ordered execution of instructions—is a fundamental feature of a computer program. As we discovered in the last chapter, the state machine architecture is one tool available to a LabVIEW programmer for assuring that the correct instructions are executed in the correct order. In this chapter, we will explore two other sequencing tools provided by LabVIEW, *data dependency* and the *Sequence Structure*.

LabVIEW is a *dataflow* computer language. This approach to programming is contrary to the *control flow* mode in which most other programming languages operate. In control flow systems, the program elements execute one at a time in an order that is coded explicitly within the program. The series of sentence-like statements in, for example, C and Python programs, which describe the sequential execution of Procedure *A* followed by Procedure *B* followed by Procedure *C*, are manifestations of the inherent control flow fashion in which these text-based languages are structured. In contrast, LabVIEW program elements abide by the principle of dataflow execution. Obeying a condition termed *data dependency*, a given object on the block diagram (called a *node*) will begin execution at the moment *all* of its input data become available. After completing its internal operations, this node will then present processed results at its output terminals. The interesting advantage of dataflow programming is that several nodes (for example, *A*, *B*, and *C*) can, through LabVIEW's *multithreading* ability coupled with a multicore processor, execute in parallel. Such parallel execution will occur whenever the receipt of node *A*'s inputs overlaps in time with the execution sequence of node *B* and/or *C*. This multithreading ability has the potential to enhance system throughput, especially in situations (not uncommon to data acquisition operations) where a portion of a particular node's execution time involves waiting. In LabVIEW, useful parallel processes can be executed while one node waits for an event, whereas in a control flow system, such waiting periods are simply dead time, putting the entire execution sequence on hold.

Despite the impression that you might glean from the previous paragraph, LabVIEW programs can, if desired, be written with a guaranteed one-step-at-a-time sequence of node execution. A proficient LabVIEW programmer may exploit the data-dependency maxim by coding a block diagram where the output of node *A* is used as a source of input data to node *B*. Such a diagram will mimic control flow execution in that *A* must fully execute to enable the execution of *B*. The wiring configurations just described will many times provide an elegant solution to the need for ordered execution within a LabVIEW program. However, to create the correct configurations requires some skill on the programmer's part.

If an elegant data-dependent wiring configuration isn't forthcoming, LabVIEW provides the *Sequence Structure*, an explicit and foolproof way to obtain control flow within a program. The Sequence Structure comes in two styles—the *Flat Sequence Structure* and the *Stacked Sequence Structure*—both of which are found in **Functions>>Programming>>Structures**. In the developmental history of LabVIEW, the Flat form of the Sequence Structure was introduced later than the Stacked form, and most LabVIEW aficionados now agree that the Flat Sequence Structure produces the most readable code. Hence, we will only use the Flat Sequence Structure in the programs we develop in this chapter.

When it is initially placed on a block diagram, the **Flat Sequence Structure** looks like a single frame of old-style movie film, as shown in the following illustration.

You can add a second frame, however, by popping up somewhere on the structure's border and selecting **Add Frame After**.

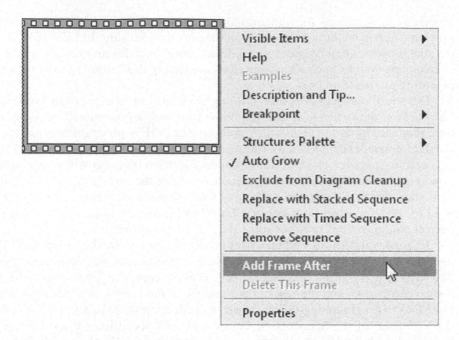

Then the original and the newly created frame will appear side by side, as below. The area enclosed by each frame can be resized using the ⤡.

By repeating this procedure, you can add any number of new frames. For example, in the next illustration, a Flat Sequence Structure with three frames has been created, with its frames labeled sequentially from left to right.

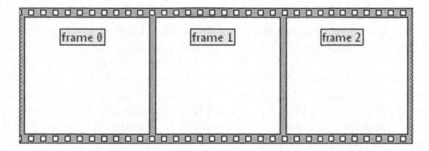

A Flat Sequence Structure executes sequentially from left to right. Hence, in a structure similar to that shown above, **frame 0** executes first, followed by **frame 1**, then **frame 2**, and so on until the last frame executes. Thus, by placing objects within different frames, you can use the Sequence Structure to control the order of execution among nodes that are not naturally coupled through data dependency. In addition, if a quantity produced in one frame is needed in a subsequent frame, that quantity can be passed between frames through a *tunnel*. For example, if one wanted to read the value of a front-panel Numeric Control, wait one-half second, and then display the value on a front-panel Numeric Indicator, that three-step sequential process can be accomplished by the following code.

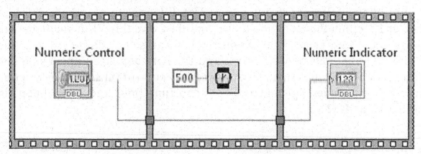

Finally, data can be input to any frame of the Flat Sequence Structure through a tunnel. Each frame can also emit output data through a tunnel, but, consistent with dataflow principles, these data will not be output from the structure until its last frame completes execution. Thus, the three-step sequential process described above can also be coded as shown next.

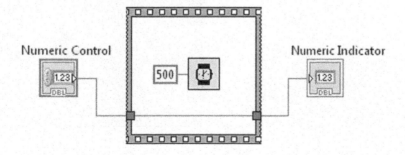

399

In this chapter, our goal is to program a multistep task in which the set of steps must proceed in a specific order. In building such a VI, it is perfectly natural to employ a Flat Sequence Structure on the block diagram and so that is what we will do in the following section. Then, in the section after that, we will demonstrate how to take advantage of LabVIEW's data-dependent mode of execution as we program the same task without the use of a multiframe Flat Sequence Structure. Along the way, we will learn how LabVIEW manages array storage in your computer's random access memory, and we will be introduced to a handy debugging tool called *Highlight Execution*.

10.2 EVENT TIMER USING A SEQUENCE STRUCTURE

A timer, which measures the duration of a given event, provides an example of an instrument in which operations must proceed in a specific order—a START initiates the timing action, followed by a STOP that terminates it. Let's build a timer VI that properly orders the required timing operations through the use of a Flat Sequence Structure. We will design our timer VI to do something interesting— compare the time it takes each of LabVIEW's two repetitive-operation structures, the For Loop and the While Loop, to build a given array of data. Our findings may surprise you!

Build the following front panel. Use **File>>Save** to create a folder called **Chapter 10** within the **YourName** folder, and then save this VI in **YourName\Chapter 10** under the name **Loop Timer (Sequence)**. Change the data-type representation of the **Number of Elements** control and the **For Loop Array** and **While Loop Array** indicators to **I32**, and then resize these objects so that they may accommodate large integer values. Format the **For Loop Time (ms)** and **While Loop Time (ms)** indicators as **U32**.

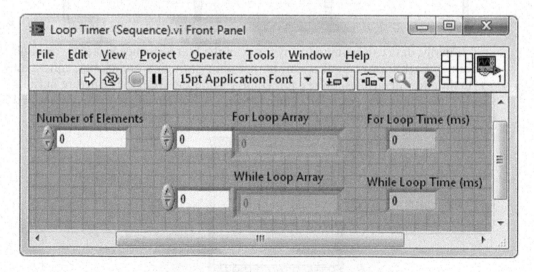

Switch to the block diagram. The basis of our timer will be **Tick Count (ms)**. This icon is found in **Functions>>Programming>>Timing** and its Help Window is reproduced next. **Tick Count (ms)** outputs an unsigned 32-bit (**U32**) integer that denotes the number of milliseconds that have elapsed since your computer was powered on.

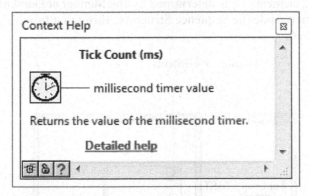

Our strategy for measuring elapsed time will be this: Call **Tick Count (ms)** at the START of an event and call it at the event's STOP. Then, by subtracting the two **Tick Count (ms)** output values, the event's elapsed time will be determined.

Put a **Flat Sequence Structure** on the block diagram. Inside the Sequence Structure's frame, place a **Tick Count (ms)** icon and wire its **millisecond timer value** output to a tunnel on the frame's border. This wire will contain your computer's internal clock value at the START of the For Loop, which will be contained in the next frame.

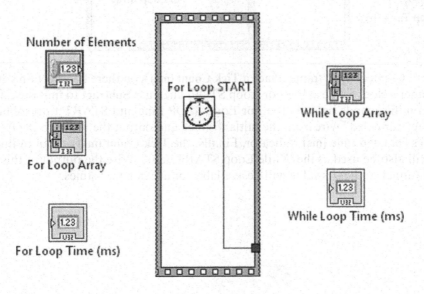

Create a second frame by selecting **Add Frame After** from the Sequence Structure's pop-up menu. Inside this frame, code a For Loop that creates an *N*-element array, where each array element has a numeric value equal to the value of its index. Output this array to the **For Loop Array** indicator on the front panel so that you may view its elements. *N* is determined by the **Number of Elements** control, which is wired from outside the Sequence Structure, through a tunnel, to the For Loop's count terminal.

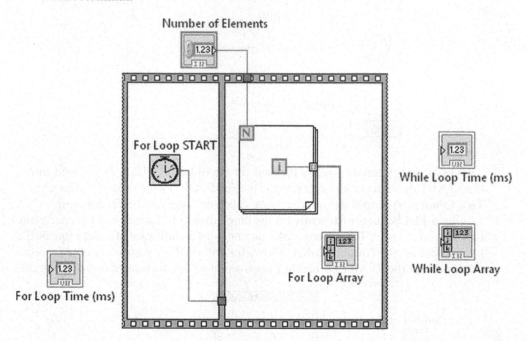

Create a third frame. Place a **Tick Count (ms)** icon there to ascertain your computer's clock value at the For Loop STOP. Then use **Subtract** to find the difference in milliseconds between the For Loop STOP time and START time (obtained in the "tunneled" wire from the initial frame), and output the result to the front panel's **For Loop Time (ms)** indicator. Finally, the **Tick Count (ms)** output in this frame will also be used as the While Loop START value. Wire the output of this icon to a tunnel so that its value will be available for use in later frames.

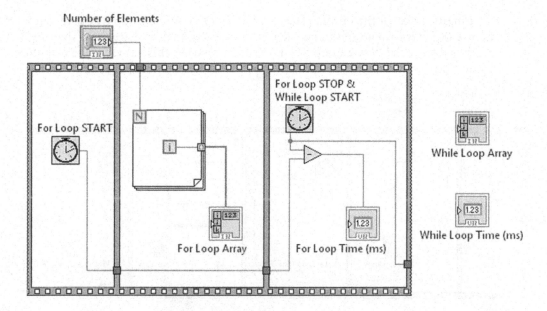

Create a fourth frame and enclose within it the While Loop-based code shown next. As we learned in Chapter 3, by wiring $N - 1$ to the lower terminal of the **Equal?** icon, the While Loop will create an N-element array. Remember to select **Tunnel Mode>>Indexing** (or, in older LabVIEW versions, **Enable Indexing**) at the While Loop tunnel; the While Loop default is **Tunnel Mode>>Last Value** (or **Disable Indexing**).

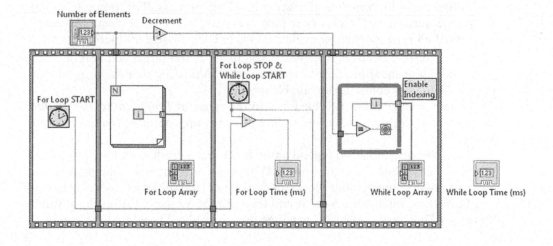

Finally, create a fifth frame. Here, determine the While Loop STOP through the use of a **Tick Count (ms)**, and then (using the "tunneled" wire from the third frame) subtract the While Loop START from it. Output this value for the elapsed While Loop execution time to the front panel's **While Loop Time (ms)** indicator.

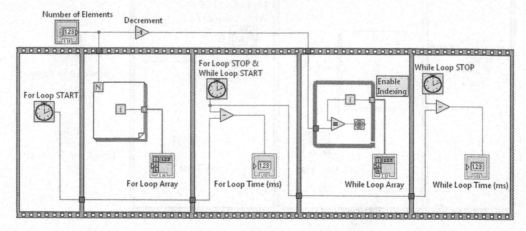

The VI is now complete, so return to the front panel and save your work. We will use **Loop Timer (Sequence)** to place the For Loop and While Loop in head-to-head competition to see if either provides superior performance in the task of assembling an N-element array.

Prior to running this program, you will need to input an appropriate value for N in the **Number of Elements** control. The proper choice for N will depend on the speed and memory capabilities of your computer. You should make N large enough so that it takes each of the two loops at least a sizable number of (at least 10) milliseconds to complete the array-building process. For times less than 10 ms, the accuracy of **Tick Count (ms)** becomes an issue, as discussed below. Since the data type for **Number of Elements** is **I32**, the largest allowed value for N is $2^{31} - 1 = 2,147,483,647$. However, if you make N too large, you will find that **Loop Timer (Sequence)** produces a runtime error indicating that the program has overwhelmed your computer's available memory.

To get started, choose a value for N in the range of *1,000,000* to *10,000,000*. Then, through an iterative process, determine an appropriate N-value for your system.

Once you've established a good value for N, get a fresh start by closing **Loop Timer (Sequence)** and then reopening it. Also, close as many other application programs as possible so that your processor can devote the bulk of its resources to running LabVIEW. Input your N-value and then run the VI. Note the resultant values of **For Loop Time (ms)** and **While Loop Time (ms)**. You should find that the For Loop produces the array significantly faster than the While Loop does. You may wish to peruse the array indicators to verify that the exact same array is being produced by the For Loop and While Loop methods.

Without closing the VI, try rerunning it a second, third, fourth, and more times and noting **For Loop Time (ms)** and **While Loop Time (ms)** for each run. You should find that the second run produces output times noticeably shorter than the initial program execution (i.e., the first run after opening the VI). Subsequent runs then reproduce nearly the same times, with only a small fluctuation on the order of a few milliseconds.

Why does the While Loop take significantly more time to produce an array than the For Loop? In comparing the array-producing code for each loop, note that the While Loop requires an **Equal?** icon to determine whether further iterations are required, but the For Loop has no such requirement. Perhaps then it is the extra execution times required for the **Equal?** operation during each iteration that account for the slower While Loop speed. To test this idea, make the loops equivalent (in regard to the **Equal?** icon) by adding an **Equal?** within the second frame's For Loop as shown below.

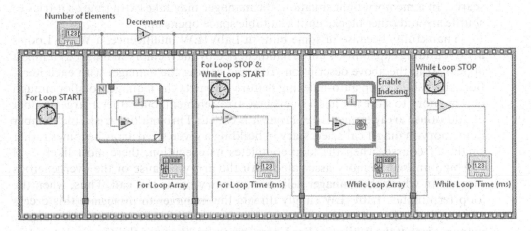

Run the VI. How do **For Loop Time (ms)** and **While Loop Time (ms)** compare once this addition is made to the VI? Does the extra time required for executing the **Equal?** icon significantly slow down the For Loop-based array-building code?

To understand the true reason why a While Loop takes more time to produce an array than a For Loop, you must first understand how LabVIEW uses memory. When LabVIEW launches, a single block of memory (either physical RAM or virtual memory) is allocated for all of the program's editing, compiling, and execution operations. During runtime, a memory manager allocates memory for tasks as needed, with the constraint that arrays and strings must be stored in contiguous blocks of memory. If the memory manager is unable to find a block of unused memory that is large enough for a particular string or array, a dialog box appears to indicate that LabVIEW was not able to allocate the required memory.

Now let's consider what happens when using the auto-indexing feature of a looping structure to build a data array. In the case of a For Loop, LabVIEW can predetermine the size of the array to be built based on the value wired to the Loop's count terminal. Thus, the memory manager is only called once as the loop initiates execution and at that time allocates the appropriate block of memory necessary to hold the array.

With a While Loop, however, the final array size cannot be known in advance. A new element is appended to the existing array with each loop iteration and this process continues until a TRUE Boolean value at the conditional terminal causes the loop to complete its execution. Because the array is constantly increasing in size as the loop iterates, the memory manager must be called continually to find an appropriately sized chunk of RAM to hold the ever-growing data array. These repeated calls to the manager take time, especially if memory starts to become scarce. In a memory-tight situation, the manager may take extra time as it tries to shuffle around other blocks until a suitable space opens up.

Thankfully, because of some built-in LabVIEW intelligence, a While Loop's array-building capability is not as handicapped by the memory manager as it might appear from the above description. To avoid calling the manager with each iteration, the While Loop auto-indexing feature instructs the manager to allot enough new memory to store not just a single new array element, but rather a large number of additional array elements each time it is called. Through this trick, the number of memory-manager calls necessary in building a given sized array becomes rather small. Of course, when the loop completes its execution, there most likely will be some unused memory associated with the array because of the overgenerous manner in which the manager delivered memory on its last call. Thus, when the loop terminates, LabVIEW simply directs the manager to dissociate this excess memory from the now-complete array. The result of this shrewd use of the memory manager is that the While and For Loops' array-building capabilities are not widely divergent (say, by powers of 10) in their performance, as you discovered through **Loop Timer (Sequence)**.

Above, we noted that the **Loop Timer (Sequence)** VI runs more slowly during its first run than it does during its subsequent executions. This observation can be traced to use of the memory manger. During the program's first run, the manager works out the memory allotments peculiar to the particular need of that VI. During subsequent runs, many of the first-run memory assignments are simply reused, diminishing the time-consuming use of the manager.

Finally, what causes the small fluctuations in execution times during the second, third, fourth,...runs? Assuming that your processor is not being asked to multitask between many open applications, these fluctuations are caused by the intrinsic accuracy of LabVIEW's built-in timing functions. These icons mark time by counting interrupts to your system's CPU that occur every 1 ms on a Windows system. It is this one-millisecond resolution in the **Tick Count (ms)** icon that accounts for the observed timing fluctuations. The lesson to be learned here is that LabVIEW's timing VIs provide a simple and effective method of measuring the duration of an event with 1 ms resolution. In situations that demand submillisecond timing (such are common in acquiring a sequence of analog-to-digital conversions), however, these icons are totally inadequate. For these high-accuracy timing applications, the submicrosecond-resolution clock on a National Instrument's data acquisition (DAQ) board can be used.

10.3 EVENT TIMER USING DATA DEPENDENCY

A skillful programmer can exploit LabVIEW's data-dependent mode of execution to force two or more nodes to execute sequentially. Thus, it is almost always possible to impose sequential execution on a block diagram without the use of a multiframe Sequence Structure.

In some cases, it is natural for node *A*'s output to be used as the input of node *B*. Then these "chained together" nodes will execute in the order *A-B*. LabVIEW's File I/O icons provide such an example. Let *A* and *B* be the two file-related icons **Open/Create/Replace File** and **Write To Text File**. Typical of all File I/O VIs, each of these icons has an input and output called **file path** (or **file**) and **refnum out**, respectively. If *A* is (somehow) provided an input **file path**, when it completes its execution, **refnum out** is output. By wiring this output to *B*'s **file** input, one guarantees that *A* executes fully before *B*. Look back at your work in Section 7.8. We implemented this method in **Spreadsheet Storage (OpenWriteClose)**. Additionally, you will find that LabVIEW icons with an associated purpose (e.g., file I/O, DAQ) have **error in** input and **error out** output terminals that can be used to chain together several related nodes to ensure sequential execution.

Many times, however, the output of *A* does not constitute a natural input for *B*. In such situations, ordered execution can still be obtained by an alternate method called *artificial data dependency*. In this scheme, it is the mere arrival of data, with no regard to their actual value, that triggers the execution of a node. An example of artificial data dependency will be constructed on the following pages.

Let's try to write a timer diagram, whose ordered execution is controlled by artificial data dependency. First, open **Loop Timer (Sequence)** and use **Save As . . .** to create a new VI called **Loop Time (Data Dependency)** in YourName\Chapter 10. Keep the front panel unchanged as shown.

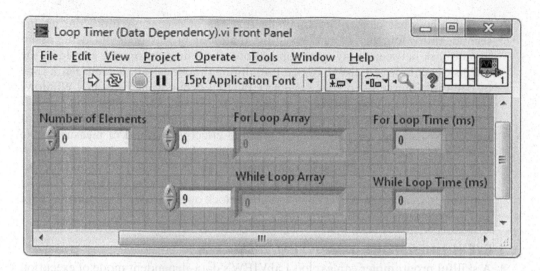

Switch to the block diagram. Eliminate the **Flat Sequence Structure** by popping up on its border and selecting **Remove Sequence**. Code the For Loop shown at the left in the next diagram, which creates an *N*-element array with the numeric value of each element equal to its index (the right portion of this diagram has not been changed yet). Wire the output of a **Tick Count (ms)** icon to the border of the For Loop. **Tick Count (ms)**'s output then is an input to the For Loop, meaning that the loop cannot initiate execution until the Tick Count (ms) icon produces a value. This clock value will be used as the For Loop START. The START value is never used within the For Loop, so it is simply the arrival of Tick Count's output that triggers the loop execution—an example of artificial data dependency.

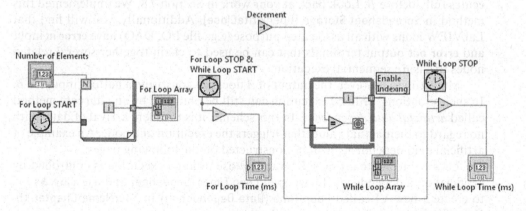

After the For Loop ceases execution, we want to obtain the For Loop STOP value using a **Tick Count (ms)** icon and then calculate the elapsed execution time in milliseconds. The trick for assuring this sequence of operations is to enclose the "STOP-value" code within a single-frame **Flat Sequence Structure** as shown below. The code within this Sequence Structure will execute upon receipt of a value at the structure's input tunnel. Wire the Start value, through the For Loop, to an output tunnel (disable the loop's auto-indexing feature by selecting **Tunnel Mode>>Last Value** or **Disable Indexing**). Then wire this For Loop output over to the Flat Sequence Structure input. The completion of the For Loop will then trigger the execution of the Sequence Structure's code, as desired.

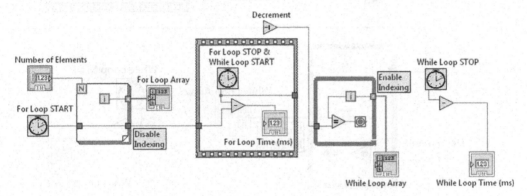

Also, when coding the Sequence Structure above, the enclosed **Tick Count (ms)**'s output is wired to an output tunnel on the structure's border. We will next turn our attention to a While Loop, which creates an N-element array. This tunnel outputs the While Loop START value and the arrival of this value at the While Loop will trigger the loop's execution — another example of artificial data dependency.

Code the array-building While Loop, whose execution is triggered by the arrival of the While Loop START value, as shown next. For array building, you will have to enable auto-indexing at the While Loop's border.

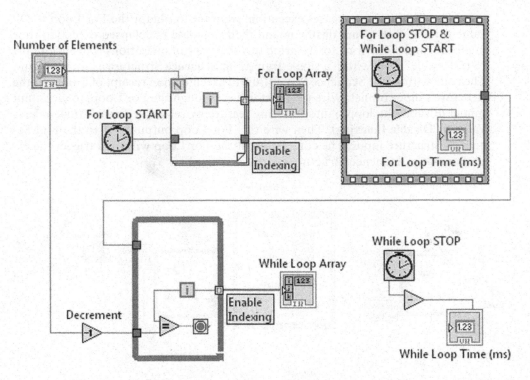

Complete the diagram as follows to calculate the time it takes for While Loop-based array building.

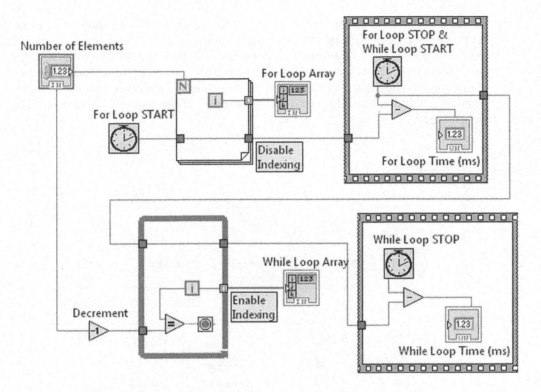

Return to the front panel and save your work. Run the VI with your optimized choice for **Number of Elements**. Check to see that the expected *N*-element array is built by both the For Loop and the While Loop. If your diagram is correct, the performance of **Loop Time (Data Dependency)** should be equivalent to that of **Loop Timer (Sequence)**.

10.4 HIGHLIGHT EXECUTION

To cement your understanding of the manner in which this VI executes, let's implement one of LabVIEW's handiest (and coolest!) debugging tools called *Highlight Execution*. Input something small (such as *10*) for **Number of Elements** on the front panel, and then switch to the block diagram. In the toolbar, enable **Highlight Execution** by clicking on the button containing the light bulb.

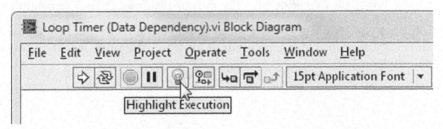

Then click on the **Run** button. The VI will execute in slow-motion animation, with the passage of data marked by bubbles moving along the wires and important data values given in automatic pop-up probes. Because the diagram executes in slow motion, the values for **For Loop Time (ms)** and **While Loop Time (ms)** obtained under Highlight Execution will be meaningless. However, this mode of execution provides a beautiful visual demonstration of the concept of artificial data dependency.

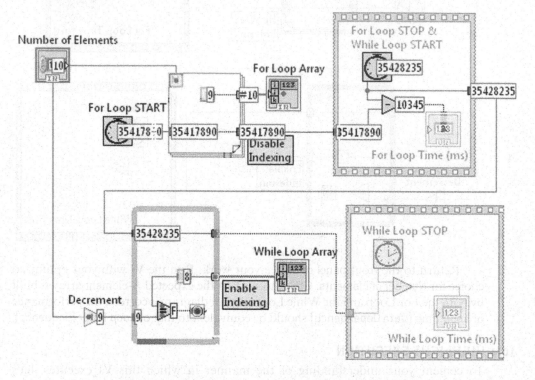

If interested, try running **Moving Average (Four Traces)** and **Moving Average (Multiple Traces)** in **YourName\Chapter 8** under **Highlight Execution** to deepen your understanding of how the shift registers are populated and arrays are built during the first several While Loop iterations.

Use **Highlight Execution** liberally in your future work (but remember to turn it off once you are finished using it by clicking again on the ⬜). This LabVIEW feature is an invaluable debugging tool for troubleshooting VIs that are causing you problems.

DO IT YOURSELF

Reaction Time.vi Write a VI that measures a user's reaction time, defined as the elapsed time from when **Square LED** is lit until the user presses the **Stop Button**. The front panel for this program includes a **Stop Button** and **Square LED** Boolean indicator suitably enlarged, as well as a **Numeric Indicator** labeled **Reaction Time (ms)**, as shown below.

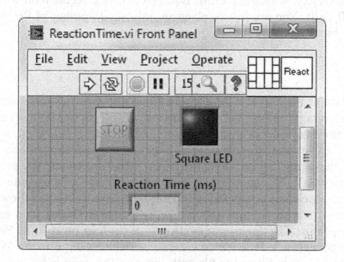

Your VI should execute in the following sequence of steps: (1) The user presses the **Run** button and then quickly moves the mouse cursor over to the **Stop Button**; when the **Run** button is pressed, **Square LED** and **Reaction Time (ms)** are initialized as *unlit* and *0*, respectively; (2) at a random time (within the range of *3* to *8* seconds) after the **Run** button has been pressed, **Square LED** becomes lit; (3) on the **Reaction Time (ms)** indicator, the value of the elapsed time in milliseconds since **Square LED** was lit is displayed continuously, and then the displayed value is halted at the moment that the user presses the **Stop Button**; and (4) **Square LED** becomes unlit.

A few helpful tips follow.

1. A random number in the range from 0 up to 1 can be obtained using the **Random Number (0–1)** icon.
2. Think carefully about the appropriate choice for the **Mechanical Action** option in the Stop Button's pop-up window.

3. Properties of a front-panel object (e.g., its color, visibility, position) can be controlled from the block diagram using a **Property Node**. To create a **Property Node** that lights **Square LED**, pop up on its block-diagram icon terminal and select **Create>>Property Node>>Value**. Once configured in its write mode (by popping up on it and selecting **Change To Write**), wiring a TRUE Boolean constant to this icon will cause the front-panel **Square LED** indicator to be lit. Conversely, wiring a FALSE Boolean constant to the icon will make **Square LED** unlit. You are free to place an unlimited number of Property Nodes on the block diagram for any given front-panel object.

USE IT!

Saving Data at the End of a Data-Taking Run The following program illustrates an easy-to-program method for providing a user with the option of saving acquired data at the end of a data-taking run. Here, we have borrowed the Use It! example of Chapter 7 to simulate the data-taking portion of a program. Since these acquired data are the input for the single Sequence Structure frame, the principle of data dependency ensures that the code within this frame will execute only after the data-taking loop has competed its operation. Hence, this VI follows a sequenced execution—data are acquired, then the **Two Button Dialog** icon causes a dialog box to appear asking the user "**Save Data?**" If the user presses the **Yes** button, then the Case Structure's **True** window executes, which contains the **Write Delimited Spreadsheet.vi** icon. Since the **file path** input of this icon is left unwired, a dialog window appears requesting a file path and the data are stored at that location in the spreadsheet format.

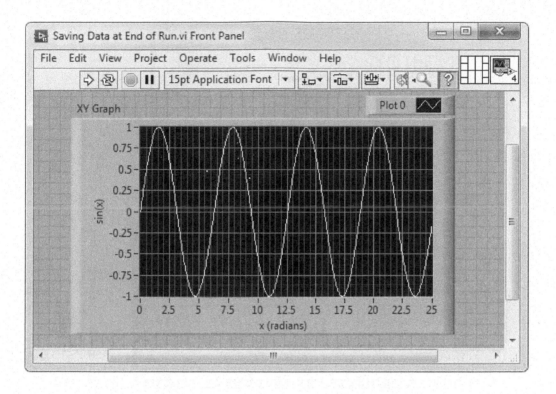

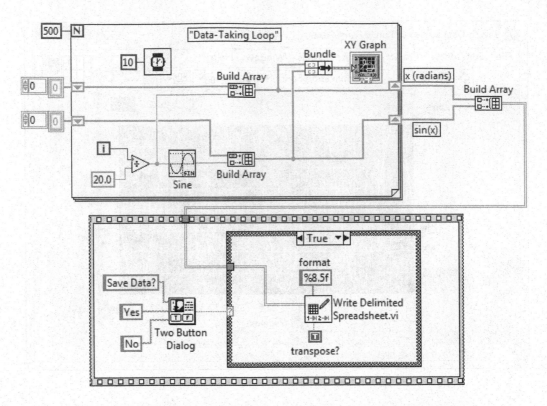

The Case Structure's **False** window is left empty as shown next.

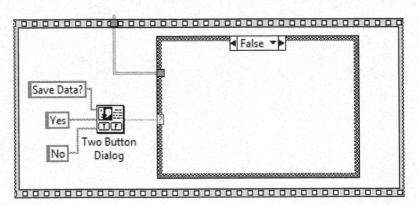

PROBLEMS

For a problem that involves writing a new program, the problem statement begins with a suggested descriptive name (including the **.vi** extension) for the VI that you will write; icons needed for the VI may be found with the aid of **Quick Drop**. For a problem that involves the use of a VI already written as part of the chapter text, the problem's topic is given in bold at the beginning of the problem statement.

1. **Five Blinking Lights (Sequence Structure).vi** In Chapter 9's Do It Yourself project, the state machine architecture was used to write a VI that sequentially illuminates five LEDs on its front panel. Using a **Flat Sequence Structure**, develop a program with the front panel shown below that carries out this same task, namely, lights the LEDs one at a time in the order 0-1-2-3-4. Make each LED remain lit for 0.2 second and have the 0-1-2-3-4 lighting pattern repeat continuously until a front-panel **Stop Button** is pressed (when the **Stop Button** is pressed, the current 0-1-2-3-4 sequence should be completed before the program halts execution). To control a LED from multiple locations on the block diagram, use **Value** Property Nodes, which can be created by popping up on the LED's icon terminal and selecting **Create>>Property Node>>Value**.

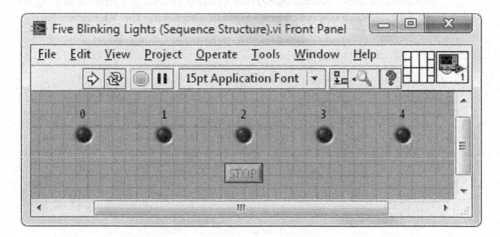

[You should find that the state machine architecture provides a more elegant solution to the Five Blinking Lights problem than a Sequence Structure (e.g., no Property Nodes are required for the state machine).]

2. **Magic Stop Button.vi** Write a program whose front panel initially appears blank as shown in the next illustration.

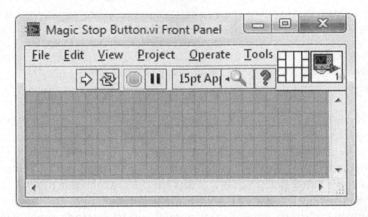

Then, when this VI is run, as shown below a blinking Stop Button appears on the front panel until the user presses it. The Stop Button then disappears and the program stops. To write this program, first place a Stop Button on the front panel, pop up on it and select **Visible>>Caption**, and then make the caption read *Press this button to stop the VI*. Next, pop up on the Stop Button and select **Advanced>>Hide Control** and secure this choice by selecting **Edit>>Make Current Values Default** and saving the VI.

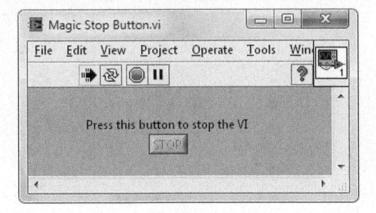

Code the block diagram so that when the program is run, first the Stop Button becomes visible, then it blinks until it is clicked, and finally the Stop Button is hidden (i.e., made not visible). The features of the Stop Button (e.g., its visibility, blinking) can be controlled by creating appropriate Property Nodes (see the discussion in this chapter's Do It Yourself project).

3. **Execution Order of Two Charts** The subdiagram shown below generates and plots 100 data samples on a Waveform Chart.

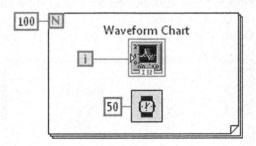

(a) **Two Charts (Simultaneous).vi** Write a VI that performs two such 100-sample plots simultaneously. That is, place two Waveform Charts on the front panel labeled **Chart 1** and **Chart 2**, and program the block diagram so that the 100-sample plots on **Chart 1** and **Chart 2** are performed over the same time interval.

(b) **Two Charts (Data Dependency).vi** Write a VI that first completes the 100-sample plot on **Chart 1** and then performs the 100-sample plot on **Chart 2**. Use artificial data dependency on the block diagram to accomplish this feat (i.e., do not use a Sequence Structure).

(c) Finally, add code to the VI you built in part (b) so that the two Waveform Charts are cleared at the end of a run. To accomplish this task, place a single-frame **Flat Sequence Structure** on the block diagram and use artificial data dependency to assure that this block-diagram object is the last item to execute before the VI completes a run. Within the **Flat Sequence Structure**, place two plot-clearing Property Nodes, one associated with each Waveform Chart. The appropriate Property Node is made by popping up on the Waveform Chart's terminal and selecting **Create>>Property Node>>History Data**. After changing this Property Node to its write mode by selecting **Change to Write** in its pop-up menu, **Create>>Constant** will produce the needed input called an **Empty Array**, as shown next.

Waveform Chart

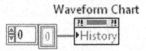

4. **Parallel While Loops with Reset.vi** Create a program in which two While Loops execute in parallel on the block diagram, simultaneously producing independent plots of random numbers (in the range 0 to 1) on two front-panel Waveform Charts until stopped by the click of a single front-panel Stop Button. Label one of the Waveform Charts **Chart 1** and the other **Chart 2**.

419

(a) First, try coding your VI using the following block diagram to accomplish the intended goal (i.e., halt simultaneously executing While Loops with a single Stop Button). Run this program and show that it does not produce two simultaneous plots. Describe briefly what this diagram does do instead and explain why.

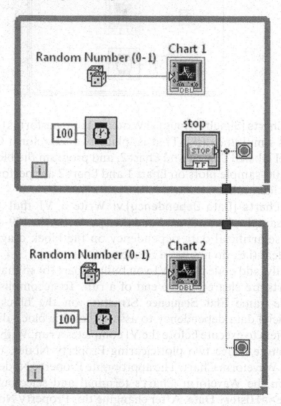

(b) For the correct diagram, one While Loop is stopped by wiring directly to the Stop Button's terminal and the other by wiring to a **Value** Property Node associated with this terminal. To produce the needed Property Node, pop up on the Stop Button's terminal and select **Create>>Property Node>>Value**. When wiring this Property Node, you will get a broken wire until the Stop Button's **Mechanical Action** is changed to something other than a "latch" mode (e.g., select **Mechanical Action>>Switch When Pressed**). Complete the diagram for your program and verify that it runs as intended.

(c) When associated with a **Value** Property Node, the Stop Button must be in a "change value" (rather than a "latch") mode. Thus, when the Stop Button is pressed to stop the VI, the button will be left in its TRUE value as the VI completes execution, which is inconvenient for subsequent uses of this program. To remedy this problem, output a quantity from each While Loop and use artificial data dependency to reset the value of the Stop Button back to FALSE before the VI completes its execution. Run this final version of your VI and verify that it executes as intended.

5. **Alternative Approaches for Loop Timer** Explore the performance of the following three array-building diagrams. These diagrams purposely avoid the use of a loop's auto-indexing feature in an effort to make explicit what this feature does automatically. Use Help Windows to understand the functioning of the unfamiliar array-related icons (found in **Functions>>Programming>>Array**).

(a) **Loop Timer (Alternative#1).vi** Code the For Loop-based block diagram given below, which initializes an appropriate-sized array (with each element defined as zero) prior to the loop structure and then avoids further calls to the memory manager through the use of **Replace Array Subset** within the loop itself. Run this VI and then **Loop Timer (Sequence)** to compare the time it takes each program's For Loop to create the same N-element array of integers. Are the times comparable or is one program significantly faster?

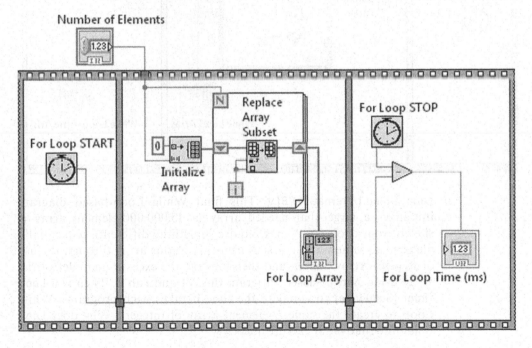

(b) **Loop Timer (Alternative#2).vi** Next, program the following While Loop-based diagram, which initializes the shift register with an empty array and then appends elements to it with each loop iteration. Run this VI and then **Loop Timer (Sequence)** to compare the time it takes each program's While Loop to create the same *N*-element array of integers. Why is **Loop Timer (Alternative#2)** such a poor performer? Although not time-efficient, in nondemanding applications (e.g., involving small-sized array), this diagram is commonly used to perform "real-time" graphing (with a calibrated *x*-axis) inside a While Loop by feeding the **Build Array** output to the terminal of a **Waveform Graph** (see Problem 6 in Chapter 8).

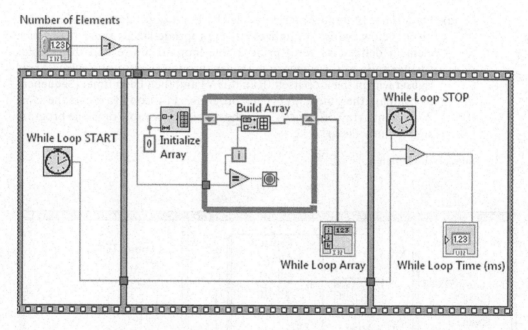

(c) **Loop Timer (Alternative#3).vi** This final While Loop-based diagram initializes a larger-than-needed array (a 15,000,000-element array is shown; your choice of *N* may require something different), sequentially places data values in the first *N* elements of this array through the use of **Replace Array Subset**, and then lops off the excess remainder of the array using **Array Subset**. Program this VI and run it. Then run **Loop Timer (Sequence)** to compare the time it takes each program's While Loop to create the same *N*-element array of integers. Why does **Loop Timer (Alternative#3)** execute quickly?

Number of Elements

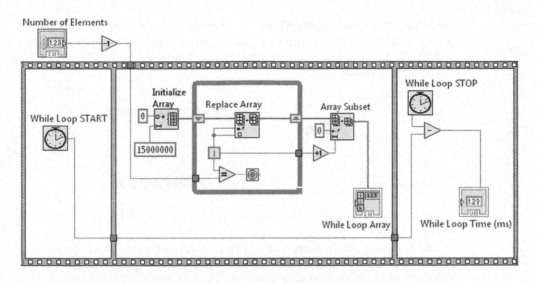

6. **Time To Press OK.vi** The **error in** and **error out** terminals present on many LabVIEW icons can be used to sequence the execution of a collection of these icons via data degeneracy. As an example of this technique, use two **Elapsed Time** and one **Prompt User for Input** Express VIs to construct a program that times how long it takes a user to click on a dialog-box **OK** button. Configure the Express VIs as shown next, and then wire them together so that data dependency resulting from the error terminals causes them to execute sequentially from left to right. The **Present (s)** terminal of Elapsed Time outputs the (universal) time in seconds at which this icon executes, so the difference of this value from the two Elapsed Time icons gives the time elapsed from when the VI was started until the user presses the dialog-box **OK** button. Add code that calculates and displays this elapsed time (in seconds) on a front-panel **Numeric Indicator** named **Time To Press OK (s)**.

7. **For Loop Timer (Icon vs. MathScript or Formula Node)** For MathScript Node users, write a VI that compares the required array-building time T of a LabVIEW For Loop icon and a MathScript Node. To build the arrays, use the diagrams shown next.

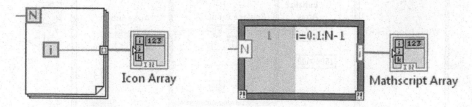

Icon Array   Mathscript Array

For various values of N, what is T for each diagram? Which diagram is the best performer?

For Formula Node users, the array size cannot be defined during run-time in a Formula Node, so its size must be explicitly given in the code. The Formula Node array-building code (with $N = 1,000,000$) for this problem is shown next.

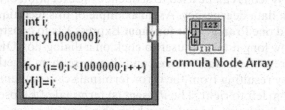

Formula Node Array

CHAPTER 11

Analysis VIs: Curve Fitting

In this chapter's exercises, you will explore the use of a *thermistor* as a temperature sensor. A thermistor has the useful characteristic that its resistance vs. temperature ($R - T$) curve can be fit to a well-known analytic function, facilitating its use as a calibrated thermometer. In Table 11.1, the $R - T$ data for a particular thermistor (Epcos Model B57863S0103F040) are given. If you will be using a different thermistor in the Do It Yourself project at the end of this chapter, please use the $R - T$ data for the thermistor that will be employed in your particular experimental setup.

11.1 THERMISTOR RESISTANCE-TEMPERATURE DATA FILE

Over the course of this chapter, you will learn how to calibrate a thermistor by fitting its $R - T$ data to a formula known as the *Steinhart–Hart Equation*. This accomplishment is contingent on the ability to read the calibration data into your fitting program. One method we will explore for inputting data into a VI is through reading a computer file that contains the relevant information. Thus, prior to working on this chapter's exercises, you will need to create and save a spreadsheet-formatted file containing a thermistor's $R - T$ data using either the data provided in Table 11.1 or the data for your own device. In Chapter 7, you learned that in the spreadsheet format, tabs separate columns and end-of-line (EOL) characters separate rows. You can use either a spreadsheet data analysis program or a word processing program to create the desired file, and store it under a descriptive name like **Thermistor R-T Data.txt**. For example, I used Microsoft Word to construct the ASCII spreadsheet file shown in the following illustration of the thermistor resistance (in kilohms) vs. temperature (in degrees Celsius) data given in Table 11.1. When saving the file using a word processing program, it is important to use the **Save** or **Save As**... command, and then select

TABLE 11.1 $R - T$ Data for Epcos Model B57863S0103F040 Thermistor

Temperature (°C)	Resistance (kΩ)
−50.00	670.1
−45.00	471.7
−40.00	336.5
−35.00	242.6
−30.00	177.0
−25.00	130.4
−20.00	97.07
−15.00	72.93
−10.00	55.33
−5.00	42.32
0.00	32.65
5.00	25.39
10.00	19.90
15.00	15.71
20.00	12.49
25.00	10.00
30.00	8.057
35.00	6.531
40.00	5.327
45.00	4.369
50.00	3.603
55.00	2.986
60.00	2.488
65.00	2.083
70.00	1.752
75.00	1.481
80.00	1.258
85.00	1.072
90.00	0.9177
95.00	0.7885
100.00	0.6800
105.00	0.5886
110.00	0.5112

Plain Text (*.txt), as opposed to, for example, **Word Document (*.docx)**, in the **Format:** selection. This procedure creates an ASCII text file devoid of any word-processor formatting characters. Take care not to inadvertently add an extra EOL character at the end of the file. Also make sure to use minus signs (i.e., keyboard hyphens, ASCII code 45), not en dashes or em dashes.

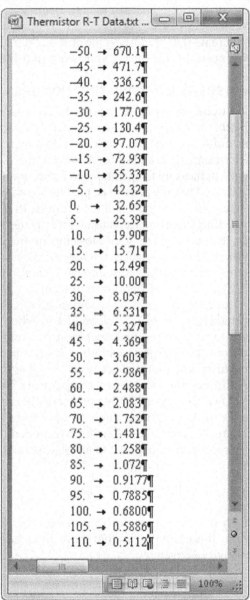

Alternatively, if you use a spreadsheet-based program to create the file, you will want to save it as a **Text (Tab delimited) (.txt)** file. This option may be listed under a **Save, Save As...**, or **Export** menu item, depending on the program you use. Make sure that you save this file in the correct manner to avoid including a bunch of program-dependent formatting characters in the file. Perhaps the best way to check that you've produced the desired file is to open it using a word processing program to see whether it appears as above.

Take a minute to create the spreadsheet file of $R - T$ data and name it **Thermistor R-T Data.txt**. You will need this file during your work in this chapter.

11.2   TEMPERATURE MEASUREMENT USING THERMISTORS

Among the many electronic temperature sensors available—including platinum resistance thermometers, thermocouples, and semiconductor devices such as the Analog Devices AD590—thermistors are used in applications that require high sensitivity, small size, ruggedness, fast response time, and low cost. When utilized with the proper support circuitry, a thermistor can measure temperatures to an accuracy of 0.01 °C over a fairly wide range (about 100 °C). The only price to be paid for the thermistor's high sensitivity is that its resistance is a nonlinear function of temperature. In the past, the nonlinearity of thermistors was a distinct disadvantage, causing users to employ resistor-based linearizing networks or less-than-accurate resistance-to-temperature conversion methods such as the Beta formula (see below). Fortunately, the widespread availability of microprocessors has enabled the use of accurate (and somewhat complex) temperature-calibration models, greatly simplifying the use of thermistors in accurate temperature measurement.

The active layer of most thermistors is composed of a semiconducting metal-oxide alloy. As is characteristic of a semiconductor, when the thermistor's temperature is elevated, loosely bound (valence) electrons within its active layer are thermally released from their respective binding sites to become conduction electrons. By this mechanism, the thermistor's resistance decreases with temperature and results in the utility of this device as a temperature sensor. Since the bound electronic state is of lower energy, an activation energy barrier E separates the conducting state from the insulating state. Thus, at a given temperature T (in degrees Kelvin), we expect that the probability of such activation, and therefore the density n of conduction electrons, is proportional to a Boltzmann factor. That is,

$$n = n_0 \exp\left[-\frac{E}{kT}\right] \qquad [11.1]$$

where n_0 is the density of conduction electrons at very high temperatures, and k is Boltzmann's constant. It is easily shown that these conduction electrons imbue the material with a resistivity ρ given by

$$\rho = \frac{m}{ne^2\tau} \qquad [11.2]$$

where e and m are the charge and mass of an electron, respectively. The scattering time τ is the average time traveled by a conduction electron until it is scattered by a lattice vibration, an impurity, or some other scattering mechanism peculiar to the material.

Thus, assuming that e, m, and τ are independent of temperature, we expect that the temperature dependence of ρ is entirely caused by the Boltzmann factor related to n. Putting Eq. [11.1] into Eq. [11.2], we predict that

$$\rho = \rho_0 \exp\left[\frac{E}{kT}\right] \qquad [11.3]$$

where ρ_0 is the temperature-independent constant given by

$$\rho_0 = \frac{m}{n_0 e^2 \tau} \qquad [11.4]$$

The macroscopic resistance R of a sample is determined by

$$R = \rho \frac{L}{A} \qquad [11.5]$$

where L and A are the sample's length and cross-sectional area, respectively.

Because a solid sample's geometric shape is little changed with temperature, we expect that the temperature dependence of R will simply mimic that of ρ. Thus,

$$R = R_0 \exp\left[\frac{E}{kT}\right] \qquad [11.6]$$

where R_0 is the (assumed) temperature-independent constant

$$R_0 = \frac{mL}{An_0 e^2 \tau} \qquad [11.7]$$

If all of our logic is correct, we predict that the temperature dependence of a thermistor's resistance should be described by Eq. [11.6]. By taking the logarithm of both sides, this relation can be rewritten as the following linear relation:

$$\frac{1}{T} = -\frac{k}{E} \ln R_0 + \frac{k}{E} \ln R \qquad [11.8]$$

or

$$\frac{1}{T} = A + B \ln R \qquad [11.9]$$

where A and B are constants defined to be $A \equiv -k \ln R_0/E$ and $B \equiv k/E$. Equation [11.9] is the so-called *Beta formula* (B is sometimes written as β) and it predicts that when a thermistor's temperature-dependent resistance data are plotted as $1/T$ (y-axis) vs. $\ln R$ (x-axis), a straight line will result with the y-intercept and slope equal to A and B, respectively.

Despite this expectation, it is found experimentally that a real thermistor's $R - T$ data exhibit some nonlinearity when plotted as $1/T$ vs. $\ln R$. Rather than being constant, the slope B decreases with decreasing temperature. Over a narrow temperature range (say, of 10 °C), B varies so slightly that the Beta formula can predict the thermistor's behavior with a temperature uncertainty of about 0.01 °C. However, for the much larger temperature span over which a thermistor is useful, the Beta formula inadequately models the data.

Something, then, must be missing in our theory. In reviewing the above derivation, one might suspect it is our assumption of a temperature-independent scattering time. It is easy to imagine that as temperature increases, the conduction electrons move faster and/or lattice vibrations become more pronounced and consequently shorten the scattering time τ. While the exponential Boltzmann factor activation of conduction electron density will still dominate the resistance's temperature dependence, the more mild temperature-dependent scattering time can possibly account for the observed slight nonlinearity of a $1/T$ vs. $\ln R$ plot.

How, then, can we more accurately model the $R - T$ data? Unfortunately, the theory for electrical conduction in metal oxide thermistors is still incomplete and cannot offer us a more refined relation to replace Eq. [11.9]. (In fact, the semiconductor energy band model advocated above is not accepted by all researchers. Some investigators, instead, hold that electric conduction in these materials is caused by the "hopping" of charge carriers from one ionic site to the next.) In light of this theoretical vacuum, the most recent literature on thermistors accounts for the nonlinearity of the $1/T$ vs. $\ln R$ curve using the standard curve-fitting technique of considering $1/T$ to be a polynomial of $\ln R$. That is, the inverse temperature is written as the following nth-order polynomial:

$$\frac{1}{T} = A + B \ln R + C \left(\ln R\right)^2 + \ldots + Q \left(\ln R\right)^n \qquad [11.10]$$

From this point of view, Eq. [11.9] is such a polynomial that has been truncated at the first-order term. One then is led to the conclusion that by retaining higher-order terms, the equation will remain valid over a wider range of temperatures.

Starting with this idea, researchers have found that the 100 °C temperature span over which a typical thermistor is active can be accurately modeled by a third-order polynomial:

$$\frac{1}{T} = A + B \ln R + C (\ln R)^2 + D (\ln R)^3 \qquad [11.11]$$

This relation can be used to convert resistance to temperature with a precision equal to that of the original $R - T$ data (typically, 0.005 to 0.01 °C).

In 1968, the two oceanographers Steinhart and Hart discovered that, when modeling up to 200 °C spans contained within the temperature range of -70 °C to 135 °C (they were especially interested in the oceanographic range of -2 °C to $+30$ °C), the second-order term in the above relation could be neglected without significant loss of accuracy. Then Eq. [11.11] reduces to

$$\frac{1}{T} = A + B \ln R + D (\ln R)^3 \qquad [11.12]$$

Equation [11.12] is called the *Steinhart–Hart Equation* and is widely used in thermistor calibration. For a typical thermistor, the constants A, B, and D have order of magnitude values of 10^{-3}, 10^{-4}, and 10^{-7}, respectively, when resistance is measured in ohms. The temperature must, of course, be on the Kelvin scale.

11.3 THE LINEAR LEAST-SQUARES METHOD

The process of scientific inquiry many times proceeds as follows: Researchers perform an experiment to investigate a particular physical phenomenon and to obtain a set of N data samples (x_j, y_j). After shutting off their instruments, the investigators apply appropriate curve-fitting techniques to determine the functional relationship $y = f(x)$ manifest in their data. Once armed with this $f(x)$, the scientific community can then judge the veracity of theoretical models posited to explain the phenomenon under investigation by testing each model's capacity for correctly predicting the experimentally observed $f(x)$. In this section, we will explore a curve-fitting method that allows one to extract an accurate $y = f(x)$ from a given data set (x_j, y_j), a skill that plays a crucial role in the above-described process. We will put this skill to immediate use in calibrating a temperature-sensing thermistor.

Quite often, it is convenient to model a functional relationship $y = f(x)$ as the linear combination of a set of basis functions b_k. The general form of this kind of model is

$$y = f(x) = \sum_{k=0}^{M-1} a_k b_k(x) = a_0 b_0 + a_1 b_1 + \dots + a_{M-1} b_{M-1} \qquad [11.13]$$

where the basis functions $b_k(x)$ are known functions of x and the a_k are the linear expansion coefficients. For example, if $b_k = x^k$, then $f(x) = a_0 + a_1 x + a_2 x^2 + \dots + a_{M-1} x^{M-1}$ is a polynomial of order $M-1$. Alternately, if $b_k = \cos(kx)$, then $f(x)$ is a function described by a Fourier series. Note that the functions $b_k(x)$ can be nonlinear functions of x. However, the coefficients a_k appear in a linear fashion in Eq. [11.13].

Once the decision is made to model $f(x)$ as a linear combination of a set of basis functions, one needs some method to determine the appropriate choice of the a_k in Eq. [11.13] so that a given data set is accurately described. The theory of data analysis provides just such a method. Because all data-taking processes are subject to random errors, the repeated acquisition of a particular experimental quantity produces an array of values that is distributed as a Gaussian ("bell curve") distribution. By carefully considering the ramification of this fact, it can be shown that Eq. [11.13] will best describe a given N data samples (x_j, y_j) when the a_k are chosen such that they minimize an error function e defined as

$$e \equiv \sum_{j=0}^{N-1} \left[y_j - f(x_j) \right]^2 = \sum_{j=0}^{N-1} \left[y_j - \sum_{k=0}^{M-1} a_k b_k(x_j) \right]^2 \qquad [11.14]$$

Note that e is a measure of the deviation between the experimentally determined y_j and the y obtained from the fitting function $y = f(x_j)$, which depends on the values chosen for the coefficients a_k. Various algorithms have been proposed to identify the a_k that minimizes e for a given data set, all of which are classified under the rubric of *linear least-squares* methods. In LabVIEW, the available methods are the Singular Value Decomposition (SVD), Givens, Householder, Lower Triangular–Upper Triangular (LU) Decomposition, and Cholesky algorithms.

LabVIEW provides built-in VIs that will implement a linear least-squares fit of your data to the functional form of Eq. [11.13]. For applications that require high performance (e.g., fastest possible execution time, smallest possible files) or detailed control over the fitting process (e.g., choice of least-squares algorithm), the low-level curve-fitting icons found in **Functions>>Mathematics>>Fitting** should be employed. In less demanding situations, however, the high-level **Curve Fitting** Express VI will provide the same results as code written with low-level icons, with much more programming ease.

In the following pages, you will write a program that fits a thermistor's $R - T$ data to the Steinhart–Hart Equation. For this task, you will employ the **Curve Fitting** Express VI to carry out the linear least-squares fitting algorithm. In the chapter's final section, you will investigate how **Curve Fitting** can also be used to perform a nonlinear least-squares fit.

11.4 INPUTTING DATA TO A VI USING A FRONT-PANEL ARRAY CONTROL

To write your curve-fitting VI, you must first understand how to input the given array of $R - T$ data into a LabVIEW program. The most straightforward method for inputting an array of values is from a front-panel Array Control. Let's write a VI that works this way. Construct the front panel shown here, which is designed to read in and then plot a set of (x, y) data samples. Place two Array Controls (select an **Array** shell first and then place a **Numeric Control** inside) and label one **X** and the other **Y**. Next, add an **XY Graph** with its x- and y-axes labeled **X** and **Y**, respectively. Using **File>>Save**, create a folder named **Chapter 11** within the **YourName** folder, and then store this VI as **Read Array Control** in **YourName\Chapter 11**.

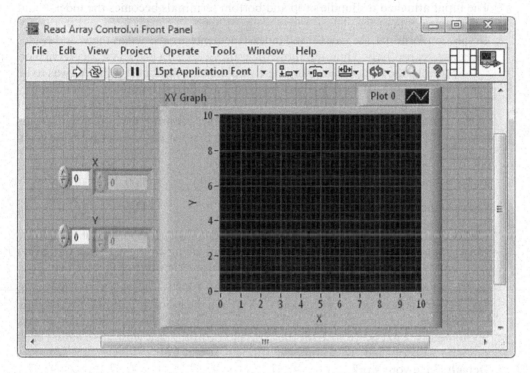

Switch to the block diagram. Write the following code that receives the input arrays, bundles them into a cluster, and then passes this cluster to the XY Graph for plotting.

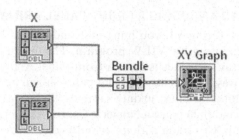

The **Bundle** icon in this diagram creates a cluster consisting of two elements. The input attached to Bundle's top and bottom terminals becomes the index-0 and index-1 elements of the cluster, respectively. When this cluster is passed to the **XY Graph** icon terminal, the index-0 and index-1 elements are plotted as the x- and y-axes variables, respectively. Thus, the above diagram yields a Y (y-axis) vs. X (x-axis) plot.

Return to the front panel. Okay, here would be the hard part if this VI was to be used to read in the values from our thermistor calibration data table. You manually type all of the temperature values from the table into the **X** control, and then type all of the resistance values from the data table into the **Y** control. Demonstrate to yourself how this procedure works by just entering the first five data values from the given thermistor calibration table:

X (Temperature):	-50.00	-45.00	-40.00	-35.00	-30.00
Y (Resistance):	670.1	471.7	336.5	242.6	177.0

Remember that the small left-hand box portion of the Array Control (called the *index display*) indicates the index of the element being displayed, and the larger right-hand box (the *element display*) is the actual numeric value of that element. You increment or decrement the index display using the 🖑. Also use this tool to highlight the interior of the element display, and then type in the numeric value that you desire.

Now, let's assume that we want to store the information in these two arrays permanently. That is, we wish for the arrays to be initialized with these values each time the VI is opened. To accomplish this feat, select **Edit>>Make Current Values Default**. Save your work.

Run the VI and you should see a graph of the thermistor's resistance as a function of temperature over the range from $-50\,°C$ to $-30\,°C$. Try closing and then opening the VI. Repeat this process a few times. Hopefully, you'll find the two front-panel arrays are initialized with the thermistor data each time the program is loaded.

For future reference, you can also input data into a program by placing an array directly on the block diagram using the **Array Constant** icon found in **Functions>>Programming>>Array**. The **Array Constant**, just like the front-panel **Array** shell, comes equipped with an index display as well as an element display receptacle that you stuff with a constant of desired data type (numeric, Boolean, string, or cluster). In this approach, you first position an **Array Constant** on your block diagram, and then place a **Numeric Constant** from **Functions>>Programming>>Numeric** (with Representation **DBL**) within its element display receptacle. You then program the Array Constant with the relevant sequence of data values. The resulting block diagram would appear as below.

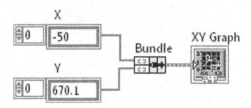

It is considered good LabVIEW style to group related programming objects in a cluster. This approach leads to well-organized front panels, reduces the quantity of block-diagram wires (termed *wire clutter*), and, coupled with the use of **Unbundle By Name**, produces a self-documented VI. Let's reform **Read Array Control** in this manner. Place a **Cluster** shell (found in **Controls>>Modern>>Array**, **Matrix & Cluster**) on a blank region of your VI's front panel and label it **XY Cluster**. Using the ⬉ , first drag **X** and then **Y** into the **Cluster** shell. The cluster can then be automatically resized by popping up on its border and selecting **AutoSizing>>Size to Fit**.

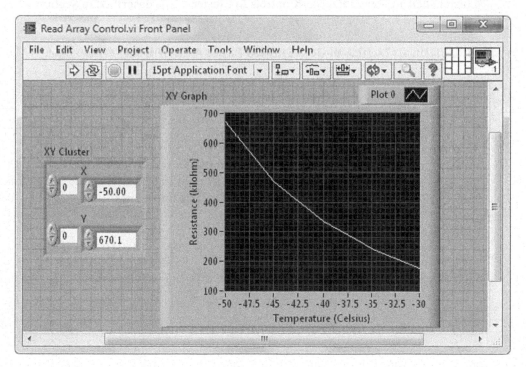

Switch to the block diagram. Remove the broken wires as well as the **Bundle** icon. Then, using the ✍, connect the **XY Cluster** and **XY Graph** icon terminals with a cluster wire.

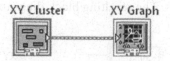

XY Cluster   XY Graph

We're finished. That's our program. Return to the front panel and save your work. Run the VI to verify that it is working properly.

There is one subtle point that is important to understand so that you obtain a valid resistance (y-axis) vs. temperature (x-axis) plot. Namely, the elements in a cluster control are ordered. This ordering is not based on the position of the elements within the cluster shell, but initially is determined by the order in which the elements were placed in the Cluster shell during programming. The first element placed in the Cluster shell is indexed 0, the second element placed there is indexed 1, and so on. Given a two-element cluster as input, an XY Graph associates the index-0 and index-1 elements with the x- and y-axes, respectively. Thus, in our present program, we want the X and Y arrays within the Cluster shell to be indexed 0 and 1, respectively. To verify this ordering, pop up on the border of the Cluster shell and select **Reorder Controls In Cluster**..., as described in Section 4.10, "Control and Indicator Clusters."

11.5 INPUTTING DATA TO A VI BY READING FROM A COMPUTER FILE

> If you already built **Read Spreadsheet.vi** as Chapter 7's Do It Yourself project, you may skip Sections 11.5 and 11.6

An alternate method for inputting data into a VI is by reading them in from a previously created computer file. Let's write a LabVIEW program that can receive and plot the thermistor spreadsheet data that you stored in a computer file in Section 11.1. We'll write this VI generically so that it can be reused in future programs to read spreadsheet data. If you haven't created the thermistor data file yet, now is the time to do it.

Construct this front panel, which simply contains an **XY Graph** with x- and y-axes labeled generically as X and Y, respectively. Design an icon for this VI, and then save it in **YourName\Chapter 11** under the name **Read Spreadsheet**.

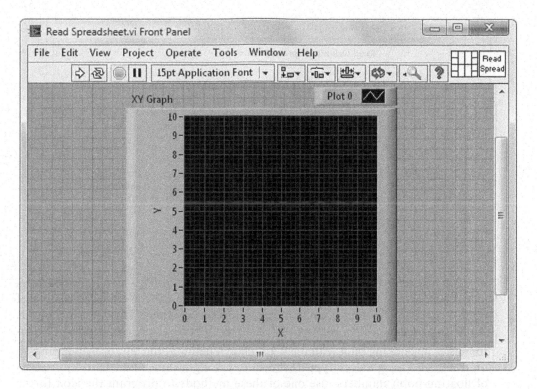

Switch to the block diagram. An ASCII spreadsheet file can be read into your VI using the **Read Delimited Spreadsheet.vi** icon found in **Functions>>Programming>> File I/O** (earlier versions of LabVIEW have an equivalent icon named **Read From Spreadsheet File.vi**). The Help Window for this icon (when configured for double-precision floating-point numbers as explained below) is shown in the following illustration. As with all of the more sophisticated LabVIEW functions, **Read Delimited Spreadsheet.vi** can be used for a lot of specialized purposes and therefore has many input and output connections. Default values for its inputs are shown in parentheses. On its Help Window, an icon's inputs are labeled in boldface, plain, or dimmed text to denote whether each input is *required*, *recommended*, or *optional*, respectively. Required inputs (**Read Delimited Spreadsheet.vi** has none of these) must be wired, or else the icon will not execute. Recommended and optional inputs control features available for your use, if desired. Optional inputs are less commonly used and so their labeling does not appear (as in the given Help Window) unless you click on the **Show/Hide Optional Terminals and Full Path** button in the Help Window's bottom left-hand corner.

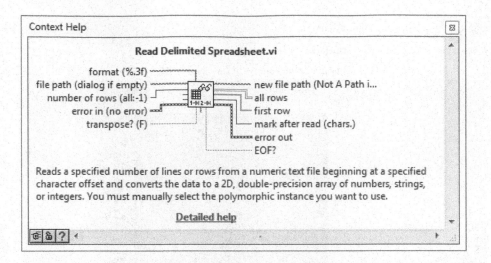

Place **Read Delimited Spreadsheet.vi** on your diagram. This icon is termed *polymorphic*, meaning that it can be configured to read spreadsheet files consisting of double-precision floating-point numbers, integers, or strings. This format choice is made using the icon's *Polymorphic VI Selector*. Operate the Polymorphic VI Selector either by clicking on it with the 🖑 as shown below at the left, or by popping up on it and choosing **Select Type** as shown to the right. Since our data file consists of floating-point numbers, use one of these methods to program the icon for its **Double** mode.

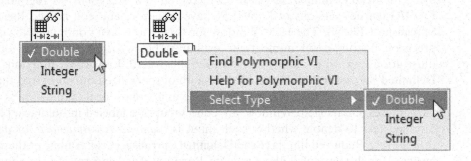

Pop up on Read Delimited Spreadsheet.vi's **file path** input and use **Create>> Control** to create a **File Path Control**. When the program is completed, a user will use this front-panel control to designate the path of the spreadsheet file wished to be read. Leave the **format** input unwired so that it takes its default value of *%.3f*

(above, when we made the polymorphic choice of **Double**, the icon was programmed with this default value; when executed, **Read Delimited Spreadsheet.vi** ignores the *.3*, while the *f* instructs the icon to format the **all rows** output as floating-point numerics). Also, since we desire to read the entire file, which is the default value for the **number of rows** input, leave this input unwired.

file path (dialog if empty)

XY Graph

11.6 SLICING UP A MULTIDIMENSIONAL ARRAY

Read Delimited Spreadsheet.vi will present your data at its **all rows** output in the form of a 2D array. This array's two columns then will have to be separated ("sliced off") into two 1D arrays to facilitate an XY plot. Separating out the columns is done using **Index Array** (found in **Functions>>Programming>>Array**), an icon we encountered previously in Section 8.7. Its Help Window is shown next.

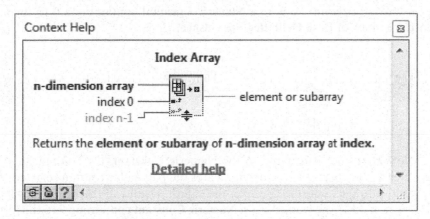

We'll explain how to implement **Index Array** for slicing off columns of a 2D array as we build the following code. Place this icon on your diagram. Note that it originates with just one **index** input.

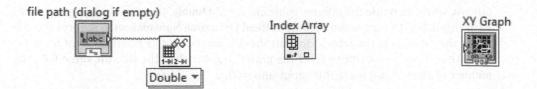

Next, wire Read Delimited Spreadsheet.vi's **all rows** output to the **n-dimension array** input of Index Array. The 2D array wire will appear, and **Index Array** responds by automatically producing a second **index** input.

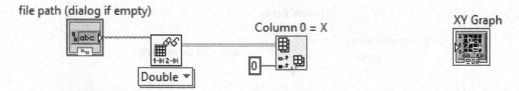

The top and bottom index inputs, which appear as small filled and hollowed-out black boxes, are the row and column indices, respectively. First, let's retrieve the X-values from the 2D array. We know that the X-values are all contained in the index-0 (first) column. So wire a **Numeric Constant** containing a *0* to the column (bottom) index of **Read Delimited Spreadsheet.vi**.

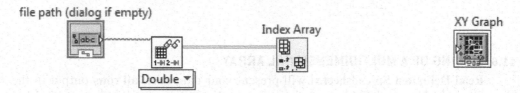

Note that once it is wired to the **Numeric Constant**, the column index input becomes a filled black box, which indicates the particular column selected. Now if we wanted to pick out, for example, the eighth X-value in this column, which has the row index of 7, we would wire a **Numeric Constant** containing *7* to the top index input. **Index Array** would then output this single array element.

However, instead of a single X-value, we want the entire column of values. To instruct **Index Array** to "slice off" the entire column, we leave the row index input unwired. When unwired, row indexing is disabled (i.e., no particular row is selected), and the icon will output all rows in the selected column in the form of a 1D array. Note that the unwired row index input appears as a hollowed-out box, indicating that indexing at this input has been toggled off.

Repeat the above procedure to slice off the Y data, which is in the index-1 (second) column.

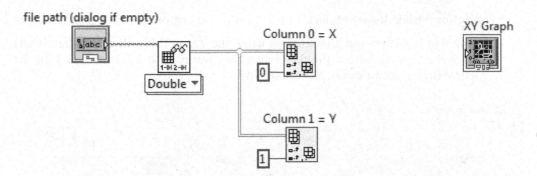

The **Index Array** icons below will each output a 1D array. Bundle these two 1D arrays together to form an *xy* cluster and wire this cluster to the XY Graph's terminal.

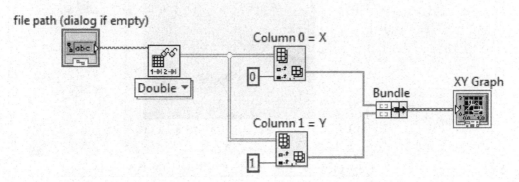

Finally, pop up on the cluster wire emanating from the **Bundle** icon and create a front-panel Indicator Cluster labeled **XY Cluster**.

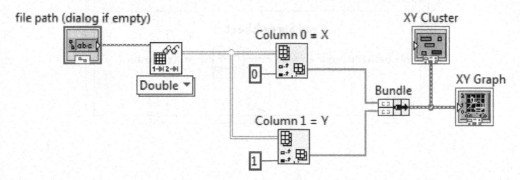

Return to the front panel and position the objects as you wish. Make the owned label of each array within **XY Cluster** visible by popping up on its index display and selecting **Visible Items>>Label**. Label the top and bottom arrays **X** and **Y**, respectively. Do not carry out the labeling using the 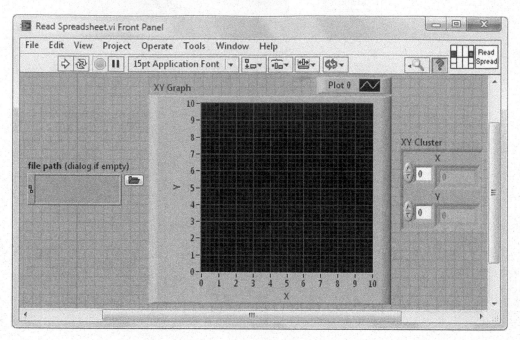 as this will create free labels rather than owned labels. Finally, assign the terminals to be consistent with the Help Window shown below. Save your work.

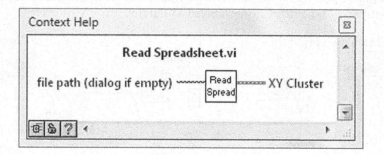

11.7 RUNNING THE VI

Your program is now complete. Take it for a test-drive by having it read and plot the spreadsheet file **Thermistor R-T Data.txt**. You have two easy methods for entering the correct path for this file into the **File Path Control**: (1) Leave the **File Path Control** empty and then press the ⮕. When the VI begins execution, a dialog window will appear, within which you can navigate your computer's directory system to the desired file; or (2) before running the VI, use the 🖑 to click on the **Browse Button** 📂 as shown below (if this button is not visible, pop up on the **File Path Control** and select **Visible Items>>Browse Button**). A dialog window will appear, within which you can navigate the directory system to the desired file's location. When you double-click on the filename, the correct path will be written in the **File Path Control**. Press the ⮕ to start the program.

Run your VI to read and plot **Thermistor R-T Data.txt**. You should be presented with a beautiful graph of your thermistor's resistance (y-axis) vs. temperature (x-axis). Note the exponential-looking decay of resistance with increasing temperature. This is a signature of the thermally activated process taking place within the material of which the thermistor is composed.

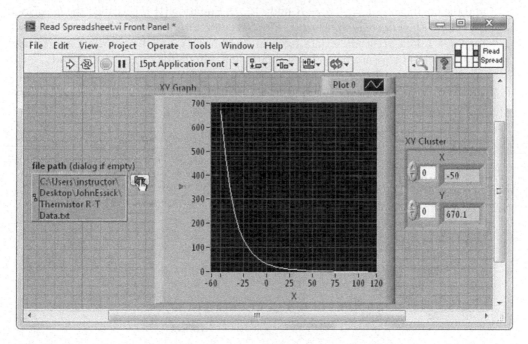

If you encounter an error when your VI tries to open the data file, carefully check to make sure that your data file is correctly constructed. An extraneous EOL character at the end of the data file is a common problem. Also make sure your negative temperatures have minus signs and not, say, en dashes. Examine the array values within **XY Cluster** to make sure that all of the temperatures and resistances are properly represented. For example, make sure that there is not an extra zero as the last element in an array (if an extra zero is present, you need to correct your **Thermistor R-T Data** file). Then save your work as you close this VI.

11.8 CURVE FITTING USING THE LINEAR LEAST-SQUARES METHOD

Now that you know how to input your thermistor data into a program, let's write a VI that uses the linear least-squares method to fit these data to the Steinhart–Hart Equation (Eq. [11.12]). At first glance, the Steinhart–Hart Equation does not appear to follow the form assumed by the linear least-squares theory (Eq. [11.13]). However, if we define $y \equiv 1/T$ and $x \equiv \ln(R)$, the Steinhart–Hart Equation becomes

$$y = A + Bx + Dx^3 \qquad [11.15]$$

which is of the form given in Eq. [11.13] with the basis functions $b_0 = 1$, $b_1 = x$, and $b_2 = x^3$ and the coefficients $a_0 = A$, $a_1 = B$, and $a_2 = D$. Equation [11.15] is called the *linearized Steinhart–Hart Equation*. The strategy for our curve-fitting VI is this: Input the thermistor's $R - T$ data, recast these data as $x = \ln R$ and $y = 1/T$, and then use the **Curve Fitting** Express VI to perform a linear least-squares fit to obtain the values of A, B, and D.

The **Curve Fitting** Express VI is found in **Functions>>Express>>Signal Analysis**. The Help Window for this icon (when configured for linear least-squares method) is shown in the next illustration.

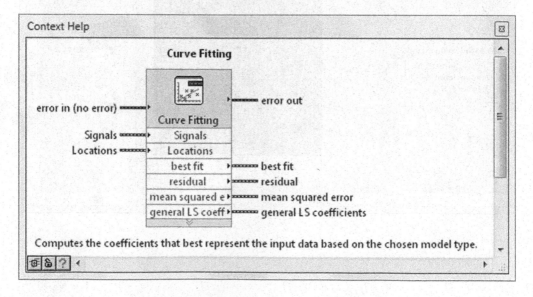

Here, it is assumed that an experimentalist has acquired N data samples (x_j, y_j). These data are input to the Express VI in the form of an N-element x-array and N-element y-array at its **Locations** and **Signals** input, respectively. After programming it with M basis functions $b_k(x)$, the VI will find the particular linear combination of the given $b_k(x)$ that best describes the data. That is, the data input to this icon are fit to Eq. [11.13] and the optimized values of the coefficients a_k are output at the **general LS coefficients** terminal. Beside the optimized values for a_k, this icon outputs the array of "best-fit" y-values, where

$$y_j^{\text{best fit}} \equiv f(x_j) = \sum_{k=0}^{M-1} a_k b_k(x_j) \qquad [11.16]$$

Also output are the N-element residual array R, where this array's jth element is defined as

$$R_j \equiv y_j - y_j^{\text{best fit}} \qquad [11.17]$$

and the *mean square error* (mse), which is defined as

$$\text{mse} \equiv \frac{1}{N} \sum_{j=0}^{N-1} R_j^{\,2} \qquad [11.18]$$

The residual array as well as the mse can be taken as a measure of the error in your fitted curve.

Build the front panel given below. To compare the resulting best-fit y-values with the original data, include an **XY Graph** with two differentiated plots called **Data** and **Best Fit** (e.g., select different colors or point styles and disable interpolation for one curve in the **Plot Legend**; for a review of how to prepare the Plot Legend for multiple plots, see Section 8.4). Also, display the determined values of a_k in an Array Indicator labeled **Best-Fit Coefficients**. Pop up on this indicator and format it to display numerics in scientific notation (using **Display Format...**). Save this VI as **R-T Fit and Plot** in YourName\Chapter 11.

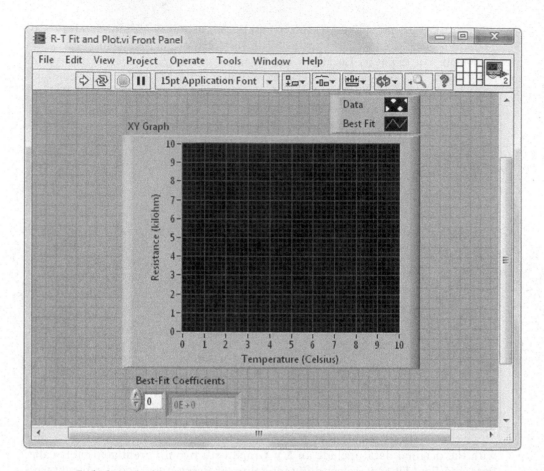

Switch to the block diagram and place the **Curve Fitting** Express VI on it. When first placed on the block diagram, the Express VI will open the dialog window shown next, where you will configure its operation. From among the several possible fitting options, select **General least squares linear**. The box labeled **Models** will become activated, and it is here that you define the basis functions to be used in the linear least-squares fitting operation. We will be fitting our data to the linearized Steinhart–Hart Equation, whose basis functions are $b_0 = 1$, $b_1 = x$, and $b_2 = x^3$, so enter *1*, *x*, and *x^3* on the first, second, and third lines of the **Models** box, respectively (within this dialog window, the exponentiation operator is ^).

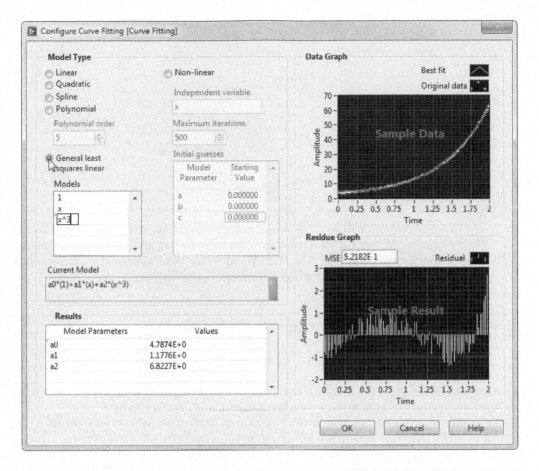

Click on the **OK** button. The dialog window will close, and the **Curve Fitting** icon will appear on your block diagram with its input and output terminals expanded (except for **error in** and **error out**).

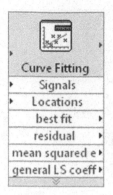

In a moment, we will have to add a few short single-variable calculations to the block diagram. When such a need arises, the easy-to-use and compact **Expression Node** offers a handy solution. The Help Window for the **Expression Node**, which is found in **Functions>>Programming>>Numeric**, is shown next.

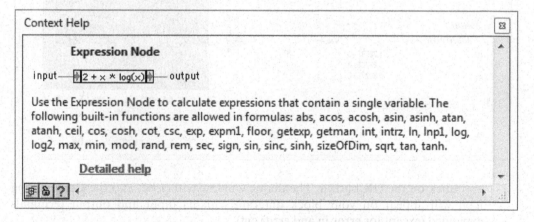

As can be seen, this icon has one input and one output terminal. When an **Expression Node** is first placed on the block diagram, its central region is blacked out . By clicking on this central region with the (or), a text-based formula can then be entered that will define the desired single-variable calculation to be carried out. You can choose any name you like for the variable within the formula. When the **Expression Node** executes, this variable assumes the numeric value at **input**, the formula is evaluated, and the result appears at **output**. If an array is wired to **input**, this process occurs for each element in the array. As an example, the **Expression Node** will evaluate the formula $5x + x^2$ (within an Expression Node, the exponentiation operator is **).

Using **Functions>>Select a VI...**, place **Read Spreadsheet** on the block diagram as shown in the next illustration. Pop up on its **file path** input and select **Create>>Constant**. In the created **File Path Constant**, enter the path for **Thermistor R-T Data.txt**. Carefully check that the file's path is perfectly correct. To find a Windows file's exact path, open the Windows folder that contains this file, then right-click on the file's icon or name and select **Properties**.

Complete the rest of the code shown, which delivers the $y = 1/T$ and $x = \ln(R)$ arrays at Curve Fitting's **Signals** and **Locations** input, respectively. The **Expression Node** $\boxed{1/(t+273.15)}$ first converts the Celsius temperature t to Kelvin temperature T (by adding the constant 273.15), and then determines the inverse Kelvin temperature. Similarly, $\boxed{\ln(R*1000)}$ converts resistance R from kilohms to ohms and then calculates the natural logarithm of this ohms resistance. The **Convert to Dynamic Data** icon is explained below.

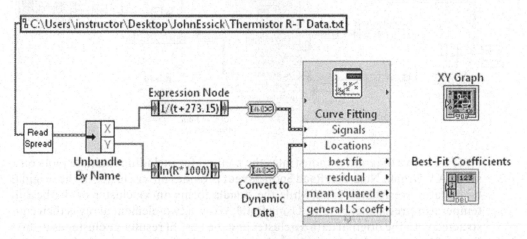

Express VIs accept and return information in a format called the *Dynamic Data Type (DDT)*, which is rendered as a dark-blue banded wire. The DDT bundles related information about the data it carries. For example, when carrying an array of numeric values, the DDT also includes the data's label name as well as the date and time the data were acquired. When you wire the output of $\boxed{1/(t+273.15)}$ to Curve Fitting's **Signals** input, the **Convert to Dynamic Data** icon will be inserted into the wire automatically. This icon converts the 1D floating-point numeric array produced by **Expression Node** to the dynamic data type used by Express VIs such as Curve Fitting. Similarly, the **Convert to Dynamic Data** icon will be inserted automatically when wiring the output of $\boxed{\ln(R*1000)}$ to Curve Fitting's **Locations** input. If you ever need to place **Convert to Dynamic Data** on a block diagram manually, it may be found in **Functions>>Express>>Signal Manipulation**.

Finally, add the following code to your block diagram that plots the original data and the best-fit results for comparison. Here, you will have to place the **Convert from Dynamic Data** icons (found in **Functions>>Express>>Signal Manipulation**) on the diagram manually. When **Convert from Dynamic Data** is first placed on the block diagram, its dialog window will open; select **Conversion>>1D array of scalars-automatic** and **Scalar data type>>Floating point numbers (double)**, and then press the **OK** button. Curve Fitting's **best fit** output contains the array of $y_j^{best fit}$ values. Since $y = 1/T$, where T is in degrees Kelvin, this array must be converted to Celsius temperatures prior to the comparison plot using the Expression Node $\boxed{1/(t+273.15)}$.

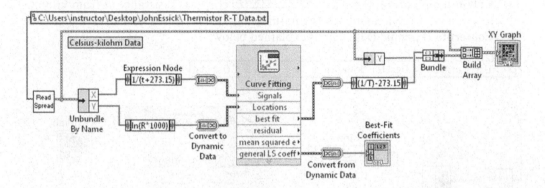

The above diagram demonstrates the method for including multiple plots on a single **XY Graph**. Note that **Read Spreadsheet** provides an xy cluster of the original temperature-resistance data, while the **Bundle** forms an xy cluster of the best-fit temperature-resistance results. Using **Build Array**, a two-element array is then constructed with the original data xy cluster and the best-fit results xy cluster as its first (index-0) and second (index-1) elements, respectively. Such an *array of clusters* is the required input for multiplots on an **XY Graph**.

Return to the front panel and run your VI. Upon completion, the **Best-Fit Coefficients** array indicator will display the Steinhart–Hart's A, B, and D constants at its 0, 1, and 2 indices, respectively.

For convenient reading, you can make all three array elements visible at once, if you like, by the following procedure. First, place the ⬧ on the side of the **Best-Fit Coefficients** indicator on the front panel so that it morphs into a Resizing Cursor.

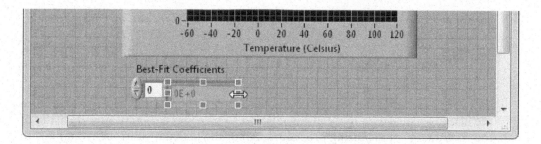

Then drag the Resizing Cursor to the right, until it has created two additional indicators.

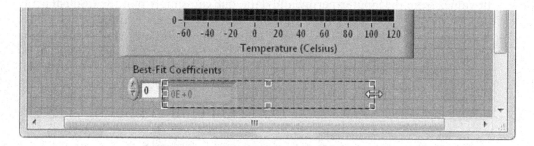

When you release the mouse button, indicators for array elements 0, 1, and 2 will then be visible. The index display provides the index for the element displayed in the leftmost indicator.

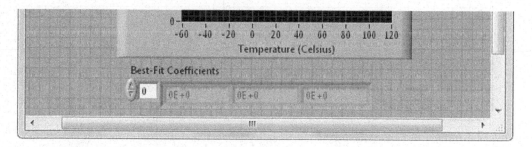

Record the resulting values of *A, B,* and *D* for your thermistor for future use. Do your *A, B,* and *D* have the order of magnitude values of 10^{-3}, 10^{-4}, and 10^{-7}, respectively, which are typical of most thermistors? Does the **XY Graph** indicate that the fitted Steinhart–Hart Equation accurately describes the temperature-dependent resistance of your thermistor?

11.9 RESIDUAL PLOT

Is there a quantitative method for judging how well the Steinhart–Hart Equation describes the thermistor's $R - T$ data? Yes, there is, and the Curve Fitting Express VI provides us with such a method. After it is has determined the array of best-fit y-values, Curve Fitting calculates the residual defined as $R_j \equiv y_j - y_j^{\text{best fit}}$ for each index j. We see that the residual is simply the deviation of the best-fit value from the actual y-value data input at **Signals** and so provides a quantitative measure of the "goodness" of the fitting procedure. All of the residual values R_j from $j = 0$ to $j = N - 1$ are output as an N-element array (contained within a DDT wire) at the Curve Fitting's **residual** output.

In general, for a quantitative display of the deviation of Curve Fitting's best-fit values from the input experimental data, one can simply plot the **residual** array output versus some appropriate quantity (such as the x-array input at **Positions**). However, in our present problem, we linearized the Steinhart–Hart Equation via $y = 1/T$, and so the residual calculated by Curve Fitting is the deviation of the inverse Kelvin temperature. That is, the **residual** output R is given by

$$R \equiv y - y^{\text{best fit}} = \frac{1}{T} - \frac{1}{T^{\text{best fit}}} \qquad [11.19]$$

where T is the experimental Kelvin temperature and T^{bestfit} is Curve Fitting's best-fit Kelvin temperature. Defining $\Delta y \equiv y - y^{\text{bestfit}}$ and $\Delta(1/T) \equiv 1/T - 1/T^{\text{bestfit}}$, we see that

$$R \equiv \Delta y = \Delta\left(\frac{1}{T}\right) \qquad [11.20]$$

To obtain a more intuitive measure, let's define the residual of Kelvin temperature as the "temperature residual" $R^T \equiv T - T^{\text{bestfit}} = \Delta T$. Then, if $y = 1/T$, we know that for small ΔT (in comparison to T itself),

$$\Delta y \approx \frac{dy}{dT}\Delta T = \frac{d}{dt}\left(\frac{1}{T}\right)\Delta T = \left(-\frac{1}{T^2}\right)\Delta T \qquad [11.21]$$

Since $R = \Delta y$ and $R^T = \Delta T$, Eq. [11.21] shows that the temperature residual can be determined from Curve Fitting's residual output using the relation

$$R^T \approx -T^2\, R \qquad [11.22]$$

Since R^T measures a difference of Kelvin temperatures, its units can be in either Kelvin or Celsius degrees because the degree interval is of the same size on both of these temperature scales.

Place a second **XY Graph** on the front panel of **R-T Fit and Plot**, and label its *x*- and *y*-axes **Temperature (Celsius)** and **Temperature Residual (Celsius)**, respectively.

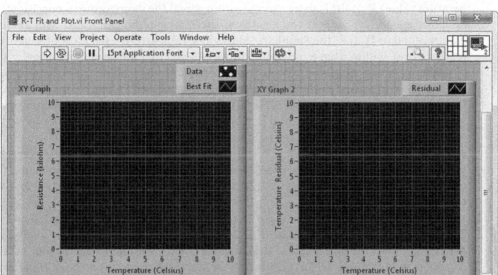

Switch to the block diagram and add the code shown below, which calculates the temperature residual and sends it to the front panel for display. The Expression Node $1/(t+273.15)$ has Celsius temperature *t* as input and $-T^2$ as output, where *T* is Kelvin temperature.

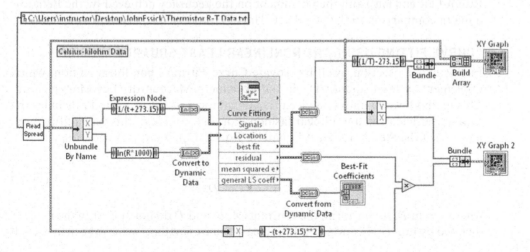

Return to the front panel and save your work. Then run your VI. Is there a temperature range within which the Steinhart–Hart consistently overestimates the temperature value ($R^T < 0$)? Is there a range within which the temperature is consistently underestimated ($R^T > 0$)? If you need your thermistor to be calibrated with an accuracy of ± 0.05 °C (i.e., the Steinhart–Hart value falls within ± 0.05 °C of the actual temperature), within what temperature range can you operate your thermistor?

When curve fitting, one expects that a better fit can be achieved if the number of basis functions used in the fitting algorithm is increased. Thus, we would expect that a third-order polynomial (Eq. [11.11]) should fit our thermistor data better than the Steinhart–Hart Equation (Eq. [11.12]), which deletes the second-order term.

Let's see if this conjecture is true. Switch to the block diagram and double-click on the **Curve Fitting** Express VI. When the dialog window opens, select **General least squares linear**, and then, in the **Models** box, program the VI to perform a fit using the four basis functions $b_0 = 1$, $b_1 = x$, $b_2 = x^2$, and $b_3 = x^3$. Then click on the **OK** button.

With this modification in place, return to the front panel. Expand **Best-Fit Coefficients** to include four indicators, and then run the VI. How has the accuracy (as determined from the Temperature Residual plot) of your fit improved? Is your result consistent with Steinhart–Hart's claim that the second-order term can be deleted from the fitting equation without significant loss of accuracy? Remember that this claim comes to us from an era when each additional term would have been calculated manually or with additional analog circuitry, and so any labor-saving tip would have been welcome.

Note that, by adding the $(\ln R)^2$ term, the new coefficients for the 1, $\ln R$, and $(\ln R)^3$ terms become different from those derived in the Steinhart–Hart case. This phenomenon results from the fact that the functions x^k that we are using as a basis are not *orthogonal*. Take a linear algebra course for more details.

Finally, try fitting the thermistor data to the Beta formula (Eq. [11.9]). In this first-order polynomial fit, only two basis functions are required: $b_0 = 1$ and $b_1 = x$. Run **R-T Fit and Plot**, and then comment on the accuracy achieved by the Beta formula in comparison to the Steinhart–Hart Equation.

11.10 CURVE FITTING USING THE NONLINEAR LEAST-SQUARES METHOD

In this closing section, we'll investigate Curve Fitting's **non-linear** option, which implements a least-squares fit via the Levenberg-Marquardt (Lev-Mar) method. This method can fit data to a given functional form $y = f(x)$, even if f includes the fitting coefficients in a nonlinear fashion. For an example of this type of function, let's revisit the Steinhart–Hart Equation (Eq. [11.12]), but this time written as

$$T = \frac{1}{A + Bx + Dx^3} \qquad [11.23]$$

where $x = \ln R$. In this form, the constants A, B, and D do not appear in the manner dictated by Eq. [11.13], and so the linear least-squares method we have been using will fail with this fitting function. However, as we will see, by having the Lev-Mar method in our bag of tricks, we'll be able to fit data to a function like Eq. [11.23] handily.

Starting with **R-T Fit and Plot**, use **File>>Save As**... to create a new VI named **R-T Fit and Plot (Nonlinear)**. Switch to the block diagram and double-click on the **Curve Fitting** Express VI so that its dialog window opens. Within the dialog window, select the **Model Type>>Non-linear** option as shown below. Then enter x as the **Independent variable** and program the **Non-linear model** box with Eq. [11.23] as follows: $1/(a+b*x+d*(x^2))$. The names of the fitting coefficients must be lowercase. Additionally, in the **Initial guesses** box, you will have to supply **Starting Values** for the **Model Parameters** a, b, and d. For the nonlinear Lev-Mar algorithm implemented by Curve Fitting to function properly, these initial guesses need to be somewhat close to the final best-fit values. For a typical thermistor, we know that the constants A, B, and D in Eq. [11.23] have order of magnitude values of 10^{-3}, 10^{-4}, and 10^{-7}, respectively, when resistance is measured in ohms. Thus, choose the **Starting Values** for a, b, and d (which are formatted to only six decimal places) to be 0.001000, 0.000100, and 0.000000, respectively. When finished, close the dialog window by pressing the **OK** button.

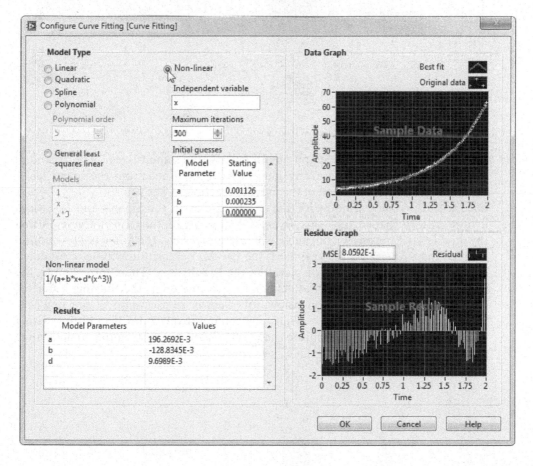

When you are returned to the block diagram, modify it as shown below. In fitting Eq. [11.23], our input variables are temperature T (in degrees Kelvin) and logarithm of resistance ln R (in ohms) so the Expression Node wired to **Signals** needs to reprogrammed as $\boxed{t+273.15}$, while the $\boxed{\ln(R^*1000)}$ wired to **Positions** can remain unchanged. The **best fit** output is in terms of Kelvin temperature T, so the Expression Node wired to it should read $\boxed{T-273.15}$ because we require only a simple conversion to Celsius temperature. Finally, the **residual** output now gives the temperature residual $R^T \equiv T - T^{\text{best fit}}$, so this quantity can be sent directly to the plot as shown, and the code related to Eq. [11.22] can be deleted. Save your work.

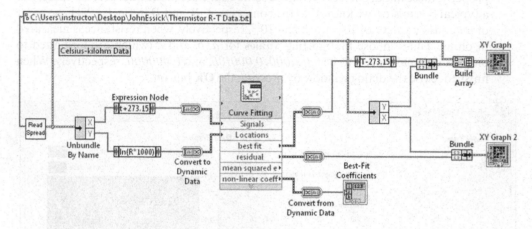

Run your VI. Verify that you obtain (nearly) the same values for A, B, and D (as well as for the temperature residual) as you obtained from the linear least-squares method implemented in the previous section. After you run your program once with the dialog window for **Curve Fitting** closed, the **Signals** and **Locations** arrays will be loaded into this Express VI. You can then open **Curve Fitting** (by double-clicking on its icon) and run it interactively via the following procedure: By pressing *<Enter>* on the keyboard, the Lev-Mar algorithm will be re-executed and the new findings will be displayed in the **Results** box. Then, for example, the fitted value of a given in the **Results** box can be entered as the **Starting Value** for a, as shown in the illustration below, followed by re-execution of the Lev-Mar algorithm by pressing *<Enter>*. Thus, you can iteratively run **Curve Fitting** with a succession of increasingly accurate choices for **Starting Values** until you determine a choice that yields a successful (i.e., self-consistent) fit.

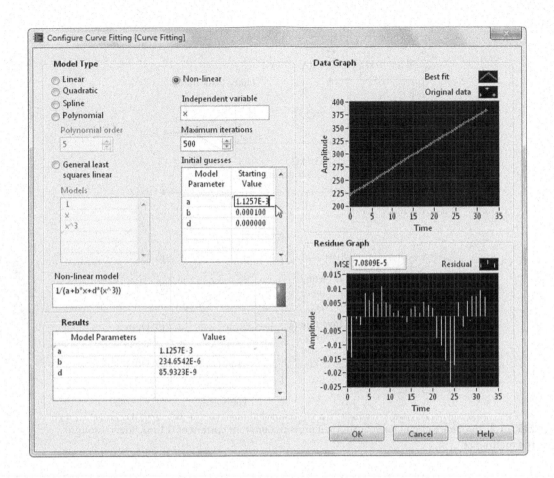

DO IT YOURSELF

Digital Thermometer.vi Build a computer-based digital thermometer that uses the thermistor that you have calibrated in this chapter as its temperature sensor.

Construct this instrument in the following steps:

1. Build the circuit shown in Figure 11.1, which produces a constant current of 0.1 mA through the thermistor. You may wish to include the high input-impedance unity-gain op-amp buffer to isolate the thermistor circuit from the voltage-sensing circuitry. If available, use an ammeter to determine the current through your thermistor precisely. It probably will differ slightly from 0.1 mA because of resistor tolerances. With a voltmeter, check that the voltage difference between V_{out} and GND is within the expected range.

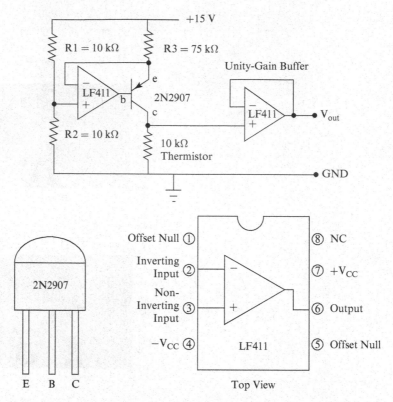

FIG. 11.1 Hardware circuit for Do It Yourself project. Constant current of 0.1 mA flows through the thermistor.

For example, the voltage difference across a thermistor whose resistance is about 10 kΩ near room temperature should be $V_{out} = IR \approx (0.1 \text{ mA})(10 \text{ k}\Omega) = 1 \text{ V}$.

2. Connect the positive and negative pins of a DAQ device's differential analog input channel to V_{out} and GND, respectively.

3. Write a VI whose front panel has a **Numeric Indicator** labeled Temperature [deg C] and a **Stop Button** as shown next. Save this VI in the **YourName\ Chapter 11** folder.

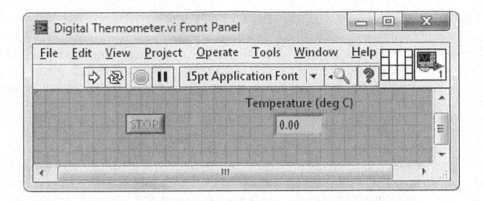

On the block diagram, place a **DAQ Assistant** icon. When the **Create New**...dialog window appears, configure it to perform an Analog Input operation that outputs temperature from a current-excited thermistor (**Acquire Signals>>Analog Input>>Temperature>>Iex Thermistor**) on the particular AI channel that you have connected to the circuit (e.g., *ai0*). When the **DAQ Assistant** dialog window then appears, configure it with the Steinhart–Hart values appropriate for your thermistor. The three Steinhart–Hart coefficients, which we called *A*, *B*, and *D* in this chapter, are respectively labeled *A*, *B*, and *C* in the DAQ Assistant dialog window. Also choose **Selected Units>>deg C**, **Iex Source>>External**, **Iex Value>>100 u** (which means 100 μA = 0.1 mA; use your measured value for the current, if available), **Configuration>>4-Wire**, and **Acquisition Mode>>1 Sample (On Demand)**. Then close the dialog window by clicking the **OK** button. When the DAQ Assistant icon appears on your block diagram, enclose it within a While Loop and complete the required extra code so that temperature readings are produced and displayed on a front-panel **Temperature (deg C)** indicator once every, say, 0.5 or 1 second, and this process repeats continually until a front-panel **Stop Button** is pressed.

For myDAQ users: A myDAQ does not support the **Temperature** option of DAQ Assistant. Thus, configure DAQ Assistant to acquire the voltage across the thermistor, and then output this value to a MathScript (or Formula) Node where Ohm's Law and the Steinhart–Hart Equation are used to determine the thermistor's resistance and temperature, respectively.

Once completed, run **Digital Thermometer** and use it to measure the room's temperature and the temperature of your skin.

USE IT!

Signal Conditioning for Thermocouple Temperature Measurement Because of their many favorable properties—including wide operating range and low cost—thermocouples are the most widely used temperature sensors. A thermocouple is constructed by joining the ends of two dissimilar metals and the resulting junction produces a well-characterized temperature-dependent voltage. However, thermocouples also have several unfavorable properties that pose challenges for the LabVIEW user. First, the voltage produced by a thermocouple is in the millivolt range and typically requires amplification before being digitized in order to resolve small temperature changes. Second, a thermocouple's voltage vs. temperature $(V–T)$ curve is nonlinear and is not predicted by a simple theoretical expression. As a result, the thermocouple's calibration equation $T(V)$—that is, temperature as a function of voltage—comes in the form of a high-order polynomial. As explored in Problem 2 of this chapter, this calibration equation has been established by the National Institute of Standards and Technology (NIST) through a curve-fitting procedure applied to the empirical $V–T$ data. Finally, a process called *cold-junction compensation* is required to account for the dissimilar-metal junctions formed at the points where the thermocouple wire connects to the supporting measurement circuitry.

Signal conditioning using a low-cost integrated circuit removes many of the difficulties of thermocouple measurement. For example, the Analog Devices AD8495 chip both amplifies the small thermocouple voltage produced by a type K thermocouple and performs cold-junction compensation. In addition, the chip produces a linear 5 mV/°C output, which provides a fairly good approximation (± 1 °C near room temperature) to the true nonlinear calibration curve. The AD8495 chip, conveniently mounted on a small printed circuit board along with its required supporting circuitry, can be purchased from Adafruit Industries (Product ID 1778). This "breakout board," when powered by the +5 V and D GND pins of a DAQ device, shifts the Type K thermocouple voltage by +1.25 V so that both positive and negative Celsius temperatures can be measured. The required connections between the breakout board and the DAQ device are shown in Figure 11.2.

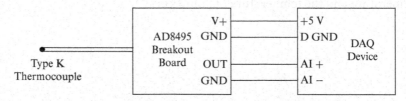

FIG. 11.2 Hardware circuit for Use It! example

The following LabVIEW program uses a DAQ device to measure the temperature of a Type K thermocouple in Celsius. Here, **DAQ Assistant** reads the output voltage from the Adafruit breakout board with the attached thermocouple.

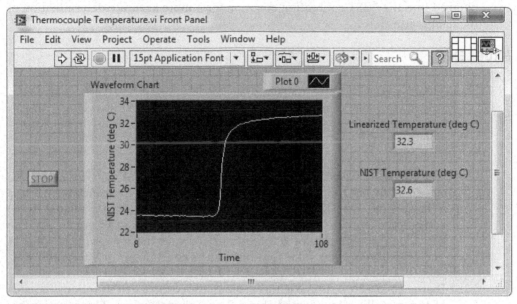

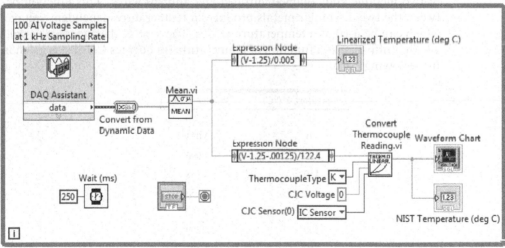

To suppress noise, **DAQ Assistant** reads an array of 100 voltage samples over a time interval of 0.1 second and the mean V of this array is computed. The temperature T (in Celsius) is then determined by the "linearized" approximation of the calibration equation $T = (V - 1.25\ V)/(0.005\ \text{V/°C})$. Alternatively, the actual thermocouple voltage V_{therm} can be calculated from $V_{therm} = (V - 1.25\ \text{V} - 0.00125\ \text{V})/122.4$ (see Analog Devices Application Note 1087). Then, using **Convert Thermocouple Reading.vi** (found in **Programming>>Numeric>>Scaling**), this voltage can be plugged into the NIST calibration equation for a more accurate determination of the temperature.

PROBLEMS

> For a problem that involves writing a new program, the problem statement begins with a suggested descriptive name (including the **.vi** extension) for the VI that you will write; icons needed for the VI may be found with the aid of **Quick Drop**. For some problems, the problem's topic is given in bold at the beginning of the problem statement.

1. **Energy Barrier Calculation** According to the theoretical considerations discussed at the beginning of this chapter, the value of B determined from fitting a thermistor's $R - T$ data to the Steinhart–Hart Equation is related to the activation energy barrier E for electronic conduction in the thermistor. Use the value you obtained for B to obtain E for your thermistor in the units of electronvolts (eV). The value for Boltzmann's constant is $k = 8.6 \times 10^{-5}$ eV/K.

2. **Thermocouple Fit.vi** A type K thermocouple is constructed by joining the ends of two metallic alloy (named chromel and alumel) wires. This junction between the two dissimilar metals produces a temperature-dependent voltage, which can be used as a temperature sensor. The type K thermocouple's voltage (in millivolts) as a function of temperature (in degrees Celsius) is given in the following table:

Temperature T (°C)	Voltage V (mV)
0	0.000
50	2.023
100	4.096
150	6.138
200	8.138
250	10.153
300	12.209
350	14.293
400	16.397
450	18.516
500	20.644

The calibration equation for the type K thermocouple has been established by the National Institute of Standards and Technology (NIST) by fitting its $V - T$ data to the following equation:

$$T = a_1V + a_2V^2 + a_3V^3 + a_4V^4 + a_5V^5 + a_6V^6 + a_7V^7 + a_8V^8 + a_9V^9 \quad [11.24]$$

Write a program that uses the **Curve Fitting** Express VI and the $V - T$ data given in the above table to determine the coefficients a_1 through a_9 in Eq. [11.24]. Consider entering the given $V - T$ data into a spreadsheet and then using **Read Spreadsheet** to input the data to your program. Alternatively, you can use Array Controls to input the data.

You might be interested in comparing your values for a_1 through a_9 with those established by NIST (see the Type K table in the online *NIST ITS-90 Thermo-couple Database*). The NIST values are derived from a more complete $V - T$ data set than the one used in this problem and so will deviate a bit from your values.

3. **Wait Until Next ms Fit.vi** The following subdiagram implements **Wait Until Next ms Multiple** to produce a y-array of millisecond timer values and a corresponding x-array of integers that index the y-values.

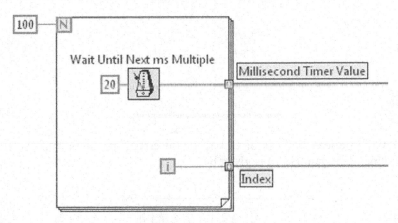

(a) Refer to **Wait Until Next ms Multiple**'s Help Window and then explain why the (x, y) data this subdiagram produces should obey the equation of a straight line $y = mx + b$, where m and b are the line's slope and y-intercept, respectively. What is the expected value for m of this line?

(b) Place the above subdiagram on the block diagram of a new VI. On this same block diagram, place a **Curve Fitting** Express VI, and program it to perform a straight-line fit on the given (x, y) data and output the resulting value for the slope to the front panel. When you run your completed VI, do you obtain the expected value for m?

4. **Resonant Circuit.vi** A resonant circuit composed of an inductor, a resistor, and a capacitor is powered by an AC input voltage whose amplitude V_{in} is constant but whose frequency f can be varied. An experimentalist measures the circuit's output voltage amplitude V_{out} at several frequencies f and then calculates the circuit's gain at each frequency, where gain is defined as the ratio of the output to input voltage amplitudes (i.e., $G \equiv V_{out}/V_{in}$). The results from this experiment are shown in the following data table:

Frequency f (Hz)	Gain G
1500	0.070
1600	0.084
1700	0.102
1800	0.128
1900	0.168
2000	0.235
2100	0.374
2200	0.740
2300	0.861
2400	0.442
2500	0.281
2600	0.206
2700	0.164
2800	0.136
2900	0.117
3000	0.103

A theoretical analysis of this resonant circuit predicts that the frequency-dependent gain $G(f)$ is given by

$$G = \frac{1}{\sqrt{1 + \left[\frac{1}{\Delta f}\left(f - \frac{f_o^2}{f}\right)\right]^2}} \qquad [11.25]$$

where the constants f_o and Δf are the resonant frequency and full width at half-power maximum, respectively (there is one frequency below f_o and one frequency above f_o where $G = 1/\sqrt{2}$; the full width is defined as the difference between these frequencies).

Write a VI that fits the given experimental $G - f$ data to the theoretically predicted relation for $G(f)$ (Eq. [11.25]). On the VI's front panel, compare the data and the best-fit curves by plotting both on the same **XY Graph**, and also display the best-fit values for constants f_o and Δf in an Array Indicator. Consider entering the given $G - f$ data into a spreadsheet and then using **Read Spreadsheet** to input the data to your program. Alternatively, you can use Array Controls to input the data.

5. **Noisy Sine Fit.vi** Write a program that fits a sine function to a noisy sine-wave input.

   (a) First, place a **Simulate Signal** Express VI on the block diagram. When its dialog window opens, program this Express VI to produce 100 samples of a sine wave with **Gaussian White Noise** added at a sampling frequency of 10000 Hz. Close the window and configure the **Simulate Signal** icon so that you have front-panel control of the sine-wave frequency and amplitude, as well as the standard deviation of the noise as shown next.

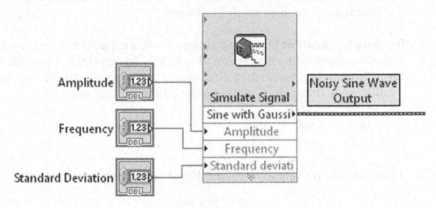

   Place a Waveform Graph on the front panel and plot the noisy sine-wave output from **Simulate Signal**. Run the VI with **Amplitude, Frequency,** and **Standard Deviation** set to *1, 250,* and *0.6*, respectively, to verify that your program functions correctly.

   (b) Put an Array Indicator on the front panel and label it **Best-Fit Amplitude and Frequency**. Then on the block diagram, place a **Curve Fitting** icon and program it to perform a nonlinear fit to the function $a* \sin (2* 3.14* b* x)$. You will also need to supply initial guesses for the parameters a and b. After closing the dialog window, connect the output of Simulate Signal to the **Signals** input of Curve Fitting. The resulting dynamic data type wire will automatically create the needed independent variable x for Curve Fitting (i.e., there is no need to explicitly wire anything to the **Locations** input). Wire the nonlinear coefficients output to the front-panel array indicator. Finally, if interested, simply join a wire from Curve Fitting's **best fit** output to the noisy sine-wave wire emanating from **Simulate Signal**. A **Merge Signals** Express VI will automatically appear, causing both curves to be plotted on the Waveform Graph.

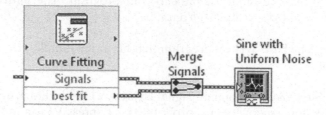

(c) Run your completed VI with various choices for **Amplitude, Frequency,** and **Standard Deviation**. To achieve a good fit of the sine function to the noisy sine wave, you will most likely have to experiment with the choice of initial guesses to the fitting parameters. For which parameter must the initial guess be most carefully chosen?

6. **Discharging Capacitor** Beginning at time $t = 0$, a capacitor of unknown value C is discharged through a resistor $R = 47$ kΩ. To determine C, as the capacitor is discharging, the voltage difference V between its plates is measured at 5 millisecond intervals, resulting in the following data set:

t (s)	0.005	0.010	0.015	0.020	0.025	0.030
V (V)	8.7	6.3	4.5	3.3	2.4	1.7

Theoretically, the relation between V and t is given by

$$V = V_0 e^{-t/RC} \qquad \text{[11.26]}$$

where V_0 is the voltage across the capacitor at $t = 0$.

(a) **Discharging Capacitor (Linear).vi** By taking the logarithm of both sides, rewrite Eq. [11.26] in a linearized form, that is, in the form $y = mx + b$. Then write a program that determines C (in μF) and also V_0 from the given V vs. t data. Build this VI so that the data set is input to its block diagram, where the data are used to construct appropriate x and y arrays, which are then input to a **Curve Fitting** icon. Program **Curve Fitting** to fit the input arrays to the equation of a straight line, use the resulting best-fit parameters m and b to calculate C and V_0, and then display these values on the front panel. Suggested icons: **Natural Logarithm, Exponential, Reciprocal**.

(b) **Discharging Capacitor (Nonlinear).vi** Write another program that also determines C as well as V_0 from the given V vs. t data. Build this VI so that the data set is input to its block diagram, where a **Curve Fitting** icon is programmed to perform a nonlinear fit of the input arrays to the equation $y = ae^{-x/b}$. Use the resulting best-fit parameters a and b to calculate C (in μF) and V_0, and then display these values on the front panel. Suggested icon: **Index Array**.

(c) Run **Discharging Capacitor (Linear)** and **Discharging Capacitor (Nonlinear)**. What values do you obtain for C and V_0 from each VI? Do the two approaches yield the same result?

7. **Discharging Capacitor (MathScript).vi** (MathScript Only) Write a MathScript-based program that analyzes the discharging capacitor experiment described in Problem 6. In particular, use the MathScript command *[p, s] = polyfit (x, y, n)* to perform a fit of the *Voltage* vs. *Time* data to a linearized form of Eq. [11.26]. A description of this command can be found by typing *help polyfit* in the LabVIEW MathScript Window. Construct this VI in the following steps:

(a) Enter the given V vs. t data as row vectors in a MathScript Node. The V row vector is created with the command $V = [8.7\ 6.3\ 4.5\ 3.3\ 2.4\ 1.7]$. Then plot V vs. t on a front-panel XY Graph with the data points as solid dots without interpolation.

(b) By taking the logarithm of both its sides and defining $y = \ln(V)$ and $x = t$, Eq. [11.26] can be linearized in the form of the first-order polynomial $y = a_0 + a_1 x$, where a_0 and a_1 are constants. Add the required code within your MathScript Node to form the y array (use *help basic* to find the MathScript command for the natural logarithm), and then use *[p, s] = polyfit(x, y, 1)* to determine the constants a_1 and a_0. These constants will be given as the first and second elements of the P row vector, that is, $p(1) = a_1$ and $p(2) = a_0$ (unlike LabVIEW arrays, MathScript-array indexing begins with *1*, rather than *0*). Using these best-fit a_1 and a_0, calculate C (in μF) and V_0, and display these values in front-panel indicators.

(c) Finally, generate an array of time values called *tfit* from $t = 0$ to 0.030 s in 0.001 s increments. Then create another array called *Vfit*, which is produced by evaluating the best-fit form of Eq. [11.26], that is, $Vfit = V_0 \exp(-tfit/RC)$, at all of the *tfit* values. Plot *Vfit* vs. *tfit* as an interpolated line on the same XY Graph used to plot the V vs. t experimental data.

(d) Run you VI. What values are obtained for C and V_0? Does the best-fit curve accurately describe the V vs. t data?

8. **Polynomial Fit.vi** Curve fitting of data must be guided by a theory describing the phenomenon under investigation. Consider the following hypothetical situation: An experimenter, investigating the relation between two quantities y and x, uses an experimental system to acquire seven samples of Y as a function of x, with the results given in the following table. These data, of course, reflect the true mathematical function $y(x)$ that relates the two quantities, but this relation is somewhat obscured by noise in the experimental system. The experimenter knows that a theory of the phenomenon being studied suggests that the measured quantities should be related by a second-order polynomial $y(x) = a_0 + a_1 x + a_2 x^2$, where $a_0 = 6.0$, $a_1 = 4.0$, and $a_2 = 2.0$ in appropriate units.

x	1.0	2.0	3.0	4.0	5.0	6.0	7.0
y	10	24	34	54	75	101	133

To compare the experimentalist's results to the theoretically predicted relation, analyze the data in the following ways:

(a) Write a program that plots the given data on an XY Graph and also inputs these data to a **Curve Fitting** Express VI. Program the Express VI to fit the data to a second-order polynomial and display the resulting **polynomial coefficients** a_0, a_1, and a_2 in an indicator on the front panel. Also, on the block diagram, include the code shown next to generate a well-resolved representation of the best-fit polynomial y_{fit} and plot this curve, along with the given data, on the front-panel XY Graph (the icon **Reverse 1D Array** is needed because the MathScript function *polyval* requires the polynomial coefficients to be in reverse order of that provided by **Curve Fitting**).

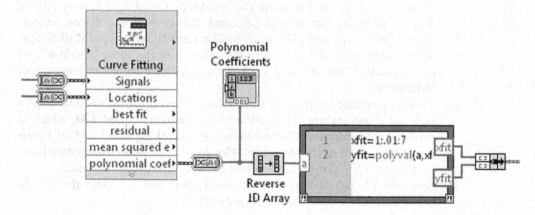

The Formula Node version does not require the **Reverse 1D Array** icon and is as shown below.

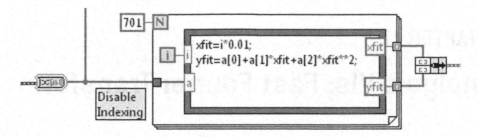

How do your fitted values of a_0, a_1, and a_2 compare with the theoretically expected values? Why do you suppose that the experimentally determined values for these coefficients don't perfectly match the theoretical values? Characterize qualitatively (in a sentence or two) how well the best-fit polynomial describes the measured data points.

(b) A better fit to the acquired data can be obtained using a higher-order polynomial. To demonstrate this fact, reprogram **Curve Fitting** so that it fits the given data to a fourth-order polynomial. From the XY Graph, does y_{fit} from this fourth-order fit agree with the experimental data more favorably than the second-order fit carried out in part (a)? What values do you obtain for a_0, a_1, a_2, a_3, and a_4? How do the first three of these values compare with the values expected from theory? Explain why, even though the fourth-order polynomial provides a closer match to the experimental data, the second-order fit carried out in part (a) provides the best comparison between the experimental findings and the given theory.

(c) Interestingly, a fitted polynomial of order $N - 1$ will perfectly pass through N given data samples. Demonstrate this fact using your program. Here, we are magnifying the effect studied in part (b). By providing too much flexibility in the fitting function, we are fitting experimental noise, at the expense of uncovering the sought-after relation $y(x)$.

Analysis VIs: Fast Fourier Transform

12.1 QUICK FAST FOURIER TRANSFORM EXAMPLE

In this chapter, we will explore the use of the *fast Fourier transform* (FFT), a signal-processing technique used by scientists and engineers in countless applications. In short, the FFT allows one to determine the amplitudes of the component sinusoidal oscillations that add together to produce a given digitized signal. As a quick introduction to the FFT, let's familiarize ourselves with a few of its properties through an Express VI-based program.

Open a blank VI. Use **File>>Save** to create a folder called Chapter 12 in the YourName folder, and then save this VI under the name FFT (Express) in YourName\ Chapter 12.

Switch to the block diagram and place a **Simulate Signal** Express VI there (found in **Functions>>Express>>Input**). When its dialog window opens, program the **Signal** box to produce a 250 Hz sine wave with an amplitude of 4 by selecting **Signal type**, **Frequency (Hz)**, and **Amplitude** as *Sine*, *250*, and *4*, respectively, as shown below. Also, in the **Timing** box, choose **Samples per second (Hz)** to be *2000* and **Number of samples** as *1024* with the **Automatic** box unchecked. When done, press the **OK** button.

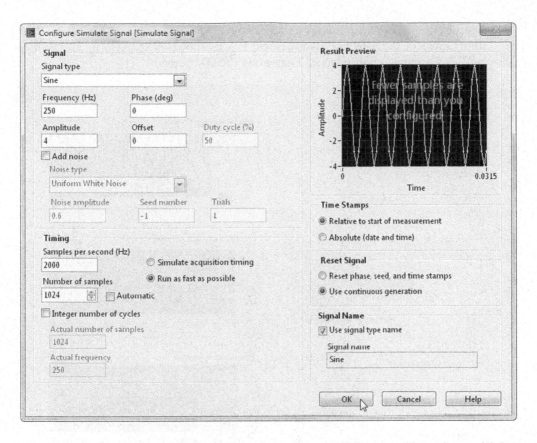

Upon return to the block diagram, configure the **Simulate Signal** icon with just two terminals—a **Frequency** input and a **Sine** output. Pop up on the **Frequency** terminal and select **Create>>Control** to create a front-panel Numeric Control named Frequency.

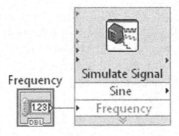

Next, place a **Spectral Measurements** Express VI (found in **Functions>>Express>>Signal Analysis**) on your block diagram. When its dialog window opens, program this icon to carry out a FFT by selecting **Magnitude (Peak)** and **Linear** in the **Selected Measurement** box as shown next. Also, in the **Window** pull-down menu, choose **None**. Then press the **OK** button.

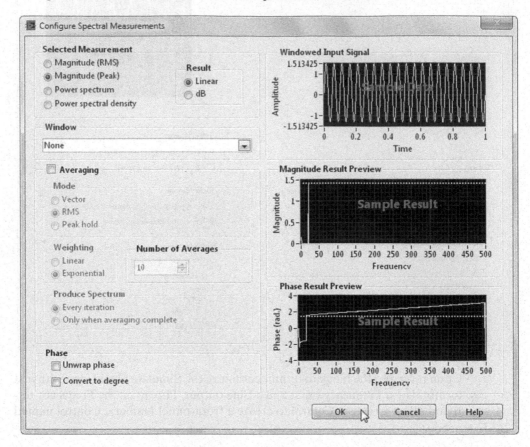

Back on the block diagram, configure the **Spectral Measurements** icon with just two terminals—a **Signals** input and a **FFT-(Peak)** output. Then wire the Simulate Signal's **Sine** output to Spectral Measurements's **Signals** input. The resulting dynamic data type (DDT) wire will pass the sine-wave signal created by **Simulate Signal** to the **Spectral Measurements** icon for FFT analysis.

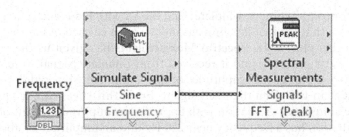

Finally, pop up on the **FFT-(Peak)** terminal and select **Create>>Graph Indicator**. This action will create a front-panel Waveform Graph, which will plot the FFT result.

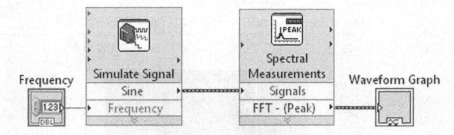

Switch to the front panel and organize it as you wish. There is no need for manual labeling of the Waveform Graph's features. When you first run this VI, the DDT wires on the block diagram will automatically replace **Plot 0** in the **Plot Legend** with an appropriate name, as well as correctly label the *x*-axis and *y*-axis.

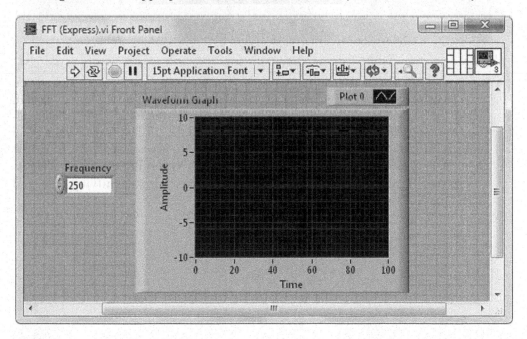

Your program is now complete. Run the VI with **Frequency** equal to *250*. First, briefly note that the plot labeling has indeed been carried out automatically. Then examine this plot. Here, **Spectral Measurements** has given us the *frequency spectrum* of the digitized signal it received from **Simulate Signal**, where a frequency spectrum is a plot of the amplitudes of the component sinusoidal oscillations that are present in the given digitized signal. From this spectrum, it appears there is only one sinusoidal oscillation with appreciable amplitude within our signal, and this oscillation has a frequency near 250 Hz and an amplitude of about 4.

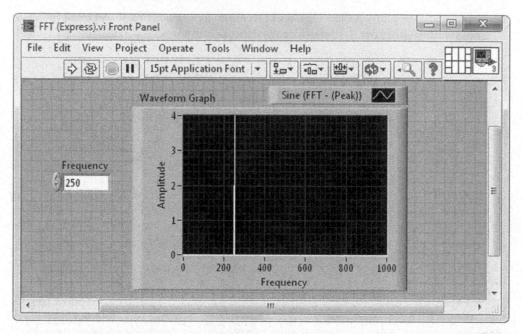

To inspect the single peak in this spectrum more closely, use the **Plot Legend** to make the individual plotted points visible by changing the **Point Style**. Then magnify the spectral peak by changing the *x*-axis endpoints. An easy way to zoom in on a peak is to use the Waveform Graph's *Zoom Tool*. To access the Zoom Tool, first pop up on the Waveform Graph's interior and select **Visible Items>>Graph Palette**. Using the ☝, click on the Zoom Tool 🔍, which looks like a magnifying glass, and select the **X-Axis Zoom** option.

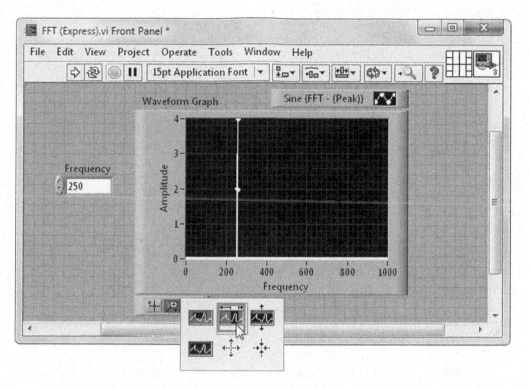

The mouse cursor will become a small magnifying glass when positioned over the plot. Click and drag it to select the region that you would like to zoom in on.

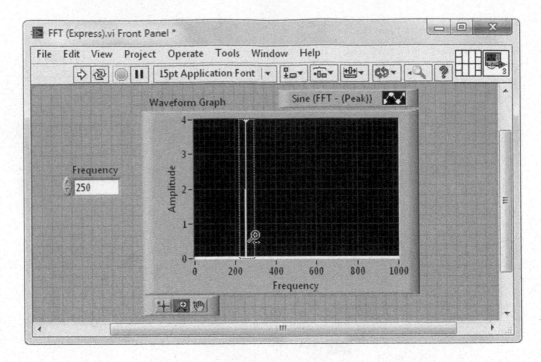

When you release the mouse button, the *x*-axis will be rescaled in the desired manner. The scaling can then be fine-tuned manually using the 🖑.

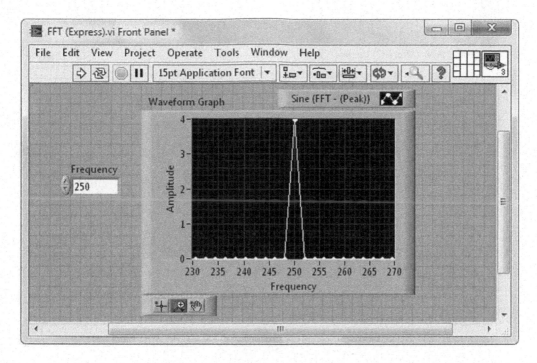

If you want to return to the original plot, select **Zoom to Fit** from the Zoom Tool menu, which will cause all of the data to be displayed again by autoscaling both axes.

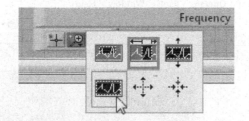

The spectrum looks perfect, doesn't it? The amplitude is equal to zero everywhere, except at the frequency of 250 Hz, where it equals +4.0, just as we expected based on the sine-wave signal we created using **Simulate Signal**. That's good news! The FFT is easy to implement and its results are easy to understand.

Unfortunately, now I have some bad news for you. Generally, the FFT is not as simple to interpret as has been suggested so far. Try repeating the above procedure with **Frequency** equal to *249*. Surprisingly, the resulting spectrum will no longer appear as a "delta-function" spike with a height of 4.0, but instead will be a broadened peak of maximum height less than 4.0, as shown.

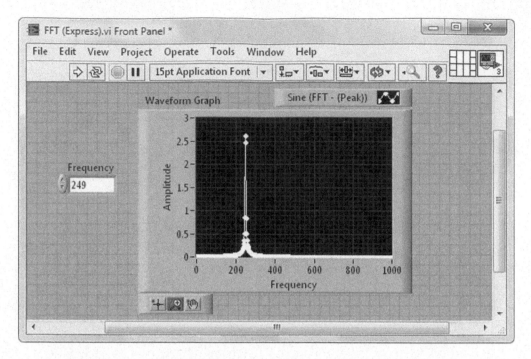

Zoom in on the peak to get a closer look.

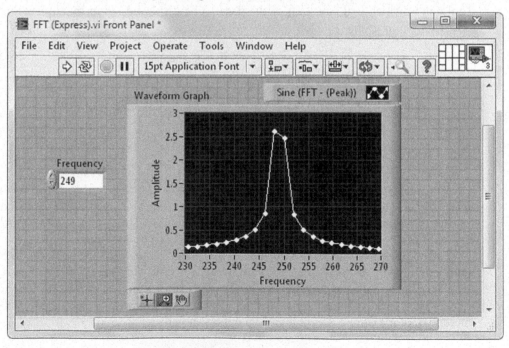

At face value, this spectrum seems to be telling us that the input signal is composed of many sinusoidal oscillations, each with a different frequency in the approximate range of 230 to 270 Hz. However, we know that this is simply untrue. We input a pure sine wave with the exact frequency of 249 Hz.

We are thus faced with a mystery. Why does our FFT-based VI work perfectly for an input frequency of 250 Hz, but fails when that frequency is changed to 249 Hz? That's one of the questions we will answer in the course of this chapter.

To close this quick introduction to the FFT, I'll demonstrate a method for mitigating the mysterious problem we just observed. Switch to the block diagram and double-click on the **Spectral Measurements** icon. When its dialog window opens, select **Window>>Hanning** as shown. Then close the dialog window by pressing the **OK** button.

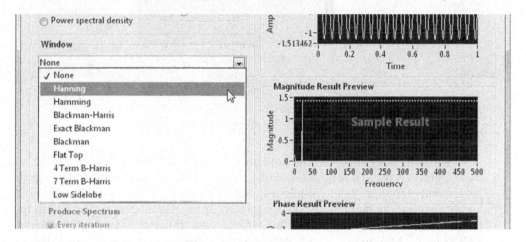

Switch to the front panel and run your VI again with **Frequency** equal to *249*. You will find that the spectral peak is still "incorrectly" spread among several frequencies in the neighborhood of 249 Hz, but the spread is now much narrower and the peak amplitude much closer to 4 than when **Window>>None** was selected.

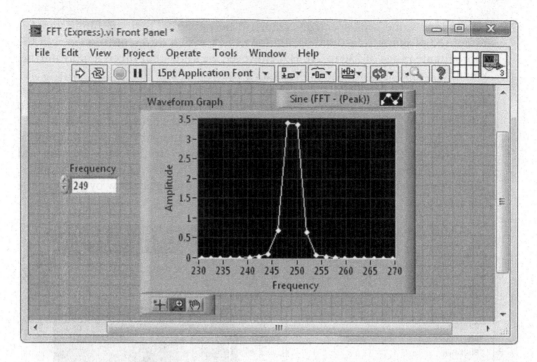

From this brief introduction, hopefully it's apparent that use of the FFT is not as straightforward as one might initially expect. In this chapter, we will develop the basic competence needed to implement and interpret FFT analysis properly. First, we will review the mathematical background of this signal-processing technique; then we will go on to demonstrate how to carry out this method using LabVIEW's **FFT.vi** icon. Along the way, we will develop a deeper understanding of the two effects (called *leakage* and *windowing*) that we encountered in this section.

12.2 THE FOURIER TRANSFORM

Imagine you are investigating a physical phenomenon by measuring the analog quantity x that varies continuously with time t, that is, $x = x(t)$. Based on our discussion at the beginning of the previous chapter, we expect that $x(t)$ can be modeled as the linear combination of a set of M basis functions $b(t)$ such that

$$x(t) = \sum_{k=0}^{M-1} a_k b_k(t) \qquad [12.1]$$

The Fourier transform is an example of just such a model. The pioneering work of nineteenth-century mathematician Baron Jean Baptiste Joseph Fourier, with subsequent elaboration by others, showed that the infinite set of complex exponentials

$\exp(i2\pi ft)$, where f includes all of the frequencies from $-\infty$ to $+\infty$ and $i = \sqrt{-1}$, is a complete basis set with which one can form linear combinations to model any arbitrary function (a complex exponential is a form of sinusoidal oscillation, as seen from Euler's formula $e^{ix} = \cos x + i \sin x$). Since an infinite number of exponentials with infinitesimally close frequency spacing is required to model an arbitrary function, the summation in Eq. [12.1] becomes an integral, yielding

$$x(t) = \int_{-\infty}^{+\infty} X(f) e^{i2\pi ft} \, df \qquad [12.2]$$

where the *Fourier component* $X(f)$ is the amplitude (I'll be much more precise in my use of the word "amplitude" shortly) of the exponential with frequency f. Generally, $X(f)$ is a complex-valued function, carrying both *magnitude* and *phase* information, a point we will carefully consider in this chapter.

Conversely, the Fourier components $X(f)$ may be obtained by the following equation:

$$X(f) = \int_{-\infty}^{+\infty} x(t) e^{-i2\pi ft} \, dt \qquad [12.3]$$

Equations [12.3] and [12.2] are called the *Fourier transform* and *inverse Fourier transform*, respectively.

12.3 DISCRETE SAMPLING AND THE NYQUIST FREQUENCY

In a real experiment, one cannot sample the quantity x continuously but must instead settle for discretely sampling it N times. If the samples are taken at equally spaced times, where the time difference between adjacent samples is Δt, then the sampling frequency f_s is given by

$$f_s = \frac{1}{\Delta t} \qquad [12.4]$$

and the N consecutive sampling times are

$$t_j = j\Delta t \quad j = 0, 1, 2, \ldots, N-1 \qquad [12.5]$$

Given such a sampling scheme, there is a maximum-frequency sine wave that we can expect to detect. This limitation is simply due to the fact that to observe a sine signal, we must at the very least, for example, first sample its positive peak, then its negative trough at the next sample, followed by its positive peak at the third sample, and so on. Thus, the minimal sampling of a sine wave is two sample points per cycle. In an experiment, then, where your instruments are acquiring data every Δt seconds (at a sampling rate $f_s = 1/\Delta t$), the maximum-frequency sinusoidal signal you can detect has a period $T = 2\Delta t$. This detection limit is called the *Nyquist frequency* $f_{nyquist}$ and is given by

$$f_{nyquist} = \frac{1}{2\Delta t} = \frac{f_s}{2} \tag{12.6}$$

Additionally, by sampling a waveform $x(t)$ at N equally spaced times t_j, one can only expect to determine the Fourier components $X(f)$ of N equally spaced frequencies f_k in the Fourier transform. This statement follows from the tacit assumption that $x(t)$ is periodic such that $x(t) = x(t + N\Delta t)$—a criterion that forces the waveform $x(t)$ to be composed only of component frequencies that fit an integer number of cycles into the sequence of N samples. The constant (zero-frequency) function obviously meets this criterion. The next lowest acceptable frequency has a period of $N\Delta t$ and thus a frequency of $f_1 = 1/N\Delta t = f_s/N$. Within the acceptable range of $-f_{nyquist}$ to $+f_{nyquist}$, the rest of the frequencies f_k are the integer multiples of f_1. Thus, defining the spacing between adjacent frequencies to be $\Delta f = f_s/N$, the N equally spaced frequencies f_k determined by discretely sampling data are

$$f_k = k\left(\frac{f_s}{N}\right) = k\Delta f \quad k = -\frac{N}{2} + 1, \ldots, 0, \ldots, +\frac{N}{2} \tag{12.7}$$

Although Eq. [12.7] appears deficient in its neglect of the frequency $-f_{nyquist}$ (given by $k = -N/2$), the two extreme frequencies $\pm f_{nyquist} = \pm f_s/2$ actually describe the same basis function due to the periodic nature of $x(t)$. Thus, the range of k-values in the above relation rightfully runs from $-N/2 + 1$ to $+N/2$.

12.4 THE DISCRETE FOURIER TRANSFORM

With the above-described constraints caused by discrete sampling, we then approximate the Fourier transform integral (Eq. [12.3]) by the following sum:

$$X(f_k) = \int_{-\infty}^{+\infty} x(t)e^{-i2\pi f_k t} \, dt \approx \sum_{j=0}^{N-1} x(t_j)e^{-i2\pi f_k t_j}\Delta t \tag{12.8}$$

Note that, using Eqs. [12.4], [12.5], and [12.7],

$$f_k t_j = (k\Delta f)(j\Delta t) = kj\frac{f_s}{N}\Delta t = \frac{kj}{N} \tag{12.9}$$

Putting Eq. [12.9] into Eq. [12.8], and realizing that Δt is a constant, we find

$$X\left(f_k\right) \approx \Delta t \sum_{j=0}^{N-1} x\left(t_j\right) e^{-i2\pi jk/N} \equiv \Delta t X_k \tag{12.10}$$

where we have defined the *discrete Fourier transform* to be

$$X_k \equiv \sum_{j=0}^{N-1} x\left(t_j\right) e^{-i2\pi jk/N} \quad k = -\frac{N}{2} + 1, \ldots, 0, \ldots, +\frac{N}{2} \tag{12.11}$$

Finally, we approximate the inverse Fourier transform integral (Eq. [12.2]) by the following sum:

$$x\left(t_j\right) = \int_{-\infty}^{+\infty} X(f) e^{i2\pi ft_j}\, df \approx \sum_{k=0}^{N-1} X\left(f_k\right) e^{i2\pi f_k t_j} \Delta f \tag{12.12}$$

Putting Eq. [12.10] into Eq. [12.12], we get

$$x_j \approx \sum_{k=0}^{N-1} X_k\, \Delta t\, e^{i2\pi f_k t_j} \Delta f \tag{12.13}$$

Then using Eq. [12.9] and also noting that $(\Delta t)(\Delta f) = \left(1/f_s\right)\left(f_s/N\right) = 1/N$, Eq. [12.13] becomes the *discrete inverse Fourier transform*

$$x_j \approx \sum_{k=0}^{N-1} \frac{X_k}{N} e^{i2\pi f_k t_k} = \sum_{k=0}^{N-1} \frac{X_k}{N} e^{i2\pi k\, j/N} \equiv \sum_{k=0}^{N-1} A_k e^{i2\pi k\, j/N} \tag{12.14}$$

Equations [12.11] and [12.14] provide us with a method of performing spectral analysis on an experimental signal x. Given that x has been sampled discretely at N evenly spaced times t_j, we first calculate the N values of X_k using the definition of the discrete Fourier transform (Eq. [12.11]). Then, in the spirit of Fourier, viewing the signal x as really the composite of oscillations at N different frequencies f_k, Eq. [12.14] tells us that the amplitude of oscillation A_k at each f_k is given by

$$A_k = \frac{X_k}{N} \tag{12.15}$$

Since A_k is derived from the complex-valued X_k, in general, A_k is a complex-valued function. Let's then name A_k the *complex-amplitude*. The plot of complex-amplitude A_k vs. frequency f_k is called the *frequency spectrum* of the sampled signal x_j.

Up to now we have taken k to be all of the integers within the range of $-N/2 + 1$ to $+N/2$. However, it is easy to show that Eq. [12.11] is periodic in k, with a period of N:

$$X_{k+N} = \sum_{j=0}^{N-1} x\left(t_j\right) e^{-i2\pi j(k+N)/N} = \sum_{j=0}^{N-1} x\left(t_j\right) e^{-i2\pi j\, k/N} e^{-i2\pi j} = X_k \qquad [12.16]$$

where we have used the fact that $\exp(-i2\pi j) = 1$ because j is an integer. Based on Eq. [12.16], the negative k-values in the range of $-N/2 + 1$ to -1 are equivalent to the k-value range of $+N/2 + 1$ to $+N - 1$. Because the array indices in computer programs generally are positive integers, algorithms to evaluate Eq. [12.11] typically let k vary from 0 to $N - 1$ (which covers one complete period). This is the convention that LabVIEW's Fourier transform-based VIs follow. Thus, in the N-element array of discrete Fourier transform values X_k that such VIs output, the elements with indices from 0 to $N/2$ contain the sequence of X_0 to $X_{N/2}$ (affiliated with positive frequencies in the range of $0 \leq f \leq +f_{nyquist}$), and elements with indices from $N/2 + 1$ to $N - 1$ contain the sequence of $X_{-N/2+1}$ to X_{-1} (affiliated with negative frequencies in the range of $-f_{nyquist} < f < 0$).

12.5 THE FAST FOURIER TRANSFORM

Since the mid-1960s, a clever algorithm to calculate the discrete Fourier transform (Eq. [12.11]) in an efficient manner has been widely used. This algorithm, known as the *fast Fourier transform (FFT)*, exploits inherent symmetries in the calculation of Eq. [12.11] that arise because of the periodic way in which the original data were taken (N evenly spaced samples taken with time spacing Δt). Where the "brute-force" evaluation of Eq. [12.11] would require on the order of N^2 complex multiplication operations, the FFT manages this task with only $N \log_2 N$ such operations. The savings in time are immense, especially when N is large. As an example, consider a digital oscilloscope configured to produce a data set of 1024 points (a typical mode of operation). To analyze the frequency spectrum of this set, the FFT algorithm will be 100 times faster than the straightforward evaluation of Eq. [12.11]. For a very large data set of 10^6 points, the FFT is 50,000 times faster!

In addition to the assumption of evenly spaced data points, there is one extra restriction to the use of FFTs. In deriving a FFT algorithm, one must assume that the size of the data set is a power of 2 (i.e., $N = 2^m$, where $m = 1, 2, 3, \ldots$) to obtain maximum calculational efficiency. If N is not equal to a power of 2, algorithms (called *discrete Fourier transforms*, or *DFTs*) exist to evaluate the discrete Fourier transform, but with a speed much slower than that of the FFT.

12.6 FREQUENCY CALCULATOR VI

The LabVIEW FFT icon that you will implement observes the following convention: The discrete Fourier transform values X_k are output as an N-element array, whose elements are indexed 0 to $N-1$, which are associated with N evenly spaced frequencies (where $\Delta f = f_s/N$) in the range of $-f_{nyquist} < f \leq +f_{nyquist}$. The array elements with indices from 0 to $N/2$ contain the sequence of X_0 to $X_{N/2}$ associated with the positive frequencies from 0 to $+f_{nyquist}$, and the elements with indices from $N/2+1$ to $N-1$ contain the sequence of $X_{-N/2+1}$ to X_{-1} associated with the negative frequencies from $-f_{nyquist} + \Delta f$ to $-\Delta f$.

Let's first write a program, to be used later as a subVI, that calculates the frequency associated with each array element of the FFT output. Start by creating the following front panel. Use **Select a Control...** in the Controls Palette to obtain the **Digitizing Parameters** control cluster in **YourName\Controls** (if you haven't created this cluster yet, see Section 4.10, "Control and Indicator Clusters"). Create two Array Indicators (select an **Array** shell first and then place a **Numeric Indicator** inside) and label one **All Frequencies** and the other **Positive Frequencies**. Then place these two Array Indicators in a **Cluster** shell named **Frequency Cluster** (with **All Frequencies** and **Positive Frequencies** being the index-0 and index-1 element, respectively). Assign the connector pane's terminals consistent with the Help Window shown and design an icon. Save this VI under the name **Frequency Calculator** in **YourName\Chapter 12**.

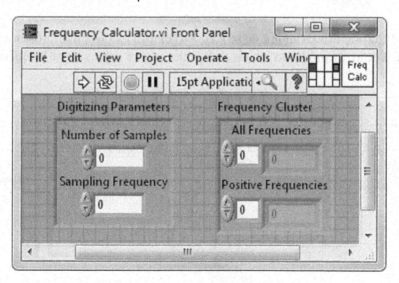

485

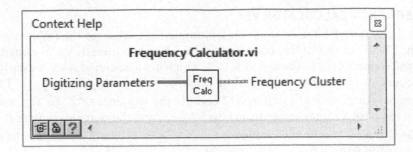

Now write a block diagram that creates an N-element array of frequencies f_k, indexed from 0 to $N - 1$. We desire the frequency at index k to be the f_k associated with the X_k at that same index in the FFT icon's output array. The separation between adjacent frequencies f_k is $\Delta f = f_s/N$, where f_s is the sampling frequency. From the above paragraphs' discussion, we see that the f_k can be calculated as follows:

$$f_k = \begin{cases} k\,\Delta f & k = 0, 1, 2, \ldots, \dfrac{N}{2} \\[2mm] (k - N)\,\Delta f & k = \dfrac{N}{2} + 1, \ldots, N - 1 \end{cases} \qquad [12.17]$$

Program this formula using a **Case Structure** as shown in the following illustration. The array produced by this code is the one we will name **All Frequencies**.

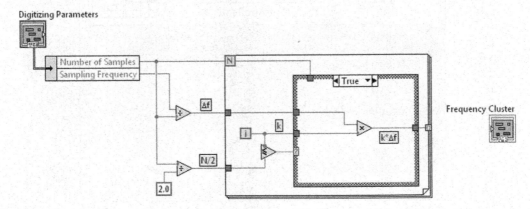

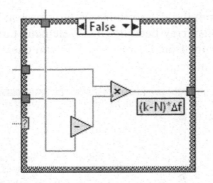

Later in the chapter, we will need an array of only the positive frequencies, including 0, but excluding the Nyquist frequency. This array, which will be called **Positive Frequencies**, is simply the first $N/2$ elements of **All Frequencies**. Use the **Array Subset** icon (found in **Functions>>Programming>>Array**) to create this array and complete the block diagram as shown below.

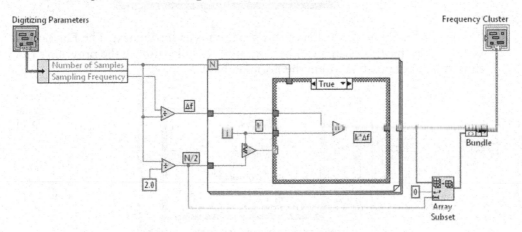

Return to the front panel. Input the values *1024* and *2000* to the **Number of Samples** and **Sampling Frequency** controls, respectively. Run the VI. If it functions properly, the positive frequencies of 0 to +1000, in increments of 2000/1024 ≅1.95, will appear in the **All Frequencies** array at the indices 0 through 512. The negative frequencies of (approximately) –998.05 to –1.95 will reside at the array elements with indices from 513 through 1023. The **Positive Frequencies** array will contain the positive frequencies of 0 to +998.05 at indices 0 through 511.

Alternatively, Eq. [12.17] can be coded using a Mathscript Node as shown below. Note that, unlike LabVIEW arrays, Mathscript-array indexing begins with *1*, rather than *0*. So, for example, within the Mathscript Node, the *N*-element

f_all array runs from index 1 to index N. However, at the f_all output of the Mathscript Node, this array becomes an N-element LabVIEW-formatted array, that is, its indexing runs from 0 to $N - 1$. You can check that this is so using the **All Frequencies** array indicator.

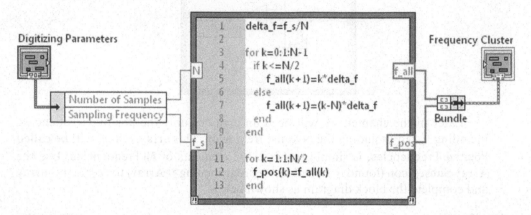

The Formula Node version of this program is as shown next. The For Loop is used so that the array size N can be a variable specified during the program's run-time, a feat not possible within the Formula Node.

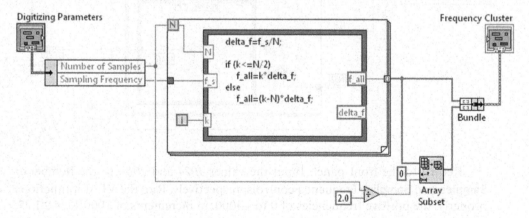

12.7 FFT OF SINUSOIDS

To get a handle on how FFT signal processing works, let's synthesize some data consisting of sinusoidal waves with known complex-amplitudes A_k and frequencies f_k, feed these data to the FFT algorithm, divide the resultant X_k values by N to produce the spectrum of A_k (see Eqs. [12.14] and [12.15]), and determine whether this spectrum matches the known input. We will use our old friend **Waveform Simulator** (which you created in Chapter 4 and saved in **YourName\Chapter 4**) to synthesize an N-element array of data and our new friend **Frequency Calculator** to produce the array of all f_k values (which is named **All Frequencies**).

Construct the front panel shown next, saving it under the name **FFT of Sinusoids** in **YourName\Chapter 12**. You can obtain the control clusters in **YourName\Controls** using **Controls>>Select a Control…** or else through the **Create>>Control** pop-menu option, once you place **Waveform Simulator** on the block diagram. We will display the complex-amplitudes A_k on the **XY Graph**, so label its x- and y-axes **Frequency** and **Complex-Amplitude**, respectively. Since the A_k are complex numbers, we will display the real and imaginary parts of the A_k via two plots. Using the **Plot Legend**, set up these two plots, and label them **Re(A)** and **Im(A)**. Distinguish these plots from each other by color (for a review of how to place multiple plots on an XY Graph, see Sections 8.4 and 11.8).

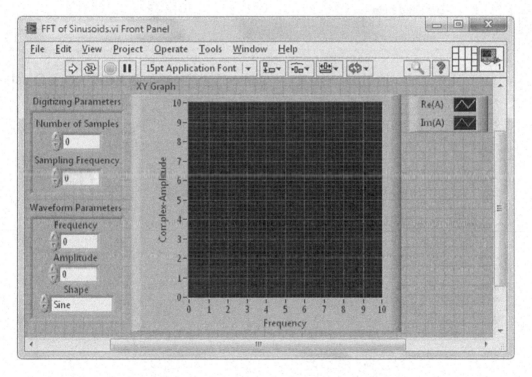

The discrete Fourier transform values X_k output by the FFT algorithm are complex numbers. LabVIEW includes a complex number data representation, so X_k is output from the **FFT.vi** icon as this data type. The Help Window for **FFT.vi** (found in **Functions>>Signal Processing>>Transforms**) is shown next. In our program, the data array we will input at **x** is purely real. Since the FFT icon is polymorphic, it will adapt itself automatically to receive this purely real input. Alternatively, if we input a complex-valued array at **x**, the icon would again automatically adapt. The complex-valued X_k are output at **FFT {x}**.

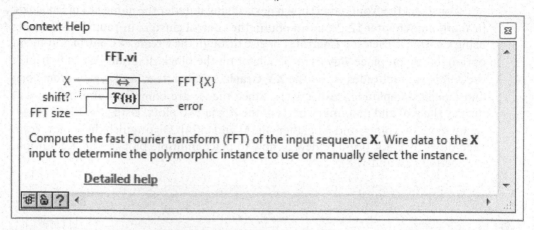

Now write the block diagram as follows. The **Complex To Re/Im** icon is found in **Functions>>Programming>>Numeric>>Complex.**

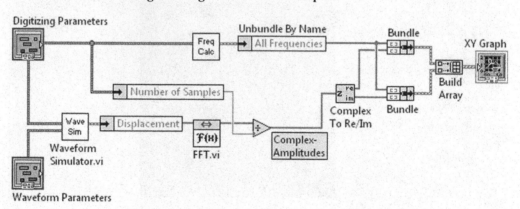

Return to the front panel of **FFT of Sinusoids** and save your work.

12.8 APPLYING THE FFT TO VARIOUS SINUSOIDAL INPUTS

Using the **Waveform Parameters** control, program your VI to produce a *Cosine* function with an amplitude of *4.0* and frequency of *250* Hz. *Note:* I will use the word "amplitude" to mean the peak-height (relative to the zero level) of a sinusoidal function. For the **Digitizing Parameters**, choose *1024* and *2000* for **Number of Samples** and **Sampling Frequency**, respectively. Remember that the FFT algorithm requires the number of samples in your input data to be a power of 2, so we have chosen $N = 2^{10} = 1024$. Run your program. The VI will find the real and imaginary parts of the complex-amplitude A_k for positive and negative frequencies f_k up to the Nyquist frequency $\pm f_{nyquist} = \pm f_s/2$. For our choice of $f_s = 2000$ Hz, the Nyquist frequency is 1000 Hz, so the 250 Hz cosine oscillation frequency that you have input will fall within the FFT's range of detection.

If all works well, the VI should output the Fourier transform of your cosine input in the following form: The real part of the complex-amplitude A is equal to zero at all frequencies f, except $A = +2$ at $f = \pm 250$ Hz. The imaginary part of A is zero at all frequencies f.

Why is this output the correct representation for the cosine input? Shouldn't the amplitude A be 4, not 2? To understand your result, plug the output back into Eq. [12.14]. Then

$$x_j \approx \sum_{k=0}^{N-1} A_k e^{i2\pi f_k t_j} = (+2)\, e^{i2\pi(250)t_j} + (+2)\, e^{i2\pi(-250)t_j}$$

$$= 4 \left[\frac{e^{i2\pi(250)t_j} + e^{i2\pi(-250)t_j}}{2} \right]$$

$$= 4\cos\left[2\pi(250)\, t_j\right]$$

where we have used the identity $\cos x = (e^{ix} + e^{-ix})/2$ for the last step. So we see that the output complex-amplitudes are indeed the correct representation for an input cosine function of amplitude 4.0.

Try a sine-wave input and see what you get. Using **Waveform Parameters**, program **FFT of Sinusoids** to produce a *Sine* function with an amplitude of *4.0* and frequency of *250* Hz, while keeping *1024* and *2000* for **Number of Samples** and **Sampling Frequency**, respectively. Now run the VI. You should find that the imaginary part of the complex-amplitude is $A = -2$ at $f = +250$ Hz, $A = +2$ at $f = -250$ Hz, and is zero elsewhere. The real part of A is zero at all frequencies. To see that this is the correct representation of your input, we again start with Eq. [12.14] and note the following:

$$x_j \approx \sum_{k=0}^{N-1} A_k e^{i2\pi f_k t_j} = (-2i) e^{i2\pi(250)\,t_j} + (+2i) e^{i2\pi(-250)\,t_j}$$

$$= 4 \left[\frac{e^{i2\pi(250)\,t_j} - e^{i2\pi(-250)\,t_j}}{2i} \right]$$

$$= 4 \sin\left[2\pi(250)\,t_j \right]$$

where we have used the facts that $i = -1/i$ and $\sin x = (e^{ix} - e^{-ix})/2i$.

There are two frequencies that behave differently from the rest: zero frequency and the Nyquist frequency. First, program **FFT of Sinusoids** with the constant *DC Level* (also called *zero-frequency*) function

$$x_j = 4.0$$

and then run the VI with **Number of Samples** and **Sampling Frequency** equal to *1024* and *2000*, respectively. Note that, rather than finding half of the amplitude at a positive and negative frequency as in the previously studied nonzero-frequency cases, the full amplitude (called the *DC component*) appears at the single frequency $f = 0$.

Now, program **FFT of Sinusoids** with the following function that oscillates at the Nyquist frequency:

$$x_j = 4.0 \cos\left[2\pi(1000)\,t_j \right]$$

and then run the VI again with **Number of Samples** and **Sampling Frequency** equal to *1024* and *2000*, respectively. If a mysterious diagonal line appears in the resulting FFT plot, remove it by popping up on the **Plot Legend** and deselecting the "connect-the-dots" mode of plotting in the **Interpolation** option. Similar to the zero-frequency case, you'll discover that the full oscillatory amplitude appears at the single frequency $f_{nyquist}$.

Here is a synopsis of our last observation: The cosine function, which is used to generate the data set, takes on its peak value when the 2000 Hz sampling process begins at time $t = 0$. The 1000 Hz Nyquist-frequency waveform then is sampled twice each cycle, producing a data set that has the following sequence—peak value, trough value, peak value, trough value, and so on. We find that by applying the

FFT algorithm of this data set, the correct value for the amplitude of oscillation (i.e., 4.0) is determined at $f_{nyquist}$.

Unfortunately, by exploring the Nyquist-frequency situation a little further, a troublesome problem emerges. Try programming **FFT of Sinusoids** to produce a Nyquist-frequency oscillation, but this time one that follows the sine function

$$x_j = 4.0 \sin\left[2\pi\,(1000)\,t_j\right]$$

and then run the VI with **Number of Samples** and **Sampling Frequency** equal to *1024* and *2000*, respectively. You'll find that the FFT detects no amplitude of oscillation at $f_{nyquist}$ in this case. Can you explain why the 1000 Hz sinusoidal oscillation is invisible in this data set?

If interested, you can try including a phase constant δ (in radians) in the sine function's argument to produce a data set that takes on neither its peak value nor its zero-value (like the cosine and sine function, respectively) at time $t = 0$. To create the desired function, open **Waveform Simulator** by double-clicking on its icon on the block diagram of **FFT of Sinusoids**. Then program Waveform Simulator's **User-Defined** function to be

$$x_j = 4.0 \sin\left[2\pi\,(1000)\,t_j + \delta\right]$$

Run **FFT of Sinusoids** with **Shape** selected as *User-Defined* in the **Waveform Parameters** control cluster. Does the FFT yield the correct amplitude of oscillation (i.e., 4.0) for any nonzero value of δ?

To summarize our Nyquist-frequency findings, in taking the FFT of a discretely sampled data set, the resultant value for the amplitude at $f_{nyquist}$ is only accurate when the Nyquist-frequency oscillation is "in phase" with the sampling process. Since one cannot guarantee that this will be the case in an actual experimental situation, the complex-amplitude at $f_{nyquist}$ should always be viewed with suspicion. Thus, in our work below, we will ignore the Fourier component determined at the Nyquist frequency.

Next, program the **User-Defined** function on **Waveform Simulator**'s block diagram to be a function that includes two simultaneous sinusoidal oscillations, along with a DC level, such as

$$x_j = 8.0 + 4.0 \sin\left[2\pi\,(250)t_j\right] + 6.0 \cos\left[2\pi\,(500)t_j\right]$$

and then run **FFT of Sinusoids**. Do you understand the output?

Finally, what happens when the input contains a sine and cosine oscillation, both at the same frequency? Run **FFT of Sinusoids** with the **User-Defined** function

$$x_j = 4.0 \sin\left[2\pi(250)t_j\right] + 6.0 \cos\left[2\pi(250)t_j\right]$$

You may have to change, for example, the **Point Style** of each plot to view the output accuracy. Do you understand the real and imaginary plots?

12.9 MAGNITUDE OF THE COMPLEX-AMPLITUDE

To this point, we have been representing a complex number as the sum of a real and an imaginary part, that is, $z = x + iy$. However, as you know, the complex number z may also be represented as a *magnitude r* times a *phase factor $e^{i\phi}$*, where $r = \sqrt{x^2 + y^2}$ and $\tan\phi = y/x$. Let's say that the oscillation of a system at frequency f is, as in the above example, the composite of a sine and cosine function with (real) amplitudes B and C, respectively:

$$x = B\sin(2\pi f t) + C\cos(2\pi f t) \qquad [12.18]$$

Then, defining $\theta = 2\pi f t$ and using the complex exponential representation of the sine and cosine functions,

$$x = B\frac{e^{i\theta} - e^{-i\theta}}{2i} + C\frac{e^{i\theta} + e^{-i\theta}}{2}$$

$$= \frac{1}{2}(C - iB)e^{i\theta} + \frac{1}{2}(C + iB)e^{-i\theta}$$

From Figure 12.1, we see that $(C + iB) = re^{+i\phi}$ and $(C - iB) = re^{-i\phi}$, where $r = \sqrt{B^2 + C^2}$ and $\phi = \tan^{-1}(B/C)$. Thus,

$$x = \frac{1}{2}\left(re^{-i\phi}\right)e^{i\theta} + \frac{1}{2}\left(re^{+i\phi}\right)e^{-i\theta}$$

$$= r\frac{e^{i(\theta-\phi)} + e^{-i(\theta-\phi)}}{2} \qquad [12.19]$$

$$= r\cos(\theta - \phi)$$

Then, equating Eq. [12.18] with Eq. [12.19] and remembering $\theta = 2\pi f t$, $r = \sqrt{B^2 + C^2}$, and $\phi = \tan^{-1}(B/C)$, we find that

$$x = B\sin(2\pi f t) + C\cos(2\pi f t) = \sqrt{B^2 + C^2}\cos(2\pi f t - \phi) \qquad [12.20]$$

Equation [12.20] tells us that a sine and cosine oscillation at the frequency f combine to produce a single phase-shifted cosine oscillation at frequency f with a net amplitude equal to the vectorial sum of the sine and cosine amplitudes.

Let's then write a VI called FFT (Magnitude Only) that, given an input data set x_j, finds the net amplitude of oscillation at the frequencies f_k, but ignores the phase information (i.e., neglects whether this oscillation is in the form of a pure sine wave or a pure cosine wave or a composite of sine plus cosine). With FFT of Sinusoids open, select File>>Save As..., and create a new VI named FFT (Magnitude Only) in YourName\Chapter 12. Use the Plot Legend to request only a single plot—labeled Mag(A)—on the XY Graph. Relabel the XY Graph's y-axis as Magnitude of Complex-Amplitude.

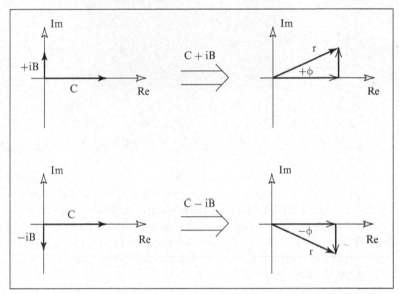

FIG. 12.1 Graphical representation of $C + iB$ and $C - iB$

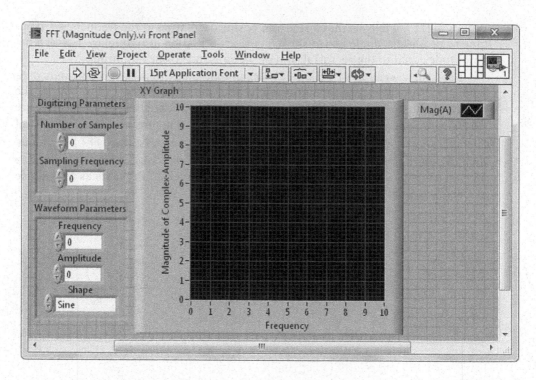

Now modify the block diagram so that, using the complex-amplitude's real and imaginary parts, its magnitude is calculated and displayed. Replace the **Complex To Re/Im** icon with **Complex To Polar** (found in **Functions>>Programming>> Numeric>>Complex**). The simple way to make this swap is by popping up on **Complex To Re/Im** and then selecting the **Replace** menu item. It will be obvious what to do from that point. Complete the modification necessary to produce the block diagram shown here.

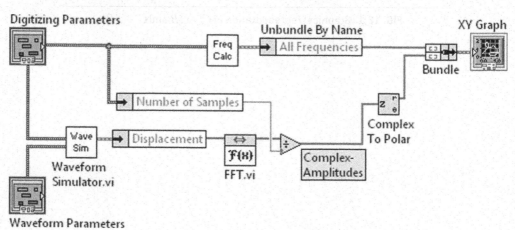

Program the **User Defined** function of **Waveform Simulator** to calculate the sum of a sine and cosine function, both having frequency $f = 250$ Hz, with amplitudes 3.0 and 4.0, respectively:

$$x = 3.0 \sin\left[2\pi(250)t\right] + 4.0 \cos\left[2\pi(250)t\right]$$

From Eq. [12.20], the above waveform should be equivalent to

$$x = \sqrt{3^2 + 4^2} \cos\left[2\pi(250)t - \tan^{-1}(3/4)\right]$$

Run **FFT (Magnitude Only)** with **Number of Samples** and **Sampling Frequency** equal to *1024* and *2000*, respectively. Is the output representative of a single 250 Hz cosine curve with net amplitude $\sqrt{3^2 + 4^2} = 5$?

The Fourier transform method represents a function as the linear combinations of the complex exponential basis set. Because $\cos(2\pi f t) = (e^{i2\pi f t} + e^{-i2\pi f t})/2$, a cosine function's amplitude is equally divided between the complex-amplitudes of the positive and negative frequency basis functions $e^{\pm i2\pi f t}$. Two exceptions to this "equal division of amplitude" occur, of course, for the basis functions at zero frequency (i.e., $e^{i2\pi(0)t} = 1$) and at the Nyquist frequency $f_{nyquist}$. When displaying the spectrum of a given input data set, this symmetry of complex-amplitudes in frequency space is commonly exploited by simply doubling the magnitude of the positive frequency's complex-amplitude and only plotting the positive-frequency axis. Such a plot displays the net amplitude of sinusoidal oscillation at each frequency $|f_k|$ directly, while phase information is completely neglected.

Modify the block diagram of **FFT (Magnitude Only)** to plot the input data's spectrum in the manner described above. Let's ignore the Fourier component at $f_{nyquist}$ because, as we saw previously, its value is sometimes suspect (that's why we excluded this frequency when constructing the **Positive Frequencies** array in the **Frequency Calculator VI**). The output of the topmost **Unbundle By Name** can be toggled from **All Frequencies** to **Positive Frequencies** by using the ✋.

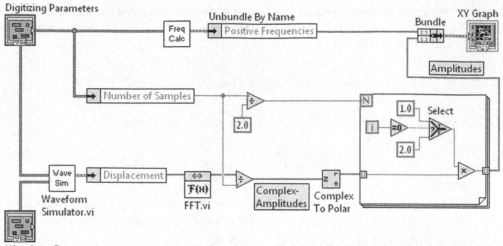

Switch to the front panel. Since the y-axis values are now equivalent to the net amplitude of sinusoidal oscillation at each (positive) frequency f_k, relabel the XY Graph's y-axis and Plot Legend as **Amplitude**. Run the VI to verify that it performs correctly.

Once satisfied that the program functions properly, return to the block diagram. The diagram is becoming fairly busy, so it's time to make a subVI. Move the **Unbundle By Name** icon for **Number of Samples** over slightly, then use the ⟨ to highlight the group of block-diagram objects in the lower-right portion of the diagram that performs the amplitude calculation.

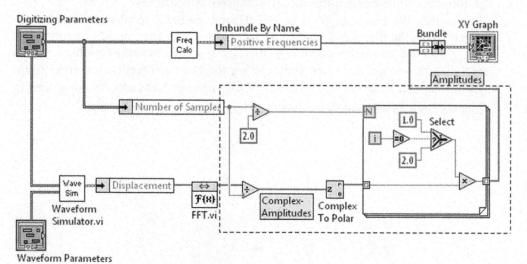

Next, use **Edit>>Create SubVI** to create a new VI. Configure its front panel as you like, then assign its terminals and create an icon consistent with the Help Window shown below. Save this VI under the name **Amplitude Calculator** in **YourName/Chapter 12** as you close it.

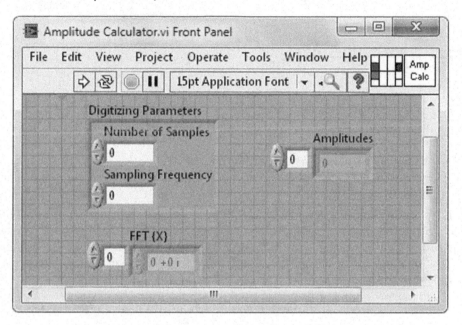

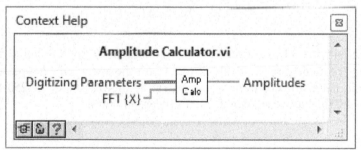

When returned to the block diagram of **FFT (Magnitude Only)**, organize the objects as shown below. Save your work.

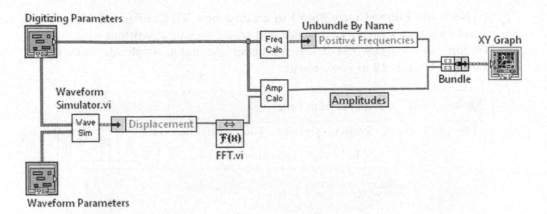

Run your VI to verify that it performs correctly.

12.10 OBSERVING LEAKAGE

Let's revisit the good news/bad news mystery we observed at the beginning of the chapter. But, first, using the front-panel **Plot Legend**, make the individual plotted points visible by changing the **Point Style**. Also, get access to the XY Graph's Zoom Tool by popping up on the XY Graph's interior and select **Visible Items>>Graph Palette**.

Okay, program **FFT (Magnitude Only)** with the function

$$x = 4.0 \cos\left[2\pi\left(250\right)t\right]$$

and then run the VI using the values *1024* and *2000* for **Number of Samples** and **Sampling Frequency**, respectively. Zoom in on the spectrum. It's perfect, with the amplitude equal to zero everywhere except at the frequency of 250 Hz, where it equals +4.0, just as we expected. That's the good news.

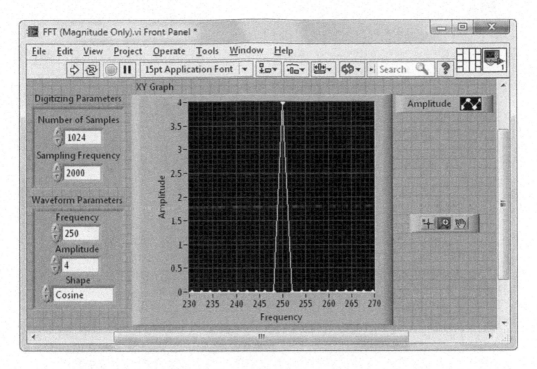

Now, the bad news. Repeat the above procedure after programming FFT (Magnitude Only) with the following function:

$$x = 4.0\cos\left[2\pi\left(249\right)t\right]$$

Before running the VI, you may wish to first turn off autoscaling on the x-axis by popping up on the XY Graph's interior and deselecting **X Scale>>AutoScale X**.

As we saw at the chapter opening, the resulting spectrum no longer appears as a "delta-function" spike with a height of 4.0, but instead is a broadened peak of maximum height less than 4.0. This spectrum seems to be telling us that the input data are oscillating at several different frequencies in the approximate range of 230 to 270 Hz. However, we know that this is incorrect. We input a pure sine wave with the exact frequency of 249 Hz.

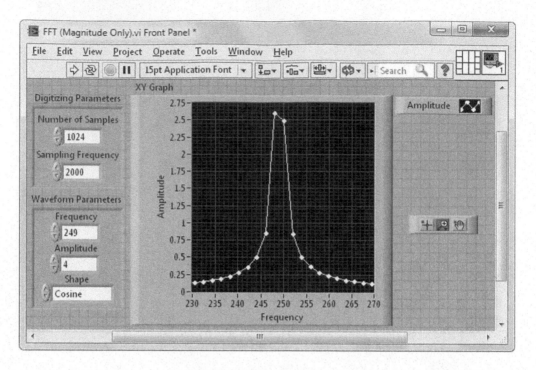

That is the mystery. Why does our FFT-based VI work perfectly for an input frequency of 250 Hz, but fails when that frequency is changed to 249 Hz? Well, here's the resolution to this paradox: For most of this chapter, I've been sneaky and always, with the exception of $f = 249$ Hz, asked you to generate input data using sine and cosine functions that oscillate at one of the N discrete frequencies f_k. For example, via Eq. [12.7], when acquiring 1024 equally spaced data samples at a sampling rate of 2000 Hz, the FFT algorithm produces complex-amplitudes A_k at the N frequencies f_k given by

$$f_k = k\left(\frac{2000\,\text{Hz}}{1024}\right) \quad k = -511, \ldots, 0, \ldots, +512$$

You can verify that 250 Hz and 500 Hz are both one of the f_k. In particular, they are f_{128} and f_{256}, respectively.

A way of summarizing our above observations is this: If you input a sinusoidal oscillation at exactly one of the frequencies f_k—for example, $f_{128} = 250.00$ Hz or $f_{256} = 500.00$ Hz—the FFT algorithm will perfectly produce the frequency spectrum of that data (i.e., a delta-function spike of the correct height at the correct frequency). However, if instead one inputs a sinusoidal oscillation at a frequency f not equal to one of the f_k—such as 249 Hz, which falls between $f_{127} = 248.05$ Hz and $f_{128} = 250.00$ Hz— the resulting spectrum is imprecise. In fact, it is as if its spectral amplitude, which rightfully should be a spike at frequency f, has diffused from this central point and distributed itself among the neighboring f_k. This smearing of spectral information—termed *leakage*—is an artifact of the finite number N of data samples contained in our discretely sampled data set. You can prove this for yourself by rerunning **FFT (Magnitude Only)** programmed to calculate a 249 Hz cosine wave, first making **Number of Samples** equal to *512*, then *1024*, then *2048*, followed by *4096*, and so on (it will be especially helpful here to have autoscaling on the *x*-axis turned off). You will find that as N increases, the spectrum becomes much more delta-function-like.

For those interested in more analytical details, a mathematical description of the leakage effect is given in Appendix B.

12.11 WINDOWING

The finite sample size of a data set (which causes leakage) can be viewed as a windowing effect. That is, when we acquire a finite number N of discretely sampled points for FFT spectral evaluation, we are in effect observing an infinite set of data d_j (where $j = -\infty, \ldots, -1, 0, +1, \ldots, +\infty$) through a rectangular viewing window in time. Defining the rectangle window function $w(t_j)$ to be zero at all times $t_j = j\Delta t$, except during the "data-viewing" time interval from $j = 0$ to $j = N - 1$, when it is equal to 1, our finite set of N sampled data points x_j is then given by the product $x_j = d_j w_j$. This idea is illustrated in Figure 12.2.

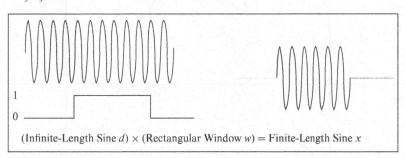

(Infinite-Length Sine d) × (Rectangular Window w) = Finite-Length Sine x

FIG. 12.2 Finite-length sine wave produced by rectangular viewing window

503

Appendix B gives a mathematical analysis of the situation illustrated in Figure 12.2, which leads to the following conclusion: The discontinuous jumps (i.e., the sharp turn-on and turn-off) at the edges of the rectangle window function are the chief cause of the broad spectral leakage we observed in the previous section.

But, as we saw at the beginning of the chapter, there is a way to mitigate this problem. Thanks to the power of your high-speed computer, you have the option of windowing your finite-length data set with some function other than a rectangle. Once the rectangular-windowed data set $(x_{rectangle})_j$ has been collected into your computer, all you have to do is construct a more desirable window w_j in software and then form the product $x_j = w_j(x_{rectangle})_j$. Then leakage can simply be suppressed by choosing a software window with a much less abrupt turn-on and turn-off than the rectangle window. For example, the *Hanning* window, which is defined to be

$$\left(w_{hanning}\right)_j = 0.5\left[1 - \cos\left(\frac{2\pi j}{N}\right)\right] \quad j = 0,1,\dots,N-1 \quad\quad [12.21]$$

is a popular choice for such a software window. The "smooth-edged" Hanning and "abrupt-edged" Rectangle windows are plotted in Figure 12.3 for comparison.

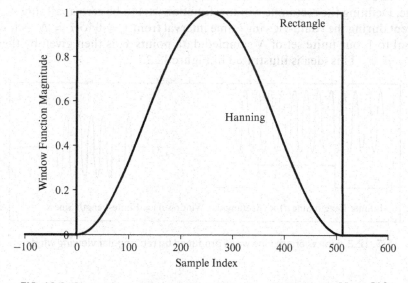

FIG. 12.3 Comparison of Rectangle and Hanning windows with $N = 512$

Let's try windowing our data set in **FFT (Magnitude Only)** and see how this improves things. To window the **Waveform Simulator**–produced data set, you will use **Scaled Time Domain Window.vi**, found in **Functions>>Signal Processing>>Windows**. Its Help Window is shown next.

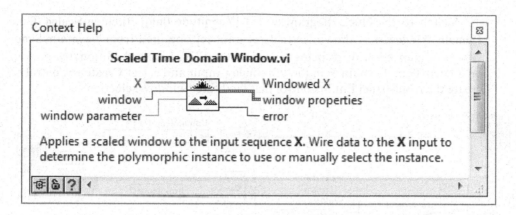

For this icon, you provide the "to-be-windowed" data array at the x input and select the desired window at **window** using an Enumerated Type ("Enum") control. The resultant array (equal to the product of the data and the window) is output at **Windowed x**. The **window properties** cluster is a two-element bundle of the selected window's *equivalent noise bandwidth (ENBW)* and *coherent gain*. These window parameters are important for use in some calculations, as will be shown in a few minutes.

When using **Scaled Time Domain Window.vi**, the user must choose one window, from among the approximately 20 available, to be applied to the input data array. Each window has its own properties such as central-lobe width and turn-on/turn-off rate, which are designed to optimize its use in particular applications. The Hanning window is a satisfactory choice in most situations. This window suppresses spectral leakage, allowing frequencies within the input signal to be well resolved. Table 12.1 lists some of the commonly used windows, along with the applications for which they are most suitable.

TABLE 12.1 Commonly Used Windows

Window	Optimal Application
Hanning	General-purpose applications
Hamming	Resolution of closely spaced sinc waves
Flat Top	Accurate amplitude measurement of isolated frequency
Kaiser*	Resolution of two closely spaced frequencies with widely differing amplitudes
Rectangle	Resolution of two closely spaced frequencies with almost equal amplitudes

*This window's central-lobe width is controlled by value input at **window parameter**.

Switch to the block diagram of FFT (Magnitude Only). Include **Scaled Time Domain Window.vi** as shown, so that the data set obtained from **Waveform Simulator** is windowed prior to being passed to the **FFT.vi** icon. Additionally, pop up on Scaled Time Domain Window.vi's **window** input and select **Create>>Control** to create the front-panel Enum control that facilitates window selection.

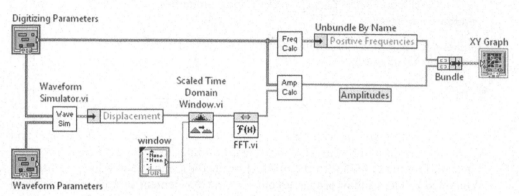

Switch to your front panel, find the Enum control labeled **window** that now appears there, position it aesthetically, and then save your work.

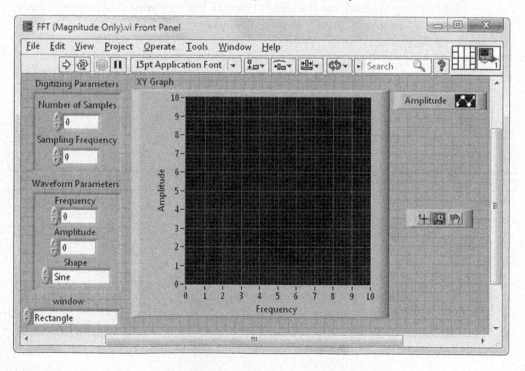

One of the nice benefits derived from having implemented **Create>>Control** to produce the front-panel **window** control is that, while creating it, LabVIEW automatically loads this Enum with the sequence of text messages that label the available options at Scaled Time Domain Window.vi's **window** input. Using the 🖑, review the sequence of selections offered by the **window** control.

Let's explore the positive effects of windowing. Assign **Number of Samples** and **Sampling Frequency** as *1024* and *2000*, respectively, and program your VI to produce the waveform $4.0\sin\left[2\pi\left(249\ \text{Hz}\right)t\right]$. Zoom in on the expected peak by scaling the axes appropriately, say, *230* to *270* for the *x*-axis and *0* to *4* for the *y*-axis, and then pop up on the XY Graph and use its pop-up menu to turn off the *x*- and *y*-axes autoscaling options. Run the VI without the benefit of windowing by selecting *Rectangle* in the **window** control.

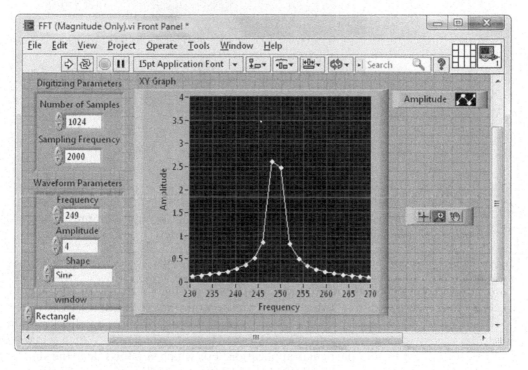

Now select the *Hanning* window and rerun the VI. Note the dramatic decrease in leakage through the use of a Hanning window.

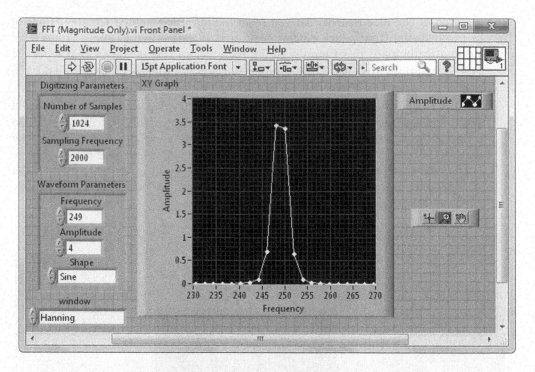

To demonstrate that certain windows perform better in a particular application, consider the challenge of resolving two sine waves of nearly the same frequency with widely differing amplitudes, a problem that might be encountered in, say, speech recognition research. Use the **User-Defined** option of **Waveform Simulator** to input the following waveform to **FFT (Magnitude Only)**:

$$x = 4.0 \sin\left[2\pi\left(249\right)t\right] + 0.1 \sin\left[2\pi\left(253\right)t\right]$$

With this input waveform (you will need to adjust the XY Graph's *y*-axis scaling appropriately), run **FFT (Magnitude Only)** several times, each time with a different choice for **window**. For the Kaiser window, you must supply a value (try *3*) at the **window parameter** input of Scaled Time Domain Window.vi. Which windows perform best for this particular application? For which windows does the smaller 253 Hz peak get obscured by the 249 Hz peak (because of large spectral leakage)?

Change the input waveform to two sine waves of nearly the same frequency with equal amplitudes:

$$x = 4.0 \sin\left[2\pi\left(249\right)t\right] + 4.0 \sin\left[2\pi\left(253\right)t\right]$$

Which window best resolves these peaks now (e.g., shows the deepest dip between the peaks)?

Finally, the *Flat Top* window is optimized to produce a peak with an amplitude that closely matches that of the input waveform. Input $4.0 \sin\left[2\pi\left(249\text{ Hz}\right)t\right]$ to **FFT (Magnitude Only)** with various choices for **window**. Is the amplitude of the spectral peak produced from this input closest to the value of 4.0 when using the Flat Top window (in comparison to when other windows are used)?

12.12 ESTIMATING FREQUENCY AND AMPLITUDE

In general, an input sinusoidal signal of frequency f and amplitude A will produce a spectral peak that is broadened over a range of f_k-values. For this situation, here is a final nifty trick for obtaining an accurate estimate for f and A. First, within the spectral peak, define the *peak power* at frequency f_k to be $P_k \equiv A_k^2$. Then the input sinusoidal frequency f that generated this finite-width spectral peak can be estimated by a weighted sum, with each observed f_k weighted by the peak power at that frequency. That is,

$$f \approx \frac{\sum\limits_k f_k P_k}{\sum\limits_k P_k} = \frac{\sum\limits_k f_k A_k^2}{\sum\limits_k A_k^2} \qquad [12.22]$$

where the sum is over the k-values that span the peak.

Also, the peak power $P \equiv A^2$ of the sinusoid with amplitude A can be estimated by

$$P \approx \frac{\sum\limits_k P_k}{\text{ENBW}} = \frac{\sum\limits_k A_k^2}{\text{ENBW}} \qquad [12.23]$$

where *ENBW* is the effective noise bandwidth of the window used in the analysis process that produced the peak. Once P is determined, then the sinusoidal amplitude is $A = \sqrt{P}$.

Write a VI called **Estimated Frequency and Amplitude** that implements Eqs. [12.22] and [12.23]. Store this program in **YourName\Chapter 12**. The front panel should look as shown next. The **Positive Frequencies** and **Amplitudes** arrays need to be the index-0 and index-1 elements, respectively, in the **Input Arrays** control cluster, to be consistent with how these arrays are bundled together on the **FTT (Magnitude Only)** block diagram. Pop up on the **Input Arrays** cluster and use **Reorder Controls In Cluster...** to make sure the two arrays are indexed properly. Assign the connector pane's terminals consistent with the Help Window shown.

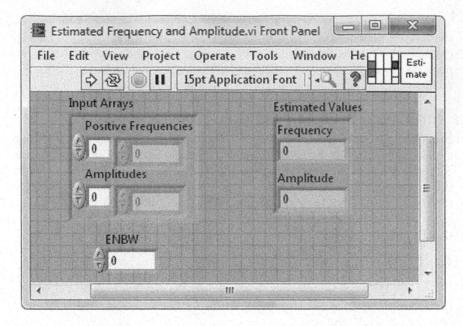

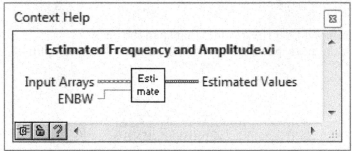

Now code the following block diagram. Because it is easy to do, we will perform the sums in Eqs. [12.22] and [12.23] over all k, but it is only necessary to include k-values for which the A_k are significantly nonzero, that is, k in the neighborhood of the peak. The MathScript Node version of the block diagram is shown next.

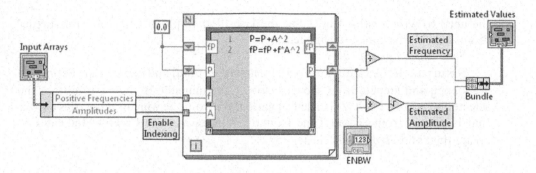

If using the Formula Node instead, the code within the For Loop is replaced by the following.

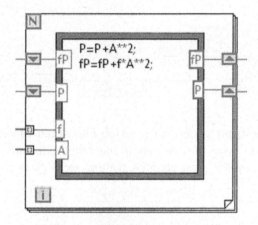

Above, we take advantage of a handy LabVIEW convenience: In addition to its ability to build an array at a loop's output, auto-indexing can be used to sequence through an array at a loop's input. Let's take the **Positive Frequencies** array as our example. Note that the wire emanating from the **Positive Frequencies** terminal is thick (denoting an array) outside the loop and then becomes thin (denoting a scalar) inside the loop. Why is this so? Auto-indexing is activated in a For Loop by default, so upon execution, the loop will sequentially input one element from **Positive Frequencies** each time it iterates. That is, on the first loop iteration when ⊞ equals 0, the element of **Positive Frequencies** with index 0 will be input; on the second iteration when ⊞ equals 1, the element of **Positive Frequencies** with index 1 will be input; and so on until the end of the array is reached. The same process happens for the **Amplitudes** array. As an added convenience, when auto-indexing is enabled on the N-element arrays entering a For Loop (note both arrays have the same size), LabVIEW automatically sets the loop's count terminal to N, thus eliminating the need to wire a value to 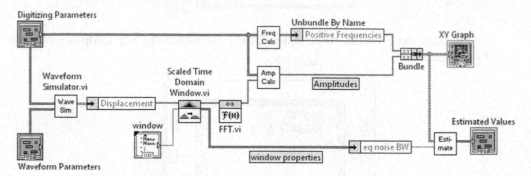. See the Use It! example of Chapter 3 for further discussion of this auto-indexing feature.

Save this VI as you close it.

Return to the block diagram of FTT (Magnitude Only) and incorporate Estimated Frequency and Amplitude as shown below. Use Unbundle By Name to obtain the selected window's ENBW (labeled eq noise BW) from the window properties cluster that is output from Scaled Time Domain Window.vi. Use Create>>Indicator to create the Estimated Values cluster.

Switch to the front panel and position Estimated Values nicely. Within this cluster, pop up on both the Frequency and Amplitude indicators and, using the Display Format... option, deselect Hide trailing zeros. Save your work.

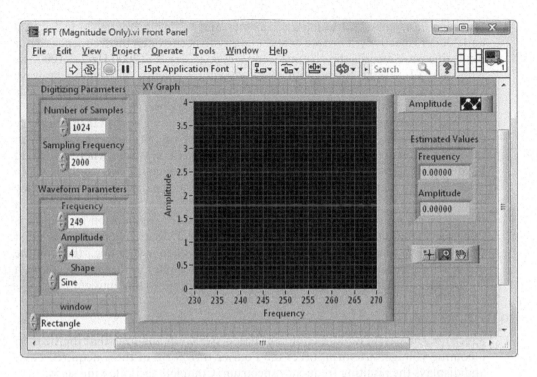

With the values for **Digitizing Parameters** and **Waveform Parameters** shown above, run the VI using various windows. I think you'll be very pleased with the estimated values of **Frequency** and **Amplitude**!

12.13 ALIASING

Finally, let's use **FFT (Magnitude Only)** to demonstrate aliasing, the phenomenon in which an incoming signal is digitally sampled too infrequently, resulting in a digitized waveform whose frequency is much lower than that of the actual signal. In Chapter 5, we found that when sampling a signal of frequency f at sampling frequency f_s, the digitized waveform will also have the frequency f only if $f \leq f_{nyquist}$, where $f_{nyquist} = f_s/2$ is the Nyquist frequency. If, instead, the frequency of the incoming signal is $f > f_{nyquist}$, then the digitized waveform will have frequency f_{alias} ($\neq f$) given by

$$f_{alias} = \left| f - nf_s \right| \quad n = 1, 2, 3, \dots \qquad [12.24]$$

where f_{alias} is in the range $0 \leq f_{alias} \leq f_{nyquist}$.

Program **FFT (Magnitude Only)** so that the input waveform is a sine wave of amplitude 4.0. Choose your favorite window, say, the general-purpose *Hanning*. In

Digitizing Parameters, set **Number of Samples** and **Sampling Frequency** to be *1024* and *2000*, respectively. Then we expect aliasing to occur when the frequency of our input sine wave exceeds the Nyquist frequency $f_{nyquist} = f_s/2 = 2000 \text{ Hz}/2 = 1000 \text{ Hz}$. Run the VI several times with the following sequence of input sine-wave frequencies: 1200 Hz, 1500 Hz, 1800 Hz, 2000 Hz, 2800 Hz, 3300 Hz, 7700 Hz. Does Eq. [12.24] correctly predict the aliased frequency you observe for each input?

In this exercise, we have the benefit of knowing the input signal's true frequency and so can use Eq. [12.24] to figure out why the digitizing process is producing an incorrect lower-frequency output. In a real experiment, however, where we would have no prior knowledge of the input signal's true frequency, we would have no other recourse than to assume that the digitized waveform has the same frequency as the incoming signal. That is, we would assume the signal is not being aliased in the digitizing process. To assure that this "no-aliasing" assumption is valid, it is imperative that an experimentalist first determine the digitizing system's Nyquist frequency (based on the known sampling rate) and then assure (e.g., through the use of a low-pass filter as in the Use It! example in Chapter 6) that no signals with frequencies greater than the Nyquist frequency are input to the digitizer.

DO IT YOURSELF

Spectrum Analyzer.vi Build a computer-based spectrum analyzer that digitizes an incoming voltage waveform, takes the Fourier transform of these acquired data, and displays the resulting frequency spectrum. Complete the following steps:

1. Write the program:
    With open FFT (Magnitude Only), use **File>>Save As...** to create a new VI and save it in the YourName\Chapter 12 folder.

    Switch to the block diagram of your VI. Delete the **Waveform Simulator** and **Waveform Parameters** icons. Place **DAQ Assistant** on your diagram and configure it to perform an analog input voltage operation on a particular AI differential channel of your DAQ device (e.g., *ai0*) with **Acquisition mode>> N Samples** selected. Include the option of digital triggering, if you wish. On your block diagram, expand **DAQ Assistant**'s terminals to include **data**, **number of samples**, and **rate**. Complete the necessary modifications so that **DAQ Assistant** receives the values of **Number of Samples** and **Sampling Rate** from the **Digitizing Parameters** cluster and then, after obtaining *N* data samples at the instructed sampling rate, passes the acquired data array to **Scaled Time Domain Window.vi**. Finally, enclose all of the block diagram code within a **While Loop** so that spectra are produced and plotted repeatedly until a front-panel **Stop Button** is pressed. Save your VI.

2. Operate the computer-based instrument:

    Using a function generator, input a sine-wave voltage with a frequency of about 100 Hz input into the analog input channel of your DAQ device that was programmed into DAQ Assistant (e.g., *ai0*). If you configured DAQ Assistant for digital triggering, also input your function generator's sync output to the DAQ device's appropriate pins. With **Number of Samples** and **Sampling Frequency** set to *1024* and *2000*, respectively, and your choice for **window**, run **Spectrum Analyzer** to see the spectral makeup of your voltage signal. How do the values of **Frequency** and **Amplitude** in **Estimated Values** compare with the known value (from the function generator's control panel) of your input signal? Try changing the input sine-wave frequency f to values within the range $0 \leq f \leq 1000$ Hz. Does the instrument perform well within this frequency range?

3. Explore interesting inputs:

    Increase the input sine wave's frequency beyond the Nyquist frequency to see the aliasing effect. Explain your observations using Eq. [12.24].

    Try inputting a square wave with a frequency of about 180 Hz and amplitude of 1 V (i.e., oscillates between –1 V and +1 V). Fourier analysis of a square wave of frequency f and unity amplitude predicts that this waveform is equivalent to the following sum of sine waves:

$$\frac{4}{\pi} \sum_{n \text{ odd}} \frac{1}{n} \sin[2\pi n f t] = \frac{4}{\pi} \left\{ \sin[2\pi f t] + \frac{1}{3} \sin[2\pi(3f)t] + \frac{1}{5} \sin[2\pi(5f)t] + ... \right\}$$

Note that this sum of sine waves only includes the odd harmonic frequencies f, $3f$, $5f$, and so on. With a square-wave input, is the output of your program consistent with this Fourier sum? Do you note any aliasing of the square wave's higher harmonics?

USE IT!

Digital Filtering In addition to the fast Fourier transform, LabVIEW offers another frequency-related signal-processing capability—digital filtering. Filtering is used to alter the frequency content of a signal. Based on the frequency range they transmit and block (called the *pass band* and *stop band*, respectively), filters are classified as the following types: *low pass*, *high pass*, *band pass*, and *band stop*. For example, a low-pass filter transmits low-frequency signals and blocks high-frequency signals and the shift between these two regimes takes place over a range of frequencies called the *transition band* (see the Use It! example in Chapter 6). The transition band begins near the filter's cutoff frequency f_c and its width is determined by the shape of the particular filter's response curve. For each filter type, there are various possible shapes of the response curve, where each shape is optimized for a particular signal-processing purpose. For example, the popular low-pass filter shapes ("topologies") are named Butterworth, Chebyshev, Elliptic, and Bessel. The following LabVIEW program allows you to explore the shapes of the response curve for various filters—that is, each filter's output amplitude vs. frequency, assuming an input signal with an amplitude of 1. This program inputs a sequence of unity-amplitude sine-wave signals with frequencies ranging from 100 Hz to 20 kHz to a filter of your choice and then records the amplitude of the sine-wave output by the filter. The result is plotted on the front-panel **XY Graph**.

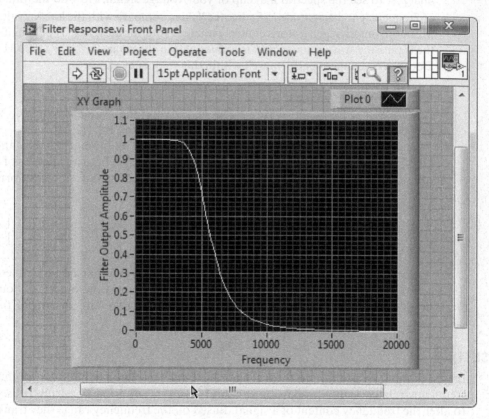

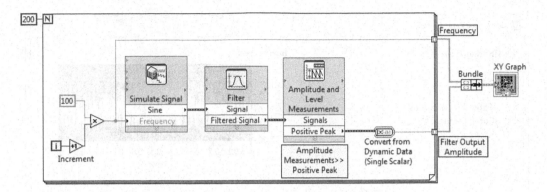

This VI implements three Express VIs, all found in subpalettes of **Functions>>Express**. The dialog window of **Simulate Signal** is programmed to produce a sine wave with an amplitude of 1 by selecting **Signal type** and **Amplitude** as *Sine* and *1*, respectively. Also, in the **Timing** box, choose **Samples per second (Hz)** to be *1000000* and, for **Number of samples**, check the **Automatic** box. In the dialog window for **Amplitude and Level Measurements**, select **Amplitude Measurements>>Positive Peak**. This icon will determine the amplitude of the sine-wave output by **Filter**. **Filter** applies a selected filter to the sequence of sine waves produced by the For Loop. Filter's dialog window is shown in the following illustration, where a Butterworth low-pass filter of order 5 with $f_c = 5000$ Hz is selected. After making this selection, Filter's dialog window was closed by pressing the **OK** button. The VI was then run, producing the filter's response curve shown on the above front panel.

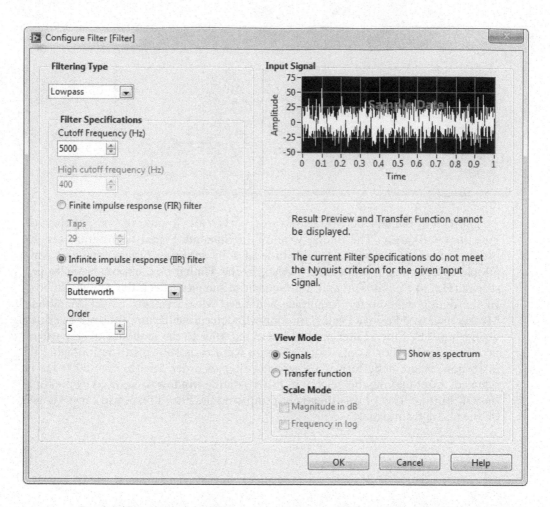

By double-clicking on **Filter**, you can change the selections within its dialog window and explore the frequency response of various filters. You will note the following features: the Butterworth low-pass filter is designed to have a "flat" plateau in its pass band (i.e., almost constant unity-amplitude output at low frequencies) and the transition region becomes narrower as the order is increased. Other filters are optimized to have a narrow transition region, but have "ripples" in their pass band and/or stop band. The Elliptic filter has the narrowest transition region of all of the available filters—the selling point of that filter.

PROBLEMS

For a problem that involves writing a new program, the problem statement begins with a suggested descriptive name (including the **.vi** extension) for the VI that you will write; icons needed for the VI may be found with the aid of **Quick Drop**. For a problem that involves the use of a VI already written as part of the chapter text, the problem's topic is given in bold at the beginning of the problem statement.

1. **Real versus Aliased Peaks** Use FFT (Magnitude Only) to investigate aliasing. Program the **User-Defined** option of **Waveform Simulator** to produce the digitized signal x of four incoming sine waves of frequency $f_1 = 25$ Hz, $f_2 = 70$ Hz, $f_3 = 160$ Hz, and $f_4 = 510$ Hz, each with an amplitude of 1 (i.e., $x = \sin[2\pi(25)t] + \sin[2\pi(70)t] + \sin[2\pi(160)t] + \sin[2\pi(510)t]$). Then, on the front panel of FFT (Magnitude Only), set **Number of Samples, Sampling Frequency, Shape**, and **window** equal to *4096, 100, User-Defined*, and *Rectangle*, respectively. Run FFT (Magnitude Only) and record the frequencies of the four peaks that you observe in the resulting spectrum. Identify each peak; that is, tell whether the peak is associated with a true frequency in the signal x or with an aliased frequency produced by the digitizing process. For each aliased frequency, use Eq. [12.24] to identify the true frequency (f_1, f_2, f_3, or f_4) that was aliased.

2. **Frequency Resolution** Use FFT (Magnitude Only) to explore the improved frequency resolution that results from long-time sampling of an input signal. Program the **User-Defined** option of **Waveform Simulator** under the assumption that the input signal x from an experiment consists of two closely spaced sine waves of frequency 399 Hz and 401 Hz, each with an amplitude of 4, that is, $x = 4\sin[2\pi(399)t] + 4\sin[2\pi(401)t]$. Then, on the front panel of FFT (Magnitude Only), set **Sampling Frequency, Shape**, and **window** equal to *2000, User-Defined*, and *Rectangle*, respectively.

   (a) Simulate sampling the input signal for longer and longer times by running FFT (Magnitude Only) with the following succession of values for **Number of Samples**—128, 256, 512, 1024, 2048, 4096—each time noting whether the 399-Hz and 401-Hz peaks are resolved in the resulting spectrum.

   (b) At what threshold value for **Number of Samples** do the two peaks begin to be resolved? From your knowledge of FFT theory (e.g., using the equations in Section 12.3), explain how this threshold value is determined by the various parameter values used in this investigation.

3. **FFT of Square Wave** Observe the Fourier transform of a square wave. Program **FFT (Magnitude Only)** with the following choices: **Number of Samples** and **Sampling Frequency** equal to *4096* and *5000*; and **Frequency, Amplitude,** and **Shape** equal to *180, 1,* and *Square*, respectively, to simulate a square wave with a frequency $f = 180$ Hz, which oscillates between 0 V and +1 V. Fourier analysis predicts that this waveform is equivalent to a DC component (of magnitude 0.5) plus a sum of sine waves as follows:

$$0.5 + \frac{2}{\pi} \sum_{n \text{ odd}} \frac{1}{n} \sin\left[2\pi n f t\right] = 0.5 + \frac{2}{\pi} \left\{ \sin\left[2\pi f t\right] + \frac{1}{3}\sin\left[2\pi(3f)t\right] + \frac{1}{5}\sin\left[2\pi(5f)t\right] + \ldots \right\}$$

Note that the sum of sine waves only includes the odd harmonic frequencies f, $3f, 5f, \ldots$, with amplitudes of $2/\pi, 2/3\pi, 2/5\pi, \ldots$. Finally, set **window** to *Flat Top* so that the Fourier peaks will have accurate amplitudes.

Run **FFT (Magnitude Only)** and record the frequency and amplitude of every peak that you observe in the resulting Fourier spectrum. Then identify the harmonic to which each peak corresponds, including those peaks that are present as a result of aliasing. Finally, comment on whether each observed peak has the amplitude expected from Fourier analysis.

4. **Frequency Doubling** Use **FFT (Magnitude Only)** to explore the *frequency doubling* (and *tripling*) phenomenon that occurs when a pure sine-wave signal is input to a nonlinear detector/amplifier (e.g., radio wave on diode detector, monochromatic light on nonlinear crystal). To investigate this effect, use the **User-Defined** option of **Waveform Simulator** to simulate the following process: A pure sinusoid $x(t) = A\sin(2\pi f t)$ is passed through a slightly nonlinear amplifier producing an output signal $a(x)$ given by

$$a(x) = \alpha x + \beta x^2$$

where α and β are constants with α significantly larger than β. If using a MathScript Node, remember that .^ is the element-wise power operator; for Formula Node users, take care to end each equation with a semicolon.

Run **FFT (Magnitude Only)** with some choice of f, α, and β ($f = 250$ Hz, $\alpha = 10$, and $\beta = 1$ might be a good place to start) to find the spectrum of $a(x)$.

(a) In terms of f, what frequencies do you observe in the spectrum?
(b) The x^2 term in $a(x)$ produces a product of sine waves that can be rewritten using the following trigonometric identity:

$$\sin\theta \sin\phi = \frac{1}{2}\left[\cos(\theta - \phi) - \cos(\theta + \phi)\right]$$

With the help of this relation, predict the frequencies that should appear in the spectrum of $a(x)$. Are these the frequencies that you observed?

(c) Explore what happens when the amplifier is even more nonlinear so that $a(x) = \alpha x + \beta x^2 + \gamma x^3$. In terms of f, what frequencies do you observe now in the spectrum? (In experimental optics, laser light of frequency f is commonly input on a nonlinear crystal to produce output light of frequency $2f$ and $3f$.)

5. **Sidebands** Observe the *sidebands* that exist in the frequency spectrum of an *amplitude modulated* (AM) wave as follows.

(a) On the block diagram of **FFT (Magnitude Only)**, replace the subVI **Waveform Simulator** (and its input control clusters) with **AM Wave** (written as Chapter 4's Do It Yourself project). Use **File>>Save As...** to create a new VI. Then, with *1024, 2000, 250,* and *50* for **Number of Samples, Sampling Frequency, Signal Frequency,** and **Modulation Frequency,** respectively, run **FFT (Magnitude Only)**. The spectrum that you observe consists of a central peak and "sidebands." In terms of f_{sig} and f_{mod}, what is the central peak's frequency and the spacing of the sidebands from the central peak?

(b) In an AM radio receiver, the AM wave received from an antenna is "demodulated" by passing it through a nonlinear diode detector. By making an appropriate modification of the block diagram of **AM Wave**, simulate this demodulation method by applying a nonlinear amplification $a(x) = x^2$ to your AM wave $x(t)$. Run your VI, and identify all of the frequencies in the resultant spectrum in terms of f_{sig} and f_{mod}. [Note that one of these frequencies is the modulation frequency f_{mod} and that by proper (low-pass) filtering it could be selected out for distribution to the radio's speaker. In optics, this technique is implemented to produce two optical laser frequencies separated by a small "radio-frequency" difference.]

6. **Lock-In Detection** Given an input consisting of the sum of many AC voltages with various frequencies, a *lock-in amplifier* can selectively measure the oscillatory amplitude of a single one of those frequencies. The theory of operation of this instrument can be simulated as follows. First, assume that the "input signal" y_{sig} to the lock-in consists of the sum of three sine waves with frequencies 100, 200, and 300 Hz and amplitudes 4, 6, and 8, respectively, that is, $y_{sig} = 4\sin\left[2\pi(100)t\right] + 6\sin\left[2\pi(200)t\right] + 8\sin\left[2\pi(300)t\right]$. Then, if a user programs the lock-in to measure the amplitude of the 200 Hz sine wave, the instrument internally produces a "reference signal" $y_{ref} = 2\sin\left[2\pi(200)t\right]$ and the product $x = y_{sig}y_{ref}$. Finally, the instrument filters x to determine the magnitude of the DC component present in x and outputs this DC value.

(a) Use **FFT (Magnitude Only)** to gain insight into the above-described theory. First, program the **User-Defined** option of **Waveform Simulator** to produce the given y_{sig} and y_{ref} and then the product $x = y_{sig} y_{ref}$. (If using the MathScript-based version of **Waveform Simulator**, remember that .* is Mathscript's element-wise multiplication operator.) Next, run **FFT (Magnitude Only)** with **Number of Samples** and **Sampling Frequency** equal to *1024* and *2000*, respectively, and **Shape** set to *User-Defined*. You will find that several frequencies, including $f = 0$, are present in x. How does the DC component compare with the amplitude of the 200 Hz sine wave present in y_{sig}?

(b) Change the reference signal to $y_{ref} = 2\sin[2\pi(300)t]$ and rerun **FFT (Magnitude Only)**. What is the resulting DC component equal to now?

(c) From the trigonometric identity $\sin\theta \sin\phi = \dfrac{1}{2}[\cos(\theta - \phi) - \cos(\theta + \phi)]$, can you explain the frequencies you observe in x?

7. **FFT Magnitude and Phase (Express).vi** With **FFT (Express)** open, use **File>>Save As...** to create a new VI that analyzes and plots both the magnitude and phase of an input sinusoidal signal. Add the second Waveform Graph to the front panel and modify the block diagram as shown below so that the two quantities are sent to the two front-panel Waveform Graphs.

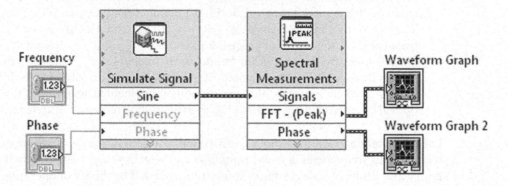

Open the dialog window for **Simulate Signal** and program the **Signal** box to produce a 250 Hz sine wave with an amplitude of 4 by selecting **Signal type**, **Frequency (Hz)**, **Phase (deg)**, and **Amplitude** as *Sine, 250, 0,* and *4,* respectively. In the **Timing** box, choose **Samples per second (Hz)** to be *2000* and **Number of samples** as *1024* with the **Automatic** box unchecked. Next, open the dialog window for **Spectral Measurements** and select **Magnitude (Peak)** and **Linear** in the **Selected Measurement** box. Also, choose **Window>>None** and **Phase>>Convert to degree**. Finally, modify the front panel as shown next so that data points are represented as dots on both plots. Choose **Frequency** equal to *250*.

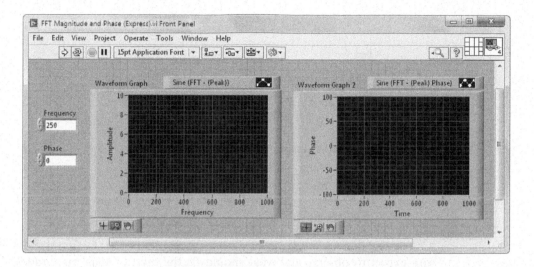

(a) Enter *0* in the **Phase** front-panel control to produce a sine signal input. Run your VI and determine the resulting values of amplitude and phase at $f = 250$ Hz for this input signal (the Zoom Tool will be handy here). Next, enter *90* in the **Phase** front-panel control to produce a cosine signal input; remember $\cos\theta = \sin(\theta + 90°)$. Rerun your VI to determine the amplitude and phase at $f = 250$ Hz for this input signal.

(b) In this chapter, you found that the complex-amplitude of a 250 Hz cosine and sine function at $f = 250$ Hz is $A = +2$ and $A = -2i$, respectively. Rewrite these two complex numbers in the form $A = |A|e^{i\theta}$ and thereby predict the phase θ expected for the FFT of a cosine and a sine function. How do these predicted values compare with the values you observed in part (a)?

(c) Enter *30* in the **Phase** front-panel control to produce an input signal $x = 4\sin[2\pi(250)t + 30°]$, and then run your VI. Is the observed value for the phase at $f = 250$ Hz in agreement with the value you expect?

8. **First-Order Anti-Aliasing Filter** Using a function generator, input a square-wave voltage with a frequency of about 180 Hz input into the first-order anti-aliasing filter (see the Use It! example of Chapter 6) shown in Figure 12.4 with $R = 15$ kΩ and $C = 0.015$ μF, whose cutoff frequency is given by $f_c = 1/2\pi RC$. Attach the output of this filter to the analog input channel of your DAQ device that was programmed into DAQ Assistant (e.g., *ai0*) on the **Spectrum Analyzer** block diagram you constructed in this chapter's Do It Yourself project. With **Number of Samples** and **Sampling Frequency** set to

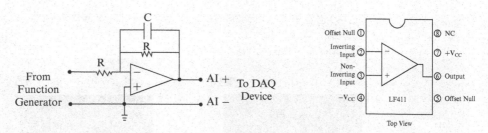

FIG. 12.4 Low-pass filter for Problem 8. Suggested op-amp is LF411.

1024 and *2000*, respectively, run **Spectrum Analyzer** to see the spectral makeup of your filtered voltage signal.

(a) Calculate f_c. Given this frequency, what harmonics of the square wave do you expect to observe and what should be the ratio of their amplitudes compared with the first harmonic (see the discussion in this chapter's Do It Yourself project)? Which harmonics do you actually observe and what are their amplitudes (in comparison with the amplitude of the first harmonic)? Comment on any deviation between your expectations and your observations.

(b) Are any of the observed peaks caused by aliasing of harmonics with frequencies higher than $f_{nyquist}$? If so, should you increase or decrease C in your filter to eliminate these peaks? (You might enjoy repeating this exercise with the eighth-order anti-aliasing filter given in the Use It! example of Chapter 6; make $f_c \approx 650$ Hz by choosing $C = 0.5$ nF.)]

CHAPTER 13

Data Acquisition and Generation Using DAQmx VIs

> Fundamental concepts of data acquisition are presented in Chapters 5 and 6. After reviewing that material, use the **Measurement & Automation Explorer (MAX)** to identify the DAQ device (e.g., model PCIe-6351 with the name *dev1*) connected to your computer. From the pinout diagram for your device, determine the pins associated with the analog input differential channel *ai0*, digital triggering (e.g., PFI 0, if your device is so equipped), analog output channel *ao0*, and relevant grounds (e.g., A GND and D GND). Have appropriate cabling on hand to connect to these pins as well as a DC voltage source, function generator, voltmeter, and oscilloscope.

13.1 DAQMX VI BASICS

Since your work in Chapter 6, you have been able to control the operation of a multifunction data acquisition (DAQ) device using programs based on LabVIEW's **DAQ Assistant**. As you know, DAQ Assistant is a sophisticated, high-level Express VI that, through the use of its dialog windows, can be configured to perform all manner of data acquisition and generation tasks (e.g., analog input, digital output). In this chapter, we will show that after DAQ Assistant has been configured for a particular task, below its glossy, easy-to-operate user interface, LabVIEW automatically generates customized code that accomplishes the requested task. This customized code is based on a collection of low-level icons called the *DAQmx VIs*. Here, you will learn how to write your own programs using the DAQmx VIs and explore the added flexibility and performance this low-level approach to programming offers in comparison to the use of DAQ Assistant.

A large assortment of data acquisition (also called *read* or *input*) operations as well as data generation (aka *write* or *output*) operations are within the capability of any multifunction DAQ device manufactured by National Instruments. When a programmer chooses to perform one such operation, that choice (e.g., analog input voltage on channel *ai0*), along with its particular timing (e.g., sampling rate) and triggering (e.g., analog trigger) selections, is termed a *task*. To program a DAQmx task, one follows a template of five sequential steps: *create and configure task, start task, read or write data, stop task, clear task.* Each of the DAQmx icons found in **Functions>>Measurement I/O>>DAQmx – Data Acquisition** is designed to execute one of these required steps and so, by properly configuring a sequence of these icons, a complete DAQmx task can be coded. Thus, most DAQmx-based data acquisition and generation programs are organized on the block diagram as shown in the next illustration.

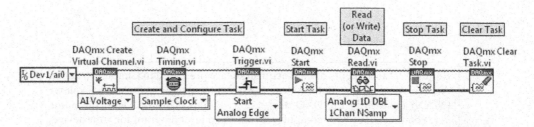

In this diagram, the task is explicitly started by **DAQmx Start Task.vi** and stopped by **DAQmx Stop Task.vi**. However, if **DAQmx Read.vi** is not preceded by **DAQmx Start Task.vi** and followed by **DAQmx Stop Task.vi**, **DAQmx Read.vi** will automatically start and, after it has completed its work, stop the task. An analogous statement applies to **DAQmx Write.vi**. Hence, a DAQmx-based diagram can often be written in the following simplified form.

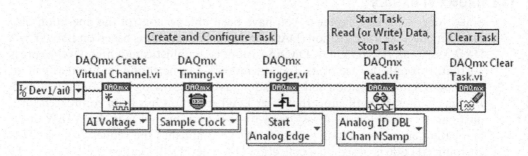

Many of the DAQmx icons are *polymorphic VIs*. When placed on a block diagram, a polymorphic VI has an associated *polymorphic VI selector* that consists of a drop-down menu of possible operational modes for the VI. For example, **DAQmx Read.vi** can be programmed to execute an analog input voltage or a digital input operation by simply making the appropriate selection on its polymorphic VI selector. To make the selection, either click on the selector with the 🖑 or pop up on the selector and use the **Select Type** option. Through this process, **DAQmx Read .vi** can be configured to operate in nearly 50 different modes. Below, we show how the 🖑 is used to select **Analog>>Single Channel>>Multiple Samples>>1D DBL**, which instructs **DAQmx Read.vi** to perform an *N*-sample Analog Input operation on one channel and output the data in the form of a 1D double-precision floating-point array.

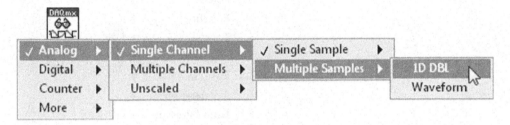

13.2 SIMPLE ANALOG INPUT OPERATION ON A DC VOLTAGE

As a step toward understanding how to program a DAQmx-based diagram, let's build a VI that turns your computer into a voltmeter. For this VI, we will use three DAQmx icons, so we begin by briefly describing the function of each of these block-diagram objects.

First, consider the polymorphic **DAQmx Create Virtual Channel.vi**. When placed on the block diagram, this VI can be used to define ("create") close to 70 different types of tasks (e.g., determine the frequency of the waveform at the counter input). The next illustration shows the selection on the polymorphic VI selector required to create a "voltmeter" task, that is, a task that digitizes voltage at a particular AI channel.

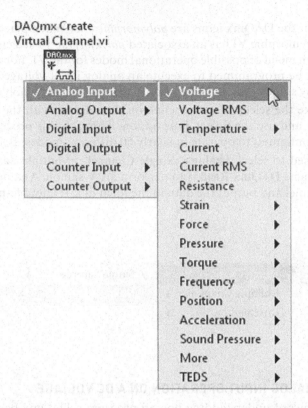

DAQmx Create
Virtual Channel.vi

The Help Window of this polymorphic VI adapts to the selection on its polymorphic VI selector. After being configured as **Analog Input>>Voltage**, DAQmx Create Virtual Channel.vi's Help Window appears as shown next. Note that the only required input (shown in boldface text) is **physical channels**; all other inputs (labeled in plain text) are recommended inputs, that is, available for your discretionary use, but may be left unwired (so that default values are used). Given the analog input channel (or channels) specified at **physical channels**, the job of this icon is to create a *virtual channel*, which consists of configurational information including the AI channel name(s) as well as the allowed input voltage range and input terminal configuration (e.g., differential mode). This virtual channel becomes part of the information that is packaged together to define a task. If a task already exists and is provided at the **task in** input, the newly created virtual channel is added to that task. However, if **task in** is left unwired, a new task is created based on the newly created virtual channel. The (modified or newly created) task is then passed out of the **task out** output for use by other DAQmx icons.

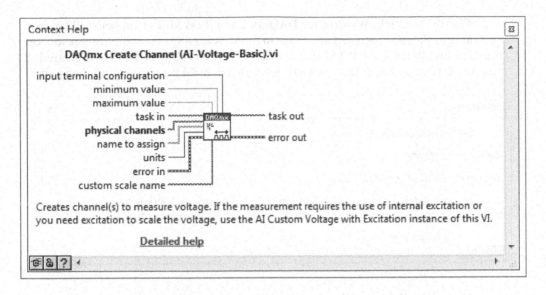

The polymorphic **DAQmx Read.vi** performs the actual data acquisition. To configure this icon to perform a single-sample analog input voltage operation on one channel and output the data in the form of a double-precision floating-point number, select **Analog>>Single Channel>>Single Sample>>DBL** on its polymorphic VI selector. After this selection has been made, the Help Window for this icon appears as shown next. Once presented with the task configurational information at its **task/channels in** input, this icon outputs the digitized voltage value as a double-precision numeric at its **data** output. Additionally, the task definition is available at the **task out** output terminal. As mentioned previously, this icon will start and stop the task if **DAQmx Start Task.vi** and **DAQmx Stop Task.vi** do not appear explicitly on the block diagram.

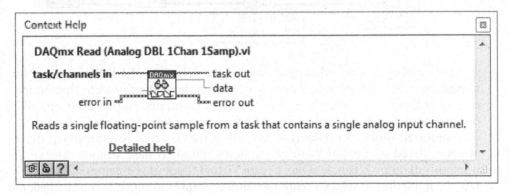

Finally, the Help Window for **DAQmx Clear Task.vi** is given below. This icon discharges the task specified at its **task in** input, releasing any computing resources such as an allotment of RAM that the task has reserved. Once a task is cleared, it cannot be run again; a new similarly defined task must be created and then run.

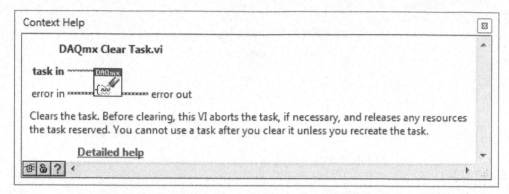

Note that all three of the above icons include error reporting via the error cluster, which appears at the **error in** and **error out** terminals.

A single voltmeter reading then can be accomplished by wiring these three icons together as follows.

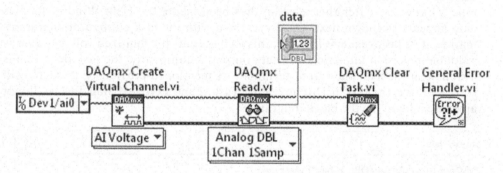

This wiring scheme takes advantage of the principle of LabVIEW programming called *data dependency*. Simply stated, data dependency means that an icon cannot execute until data are available at *all* of its inputs. In this diagram, all of the inputs to **DAQmx Create Virtual Channel.vi** are wired (**physical channel**, the only required input, is explicitly wired; all of the unwired recommended inputs are considered by LabVIEW to be implicitly wired to their default values). So, when this diagram is run, **DAQmx Create Virtual Channel.vi** executes immediately. Upon completion, **DAQmx Create Virtual Channel.vi** outputs a task at its **task out** terminal, which is passed through the wiring to DAQmx Read.vi's **task in** input.

Because of data dependency, **DAQmx Read.vi** cannot execute until it receives the task from **DAQmx Create Virtual Channel.vi**. Then, after **DAQmx Read.vi** completes its execution, it passes the task from its **task out** terminal to DAQmx Clear Task.vi's **task in** input. Only then can **DAQmx Clear Task.vi** execute. Thus, through this programming trick, we are assured that the icons will execute in the desired sequence: **DAQmx Create Virtual Channel.vi** followed by **DAQmx Read.vi** followed by **DAQmx Clear Task.vi**.

Also, the correct manner of chaining together DAQmx icons for error reporting is shown above. If an error does occur at one point in the chain, subsequent icons will not execute and the error message will be passed to the **General Error Handler.vi**, which will display the message in a dialog box. **General Error Handler.vi** is found in **Functions>>Programming>>Dialog & User Interface**. Rather than using this dialog box approach to error reporting, one could instead pass error information to a front-panel error cluster for viewing.

Okay, we are ready to write our simple voltmeter program. Open a new VI and, using **File>>Save**, first create a folder called Chapter 13 in the YourName folder, and then save this VI under the name DC Voltmeter (DAQmx) in YourName\Chapter 13.

Switch to the block diagram and write the following code, which continually reads and displays the voltage difference at the AI differential channel *ai0* every 0.5 s until the front-panel **Stop Button** is pressed or an error occurs. The three DAQmx icons are found in **Functions>>Measurement I/O>>DAQmx – Data Acquisition**. Use **Create>>Constant** in the pop-up menu to create the wired inputs of **DAQmx Create Virtual Channel.vi**. The **DAQmx Physical Channel Constant** created at the physical channels input can be programmed by clicking on its *menu button* with the . You will be presented with a list of available analog input channels on your particular DAQ device (as determined by MAX when it was last run). Otherwise, you can highlight the interior of the **DAQmx Physical Channel Constant** and manually enter the name of the desired AI channel (e.g., *Dev1/ai0*). Also note that this diagram configures the analog input channel to accept a minimum and maximum voltage difference of $V_{min} = -10$ V and $V_{max} = +10$ V, respectively. Use values appropriate for your DAQ device on your block diagram. The **Or** icon is found in **Functions>>Programming>>Boolean**.

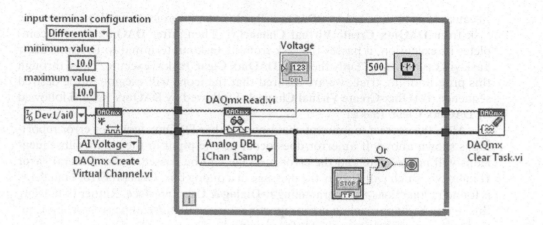

If a bad wire resulted when you wired the error cluster to the input of the **Or** icon, your (older) version of LabVIEW requires that the error cluster's status value be unbundled as shown next (see Section 6.4).

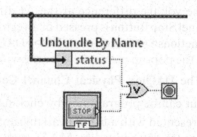

Switch to the front panel. Arrange the two objects as desired and rename the indicator **Voltage**, if you like. The voltage resolution ΔV of the n-bit analog-to-digital conversion process performed by your DAQ device is given by Eq. [5.1], i.e., $\Delta V = \left(V_{max} - V_{min}\right)/2^n$. So, for a device such as the PCIe-6351 with $n = 16$, $\Delta V = (20\text{ V})/2^{16} = 0.3\text{ mV}$. For $n = 14$ (e.g., USB-6001, USB-6009 legacy device), $\Delta V = 1\text{ mV}$. Use this information to select the proper number of **Digits of precision** on the **Voltage** indicator. For example, if $\Delta V = 0.3\text{ mV}$, **Digits of precision** should be *4*. Save your work.

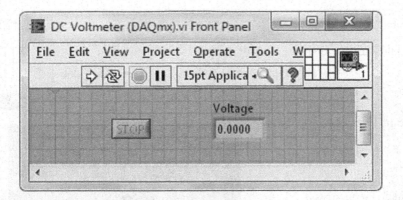

Connect a given DC voltage difference (say, 2 V) to differential channel *ai0* of your DAQ device. For example, the high-voltage and low-voltage wire from your DC source, respectively, should be connected to AI 0 and AI 8 for the PCIe-6351, AI 0 and AI 4 for the USB-6002, and AI 0^+ and AI 0^- for the myDAQ or NI ELVIS II. If your voltage source is "floating," use an extra wire (or 100 kΩ) to connect its negative terminal to the DAQ device's AI GND pin (see the discussion in Section 5.6).

Run DC Voltmeter (DAQmx). Does it read the input voltage correctly?

13.3 DIGITAL OSCILLOSCOPE

With a quick change on a polymorphic VI selector and an additional DAQmx icon or two, you can morph your DC voltmeter into a digital oscilloscope program. This new VI—named Digital Oscilloscope (DAQmx)—will acquire *N* equally spaced voltage samples of a time-varying analog input signal and then quickly plot the array of data values. By repeating this process over and over, we'll achieve a real-time display of the waveform input.

With DC Voltmeter (DAQmx) open, use File>>Save As... to create the new program called Digital Oscilloscope (DAQmx) and store it in YourName\Chapter 13. On the front panel, delete the Numeric Indicator and replace it with a Waveform Graph, whose *x*- and *y*-axes are labeled Time and Voltage, respectively. Also, using Controls>>Select a Control..., locate a Digitizing Parameter cluster in YourName\ Controls and place it on the front panel (if you haven't stored a Digitizing Parameter cluster in YourName\Controls previously, construct the cluster now as described in Section 4.10, "Control and Indicator Clusters"). Save your work.

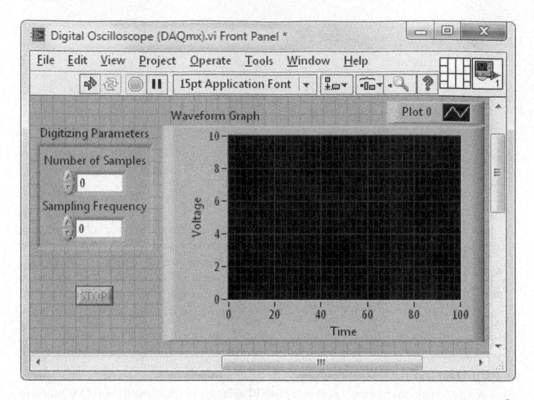

Switch to the block diagram. Remove all broken wires. Then, using the 🖐,
click on the **DAQmx Read.vi**'s polymorphic VI selector and select **Analog>>Single
Channel>>Multiple Samples>>Waveform**. The next illustration shows how the 🖐
is used to make this selection, which instructs DAQmx Read.vi to perform an *N*-
sample analog input operation on one channel and output the data in the waveform
data type.

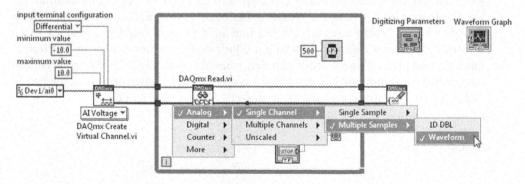

What is the *waveform* data type? Remember that previously (see Section 3.9), when using a Waveform Graph to plot an N-element 1D numeric array of data samples with a calibrated x-axis, we had to form the cluster shown below consisting of the initial x-axis value x_0, the x-axis increment Δx between neighboring data samples, and the 1D data array itself.

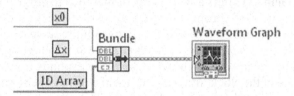

Similar to the dynamic data type used by Express VIs, the waveform data type automatically packages these items (x_0, Δx, and the 1D data array) together in a single wire. Thus, by programming DAQmx Read.vi's output to be in the waveform (rather than 1D DBL) format, its **data** output terminal can be wired directly to the Waveform Graph terminal. The resulting waveform wire will produce the same calibrated plot as would result from the "bundling" code given above (see Problem 8 in Chapter 3).

Place the **Waveform Graph** icon terminal within the While Loop and wire it to DAQmx Read.vi's **data** output. Note the banded brown wire, which denotes the waveform data type. Also, delete the **Wait (ms)** icon and its associated **Numeric Constant** within the While Loop.

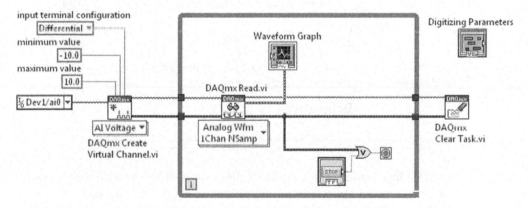

Our diagram now reads as follows. The one-time act of creating the task is done by **DAQmx Create Virtual Channel.vi**. This task is then passed into the While Loop where, during each iteration, **DAQmx Read.vi** acquires N data samples at a given sampling frequency f_s and then sends these data to the **Waveform Graph** for plotting. This data acquisition and presentation process repeats continuously until the While Loop ceases its execution in response to the **Stop Button** being pressed or an error having occurred. After exiting the While Loop, the one-time act of disposing of the task is done by **DAQmx Clear Task.vi**.

The unwired Digitizing Parameters cluster reminds us that we have a bit more work to do, namely, program the DAQ device with the desired values for N and f_s. To accomplish this feat, we use the polymorphic **DAQmx Timing.vi** icon.

DAQmx Timing.vi offers a wide array of methods to control when an incoming analog signal is digitized. From among these possibilities, we will use **Sample Clock**, a "hardware timing" method in which a digital square wave with a very precise frequency (termed the *sample clock*) controls the rate at which samples of the incoming analog signal are acquired. Each tick, which corresponds to a rising (or else, falling) digital edge of the clock, initiates the acquisition of one sample. After selecting **Sample Clock** on its polymorphic selector, the Help Window of **DAQmx Timing.vi** appears as shown below. If the **source** input is left unwired, the built-in ("on-board") sample clock on your DAQ device will be used to produce the required digital edges.

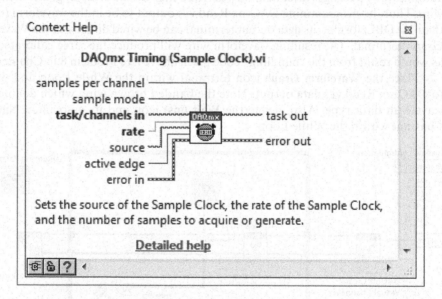

Place **DAQmx Timing.vi** on your block diagram as shown next. Since the timing information being configured by this icon becomes part of the task definition, place it outside the While Loop so that it executes one time only. Here, Number of Samples and Sampling Frequency are wired to the **samples per channel** and **rate** inputs of DAQmx Timing.vi, respectively. Using the **Create>>Constant** pop-up selection, create the ring constant wired to the **sample mode** input of DAQmx Timing .vi and then select its **Finite Samples** mode. In this mode, a finite number of samples (given by the value of **samples per channel**) is acquired each time DAQmx Read.vi executes. Save your work.

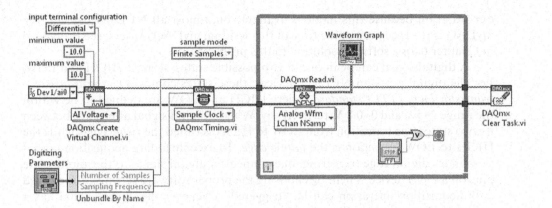

Return to the front panel and set **Number of Samples** and **Sampling Frequency** equal to *100* and *1000*, respectively. Connect the positive and negative (ground) outputs from a function generator to positive and negative pins *ai0* on your DAQ device, and then adjust the settings on the function generator so that it is outputting an analog sine wave with amplitude less than 10 V and a frequency of about 50 Hz. Run **Digital Oscilloscope (DAQmx)**. If all goes well, you will see a plot of about five cycles of the 50 Hz sine wave.

I'm sure you're impressed by your computer-based instrument, but at the same time, concerned about an obvious flaw. Most likely, the sine-wave trace you are observing does not appear stationary but instead is moving either to the right or to the left. And, by slightly changing the function generator's sine-wave frequency, the plotted sine wave can be made to move first in one direction and then the other. As explained in Chapter 6, the problem encountered here can be cured by properly *triggering* our digital oscilloscope, that is, by beginning all *N*-sample acquisitions on the same well-defined point of the incoming periodic signal's cycle. For **Digital Oscilloscope (DAQmx)** to be a useful program, we must build in this triggering capability.

In a commercially available oscilloscope, triggering is accomplished in the following manner. The input signal is monitored by an analog *"level-crossing"* circuit. The purpose of this circuit is to determine each time the incoming signal passes through a specified voltage level and then immediately trigger the scope's data acquisition process. Using knobs on the scope's front panel, the oscilloscope user sets the circuit's threshold level and specifies whether acquisition should be initiated when the level is passed through starting from above (*negative*, or *falling*, *slope*) or starting from below (*positive*, or *rising*, *slope*).

Many National Instruments DAQ devices are capable of performing the analog level-crossing triggering procedure described above. For example, the PCIe-6351 and ELVIS II have this capability; however, the lower-cost USB-6002 and myDAQ do not. So that most readers can include triggering in their **Digital Oscilloscope (DAQmx)** VI, we will implement an alternate mode—*digital edge triggering*—in our program because this mode is available on almost all NI DAQ devices (the myDAQ is the exception at the time of this writing; myDAQ owners, see Problem 6 in Chapter 6 for a software solution to this problem).

A digital signal can be in one of two possible states, termed HIGH and LOW. For the digital ports on NI DAQ devices, these states conform to the *transistor–transistor logic (TTL) standard*, where HIGH and LOW are defined to be within the range 2–5 V and 0–0.8 V, respectively. When a digital signal alternates between its two states, the transition from LOW to HIGH is called the *rising edge*, while the HIGH to LOW transition is the *falling edge*. To execute a data acquisition operation under digital edge triggering, one connects a digital signal to the appropriate pins on a DAQ device. Then, by choosing the proper software settings, the desired data acquisition operation can be "triggered" whenever the digital signal has a rising or, alternatively, falling edge.

Stop **Digital Oscilloscope (DAQmx)**, if it's running, and then switch to its block diagram. We will add triggering capability to this VI using the polymorphic **DAQ Trigger .vi** icon, which can configure a task to include digital, analog, or more novel forms of triggering. To configure this icon in its digital triggering mode, select **Start>>Digital Edge** in its polymorphic selector. After being configured in this way, the Help Window for **DAQ Trigger.vi** appears as shown below. The required **source** input specifies the DAQ-device pin (e.g., PFI 0) to which the digital triggering signal is connected.

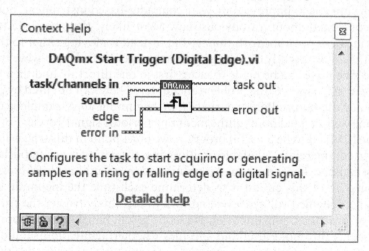

Add **DAQmx Trigger.vi** to your block diagram as shown below. Use the pop-up selection **Create>>Control** to create the **edge** front-panel control. Also use **Create>>Constant** to create a **DAQmx Terminal Constant** ⬚ wired to **source**. Clicking on its *menu button* ⬚ with the 🖑 will present you with a list of available triggering pins from which to choose. Otherwise, you can highlight the interior of the **DAQmx Terminal Constant** and manually enter the name of the desired triggering pin on your DAQ device (e.g., */Dev1/PFI0*).

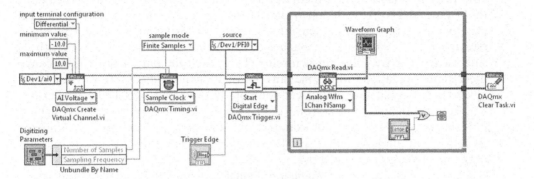

Return to the front panel. Arrange the objects to accommodate the new control and change its label from **edge** to **Trigger Edge**, if you wish. **Digital Oscilloscope (DAQmx)** is now configured for digital-edge triggering. Save your work.

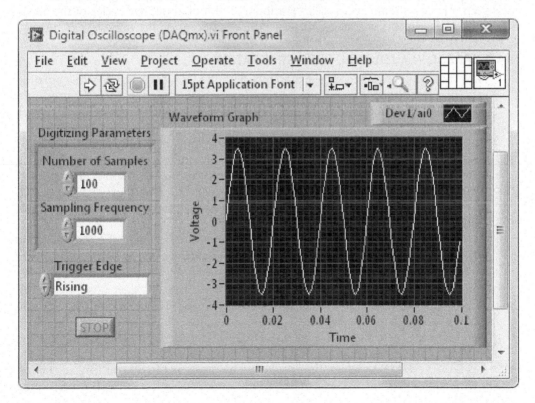

Besides producing a sine wave with frequency f at its *main* output, your function generator creates a TTL digital signal of the same frequency f, which is available at its *sync* (sometimes called *TTL* or *trigger out*) output. The digital edges of this TTL signal coincide in time with particular points on the sine wave's cycle. For example, on many function generator models, the rising edge of the TTL signal occurs when the sine wave is at its peak, while on other units it coincides with the sine wave's positive-going zero crossing. Thus, the sync output provides a convenient digital signal with which one can control the acquisition of sine-wave data (i.e., always begin at the same equivalent point on the cycle) via digital edge triggering.

Connect the sync output from your function generator to the digital triggering pins on your DAQ device. For example, sync's positive and negative (ground) terminals would connect to PFI 0 and D GND, respectively, for the software selection in the block diagram given above.

On the front panel, set **Number of Samples**, **Sampling Frequency**, and **Trigger Edge** equal to *100*, *1000*, and *Rising*, respectively, and set the sine-wave frequency on your function generator to approximately 50 Hz. Run **Digital Oscilloscope (DAQmx)**. You should see a "stationary" plot of about five cycles of the 50 Hz sine wave.

Stop **Digital Oscilloscope (DAQmx)**, and then change **Trigger Edge** to *Falling*. Run your VI. You should again see a "stationary" plot of about five cycles of the 50 Hz sine wave, but this time beginning on a different point on the sine-wave cycle.

13.4 EXPRESS VI AUTOMATIC CODE GENERATION

Now that we have gained some experience with the DAQmx VIs, we're in a position to appreciate the automatic code generation feature of Express VIs by programming **DAQ Assistant** to do the same task we have just coded on the block diagram of **Digital Oscilloscope (DAQmx)**.

Place the **DAQ Assistant** Express VI on a blank block diagram. When the **Create New...** dialog window opens, make the selections **Acquire Signals>>Analog Input>>Voltage>>ai0**, and then press the **Finish** button. You will recognize that these selections are the choices we programmed on the polymorphic VI selector and source input of **DAQmx Create Virtual Channel.vi** on the block diagram of **Digital Oscilloscope (DAQmx)**.

Next, when the **DAQ Assistant** dialog window opens, under the **Configuration** tab, in the **Signal Input Range** box, select **Max>>10** and **Min>>–10**. Also select **Terminal Configuration>>Differential** and, in the **Timing Settings** box, **Acquisition Mode>>N Samples**. Under the **Triggering** tab, in the **Start Trigger** box, select **Trigger Type>>Digital Edge**, **Trigger Source>>PFI0**, and **Edge>>Rising**. Again, you will recognize that these selections are the associated choices we programmed for the **DAQmx Create Virtual Channel.vi**, **DAQmx Timing.vi**, and **DAQmx Triggering .vi** icons on the block diagram of **Digital Oscilloscope (DAQmx)**.

Exit the **DAQ Assistant** dialog window by pressing the **OK** button. In the following moments, you will see a dialog window indicating that LabVIEW is building a DAQmx-based VI capable of carrying out the task that you have selected. After LabVIEW has created this (hidden) lower-level code, you will be returned to the Express VIs block-diagram icon.

To view the DAQmx-based code that LabVIEW created for the task that you configured in the dialog windows, pop up on the Express VI's icon and select **Open Front Panel** and then **Convert** in the dialog window that follows. When the front panel opens, switch to the block diagram. There, you will find DAQmx-based code that is very similar to the block diagram you wrote for **Digital Oscilloscope (DAQmx)**. If desired, you can modify, save, and run this VI just like you would with any other LabVIEW program.

13.5 LIMITATIONS OF EXPRESS VIS

If the DAQ Assistant Express VI can automatically write the required low-level code for any data acquisition and generation operation requested, why would one ever wish to write their own DAQmx-based code manually? The answer to this question is that the Express VI developers designed these high-level functions to carry out well-defined, commonly needed tasks. If your need fits these particular tasks exactly, then Express VIs are the (painless) way to go. However, if your needs deviate somewhat from the Express VIs' defined functionality or if your project demands the utmost in software efficiency, the Express VI approach may not be able to offer the flexibility and performance level you require.

As a simple demonstration of Express VI limitations, open **Digital Oscilloscope (Express)**, the DAQ Assistant-based digital oscilloscope program you wrote in Chapter 6 and stored in **YourName\Chapter 6**. As you may remember, this program is digitally triggered and expects an analog signal input at AI channel *ai0*, so configure a function generator to produce a (close to) 100 Hz sine wave and input the sine wave and the generator's sync output to *ai0* and PFI 0 (and D GND) pins of your DAQ device, respectively. On the VI's front panel, make **Number of Samples** and **Sampling Frequency** equal to *100* and *1000*, respectively, and then run **Digital Oscilloscope (Express)**. You will see (about) 10 cycles of the digitized sine wave.

Now, let's say that you wanted to "zoom in" on one cycle of the sine wave, which can be accomplished by simply increasing the sampling frequency by a factor of 10 while keeping the number of samples the same. With **Digital Oscilloscope (Express)** still running, change **Sampling Frequency** to *10000* (to secure this choice, you will have to click on the ✓ or press <*Enter*>). You will find that the VI does not appear to respond to this new choice for the sampling frequency.

The lack of response to an attempt at changing its sampling frequency during runtime is the result of an intrinsic feature of DAQ Assistant. DAQ Assistant is hardwired to set its various digitizing parameters (including its sampling frequency) only during its first execution after the **Run** button has been pressed. To demonstrate this feature of DAQ Assistant, halt **Digital Oscilloscope (Express)** by pressing its **Stop Button**. Then, with **Sampling Frequency** set to *10000*, press the **Run** button. You will find that DAQ Assistant has accepted the new sampling frequency and a single cycle of the digitized sine wave will be displayed.

In the interest of optimizing the performance of our VI, we programmed **Digital Oscilloscope (DAQmx)** to have the same behavior as **Digital Oscilloscope (Express)**. That is, so that the digitizing parameters were not needlessly updated with each While Loop iteration, we placed all configurational DAQmx icons outside the loop. The advantage of writing DAQmx-based code, however, is flexibility. If we now decide to make **Digital Oscilloscope (DAQmx)** capable of responding to run-time changes of its front-panel inputs, we can do so as follows. First, pop up on the While Loop and select **Remove While Loop**.

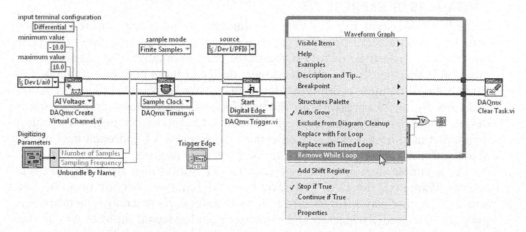

Then modify the block diagram so that **DAQmx Timing.vi** and **DAQmx Trigger .vi** are included within a newly placed While Loop as shown next. Save your work.

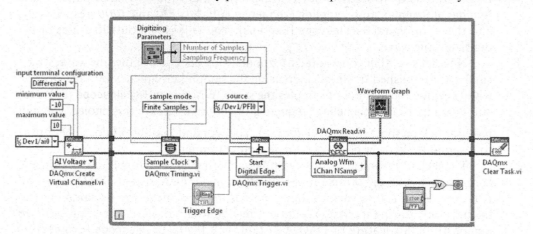

Run this modified version of Digital Oscilloscope (DAQmx). Try changing the front-panel values of Number of Samples, Sampling Frequency, and/or Trigger Edge during runtime, and enjoy the instant response to your command.

13.6 IMPROVING DIGITAL OSCILLOSCOPE USING STATE MACHINE ARCHITECTURE

In modifying Digital Oscilloscope (DAQmx), we sacrificed performance efficiency to gain runtime response. Namely, most of the DAQmx icons contained on the block diagram simply set up (**DAQmx Create Virtual Channel.vi, DAQmx Timing.vi, DAQmx Trigger.vi**) or shut down (**DAQmx Clear Task.vi**) data acquisition operations and, thus, need only be called during one (or a small subset) of the multitude of Digital Oscilloscope (DAQmx)'s While Loop iterations. These set-up and shut-down actions are collectively termed *overhead operations* and, as the program is now written, two of these overhead operations (**DAQmx Timing.vi, DAQmx Trigger.vi**) are executed each time the While Loop iterates. Since your data-taking system is not able to digitize the incoming stream of real-time data during overhead operations, needlessly including them increases the program's *dead time*, the percentage of each iteration period during which the system is not sensitive to the incoming data. With increasing dead time, more and more cycles of the incoming repetitive data will stream past your system's input undetected. In certain situations, such as when collecting a large number of data cycles with the intent of adding them together so as to average out random noise, a programming inefficiency that causes you to miss a significant percentage of the incoming data cycles can easily extend the completion time of your experiment by minutes (or sometimes even hours).

If you care to correct the above-described problem, write the following program, which only executes set-up and shut-down operations when necessary (e.g., **DAQmx Trigger.vi** executes only when a change has been made to the front-panel Trigger Edge control). With Digital Oscilloscope (DAQmx) open, use **File>>Save As...** to create a new VI named Digital Oscilloscope (DAQmx State Machine) in YourName\ Chapter 13. Consolidate all three controls in one control cluster by placing Trigger Edge in the Digitizing Parameters control cluster.

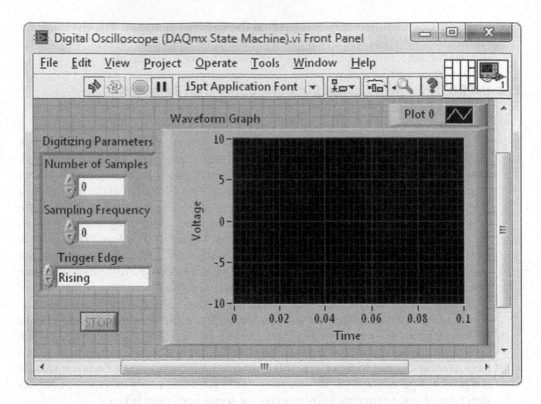

Switch to the block diagram and find a fairly large blank work area where you can build your new code. You can either delete the icons currently on the block diagram (without deleting the terminals for the front-panel objects) or stockpile these icons in some out-of-the-way place and then reuse them as you construct your new code.

The block diagram we will write is based on the *state machine* architecture that we studied in Sections 9.3 and 9.4. A state machine consists of a Case Structure nested within a While Loop. The While Loop executes continuously until its ⊙ is set to TRUE, and with each loop iteration, one of the Case Structure cases (termed a *state*) executes. The state that executes during a particular iteration performs some operation (e.g., reads data) and, in addition, selects which state will be executed during the next iteration. The block-diagram template for a state machine is shown in the next illustration. The state selection is made using an Enum Constant, which is stored in a shift register. Information that needs to be passed from one state to another is similarly stored in other shift registers.

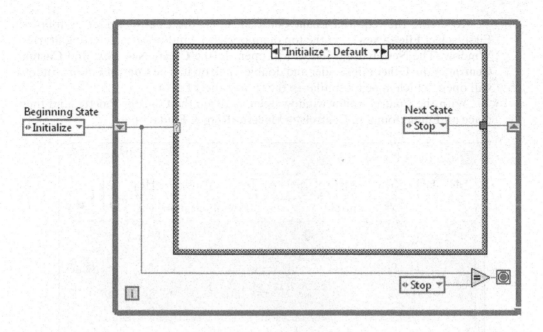

A state diagram representation of the algorithm for our digital oscilloscope state machine is shown next. Here, we see that the algorithm is carried out through five distinct states, which can be labeled *Create Task*, *Check Settings*, *Change Settings*, *Read Data*, and *Clear Task*. Thus, we need to create a block-diagram Enum Constant programmed with these five items.

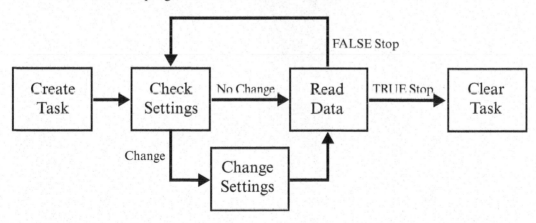

To create the required **Enum Constant**, open the **Control Editor** as follows: First, select **File>>New...** at the top of an open VI window or the Getting Started Window. The **New** dialog window will open. In the **Create New** box, find **Custom Control** in the **Other Files** folder and double-click on it. The **Control Editor** window will open, which appears similar to the front panel of a VI.

With the **Control Editor** window open, activate the **Controls Palette**, and then place an **Enum** (found in **Controls>>Modern>Ring & Enum**) on it.

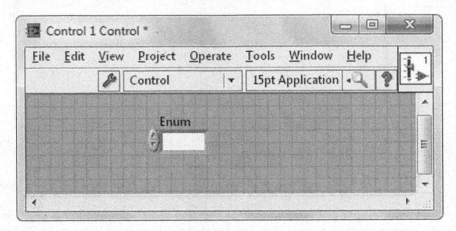

Pop up on the **Enum** and select **Edit Items...**, and then program the **Items** box with the five state names.

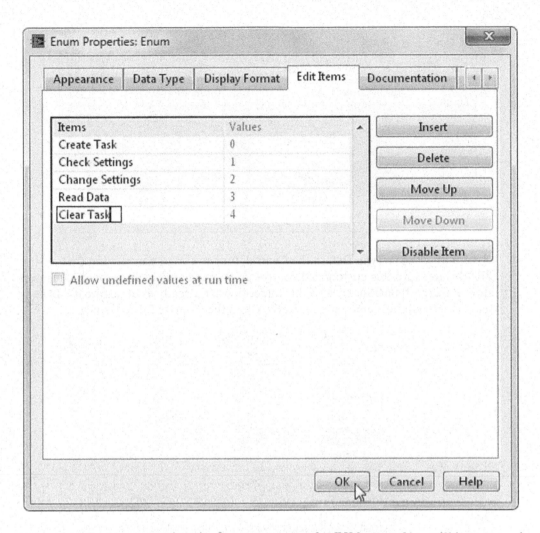

After programming the five states, press the **OK** button. You will be returned to the **Control Editor** window, which now contains the completed Enum. You may wish to resize the Enum to make all of the text within it visible.

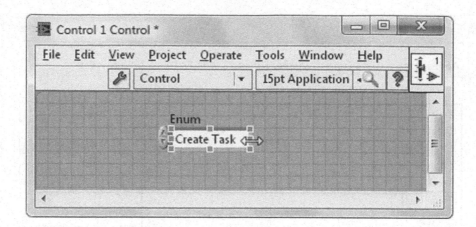

Finally, select **Type Def.** in the **Control Type** pull-down menu and then, using **File>>Save**, save this customized control under the name **Oscilloscope States** and store it in **YourName\Controls** (if this folder doesn't already exist, create it). The extension **.ctl** will be added automatically. Close the Control Editor window.

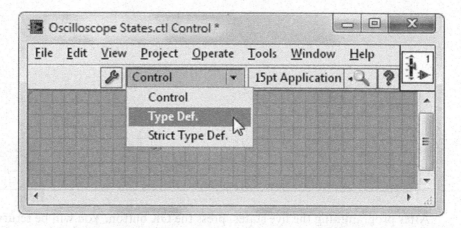

By saving **Oscilloscope States** in the **Type Definition** mode, it becomes a master file that is linked to every related Enum Constant placed on the block diagram. Then any change made to the **Oscilloscope States** file will be transmitted to every related Enum Constant. This feature can be quite labor-saving when writing a state machine and finding well into the project that an additional state must be added.

Begin coding the block diagram of **Digital Oscilloscope (DAQmx State Machine)** by constructing the **Create Task** state as shown in the next diagram. To obtain a copy of your customized Enum, click on **Select a VI...** in the Functions Palette. In the dialog window that opens, navigate to the **YourName\Controls** folder and double-click on the filename **Oscilloscope States.ctl**. You will be returned to the block diagram, where you can place the customized **Enum Constant**. The free label **State** is created using the 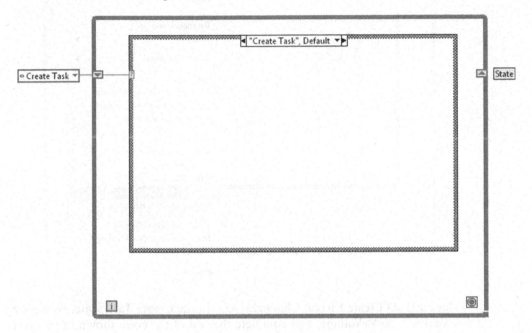.

When the **Enum Constant** is connected to the Case Structure's selector terminal, two cases will be activated that are associated with the first two items programmed on the **Enum Constant (Create Task** and **Check Settings**). Pop up anywhere on the Case Structure's border and choose **Add Case for Every Value**. The Case Structure will then create a case for each of the five states.

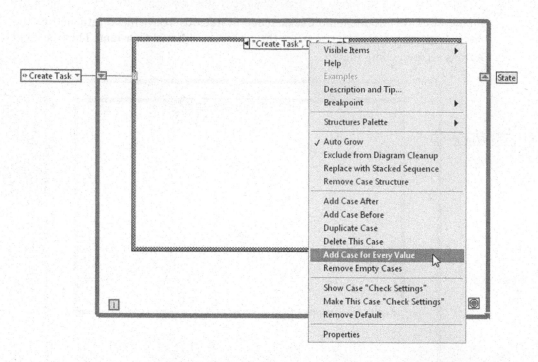

Place a **DAQ Create Virtual Channel.vi** icon in the **Create Task** state, configure it for **Analog Input>>Voltage**, and complete the rest of the code shown in the next illustration. You can create multiple copies of the **Enum Constant** by cloning (pressing <*Ctrl*> while clicking on the object with the ⬧). Once on the block diagram, the desired item of the **Enum Constant** can be selected by clicking on it with the ⬧. This state selects **Check Settings** as the next state to be executed.

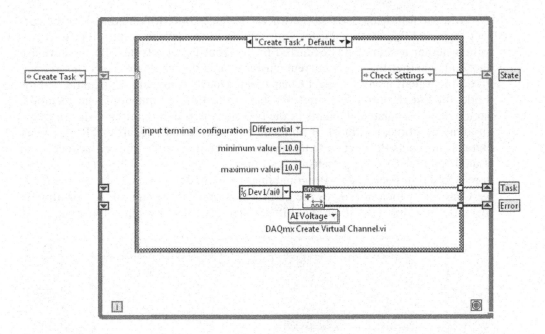

Switch to the **Check Settings** state and place the **Digitizing Parameters** icon terminal here. Complete the code shown, which reads the current cluster values and stores them in a shift register for use in subsequent iterations.

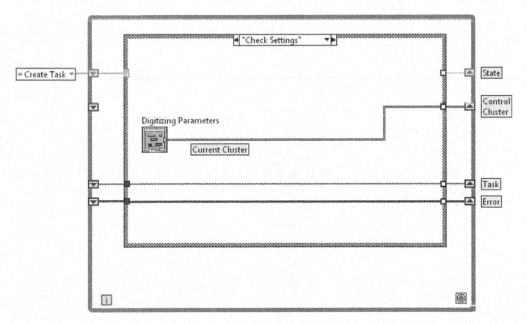

Add the following code to the **Check Settings** state, which compares the current **Digitizing Parameters** cluster with that from the previous iteration, to determine whether a change has occurred in the front-panel settings. Because we are checking whether the whole current cluster equals the whole previous cluster, pop up on the **Equal?** icon and select **Comparison Mode>>Compare Aggregates**. In this mode, the **Equal?** output is a single Boolean value (in the **Compare Elements** mode, each pair of associated elements in the two clusters is compared and the output is an array of Boolean values). If no change has occurred (**Equal?** is TRUE), **Read Data** is chosen as the next state to execute. Alternatively, if a change has occurred (**Equal?** is FALSE), **Change Settings** is chosen as the next state. The **Select** icon used here is found in **Functions>>Programming>>Comparison**. Also, to create the **Cluster Constant**, which initializes the **Control Cluster** shift register (since it is already formatted), simply pop up on the shift register and select **Create>>Constant**.

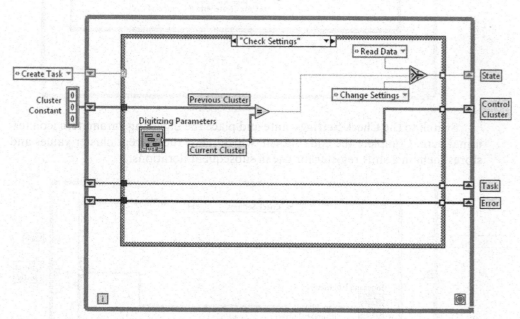

Switch to the **Change Settings** state and code it as shown. This state reconfigures the DAQ device with new Timing and Triggering parameters and selects **Read Data** as the next state to execute. The order of **Unbundle By Name**'s output terminals is chosen (using the) to avoid crossed wires.

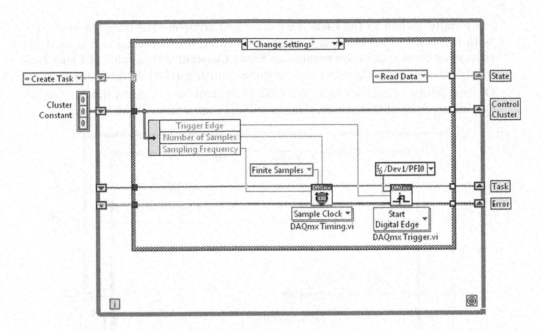

Next, code the **Read Data** state. Here, if the **Stop Button** has been pressed (so that its value is TRUE), **Clear Task** is chosen as the next state to execute. Alternatively, if the **Stop Button** has not been pressed (so that its value is FALSE), **Check Setting** is chosen as the next state (which commences another cycle of data taking).

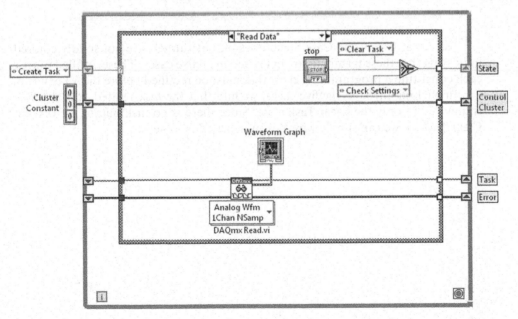

Finally, switch to the **Clear Task** state and complete the diagram as shown. With this code, after the **Clear Task** state executes, the While Loop will cease execution on its next iteration because an **Enum Constant** with a value of **Clear Task** will be passed to the **Equal?**'s upper terminal, causing a TRUE to be output to the **Or** icon, hence passing a TRUE to the ⬛. If an error occurs at any time during the execution of the state machine, this too will cease execution of the VI.

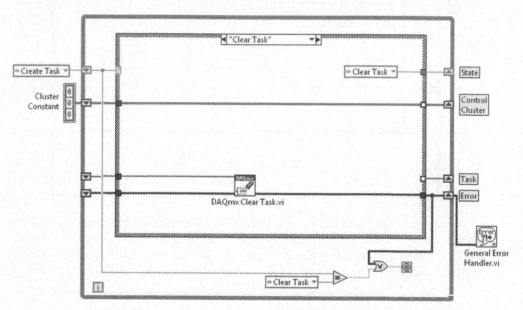

Note that two of the Case Structure's output tunnels are not solidly colored, indicating that these terminals are not wired in all five cases. These unfilled tunnels each constitute a programming error that must be rectified before the VI can run.

In clicking through the five cases, we find that the *task*-related output tunnel is unwired in only the **Clear Task** state. Since there is no task output for **DAQmx Clear Task.vi**, we can simply wire this tunnel as shown next.

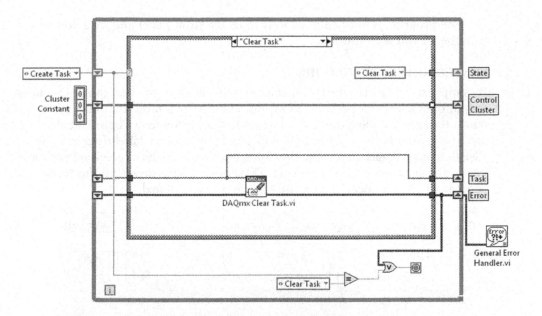

For the *cluster*-related output tunnel, the **Create Task** state is the culprit. Resolve this problem by including the wire shown.

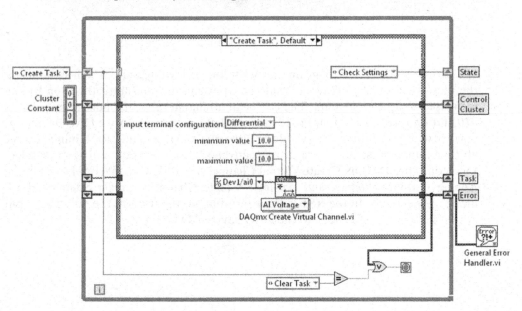

The program is finished now. Return to the front panel and save your work. Run the VI and see if it works.

13.7 ANALOG OUTPUT OPERATIONS

By simply reconfiguring the DAQmx icons with which we are now familiar, analog output operations can be carried out. Let's first write a program that simply outputs a requested voltage value. On a new VI, place a **Numeric Control** labeled Voltage and a **Stop Button**. If you like, hide the Stop Button's label (toggle it off via **Visible Items>>Label**) and make the Voltage control's **Digits of precision** appropriate for the accuracy of your particular DAQ device. Save this VI in the YourName\ Chapter 13 folder under the name DC Voltage Source (DAQmx).

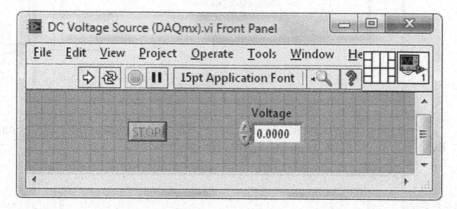

Switch to the block diagram and write the following code, which will output the requested analog voltage at analog output channel *ao0* until the **Stop Button** is pressed or an error occurs. The choice of **minimum value** and **maximum value** as *−10* and *10*, respectively, is appropriate for all of our representative DAQ devices (if you are using a USB-6009 legacy device, the appropriate values for **minimum value** and **maximum value** are *0* and *5*, respectively). Using the polymorphic VI selectors, configure **DAQmx Create Virtual Channel.vi** and **DAQmx Write.vi** to be in the **Analog Output>>Voltage** and **Analog>>Single Channel>>Single Sample>>DBL** mode, respectively. In the **RSE** output terminal configuration, the voltage output at the AO 0 pin is referenced to the DAQ device AO GND.

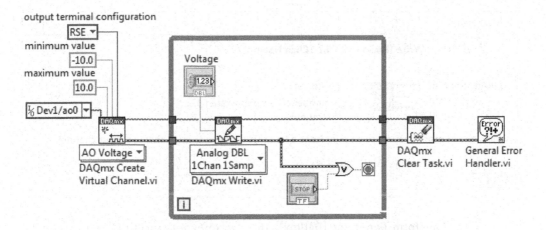

Return to the front panel and save your work. Connect the positive and negative terminals of a voltmeter or oscilloscope to the AO 0 and AO GND pins on your DAQ device. Run the VI with various values for **Voltage**. You will now find that when **Voltage** is set to outside of the range allowed by **minimum value** and **maximum value**, the program ceases operation.

When you are finished, run **DC Voltage Source (DAQmx)** one more time with **Voltage** set to *0*, so the channel *ao0* is left in the zero-voltage state.

13.8 WAVEFORM GENERATOR

If your DAQ device supports hardware-timed analog output operations (all of our representative DAQ devices do; the USB-6009 does not), try building the following hardware-timed waveform generator called **Waveform Generator (DAQmx)**. This VI is based on the **DAQmx Write.vi** icon configured in its **Analog>>Single Channel>>Multiple Samples>>1D DBL** mode. When in this mode, the Help Window for **DAQmx Write.vi** is as shown in the following illustration. Here, a 1D array of *N* double-precision floating-point numbers describing a few cycles of a periodic waveform is input at **data**. The icon then writes this *N*-element array to a memory buffer. If **autostart** is set to TRUE, the icon then instructs the DAQ device to begin outputting the sequence of analog voltage values as dictated by the 1D array within the memory buffer.

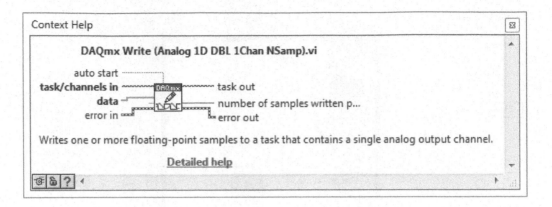

In **Waveform Generator (DAQmx)**, the "waveform-describing" array will be created using **Waveform Simulator**, whose operation is controlled by the **Digitizing Parameters** and **Waveform Parameters** control clusters. Starting with a blank VI, build the following front panel. **Digitizing Parameters** and **Waveform Parameters** are found in **YourName\Controls**. Save this VI under the name **Waveform Generator (DAQmx)** in **YourName\Chapter 13**.

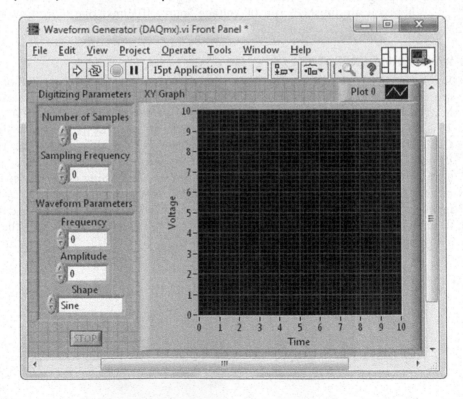

Switch to the block diagram and code it as shown next. Wire Waveform Simulator's **Displacement** array to the **data** input of DAQmx Write.vi. Also, wire **Sampling Frequency** to the **rate** input of DAQmx Timing.vi and configure this icon's **sample mode** to be **Continuous Samples**.

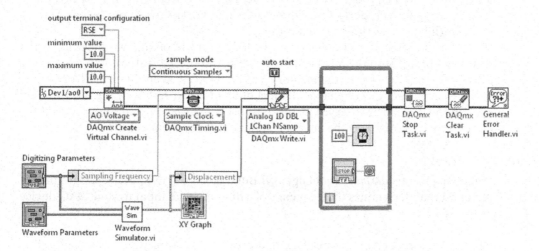

When this diagram is run, **DAQmx Write.vi** will write the N-element **Displacement** array into the memory buffer associated with the defined AO task, and the array will also be plotted on the front-panel **XY Graph** for the user to view. The continuous AO operation will then be started. In this operation, the DAQ device will output the prescribed sequence of voltages at the selected analog output channel in a circular fashion (i.e., after the last array value is output, the next output is the first array value). The rate at which the output voltage is changed ("updated") from one value to the next is controlled by **Sampling Frequency**. While this continuous analog output operation executes, the value of the front-panel **Stop Button** is checked every 100 ms. When the button is pressed, the AO task is stopped and cleared.

Return to the front panel and save your work. Then connect the positive and negative (ground) inputs of an oscilloscope to the AO 0 and AO GND pins of your DAQ device, respectively.

Program **Waveform Generator (DAQmx)** to create two cycles of a 5 V, 50 Hz sine wave, where each cycle is defined by 100 samples. That is, in the **Waveform Parameters** control cluster, set **Frequency**, **Amplitude**, and **Type** equal to *50*, *5*, and *Sine*, respectively; in the **Digitizing Parameters** control cluster, set **Number of Samples** and **Sampling Frequency** to be *200* and *5000*, respectively. Run the VI. If your oscilloscope is triggered properly, you will observe a 50 Hz sine wave produced continuously until you press the VI's **Stop Button**.

What happens if you don't write a whole number of sine-wave cycles to the memory buffer? Stop the VI. Then program it to write 1½ (rather than two) cycles of the 50 Hz sine wave to the memory buffer. That is, keep all of the front-panel settings as before, except change **Number of Samples** to *150*. Run the VI. Can you explain the waveform observed on the oscilloscope based on your knowledge of the circular fashion in which the memory buffer values are sequenced?

DO IT YOURSELF

Frequency Meter (DAQmx Count Edges).vi Build a (crude) DAQmx-based frequency meter VI that determines the frequency of a digital signal input to your DAQ device as follows.

1. Configure a counter input task that uses *counter 0* on your DAQ device (e.g., *Dev1/ctr0*) to count falling digital edges at the counter's Source input (e.g., *PFI 0*).
2. Within a While Loop, write code that executes the following procedure for each iteration: Start the counting task, use **Wait (ms)** to define the counting interval to be 100 ms, read the number of counts N accumulated during the 100 ms, calculate the frequency $f = N/(0.1s)$ and send it to the front panel for display, and then stop the task.
3. Configure the While Loop so that its operation ceases when a front-panel Stop Button is pressed or an error occurs.

After you have completed your VI, connect the sync output of a function generator to the input of counter 0 (and GROUND) on your DAQ device, and then use it to measure the frequency of this digital signal.

(OPTIONAL) **Frequency Meter (DAQmx Frequency).vi** All of our representative DAQ devices are equipped to count falling edges of digital signals; thus, everyone can carry out the above project. If you have a more advanced DAQ device (such as the PCIe-6351), construct another frequency meter VI in which the DAQmx task is configured as **Counter Input>>Frequency** (as opposed to **Counter Input>>Count Edges**) with **Measurement Method>>Low Frequency with 1 Counter**. Input the digital pulse train whose frequency you wish to measure at the counter's Gate input. Then, rather than measuring time with a **Wait (ms)** icon, the time measure can be carried out by a much more accurate clock onboard your DAQ device.

USE IT!

Serial Peripheral Interface Communication Many modern-day integrated circuits designed to perform useful laboratory tasks, such as sensor signal conditioners and digital potentiometers, communicate with a computing system using the serial peripheral interface (SPI). In this method, up to four wires ("lines") carry digital signals between the SPI-equipped chip and the computing system. Two of these lines—labeled *Chip Select (CS)* and *Clock (CLK)*—synchronize the transfer of digital data that occurs on the remaining one or two lines. When the CS line is in the LOW state, SPI communication is activated and one bit of digital data is transferred with each designated transition of the digital square-wave on the CLK line. For some SPI-equipped chips the designated transition is *falling* (HIGH-to-LOW), while for other chips it is *rising* (LOW-to-HIGH); check a particular chip's datasheet to determine which convention is chosen.

In the following example, LabVIEW-based SPI communication is demonstrated using the Maxim MAX31855 chip. This chip amplifies and performs cold-junction compensation on the voltage signal from an attached type K thermocouple, and digitizes the result. In addition, the chip uses a linearized calibration equation (see the Use It! example in Chapter 11) to convert the digitized voltage to its associated temperature and outputs this temperature as a 14-bit binary number using its SPI circuitry. Since the MAX31855 only outputs temperature data and is not equipped to receive data input, it only has one SPI-related data line (labeled *DO*). The MAX31855 chip, conveniently mounted on a small printed circuit board along with its required supporting circuitry, can be purchased from Adafruit Industries (Product ID 269). A similar "breakout board" is available from SparkFun (Part #SEN-13266).

The experimental setup consists of an Adafruit breakout board with the MAX31855 chip powered at its Vin and GND inputs from a DAQ device's +5 V and D GND pins, respectively. The three-line SPI communication is facilitated by connecting the breakout board's CS, CLK, and DO pins to three of the DAQ device's DIO lines (here, we arbitrarily used lines 0, 1, and 2 of port 0) as shown in Figure 13.1.

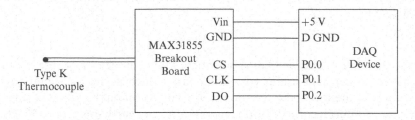

FIG. 13.1 Hardware circuit for Use It! example

The following program uses the DAQ device to read data from the MAX31855 chip using the SPI.

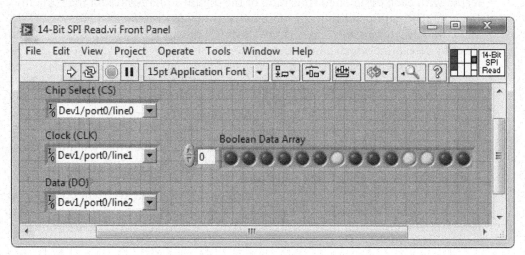

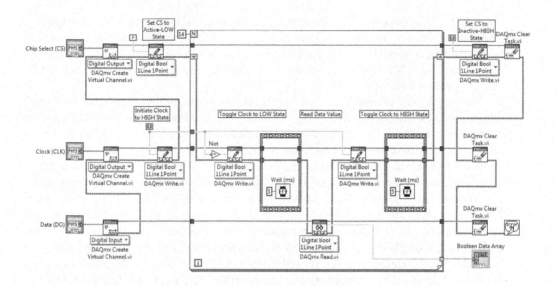

The above program is consistent with the MAX31855 datasheet, which states that the DO pin sequences out a 14-bit binary representation of the thermocouple's temperature (see Problem 4 for more information). The first bit is present at the DO pin 100 ns after the HIGH-to-LOW transition of CS (in the program, the first falling edge of the clock is made to occur near this time as well). In the above program, this data value is read into the DAQ device's Data line as a Boolean value 5 ms later. As the For Loop then iterates, the 13 remaining bits appear at DO and are sequentially read as each of the 13 falling transitions of the clock occurs (a bit is read 5 ms after the clock's falling transition). The proper execution order of the icons is ensured by the error cluster wiring and the principle of data dependency. The 14 bits are stored as an array of Boolean values using the For Loop's auto-indexing feature and output to the Array Indictor labeled **Boolean Data Array**. To create this Boolean Array Indicator, place a **Round LED** in an **Array** shell and then expand to the resulting Array Indicator to expose

14 LEDs (this procedure is described in Section 3.11 and 11.8 for a Numeric Array Indicator). The 14-element Boolean array is a two's-complement representation of a signed integer with the sign bit as the index-0 element, the most significant bit (MSB) as the index-1 element, and the least significant bit (LSB) as the index-13 element.

Using **14-Bit SPI Read.vi** as a subVI, the next program reads and plots the thermocouple's temperature every 500 ms. The **Boolean Array To Number** icon converts the 14-bit binary number obtained from the MAX31855 chip to its equivalent signed integer. This icon assumes that the array is arranged in the opposite sense to that provided by the chip (i.e., with the LSB as the index-0 element), so the array ordering is first reversed via the **Reverse 1D Array** icon. In addition, by default, **Boolean Array To Number** outputs an unsigned integer (U32). To program it to produce the required signed integer, one must pop up on this icon and select **Properties**. In the dialog box that appears, under the **Output** tab click on the U32 icon and select **Representation>>I32**. Finally, according to the MAX31855 datasheet, the signed integer is the thermocouple temperature in units of 0.25 °C. Thus, to obtain the temperature in Celsius, the signed integer is multiplied by 0.25.

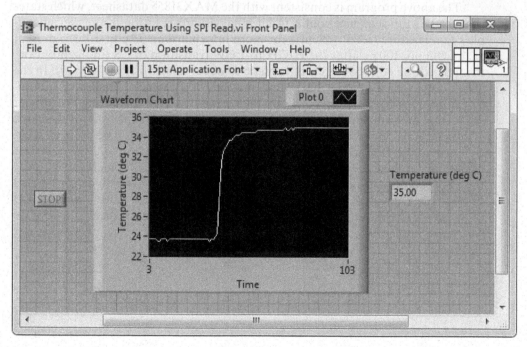

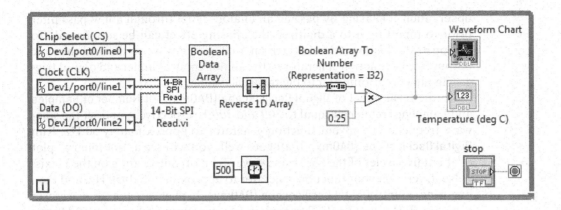

PROBLEMS

> For a problem that involves writing a new program, the problem statement begins with a suggested descriptive name (including the **.vi** extension) for the VI that you will write; icons needed for the VI may be found with the aid of **Quick Drop**. For a problem that involves the use of a VI already written as part of the chapter text, the problem's topic is given in bold at the beginning of the problem statement.

1. **Digital Resolution** Observe the resolution ΔV of an analog input signal digitized by your DAQ device and verify that your observation is consistent with the prediction of Eq. [5.1] in Chapter 5, that is, $\Delta V = V_{span}/2^n$. Attach a slowly varying analog signal (e.g., 1 Hz triangle wave) and the sync output from a function generator to the analog input channel and digital triggering pins programmed into **Digital Oscilloscope (DAQmx)**. Run **Digital Oscilloscope (DAQmx)** with a fast sampling rate and fairly small number of samples so that the input signal is effectively constant over the trace that you obtain. Stop the VI, and then by changing the y-axis scaling, zoom in on the acquired data (alternatively, you can use a **Probe** to view the numerical values of the data samples). Under "high magnification," you should find that the data samples of this "effectively constant" input signal are distributed in discrete levels (because of electronic noise and, possibly, the signal's slow time-variation). Measure the spacing ΔV between two of these adjacent levels. Does your value for ΔV agree with the prediction of Eq. [5.1]?

2. **Observation of Aliasing** By passing an analog signal through a low-pass filter prior to inputting it to a digitizer, the aliasing effect can be suppressed. To demonstrate this procedure, attach an analog sine-wave signal and the sync output from a function generator to the analog input channel and digital triggering pins programmed into **Digital Oscilloscope (DAQmx)**.

On the front panel of **Digital Oscilloscope (DAQmx)**, set **Number of Samples** and **Sampling Frequency** equal to *200* and *2000*, respectively, and set the sine-wave frequency f on your function generator to approximately 50 Hz. Run **Digital Oscilloscope (DAQmx)**. If all goes well, you will see a "stationary" plot of about five cycles of the 50 Hz sine wave. Turn off autoscaling on the y-axis.

Next, set f on your function generator to approximately 1950 Hz, and then fine-tune f until **Digital Oscilloscope (DAQmx)** displays an aliased waveform of about 50 Hz. Use Eq. [5.2] to explain why the input with $f = 1950$ Hz appears as a 50 Hz sine wave when digitized because of aliasing.

Finally, without changing the frequency setting on your function generator, pass the $f = 1950$ Hz sine-wave voltage through the low-pass op-amp filter shown in Figure 13.2 with $R = 10$ kΩ and $C = 0.1$ μF, prior to inputting it to the DAQ device. When **Digital Oscilloscope (DAQmx)** is now run, by what factor is the amplitude of the aliased signal attenuated in comparison to the unfiltered situation? What is the cutoff frequency $f_c = 1/2\pi RC$ of the low-pass filter? If you wanted the aliased signal to be attenuated even further, what component in the filter would you change? (This process can be optimized using a higher-order low-pass filter than the simple first-order circuit used here; see the Use It! example in Chapter 6.)

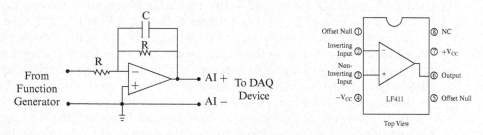

FIG. 13.2 Low-pass filter for Problem 2. Suggested op-amp is LF411.

3. **Dual Digital Oscilloscope (DAQmx).vi** Build a DAQmx-based two-channel digital oscilloscope. With **Digital Oscilloscope (DAQmx)** open, use **File>>Save As...** to create a new VI. On the block diagram, you must change the single physical channel selected (e.g., *Dev1/ai0*) to a list of two channels. Use the Help Window of **DAQmx Create Virtual Channel** to learn the proper syntax for making such a list. Besides programming this list of physical channels, there is only one other change you must make on the block diagram to complete your program.

    Once the VI is finished, attach a voltage waveform and the sync out digital signal from a function generator to the two analog input channels of your DAQ device that are programmed on the block diagram. Additionally, attach the sync out digital signal to the triggering channel (e.g., PFI 0) of your DAQ device. Run your VI and verify that it simultaneously displays traces of the two inputs.

4. **NIST Thermocouple Temperature Using SPI Read.vi** Write a program that determines the NIST-calibrated temperature of the type K thermocouple attached to a MAX31855 chip, as opposed to the approximate "linearized" temperature this chip outputs by default. According to its datasheet, the MAX31855 chip can report both the "linearized" temperature T_{therm} of its attached thermocouple (the temperature we found in this chapter's Use It! example) as well as its determination for the cold-junction compensation temperature T_{cold}. These two temperatures are contained in a 32-bit binary array transferred at the DO pin via the SPI. T_{therm} and T_{cold} are encoded as a 14-bit binary number in the array's index-0 through index-13 elements and as a 12-bit binary number in the array's index-16 through index-27 elements, respectively. The 14-bit binary number is a two's-complement representation of a signed integer that is T_{therm} in units of 0.25 °C, while the 12-bit binary number is a two's-complement representation of a signed integer that is T_{cold} in units of 0.0625 °C.

    Modify the Use It! example in this chapter to perform a 32-bit SPI read from the MAX31855 chip, then decode the resulting Boolean array to find T_{therm} and T_{cold}. Next, find the thermocouple's voltage V_{therm} using the following relation from the datasheet:

    $$V_{therm} = \left(41.276 \times 10^{-6} \text{ V/°C}\right)\left(T_{therm} - T_{cold}\right)$$

    Finally, use the **Convert Thermocouple Reading.vi** icon (see the Use It! example in Chapter 11) to plug this voltage into the NIST calibration equation for an accurate determination of the thermocouple temperature.

5. **Time Constant of RC Circuit** Use **Digital Oscilloscope (DAQmx)** to measure the time constant of an *RC* circuit. Attach a TTL digital square wave from the sync output of a function generator to the triggering channel (e.g., PFI 0) of your DAQ device as well as across the *RC* series circuit as shown in Figure 13.3 with $R = 4.7$ kΩ and $C = 0.1$ μF. When the square wave transitions from its LOW to HIGH (HIGH to LOW) state, the capacitor will be charged (discharged) through the resistor. Set the square wave frequency to a low frequency (say, 10 Hz) to allow the capacitor to fully charge and discharge after each transition. To observe the capacitor discharge, attach the analog input channel for which **Digital Oscilloscope (DAQmx)** is configured across the capacitor. Then, choosing appropriate values for set **Number of Samples**, **Sampling Frequency**, and **Trigger Edge**, observe the discharge of the capacitor and accurately measure its time constant τ. By definition, τ is the time taken for the capacitor's voltage to discharge to $e^1 = 0.37$ of its initial value. Compare the value you measure to the theoretically expected value of $\tau = RC$.

To assist in measuring voltages along the discharge curve, you may wish to activate a cursor by popping up on the Waveform Graph and selecting **Visible Items>>Cursor Legend**. When the cursor legend appears, pop up on it and select **Create Cursor>>Single Plot**. Once activated, by moving the cursor to particular data points on the plot, the *x*- and *y*-coordinates of this point of interest are displayed in the cursor legend.

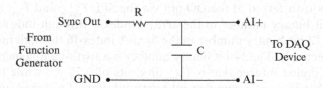

FIG. 13.3 *RC* circuit for Problem 5

6. **Waveform Generator (DAQmx State Machine).vi** Write a program that allows the parameters of the analog waveform being output by your DAQ device to be changed while the program is running. Use the state machine architecture for your block diagram. When a change in front-panel setting is detected, the task first must be stopped before changes are made to the sampling frequency.

7. **Digital Potentiometer Using SPI Write.vi** The Analog Devices AD5206BN-100 chip contains six digitally controlled potentiometers. Through SPI communication, each of these variable resistances can be placed in 256 equally

separated resistance states ("positions"), which span the range from (almost) $0 \, \Omega$ to $100 \, k\Omega$. Since this chip only receives commands and is not equipped to output data, it only has three SPI-related lines—*CS*, *CLK*, and a data line labeled *SDI*. The command sent to the chip is an 11-bit binary number, where the first 3 bits are the address of the potentiometer (addresses for potentiometers 1 to 6 run from 000_2 to 101_2) and the last 8 bits specify the desired position of the addressed potentiometer. The possible positions run from (decimal) 0 to 255. Bits are arranged from MSB to LSB.

Write the following program, which uses a DAQ device to transmit a given 11-bit Boolean array using the SPI. Here, we assume the 11-bit array represents a command being sent to an AD5206BN-100 chip. The three-line SPI communication is facilitated by connecting the chip's CS, CLK, and SDI pins to three of the DAQ device's DIO lines (here, we arbitrarily used lines 0, 1, and 2 of port 0). The DAQ device's +5 V and Digital GND pins can also power the chip at its VDD and VSS pins, respectively.

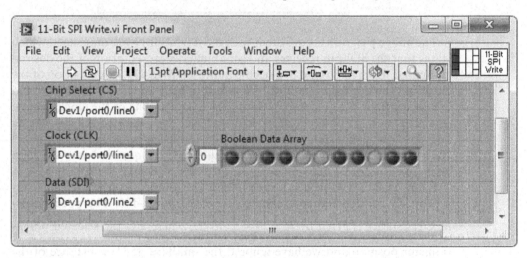

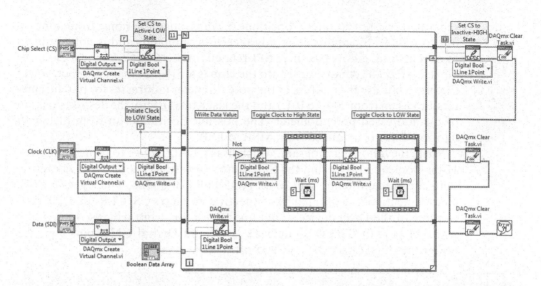

The above program is consistent with the AD5206BN-100 datasheet, which states that the SDI pin receives an 11-bit binary representation of the desired command. Each bit is read by the chip at the LOW-to-HIGH transition of the clock. As the For Loop iterates, the 14 bits appear one at a time at SDI and are sequentially read as each of the 14 rising transitions of the clock occurs. The proper execution order of the icons is ensured by the error cluster wiring and the principle of data dependency. The 11 bits originate in a front-panel Array Control labeled **Boolean Data Array** and are sequenced into the loop using the For Loop's auto-indexing feature.

Using **11-Bit SPI Write.vi** as a subVI, build the following program that sends a properly constructed command to the AD5206BN-100. The potentiometers' resistance is given by $R = \left(\dfrac{x}{256}\right)100\,\text{k}\Omega$, where x is the selected potentiometer position and we have ignored the small $(\approx 45\,\Omega)$ resistance of the chip's terminals.

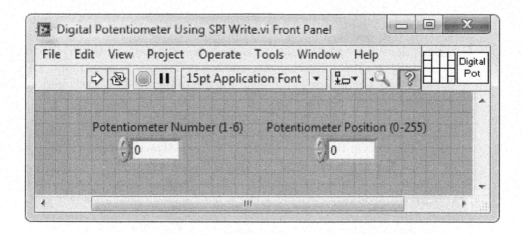

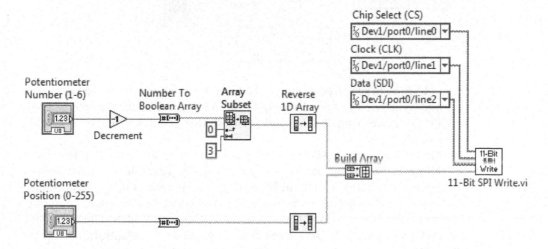

8. **Spectrum Analyzer (DAQmx).vi** Using the information given in Chapter 12's Do It Yourself project, build a spectrum analyzer with a DAQmx-based (rather than DAQ Assistant Express VI-based) program.

9. **Creating Task Using MAX** In this problem, you will create a task using the Measurement & Automation Explorer (MAX). This approach provides an alternate method to that used within the chapter text, where tasks were created on the block diagram. Use MAX to create a digital oscilloscope task similar to your **Digital Oscilloscope (DAQmx)** VI through the following steps.

   With MAX open, right-click on **Data Neighborhood** and select **Create New...>>NI-DAQmx Task**. In the dialog windows that follow, choose selections **Acquire Signals>>Analog Input>>Voltage>>ai0**, name your task

Oscilloscope, and then press the **Finish** button. Within the MAX window, you will then find the **Configuration** tab displayed for your task. There, select **Acquisition Mode>>N Samples, Samples to Read>>100**, and **Rate (Hz)>>1k**. Next, click on the **Triggering** tab, and select **Trigger Type>>Digital Edge**. Finally, click on the **Save** button near the top of the window. The creation of your task is now complete.

(a) **Digital Oscilloscope [MAX Task].vi** To implement your task, open a blank VI. On the block diagram, place a **DAQmx Task Name**, found in **Functions>>Measurement I/O>>DAQmx**. Using the 🖑, click on its menu button and select *Oscilloscope*. Then complete the following diagram and save your work.

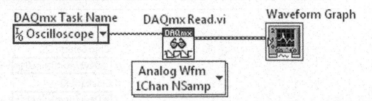

Using a function generator, attach a 100 Hz sine-wave voltage waveform and sync out the digital signal to the AI 0 and PFI 0 channels of your DAQ device. Run this VI and verify that it obtains a single $N = 100$ sample trace.

(b) **Digital Oscilloscope [MAX Task Autocode].vi** Alternatively, the DAQmx Task Name can generate code for you automatically. To implement this feature, open a new blank VI. On the block diagram, place a **DAQmx Task Name**; using the 🖑, click on its menu button and select *Oscilloscope*. Then pop up on the DAQmx Task Name and select **Generate Code>>Configuration and Example**. A **DAQmx Read.vi** icon wired to a configurational subVI will be automatically created. Open the subVI and you will find code that is familiar. Save your work. With the 100 Hz sine-wave and digital-sync signal inputs from the function generator, run your VI and verify that it obtains a single N sample trace. With some simple modifications, you can transform this VI into a program that is equivalent to **Digital Oscilloscope [DAQmx]**, if you wish.

10. **Creating Virtual Channel Using MAX** In this problem, you will create a "virtual channel" using the Measurement & Automation Explorer (MAX). This approach provides an alternative method to that used within the chapter text, where virtual channels were created on the block diagram using the **DAQmx**

Create Virtual Channel.vi icon. Use MAX to create an analog input voltage virtual channel such as that used in your **Digital Oscilloscope (DAQmx)** VI through the following steps:

(a) With MAX open, right-click on **Data Neighborhood** and select **Create New…>>NI-DAQmx Global Virtual Channel**. In the dialog windows that follow, select **Acquire Signals>>Analog Input>>Voltage>>ai0**, name your virtual channel *AI Voltage*, and then press the **Finish** button. Within the MAX window, you will then find the **Configuration** tab displayed for your virtual channel, where you can choose parameters related to the virtual channel such as the **Signal Input Range** and **Terminal Configuration**. If any changes are made from the default values, then click on the **Save** button near the top of the window. The creation of your virtual channel is now complete.

(b) **Digital Oscilloscope (MAX Virtual Channel).vi** To implement your virtual channel, open **Digital Oscilloscope (DAQmx)** and, using **File>>Save As…**, create a new VI. On the block diagram, delete the **DAQmx Create Virtual Channel.vi** icon (and the various items wired to it) and replace it with a **DAQmx Global Channel**, found in **Functions>>Measurement I/O>>DAQmx**. Using the ✋, click on its menu button and select *AI Voltage*. Then simply modify the diagram as shown below. Save your work.

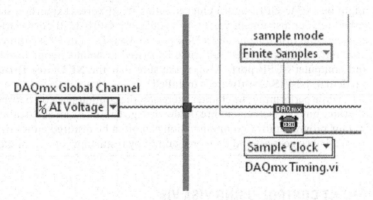

Using a function generator, attach a 100 Hz sine-wave voltage waveform and sync out the digital signal to the *ai0* and triggering (e.g., PFI 0) channels of your DAQ device. Run your VI and verify that it performs exactly like **Digital Oscilloscope (DAQmx)**.

CHAPTER 14

Control of Stand-Alone Instruments

To perform the exercises in this chapter, you must have a stand-alone instrument equipped with a General Purpose Interface Bus (GPIB) and/or Universal Serial Bus (USB) interface, ideally the Keysight/Agilent 34410A (or 34401A) Digital Multimeter. For GPIB communication, you must also have a National Instruments GPIB device connected to your computer. This device might be a PCIe-GPIB board plugged into a PCI Express expansion slot and linked to your instrument via a GPIB cable or a GPIB-USB device attached between your instruments and a USB port on your PC. For USB communication, only an appropriate USB cable is needed to connect your instrument and computer's USB port. Also, make sure that the **NI Device Drivers** (in particular, **NI-VISA**) software is installed.

If, while performing the chapter exercises, the communication between the interface bus and instrument breaks down (e.g., caused by an accidental error in your programming), communication can often be restored either by turning the instrument off and then on again or by restarting your computer.

14.1 INSTRUMENT CONTROL USING VISA VIs

In previous chapters, you have used LabVIEW software to transform a personal computer (connected to an appropriate National Instruments DAQ device) into several handy laboratory instruments. In particular, you programmed this system to become a DC voltmeter, digital oscilloscope, waveform generator, spectrum analyzer, digital thermometer, and frequency meter. Pause to consider the following tantalizing prospect: Perhaps the only instrument required in a modern-day laboratory is a DAQ device-equipped computer controlled by LabVIEW software. That is, by simply writing a collection of appropriate VIs, it might be possible for you—the cutting-edge experimentalist—to satisfy all of your laboratory instrumentation needs with this single LabVIEW-based data acquisition and generation system. This system's tremendous flexibility would then obviate the need to purchase an expensive collection of stand-alone electronic equipment such as power supplies, function generators, picoammeters, spectrum analyzers, and oscilloscopes.

The functioning VIs that you have written in previous chapters demonstrate that the above "tantalizing prospect" can, at least in certain situations, be realized. But don't discard your stand-alone instruments just yet. The timing, speed, sensitivity, and simultaneous data-taking requirements of many contemporary research experiments are beyond the capabilities of your DAQ device. For instance, although the LabVIEW-based digital oscilloscope we constructed worked well for observing audio-range frequencies (less than 20 kHz), it would prove miserably inadequate at displaying the several nanosecond-wide voltage pulses emanating from a photomultiplier tube. In this latter situation, a stand-alone digital oscilloscope with a very fast analog-to-digital converter (on the time scale of several giga-samples per second) would do the job nicely. Thus, stand-alone instruments play a central role in state-of-the-art research, and so it might not surprise you to find that they, too, fall under the scope of LabVIEW.

Over the past several decades, a message-based communications standard has evolved by which stand-alone instruments can be software-controlled using a personal computer. In this communications scheme, a particular instrument obeys an array of manufacturer-defined ASCII character commands that represent all the possible ways of manually pressing buttons, turning dials, and viewing output data on its front panel. Although the hardware conduit (called an *interface bus*) through which these ASCII messages are passed between the PC and laboratory instrument can take on various guises (including RS-232, GPIB, Ethernet, and USB), there is a single set of LabVIEW icons available to control this communication process. This icon set is named *VISA* (short for Virtual Instrument Software Architecture) and is found in **Functions>>Instrument I/O>>VISA**.

In this chapter, you will learn how to use VISA icons to control the message-based communication between a stand-alone instrument and your computer. You will explore generic features of this communication process such as the Standard Commands for Programmable Instruments (SCPI) language and various synchronization methods while writing code that controls a particular stand-alone instrument—the *Keysight/Agilent 34410A Digital Multimeter*—using two interface buses, the *General Purpose Interface Bus* and the *Universal Serial Bus*.

14.2 THE VISA SESSION

When using VISA icons to facilitate message-based communication between a computer and a particular stand-alone instrument, the instrument is termed a *VISA resource* and the communication activity is called a *VISA session*. To *query* a VISA resource (i.e., send it a command and then receive back its response), the required VISA session consists of the following four steps: *open the session, write the command message to the resource, read the response from the resource, close the session.* In **Functions>>Instrument I/O>>VISA** (and its subpalette **VISA Advanced**), the following four icons are available to perform the four given steps: **VISA Open**, **VISA Write**, **VISA Read**, and **VISA Close**. To understand how to wire these four icons together to query an instrument, we will first briefly describe the function of each individual icon. In the Help Windows that follow, remember that required inputs are given in boldface text; recommended and optional inputs are in plain and dimmed text, respectively.

The Help Window for **VISA Open** is shown in the following illustration. The job of this icon is to begin a VISA session between your computer and the resource defined at its **VISA resource name** input. The **VISA resource name** consists of text that specifies the interface type being used (e.g., GPIB or USB), followed by resource information (in the case of GPIB, the instrument's address, a number we'll discuss in a few minutes), and ending with the resource type. For our work, the resource type will be a stand-alone instrument denoted by *INSTR*. To pass the **VISA resource name** to other VISA icons, this quantity is available at the **VISA resource name out** output terminal.

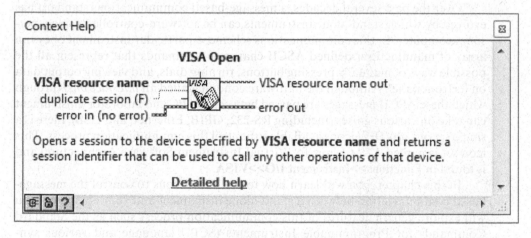

VISA Write, whose Help Window follows next, performs the actual ASCII message transfer from your computer to the stand-alone instrument. Once presented with the open session's **VISA resource name**, this icon writes the ASCII string at its **write buffer** input to the instrument. This string is one of the commands recognized by the instrument and, when received by the instrument, configures it properly for a desired data-taking measurement. Additionally, the VISA resource name is available at the **VISA resource name out** output terminal.

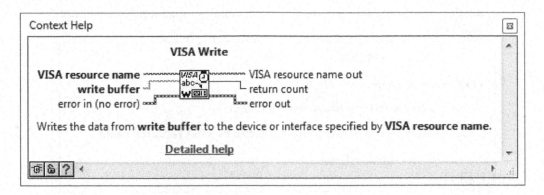

Next, the Help Window for **VISA Read** is shown. This icon transfers the response message from the stand-alone instrument into your computer's memory. When given the open session's **VISA resource name,** this icon receives the ASCII response string consisting of (a maximum of) **byte count** number of bytes and outputs this string at its **read buffer** terminal. This string typically contains the results of a data-taking measurement performed by the instrument. Additionally, the VISA resource name is available at the **VISA resource name out** output terminal.

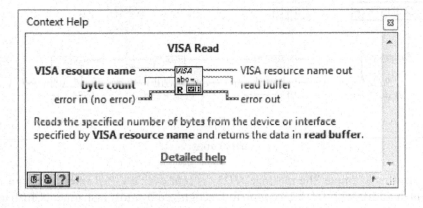

Finally, **VISA Close**'s Help Window is given below. This icon closes the VISA session specified at its **VISA resource name** input.

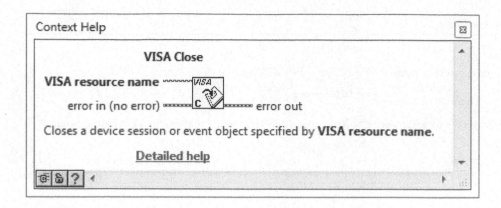

Note that all four of these VISA icons include error reporting via an error cluster, which appears at the **error in** and **error out** terminals.

The four-step VISA session to query an instrument is accomplished by wiring these four VISA icons together as follows.

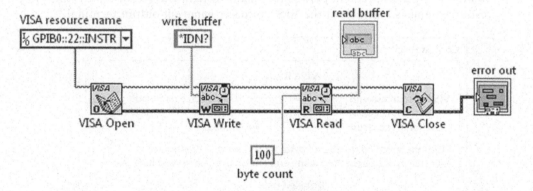

In this example, the message *IDN?* is sent over a GPIB interface to an instrument at address 22. *IDN?*, a command recognized by most instruments, instructs the instrument to identify itself. After receiving this command, the instrument's response (which is an ASCII string consisting of identification information) is received by the computer over the GPIB and displayed in the **read buffer** front-panel string indicator.

Similar to the File I/O and DAQmx icons that you have studied previously, the wiring scheme of VISA icons takes advantage of the principle of LabVIEW programming called *data dependency*. Simply stated, data dependency means that an icon cannot execute until data are available at *all* of its inputs. In the previous diagram, all of **VISA Open**'s required inputs (there is only one—**VISA resource name**) are wired. So, when this diagram is run, **VISA Open** executes immediately. Upon completion, **VISA Open** outputs a VISA resource name at its **VISA resource name**

out terminal, which is passed through the wiring to VISA Write's **VISA resource name** input. Because of data dependency, **VISA Write** cannot execute until it receives the VISA resource name from **VISA Open**. In a similar way, **VISA Read** cannot execute until **VISA Write** completes, and so on. Thus, through this programming scheme, we are assured that the icons will execute in the desired sequence: **VISA Open** followed by **VISA Write** followed by **VISA Read** followed by **VISA Close**.

Also, the correct manner of chaining together VISA icons for error reporting is shown above. If an error does occur at one point in the chain, subsequent icons will not execute and the error message will be passed to the **error out** indicator cluster.

Because VISA-based programming is so robust, you can write highly dependable data-taking programs with just the information already presented. However, with a bit more grounding in the message-based communication scheme, you'll be able to create programs in which you can have near-total confidence. The following paragraphs will take you to the next level of sophistication in stand-alone instrument control.

14.3 THE IEEE 488.2 STANDARD

When remote control of laboratory instruments first became possible, there was a chaotic period during which, more or less, each instrument manufacturer defined its own communications protocol through a unique blend of parallel and serial modes, positive and negative polarities, and assorted handshaking signals. In 1965, Hewlett Packard (which then became Agilent and now Keysight) ended this cacophony by designing a universal instrument interface called the Hewlett Packard Interface Bus (HP-IB) and offered it as the only option on all of its new computer-programmable instruments. Because of its high transfer rates, HP-IB quickly gained popularity with other instrument manufacturers and, in 1975, was accepted as an industry-wide standard known as IEEE-488 or, more commonly, the General Purpose Interface Bus (GPIB). In 1987, an improved version of this standard called IEEE-488.2 was adopted, which enhanced and strengthened message-based communication by specifically defining an instrument's minimally required communication capabilities, a protocol for message exchange, a generic set of commonly needed commands, and a status-reporting system. Today, most computer-controlled laboratory instruments are IEEE 488.2 compliant, even those that communicate over interface buses other than the GPIB (e.g., Ethernet and USB).

14.4 COMMON COMMANDS

One important innovation of the IEEE 488.2 standard was the introduction of a standardized set of *common commands* for the many generic operations that all instruments must perform. The common commands begin with asterisks to delineate them from the other device-specific commands recognized by a particular

TABLE 14.1 Common Commands for IEEE488.2 Compliant Instruments

Mandatory Common Commands	Function
*IDN?	Reports instrument identification string.
*RST	Resets instrument to known state.
*TST?	Performs self-test and reports results.
*OPC	Sets operation complete (OPC) bit in SESR upon completion of the command.
*OPC?	Returns "1" to the output buffer upon completion of the command.
*WAI	Waits until all pending operations complete execution.
*CLS	Clears status registers.
*ESE	Enables event-recording bits in SESR.
*ESE?	Reports enabled event-recording bits in SESR.
*ESR?	Reports value of SESR.
*SRE	Enables a SBR bit to assert the SRQ line.
*SRE?	Reports SBR bits that are enabled to assert the SRQ line.
*STB?	Reports the contents of the SBR.

instrument. All IEEE 488.2 compliant instruments, at the very least, are required to recognize the subset of 13 common commands given in Table 14.1. Many of these commands are related to the reporting of events using two status registers called the *Status Byte Register* (SBR) and *Standard Event Status Register* (SESR), which will be described in detail starting in the next paragraph.

14.5 STATUS REPORTING

Another IEEE 488.2 innovation is a standardized scheme for *status reporting*. This status report system is available to inform you of significant events that occur within each instrument connected to an interface bus. In this scheme, an instrument is equipped with two status registers called the *Standard Event Status Register (SESR)* and the *Status Byte Register (SBR)*. Each bit in these registers records a particular type of event that may occur while the instrument is in use, such as an execution error or the completion of an operation. When the event of a given type occurs, the instrument sets the associated status register bit to a value of 1, if that bit has previously been enabled (as described in the following paragraphs). Thus, by reading the status registers, you can tell what events have transpired.

The Standard Event Status Register, which is schematically shown here, records eight types of events that can occur within a data-taking instrument.

Standard Event Status Register (SESR)

7	6	5	4	3	2	1	0
PON	URQ	CME	EXE	DDE	QYE	RQC	OPC

The eight events associated with the 8 bits of the SESR are described in Table 14.2. In our work, the OPC bit will be most useful.

TABLE 14.2 Eight Standard Event Status Register (SESR) Events

Bit	Associated Events of SESR
7 (MSB)	**PON** (Power On): Instrument was powered off and on since the last time the event register was read or cleared.
6	**URQ** (User Request): Front-panel button was pressed.
5	**CME** (Command Error): Instrument received a command with improper syntax.
4	**EXE** (Execution Error): Error occurred while the instrument was executing a command.
3	**DDE** (Device Error): Instrument is malfunctioning.
2	**QYE** (Query Error): Attempt was made to read the instrument's output buffer when no data were present, or a new command was received before previously requested data had been read from the output buffer.
1	**RQC** (Request Control): Instrument requests to be controller.
0 (LSB)	**OPC** (Operation Complete): All commands prior to and including an *OPC command have been executed.

The SESR exists as an event-signaling tool for you to use in your programs. However, this status register completely lacks initiative and will not perform any work unless you request it to do so. Thus, when initiating communications with an instrument, one of the messages that you may wish to send is an instruction that activates the subset of event-reporting SESR bits that are of interest to you. For instruments that conform to the IEEE 488.2 standard, this activation process is accomplished via the *ESE* (Event Status Enable) command. For example, suppose you wish the QYE bit to be activated and thus record any execution errors in the SESR's bit 2. Since $00000100_2 = 4_{10}$, the QYE bit can be activated by performing a VISA Write of the ASCII command *ESE 4* to the instrument. In our work to come, we will activate the OPC bit with the command *ESE 1*.

The Status Byte Register, which is schematically shown next, records whether data are available in the instrument's output buffer, whether the instrument requests service, and whether the SESR has recorded any events.

Status Byte Register (SBR)

7	6	5	4	3	2	1	0
—	RQS	ESB	MAV	—	—	—	—

The functions of the eight bits of the SBR are described in Table 14.3. The SBR bits are studious, performing their status-reporting duties without need of a request from you.

TABLE 14.3 Functions of Eight Standard Byte Register (SBR) Bits

Bit	Function of SBR Bit
7 (MSB)	May be defined for use by instrument manufacturer.
6	**RQS** (Request Service): The instrument has asserted a SRQ because it requires service from the GPIB controller.
5	**ESB** (Event Status Bit): An event associated with an enabled SESR bit has occurred.
4	**MAV** (Message Available): Data are available in the instrument's output buffer.
3–0	May be defined for use by the instrument manufacturer.

An instrument can be configured to assert a *Service Request (SRQ)* in response to either of two events—an event detected by the Standard Event Status Register or the presence of previously requested data in the instrument's output buffer [i.e., the assertion of the Event Status Bit (ESB) or Message Available (MAV) bit, respectively]. This configuration process is accomplished on IEEE 488.2 instruments using the *SRE* (Service Request Enable) command. For example, if you wish an event detection by the SESR to trigger a request for service by the instrument, initialize the instrument by writing the ASCII command *SRE 32* to the instrument. Since $32_{10} = 00100000_2$, the setting of the SBR's fifth (ESB) bit will then be the criterion for the instrument asserting a SRQ. If, instead, you wish the presence of data in the output buffer (signaled by the MAV bit being set) to trigger a SRQ, then write *SRE 16*. Finally, *SRE 0* will disable the instrument's ability to assert a SRQ.

The relationship between the Standard Event Status Register, the instrument's output buffer, and the Status Byte Register (along with the common commands that configure and query each) is illustrated in Figure 14.1.

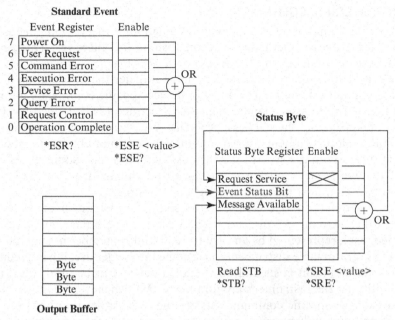

FIG. 14.1 Relationship between Standard Event Status Register (SESR) and Status Byte Register (SBR).

As an alternative to the use of the SRQ, *serial polling* is a common method for determining the status of an instrument. In a serial poll process, the interface bus queries an instrument, and the instrument responds by returning the value of the bits in its Status Byte Register. A serial poll is easily accomplished in LabVIEW using **VISA Read STB**, found in **Functions>>Instrument I/O>>VISA**. This icon's Help Window is shown next.

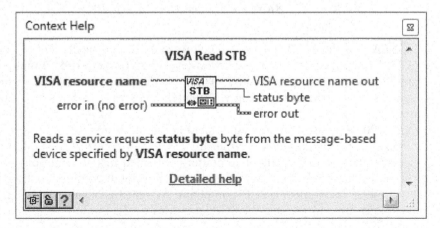

14.6 DEVICE-SPECIFIC COMMANDS

Finally, each stand-alone instrument is designed for a specialized purpose and has its own idiosyncratic methods for accomplishing its objectives. Thus, every programmable instrument comes with a set of *device-specific commands* that allow the user to control its functions remotely and to transfer the information it produces into a computer's memory. The array of device-specific commands for an instrument is listed in its user manual. This set of commands is defined by the instrument's maker ... and therein lies a problem. When surveying the user manuals for programmable instruments of varying models and manufacturers, you will find a great diversity in the style of the various command sets. Some (especially those associated with older model instruments) are an alphabetized collection of cryptic one- or two-character strings (the designers' thinking was obviously "shorter commands yield quicker and, therefore, better computer-instrument communication"). At the other extreme are the user-friendly sets, with similar commands logically grouped, each represented by an easy-to-read-and-remember mnemonic.

As programmable instruments have come into wider use, it has become apparent that development costs and unscheduled delays can be diminished markedly by simplifying the instrument programmers' task whenever possible. Thus, user-friendly device-specific command sets are the rule, rather than the exception, for instruments currently being manufactured. Commonly, these command sets are organized in a hierarchical *tree structure*, similar to the file system used in computers. Each of an instrument's major functions—such as **TRIG**ger, **SENS**e (alternatively, **MEAS**ure), **CALC**ulate, and **DISP**lay—defines a *root* and all commands associated with that root form its *subsystem*. So, for example, to configure the Keysight/ Agilent 34410A Digital Multimeter to measure a DC voltage whose value is expected to fall within the range of ± 10 V (an action within its **SENS**e subsystem), the appropriate command is as follows:

$$SENSe:VOLTage:DC:RANGe\ 10$$

Here, *SENSe* is the root keyword and colons (:) represent the descent to the lower-level *VOLTage*, then *DC*, then the lowest-level *RANGe* keywords. Finally, *10* is a parameter associated with *RANGe*. Although the full command given here can be sent to the instrument, it is only absolutely necessary to send the capitalized characters.

In 1990, a consortium of equipment manufacturers defined the *Standard Commands for Programmable Instruments (SCPI)* in an effort to standardize the device-specific command sets of computer-controlled instrumentation. Although this standard has not been universally adopted, it is not uncommon to discover that your post-1990 instrument is SCPI compliant. As a means of categorizing generally applicable command groups, the SCPI standard posits the following model for a

SIGNAL MEASUREMENT INSTRUMENT

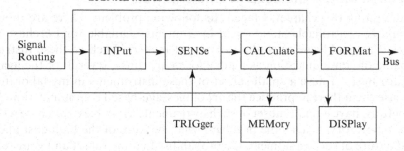

FIG. 14.2

generic programmable instrument. An instrument that performs measurements on an input signal is assumed to have the root functions shown in Figure 14.2. Here, for example, **SENSe** includes any action involved in the actual conversion of an incoming signal to internal data, such as setting the range, resolution, and integration time, while **INPut** consists of actions that condition the signal prior to its conversion, such as filtering, biasing, and attenuation. Alternatively, an instrument that generates signals is modeled by Figure 14.3.

SIGNAL GENERATION INSTRUMENT

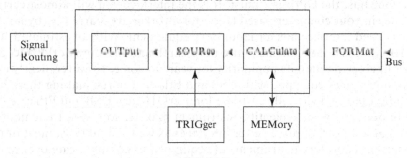

FIG. 14.3

The SCPI command set is organized in a hierarchical tree structure using the syntax illustrated above by the *SENSe:VOLTage:DC:RANGe 10* command. You'll learn more about the SCPI command syntax as you work your way through this chapter. But maybe now is a good time to dive in and actually control a stand-alone instrument.

14.7 SPECIFIC HARDWARE USED IN THIS CHAPTER

In designing this chapter, I faced the following problem. There are thousands of computer-controllable stand-alone instruments available for purchase from the myriad of worldwide scientific instrument makers. Each of these instruments is capable of communicating using one (or many times, several) of the available interface buses. I have a small subset of these instruments in my laboratory and I can use them there to practice the art of message-based communication. You also, hopefully, have a small subset of such instruments available to practice with in your own laboratory. What's the problem? Well, because of the high cost and specialized nature of such equipment, the probability that my subset and your subset have some common instrument is most likely very small. The unfortunate thing about this situation is that each stand-alone instrument is designed to take specialized measurements and understands its own unique set of ASCII commands (which are defined by its maker and are listed in its user manual). Thus, before attempting to control a particular instrument using an interface bus, the programmer must have a detailed understanding of the measurement that the instrument is designed to take, the procedure that it implements in doing its work, and the command list that it recognizes. All of these considerations greatly constrain the writing of a set of generic laboratory exercises that everyone can perform.

That said, I still was faced with the fact that I had to choose a particular instrument and interface bus to work with in this chapter's exercises. With regard to the interface bus, the GPIB is a must. It is the interface you will almost certainly encounter in your computer-based laboratory work and is currently (by far) the most widely used interface bus for laboratory equipment. With an estimated 10 million GPIB-equipped instruments in use in research and industry worldwide, the GPIB most likely will retain its popularity for years to come. However, nearly every new computer comes equipped with USB and Ethernet ports, making these interfaces both familiar and virtually cost-free for users. Hence, USB and Ethernet are gaining in popularity with scientific instrument makers, so I want to demonstrate the use of at least one of these newer interfaces as well. Thankfully, most of the latest models in laboratory instruments are equipped to communicate over multiple interface buses and so a single such device can be used to demonstrate the use of both the GPIB and, say, the USB interfaces.

For the following reasons, the Keysight/Agilent 34410A Digital Multimeter (I'll call it the *34410A DMM* from now on) has many features that make it ideal for our work and so I have chosen it as our "device to practice with" in this chapter's exercises. First, this instrument measures voltage, current, and resistance—vanilla-flavored quantities that require no specialized knowledge to understand (unlike, for instance, the control of grating angle and slit size in a spectrometer). Second, this instrument is equipped to communicate via three possible interfaces: GPIB, USB, and Ethernet (we will use the first two of these in this chapter). Third, for such a high-quality instrument standardly equipped with several interfaces, its price tag of approximately $1400 makes it affordable (at least by scientific instrument standards). Every lab should have one and many do! Fourth, this instrument is both IEEE 488.2 and SCPI compliant and so we will be able to use it to gain experience with generic features of widely adopted communication schemes such as SCPI command syntax and IEEE 488.2 status reporting. Fifth, the 34410A DMM is the successor to the longtime industry-standard Keysight/Agilent/HP 34401A DMM. Thanks to the manufacturer's interest in maintaining backward compatibility between these two instruments, the commands I have chosen to use with the newer 34410A DMM can also be used by 34401A DMM owners (of which there are many) to carry out this chapter's exercises on their instrument's GPIB interface (the 34401A DMM has no USB or Ethernet capabilities).

In the best circumstance, a 34410A (or 34401A) DMM is already available for your use, or, with a modest investment, you can purchase this worthwhile instrument. Then, without need for modification, you can straightforwardly work your way through the given exercises to learn the basics of message-based communication. If, instead, you have some other interface-equipped instrument available, try reading the following pages to understand the generic issues being investigated. Then, by consulting the user manual, it may be fairly easy, for instance, to use the interface-appropriate VISA resource name and substitute an ASCII command string here and there to adapt the exercises to your particular instrument and interface bus. If neither of the above describes your situation, simply read through the following pages. I believe you will learn some valuable features of instrument control that will serve you well in future work.

14.8 MEASUREMENT & AUTOMATION EXPLORER (MAX)

To carry out the exercises in this chapter, you must have a stand-alone instrument equipped with a GPIB and/or USB interface. To communicate over the GPIB, you must also have a National Instruments GPIB device (e.g., PCI-GPIB or PCIe-GPIB) connected properly to your computer, which is in turn connected to the stand-alone instrument by a GPIB cable. To use the USB, simply connect an appropriate USB cable between the stand-alone instrument's USB connector and a USB port on your computer. Additionally, the National Instrument's **NI Device Drivers** (in particular, **NI-VISA**) software, which is included in your LabVIEW installation package, must be correctly installed. To verify that these conditions are met, we will use the handy utility *Measurement & Automation Explorer*, which is nicknamed *MAX*.

To open MAX, either select **Tools>>Measurement & Automation Explorer…** (if you have an open VI or Getting Started Window) or double-click on MAX's desktop icon (if available). After MAX opens, double-click on **Devices and Interfaces** under the **My System** heading at the left. This action will command MAX to determine all of the National Instrument devices present within your computing system.

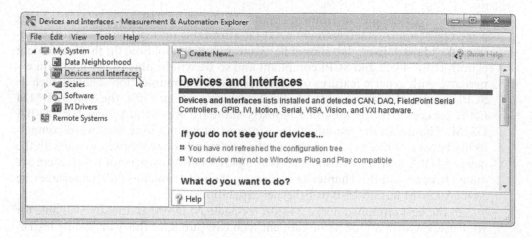

MAX will list the findings of its device survey in hierarchical tree fashion as shown next. If a GPIB device is connected correctly to your computer, an item with the shorthand label **"GPIB0"** will appear in the resulting list (your system may have a different number than *0* in the folder's label). To find all of the stand-alone instruments properly connected to this device, right-click on the **"GPIB0"** item and select the **Scan for Instruments** option. Alternatively, you can click on the **Scan for Instruments** button in the toolbar near the top of the window.

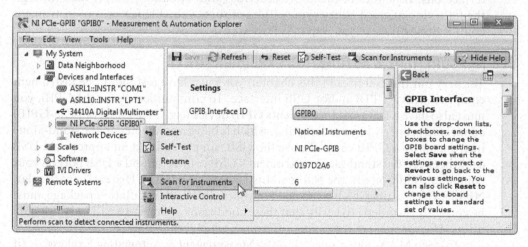

In a few moments, MAX will complete the scan. To view its results, double-click on the **GPIB0** item. For the case shown below, one stand-alone instrument named **34410A** was found.

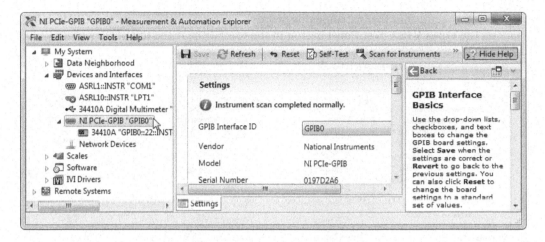

Because up to 15 instruments can be connected to a single GPIB device, each instrument has an identification number called its *GPIB address*. A GPIB address can be any integer between 0 and 30 and is typically defined via a hardware DIP switch setting within the instrument or a sequence of button-pressing and/or knob-turning on its front panel. The instrument's user manual will describe the method for setting its address.

In our example, the GPIB address of the instrument found in the **Scan for Instruments** operation is viewed by clicking on the line with the **34410A** listing. After the click, in the box associated with the **Settings** tab, we find that the instrument's GPIB address is 22 (the address of your instrument may be different) and that the correct **VISA resource name** for this instrument is *GPIB0::22::INSTR*. A text description identifying the instrument also appears.

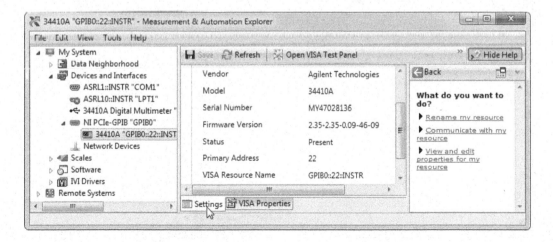

To verify that the instrument is properly communicating over the GPIB, click on the **Open VISA Test Panel** button in the toolbar near the top of the window.

An interactive dialog panel will appear. You might wish to take a few moments to explore some of the interface-related information available in this panel by selecting its various windows (e.g., **Configuration**) and their associated tabs. After exploring, select the **Input/Output** window by clicking on its name.

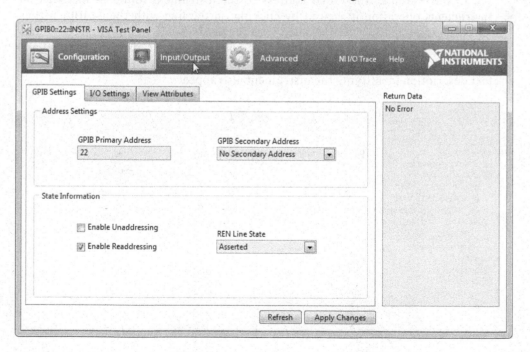

Once in the **Input/Output** window, select its **Basic I/O** tab. Here, after typing an ASCII message in the **Select or Enter Command** box, a mouse-click on the **Query** button will carry out a write-then-read action. That is, the message in **Select or Enter Command** will be written over the GPIB to the addressed instrument, and then the instrument's ASCII response will be read back over the GPIB to the computer and displayed in the large rectangular box at the window's bottom left. When the **Input/Output** window opens, its **Select or Enter Command** box is preloaded with *IDN?*, the IEEE 488.2 common command for an instrument to identify itself (the backslash message terminator \n is also included; more on that later).

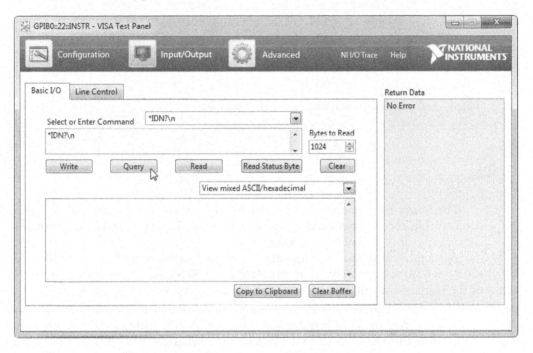

Click on the **Query** button. If your GPIB communication is configured properly, the identification string received from the instrument will appear in the large rectangular box. For the 34410A DMM used here, this identification string identifies the instrument's manufacturer and model information followed by some integers, which denote the version numbers of installed firmware that controls the multimeter's internal microprocessors. Note that this identification string is 60 bytes long.

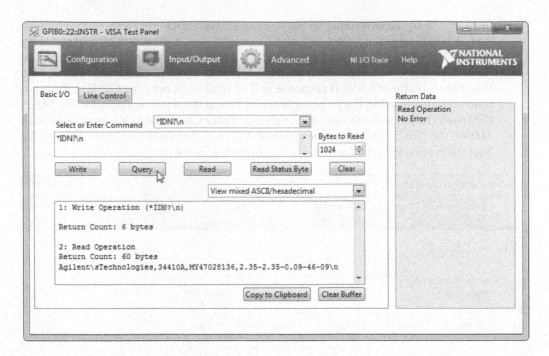

For future reference, this interactive dialog window is a handy tool for use in determining correct command syntax when developing message-based communication programs for a new instrument.

Alternatively, if your instrument is set up to communicate over the USB, after the device survey initiated by the double-click on **Devices and Interfaces**, the name of the instrument will appear in the resulting list. As seen below, included next to its name, the Neptune's Trident symbol ⚕ denotes that this instrument is communicating over the USB interface. By clicking on this listing and then selecting the **Settings** tab, the USB-related **VISA resource name** for your instrument can be found (as seen below, the DMM on my system is named *USB0::0x0957::0x0607:: MY47028136::INSTR*).

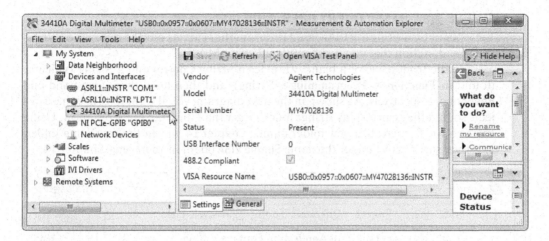

After opening a **VISA Test Panel**, try querying your instrument for its identification string over the USB. You should, of course, get the same result as when querying over the GPIB.

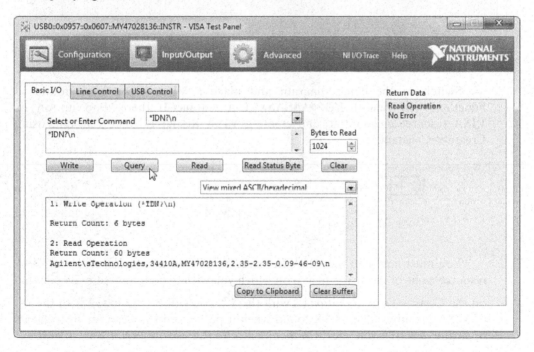

After you are finished, exit this dialog window and then close MAX.

14.9 SIMPLE VISA-BASED QUERY OPERATION

Let's begin by writing a VISA-based program that carries out the query (i.e., write-then-read) action that you just completed using MAX.

On the front panel of a blank VI, place a **String Control** and a **String Indicator** (found in **Functions>>Programming>>String**), and then label them **Command** and **Response**, respectively. As shown in the next diagram, you'll want to resize these objects so that they can display strings much larger than their default sizes allow. Using **File>>Save**, first create a new folder named **Chapter 14** within the **YourName** folder, and then save this VI under the name **Simple VISA Query** in **YourName\Chapter 14**.

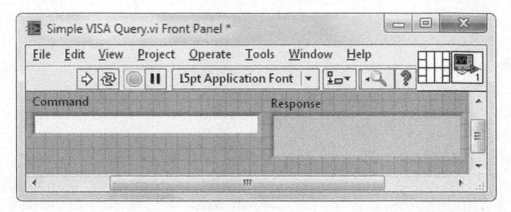

Switch to the block diagram and place a **VISA Open** icon (found in **Functions>>Instrument I/O>>VISA>>VISA Advanced**) there. Pop up on its **VISA resource name** input, and create a **VISA Resource Name Constant** using **Create>>Constant**.

You must now load the **VISA Resource Name Constant** ⌨ with the **VISA resource name** of the instrument with which you wish to communicate. By clicking on the Constant's *menu button* ⌄ with the 🖑, you will be presented with the list of VISA resources that MAX found when it performed the **Scan for Instruments** operation (or after you select the menu button's **Refresh** option), as shown next.

You can then simply choose the desired **VISA resource name** from this list. Alternatively, you can manually enter the appropriate **VISA resource name** for your instrument (as found using MAX) into the [icon]. The syntax for a **VISA resource name** is *Interface Bus Name::Resource Information::Resource Type*.

Here, from the menu button's list, I have chosen the **VISA resource name** appropriate for communicating with my DMM over the GPIB.

If, instead, I wished to communicate with my DMM over the USB, I would choose this available **VISA resource name** from the menu button's list.

Here is the almost-too-good-to-be-true fact: The VISA icons we will use to write our program are interface independent. That is, once the appropriate **VISA resource name** is wired to **VISA Open**, the rest of our VISA-based block diagram will be the same, regardless of whether we are communicating over the GPIB, the USB, or several other possible interface buses (including Ethernet, RS-232, PXI, VXI). All of the low-level details related to the specific interface bus used are taken care of "under the hood" by the VISA icons. We are left to simply direct the communication between the instrument and PC at a generic high level, for example, orchestrating what, and at what time, commands and data are written and read. For the block diagrams in the rest of this chapter, my selected **VISA resource name** will be appropriate for GPIB communication. As just described, however, it is a simple matter to change the **VISA resource name** choice to switch to communication over the USB, if desired.

Complete the block diagram as shown next using the VISA icons found in **Functions>>Instrument I/O>>VISA** (and its subpalette **VISA Advanced**). Wire the **Command** control terminal to VISA Write's **write buffer** input, and wire the **Response** indicator terminal to VISA Read's **read buffer** terminal. When executed, **VISA Read** will read up to N bytes from the selected resource, where N is equal to the integer wired to its **byte count** input. In a moment, we will read the 34410A DMM identification string. Using the VISA Test Panel a few minutes ago, we saw that this identification string is 60 characters long. Thus, wire the **byte count** input to an integer (**U32**) greater than or equal to 60 (I used *100*) as shown. Create the **error out** indicator cluster using the pop-up menu option **Create>>Indicator**.

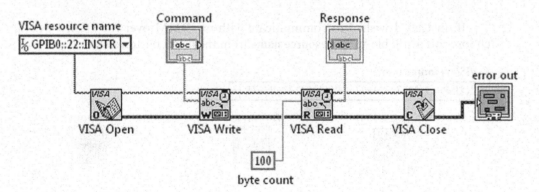

Return to the front panel, arrange the objects nicely, and then save your work.

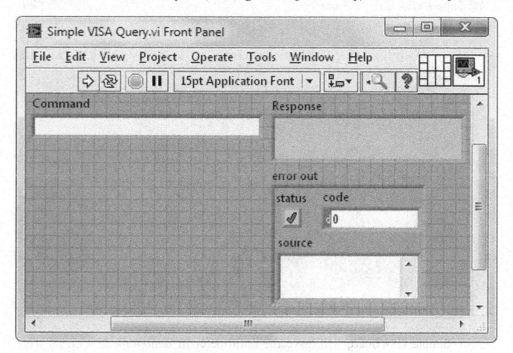

Enter *IDN? into **Command**, and then run your VI. If all goes well, **Response** will display the instrument's identification string upon completion of the VI execution, as shown.

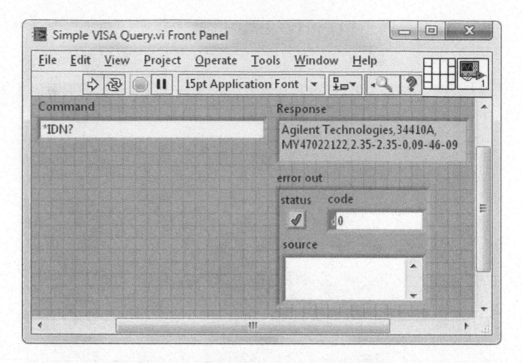

Simple VISA Query will leave the multimeter in *remote mode* with its triggering circuitry "idled." You can return to *local* mode, which continuously "triggers" measurements, by pressing the instrument's front-panel Shift/Local key.

14.10 MESSAGE TERMINATION

At the conclusion of a message-transfer process, some method must be used to signal that the complete message has been passed. The IEEE 488.2 standard appoints the ASCII LF (line feed, also called new line) as its special *end of string* (EOS) character. That is, when receiving a message string, the LF character is always interpreted by the receiver as the last byte of a message. Thus, appending LF to a command string is one method of signaling message termination in IEEE 488.2 communication. Alternatively, the IEEE 488.2 standard allows the assertion of an *end or identify* (EOI) while the last character in the string is being passed as another acceptable termination method. The EOI is a digital signal on a dedicated wire within the GPIB cable. When using VISA icons to control an IEEE 488.2 compliant instrument, message termination is taken care of automatically, allowing you to remain blissfully ignorant of this lower-level activity. If you would like to view an example of this (usually invisible) message termination activity, pop up on the **Response** indicator on the front panel of **Simple VISA Query**, and then select **'\' Code Display**. The \n character you see at the end of the identification string is the backslash code for LF (see Section 7.10). The multimeter appended this termination character to its identification string to alert the receiver (in this case, the GPIB device) that the message has ended.

To deactivate backslash coding, pop up on **Response** and select **Normal Display.**

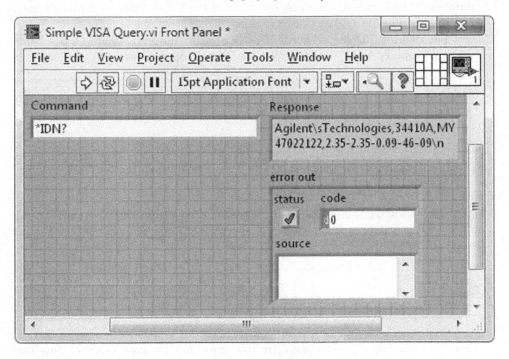

14.11 GETTING AND SETTING COMMUNICATION PROPERTIES USING A PROPERTY NODE

In addition to message termination, there are other low-level functions connected with message-based communication. Many of these low-level functions have an associated parameter setting, which is termed a *VISA property*. VISA assigns default values for these properties, and as long as the VISA-based programs that you write fall within the scope of these default settings, the VISA icons will automatically take care of these low-level functions without any programming effort needed by you (as demonstrated by the message termination example shown above). At times, however, you will most likely write programs that fall outside the range of the default VISA property settings and so will need to assign nondefault values to these quantities.

Reading ("getting") and writing ("setting") VISA property values can be done within your programs using a *Property Node* (and also can often be done in MAX).

As a concrete example of a VISA property, consider **VISA Read**'s *timeout*, which is a fail-safe feature of **VISA Read** that prevents a program from running endlessly if an error occurs. If, for example, an instrument that is being queried doesn't seem to be responsive (perhaps a nameless experimenter forgot to flip on the instrument's power switch), **VISA Read** will only wait for the instrument's response for a certain number of milliseconds (given by the value of the VISA property named **Timeout**) before aborting the read operation and issuing an error message.

The default timeout value for VISA Read on your system can be determined using a **Property Node**. The Help Window for a **Property Node** is shown next.

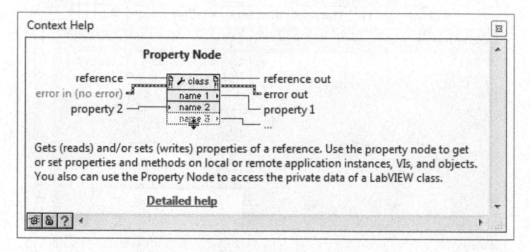

Write the following VI, which reads the current **Timeout** value on your system. Open a new VI, and save it under the name Get Timeout Value in YourName\ Chapter 14. Switch to the block diagram and place a **VISA Property Node** (found in **Functions>>Instrument I/O>>VISA>>VISA Advanced**) there. Then, using **Create>>Constant**, wire the **VISA resource name** for your instrument to the Property Node's **reference** input as shown.

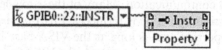

Next, using the 🖑, click on the **Property** terminal and select **General Settings>>Timeout Value**. You might explore what other Properties appear in this menu, many of which are specific to a particular interface bus.

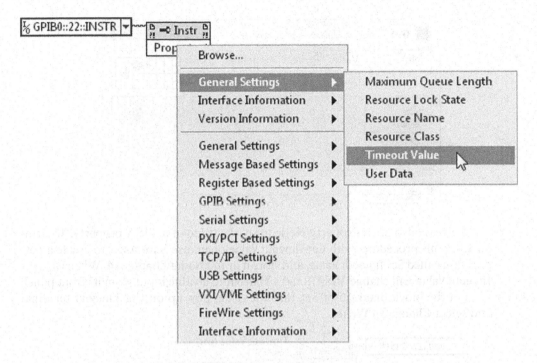

Note that, within the **Timeout** terminal, a small arrow at the right points outward from the terminal's interior. This outward-directed arrow indicates that the **Timeout** terminal is configured in its read mode; that is, it reads ("gets") the current **Timeout** value. Using **Create>>Indicator**, create a front-panel indicator to display the value of **Timeout**.

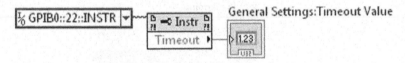

Switch to the front panel, change the indicator label to **Timeout Value (ms)**, and then save your work. Run the VI. As shown below, the (default) **Timeout** value for my system is 3000 ms = 3 s.

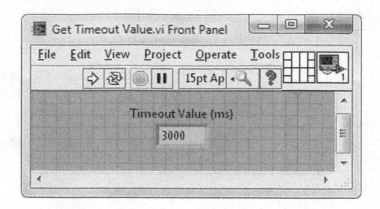

You can also use a Property Node to set the value of a VISA property. To demonstrate this procedure, with **Get Timeout Value** open, use **Save As...** to create a new program called **Set Timeout Value**, and store it in **YourName\Chapter 14**. When run, **Set Timeout Value** will change VISA Read's **Timeout** to a value input from its front panel.

On the block diagram of **Set Timeout Value**, pop up on the **Timeout** terminal and select **Change To Write**.

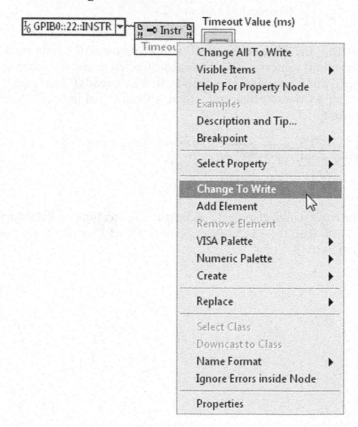

Note that, within the **Timeout** terminal, the small arrow is now at the left pointing inward toward the terminal's interior. This inward-directed arrow indicates that the **Timeout** terminal is configured in its write mode; that is, it writes ("sets") the **Timeout** value. Delete the **Timeout Value (ms)** indicator terminal and then, using **Create>>Control**, create a front-panel control labeled **Timeout Value (ms)**.

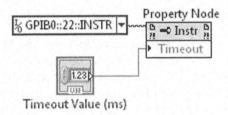

Timeout Value (ms)

Return to the front panel and save your work.

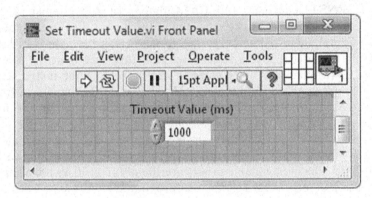

Set **Timeout Value (ms)** to be *1000*, and then run **Set Timeout Value**. Next, run **Get Timeout Value**. Is **Timeout** now equal to 1000 ms = 1 s? Try setting **Timeout Value (ms)** equal to *2500*. You will find that only certain values for **Timeout** are allowed. LabVIEW takes the value you input to **Set Timeout Value** as a suggestion (rather than an order) and sets **Timeout** to the nearest allowed value. When finished, run **Set Timeout Value** one last time, setting **Timeout** equal to *3000*.

14.12 PERFORMING A MEASUREMENT OVER THE INTERFACE BUS

Now that **Simple VISA Query** has given us a template for the VISA query process, let's try controlling a real measurement. Hook up some known DC voltage difference, say, 5 or 6 V, between the HI and LO Voltage Inputs of the 34410A (or 34401A) DMM. This instrument's user manual instructs us that delivering the following sequence of ASCII commands will result in one DC voltage sample being acquired and then loaded into the instrument's output buffer (which is part of its interface circuitry):

CONF:VOLT:DC<Space>10,0.00001

INIT

FETC?

Here is the meaning of this secret code. First, the 34410A DMM can be programmed to perform 14 different types of measurement functions, including DC voltage, AC voltage, DC current, AC current, resistance, frequency, capacitance, and temperature. Given these options, the first command instructs the instrument that we desire to take a DC voltage measurement. The full command is *CONFigure:VOLTage:DC <Space><Range>,<Resolution>* (this command actually executes a collection of commands drawn from the 34410A DMM's **INP**ut, **SENS**e, **TRIG**ger, and **CALC**ulate root subsystems). The command *CONFigure:VOLTage:DC* is constructed in the hierarchical tree structure, typical of SCPI-compliant instruments. *CONFigure* is the root-level keyword and colons (:) represent the descent to the lower-level *VOLTage* and then the lowest-level *DC* keywords. Although the full command can be sent to the instrument, it is only absolutely necessary to send the capitalized characters. Separated from the command *CONF:VOLT:DC* by a *<Space>*, the numeric values for two measurement parameters—*<Range>* and *<Resolution>*—are specified. *<Range>* selects among the instrument's five available voltage measurement scales. Each scale offers a different sensitivity, with *<Range>* giving the maximum measurable value on a particular scale. The five available ranges are 100 mV, 1 V, 10 V, 100 V, and 1000 V. In our situation of measuring a signal of approximately 5 V, the 10 V scale is appropriate. *<Resolution>* specifies the precision of the measurement, with eight levels of accuracy available on the 34410A DMM. For our programs to be compatible with both the 34410A and the older-model 34401A DMM, we will use only two resolution levels in our work. These resolutions have 5 1/2 and 6 1/2 digits of precision (where the 1/2 digit means that the most significant decimal place can only take on a value of "1" or "0"), and so, on the 10 V scale, voltages can be resolved at the level of either 0.0001 or 0.00001 V, respectively. The trade-off in requesting higher accuracy is that the measurement takes a longer time. In the command sequence above, the highest resolution of 6 1/2 digits is selected by setting *<Resolution>* equal to *0.00001* when *<Range>* equals *10*. Note that the syntax of the *CONF* command obeys the conventions of the SCPI language: a *comma* (,) separates the parameters from each other and a *<Space>* separates the command from the parameters.

Once the multimeter has been configured for the desired measurement function as described in the previous paragraph, the data-taking process is begun by sending the *INITiate* command (from the **TRIG**ger root subsystem). Upon receipt of *INIT*, the multimeter will acquire the requested voltage sample and then store this value in its internal memory. Finally, the *FETCh?* command (from the **MEM**ory root subsystem) instructs the instrument to transfer the reading in its internal memory to its interface-related output buffer.

We would like now to place this command sequence into **Simple VISA Query**. Since there are three commands to be sent, it appears that we must modify the VI to include a sequence of three successive implementations of **VISA Write**. Although you are free to do so, a much easier solution is available. The SCPI language allows the programmer to concatenate several commands together into one long multicommand string that can be sent in a single **VISA Write** statement. The syntax for this concatenation process is as follows:

- Use a semicolon (;) to separate two commands within the string.
- Begin a command with a colon (:) if it has a different root level than the command preceding it. The first command in the concatenated string and IEEE 488.2 common commands (which begin with an asterisk) do not require a leading colon.

Since each of our three commands has a different root level, applying the above rules results in the following concatenated string:

CONF:VOLT:DC<Space>10,0.00001;:INIT;:FETC?

Type this command into the **Command** control on the front panel of **Simple VISA Query** as shown next. Run the VI. Your computer will instruct the multimeter to acquire a 6 1/2 digit voltage reading on the 10 V range, retrieve this value, and then display it on the front-panel in the **Response** indicator. Cool, eh?

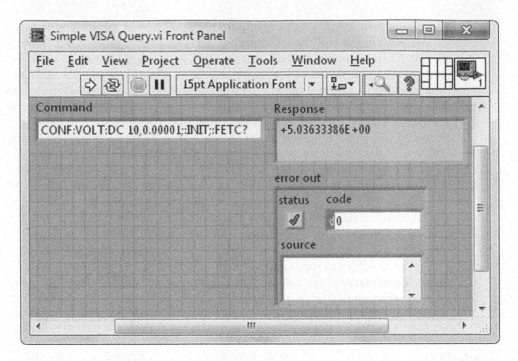

Note that, although extra digits are displayed, the value within **Response** is only accurate to the fifth decimal place.

As shown above, the 34410A DMM reports its data samples in the form of an ASCII character string using the exponential format SD.DDDDDDDDESDD, where S is a positive or negative sign, D is a numeric digit, and E is an exponent. For future reference, note that the string that represents a data sample is 15 bytes long. If you want to use this reading as input to a mathematical calculation (a common situation), you will need to convert the string representation into a numeric format. Such conversion operations can be easily accomplished in LabVIEW using the collection of conversion icons found in **Functions>>Programming>>String**. In the present case, use **Fract/Exp String To Number** in **Functions>>Programming>> String>>String/Number Conversion**. The Help Window for this icon is given next.

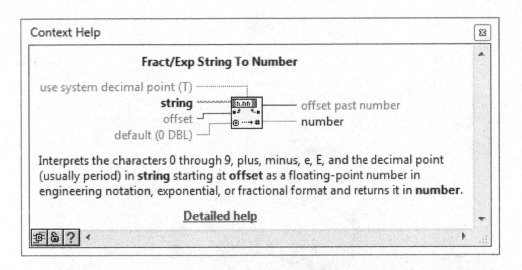

Place a **Numeric Indicator** on the front panel of Simple VISA Query and label it **Numeric Voltage**. Use **Display Format...** in this indicator's pop-up menu to make its **Digits of precision** equal to 5 and disable **Hide trailing zeros**. Then modify the block diagram as follows.

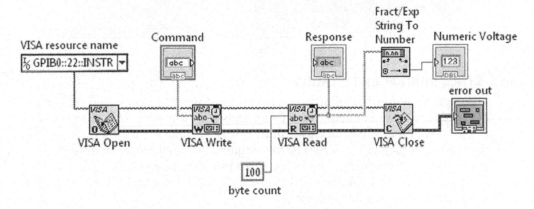

Run the VI to verify that the string-to-number conversion icon performs as expected.

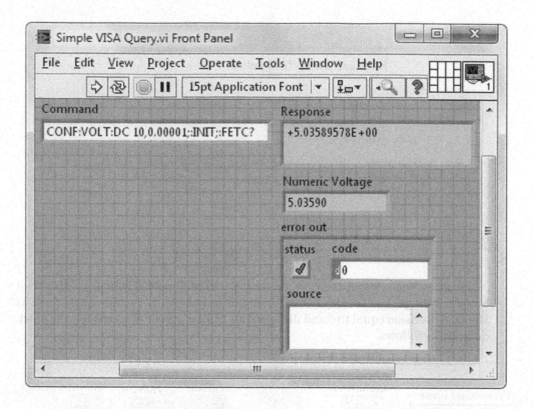

14.13 SYNCHRONIZATION METHODS

Although most ASCII commands are completed quickly after being received by a programmable instrument, some commands start a process that requires a significant amount of time (such as acquiring a large amount of data or moving an object from Point *A* to Point *B*). The time required for such processes must be taken into account when writing a data acquisition program, or else, upon execution, the program may request data before they are available, induce undesirable motion, or cause some other chaotic outcome.

As an example, in its default configuration, the 34410A DMM acquires one data sample after receipt of the *INITiate* command and then stores this measured value in its internal memory. However, through use of the *SAMPle:COUNt <Space><Value>* command, the multimeter can be instructed to take and store multiple data samples upon receiving *INITiate*. The 34410A DMM is configured to acquire 1500 DC voltage samples with 6 1/2-digit resolution via the following concatenated string of commands:

CONF:VOLT:DC<Space>10,0.00001;:SAMP:COUN<Space>1500;:INIT;:FETC?

The *FETCh?* command will load the 1500 acquired samples from the multimeter's internal memory (which, by the way, can hold up to a maximum number of 50000 measured values; the 34401A can store only 512 values) into the instrument's interface-related output buffer.

Let's write a VI that uses the given command string to gather a sequence of 1500 voltage samples. Open **Simple VISA Query**, and then use **Save As...** to create a new VI called **Simple VISA Query (Long Delay)**. Delete **Numeric Voltage** from the front panel and enlarge **Response** so that it can display a very long string (which is the concatenation of 1500 voltage values), and activate its scrollbar by selecting **Visible Items>>Vertical Scrollbar** in its pop-up menu. Type the command given above into the **Command** control (34401A DMM owners, your instrument performs more slowly, so configure it to acquire 15, rather than 1500, i.e., use *SAMP:COUN<Space>15* in your command string). Once entered into **Command**, you can keep this command permanently loaded there by selecting **Edit>>Make Current Values Default**.

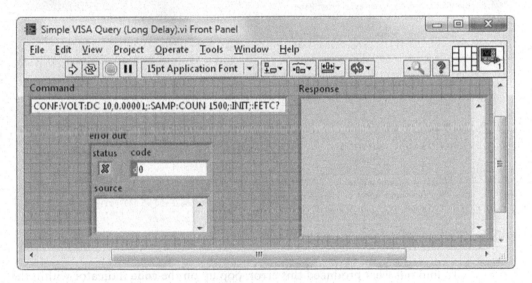

Switch to the block diagram. Delete **Fract/Exp String To Number**. After the given command string is written to the multimeter by **VISA Write**, **VISA Read** will receive a string containing 1500 voltage samples. Since each voltage sample is reported as a 15-bytes string and a (single) delimiting ASCII character will be needed to separate each sample, this 1500-sample string is expected to be about $(1500 \times 15) + 1500 = 24000$ bytes long. Input an integer larger than 24000 to **byte count** as shown.

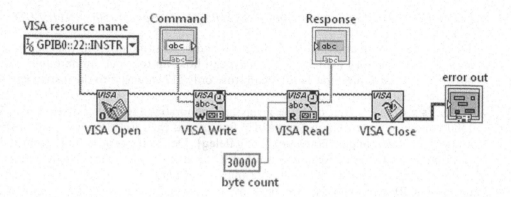

VISA resource name — Command — Response — error out

VISA Open — VISA Write — VISA Read — VISA Close

byte count
30000

Save your work, return to the front panel, and then run **Simple VISA Query (Long Delay)**. Count down the seconds, 3...2...1... Disappointingly, you will find that your VI outputs only a subset of the expected 1500 values and produces an error.

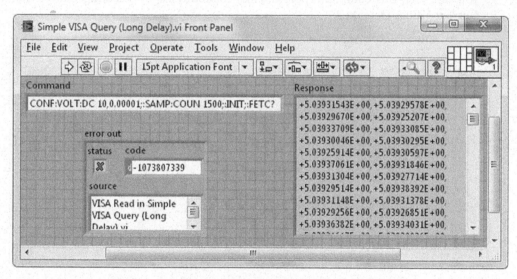

To find out what produced this error, pop up on the **code** indicator within the **error out** cluster and select **Explain Error**.

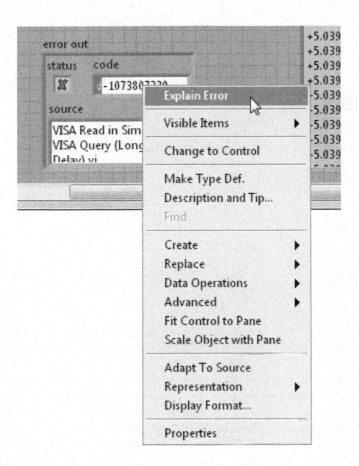

A dialog window appears, where we are told that a "timeout expired" at **VISA Read** before the requested operation (i.e., take 1500 data samples) could be completed.

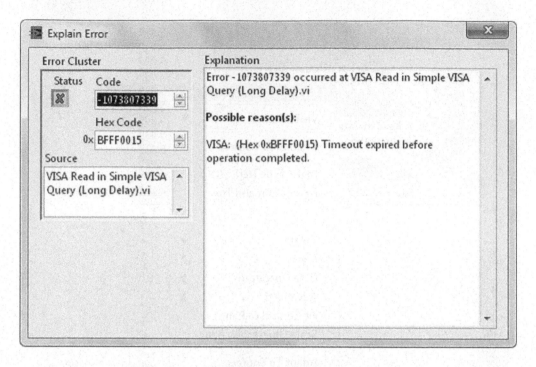

After some head scratching and checking of the 34410A DMM user manual, the following explanation then emerges for the error we observed when running **Simple VISA Query (Long Delay)**. Simply stated, voltage sampling takes time. When configured for 6 1/2-digit resolution, it takes the 34410A multimeter 0.2 power line cycles (PLC) for each voltage sample. Additionally, the 34410A DMM has an optional autozero feature, which operates as follows: After each voltage measurement, the multimeter internally disconnects the input signal and takes a zero reading. The instrument then subtracts the zero reading from the preceding measured value to prevent offset voltages in the multimeter's internal circuitry from affecting measurement accuracy. Since the zero reading also takes the same number of PLC as a regular voltage measurement, in comparison to when autozeroing is not activated, each complete voltage sample by the multimeter takes twice as long. Consulting the DMM user manual, we find that the autozeroing feature is turned off by default for the resolution we have chosen, so each voltage sample takes 0.2 (not 0.4) PLC. Assuming this instrument is plugged into a 60 Hz power source (i.e., 60 PLC per second), 1500 voltage samples will take about

$$1500 \times \left(\frac{0.2\ \text{PLC}}{60\ \text{PLC/s}} \right) = 5\ \text{s}$$

There's the problem! A few moments ago, we found that the default timeout value for **VISA Read** is 3 seconds, but the measurement we have initiated takes about 5 seconds. Thus, before all of the requested data are available, **VISA Read** terminates the execution of **Simple VISA Query (Long Delay)**.

For 34401A users, each data sample by your DMM takes 10 PLC and autozeroing is turned on by default. Thus, 15 samples take $15 \times (20 \text{ PLC}/60 \text{ PLC/s}) = 5$ s.

There are a couple of crude solutions to this dilemma. First, on the block diagram of **Simple VISA Query (Long Delay)**, you can insert a single-frame **Sequence Structure** into the VISA execution chain that simply contains a **Wait (ms)** icon, wired to produce a delay of greater than 5 seconds between the issuance of the data-taking command and the order to read the gathered data samples. The resulting diagram would appear as follows.

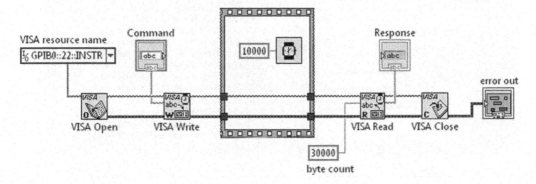

Second, for a slightly more elegant fix, you can use a **Property Node** to change the **Timeout** value for **VISA Read** from its default value (on my system) of 3000 ms to something larger than 5 seconds. Use this approach to modify the block diagram of **Simple VISA Query (Long Delay)** as shown below. Here, the **Timeout** value is chosen to be 10 seconds.

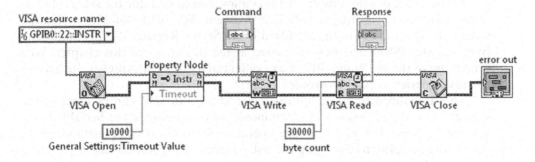

Return to the front panel of **Simple VISA Query (Long Delay)**. Turn your DMM off and then on again to clear the unread data in its output buffer from the previous unsuccessful run. Then, with the command to perform 1500 samples programmed into **Command**, run the VI (34401A DMM owners, request 15 samples, not 1500). About 5 seconds later, you should see something like the following front panel. Note that the delimiter used by the multimeter to separate neighboring data values is a comma.

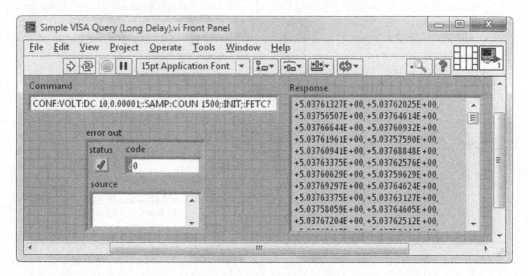

In the preceding example, we found that with a detailed knowledge of the measurement process being implemented, it was possible to troubleshoot a malfunctioning VISA-based VI. Please note that lack of communication (in particular, the interface bus not correctly knowing when the instrument's data will be available) is the root problem that led to the malfunction.

Fortunately, powerful tools exist that allow one to monitor the status of tasks being performed by a programmable instrument. For IEEE 488.2-compliant instruments, these tools are the Standard Event Status Register (SESR) and Status Byte Register (SBR) that were discussed at the beginning of this chapter. With proper use of the SESR and SBR, many potential data-taking glitches, such as the one just experienced, can be avoided.

The status-reporting capabilities of the SESR and SBR can be employed in several ways. We will explore two commonly used techniques: the Serial Poll and Service Request Methods. The core operation for both of these methods is the same—the completion of an assigned task triggers the Operation Complete (OPC) bit in the Standard Event Status Register to be set, which in turn sets the Event Status Bit (ESB) of the Status Byte Register.

In the Serial Poll Method, the setting of ESB is detected by directly checking the Status Byte Register, whose state is obtained by serial polling the instrument. The complete step-by-step process of this method is shown in Figure 14.4.

Serial Poll Method

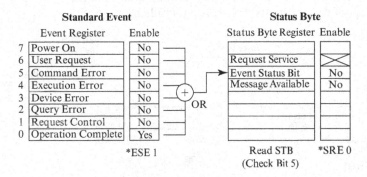

FIG. 14.4

In the Service Request Method, the Status Byte Register is configured such that, when its ESB is set, the Request Service bit is induced to be set also. This action then causes the instrument to assert a SRQ, which alerts the interface bus that the assigned operation is complete. This method is pictured in Figure 14.5.

Service Request Method

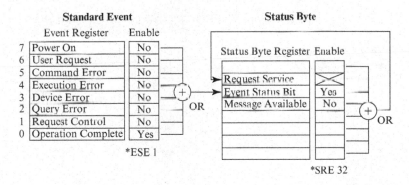

FIG. 14.5

In the following sections, we will write VIs that implement both of these approaches to status reporting.

14.14 MEASUREMENT VI BASED ON THE SERIAL POLL METHOD

Let's try the Serial Poll Method first. To configure the 34410A (or 34401A) DMM for status reporting using the Serial Poll Method, write the following VI called **Status Config (Serial Poll)** and save it in **YourName\Chapter 14**. First, code the VI's block diagram as shown next. Use the autocreation feature in pop-up menus to create all of the constants, controls, and indicators.

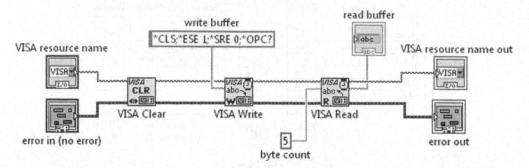

Switch to the front panel and arrange the objects logically. Design an icon and assign the connector pane's terminals consistent with the Help Window shown.

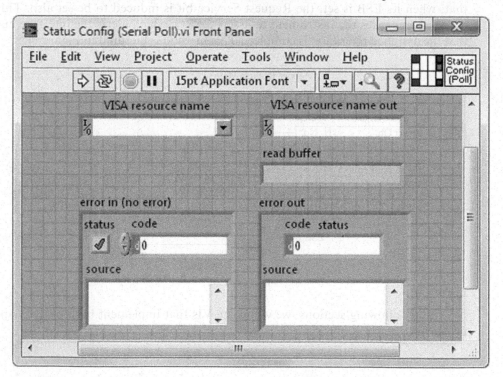

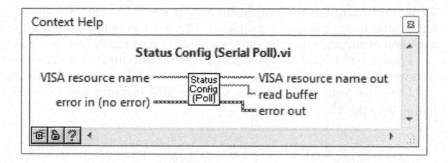

Here's how the VI works, assuming that the instrument referenced by **VISA resource name** is IEEE 488.2 compliant. Within the chain of VISA icons, **VISA Clear** (found in **Functions>>Instrument I/O>>GPIB**) executes first. The Help Window for this icon is shown below.

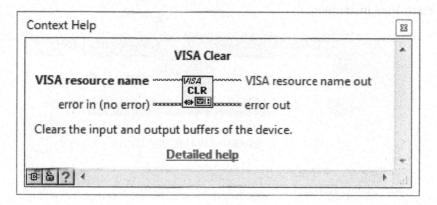

Although not an absolute necessity for inclusion in **Status Config (Serial Poll)**, this VI performs the precautionary action of "clearing" the instrument. **VISA Clear** instructs the instrument to abort all measurements in progress, disable its triggering circuitry, clear its interface-related output buffer, and prepare to accept a new command string.

Next, **VISA Write** sends the concatenated command string **CLS;*ESE 1;*SRE 0;*OPC?* to configure the instrument for status reporting using the Serial Poll Method. Note that since the component strings are all IEEE 488.2 common commands, leading colons are not required in the concatenation. In this sequence of commons, **CLS* clears the contents of the SESR and SBR. As described in the beginning of this chapter, **ESE 1* enables the SESR's OPC bit to set the ESB in the Status Byte Register and **SRE 0* disables the instrument from asserting a SRQ.

Then *OPC?* requests the instrument to return a "*1*" to the instrument's output buffer after this command is completed. This last command is included simply as a method of checking that the entire sequence of commands has been executed.

Finally, **VISA Read** reads the contents of the instrument's output buffer. If all goes well, there should be a single ASCII character "1" read into the computer.

Test-drive your VI as follows. Click on **VISA resource name** control's menu button with the 🖑.

From the list presented, select the **VISA resource name** for your computer-controlled instrument.

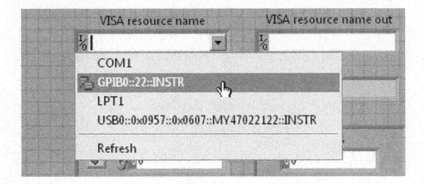

Then run **Status Config (Serial Poll)**. Upon completion, does the **Buffer Reading** string indicator display an ASCII character "1"?

Next, construct a VI called **Serial Poll**, which continuously reads the Status Byte Register of an instrument until a given bit is set. A suggested coding of **Serial Poll** is shown in the following diagrams, and explanations of the unfamiliar icons are in the subsequent paragraphs. Save **Serial Poll** in **YourName\Chapter 14**.

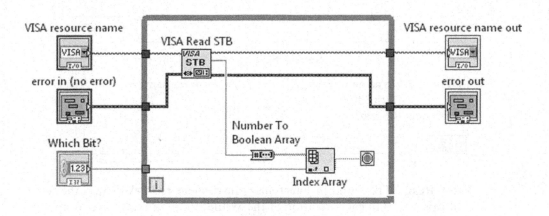

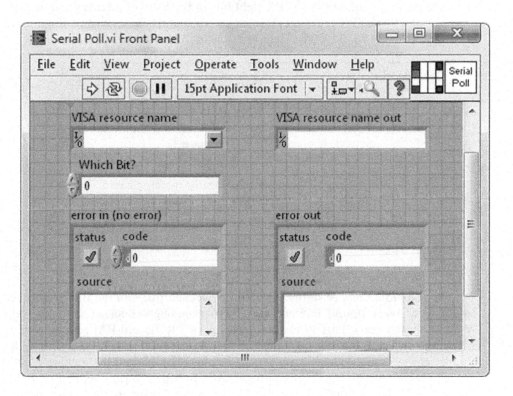

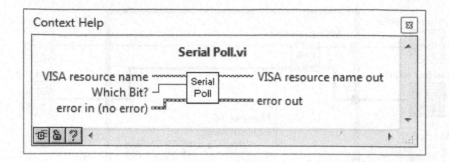

VISA Read STB, found in **Functions>>Instrument I/O>>VISA**, is the work-horse of this VI. With each iteration of the While Loop, its **status byte** output returns the current values of the SBR's eight bits in the form of an unsigned integer. For example, if the SBR's fifth bit (ESB) is set, then **status byte** outputs the integer 32, since $00100000_2 = 32_{10}$. The Help Window for **VISA Read STB** is shown next.

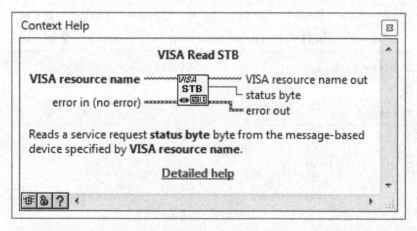

The individual bits of **status byte** can be checked through the use of **Number To Boolean Array** (found in **Functions>>Programming>>Boolean** with its Help Window shown next). This VI creates an array of TRUE and FALSE values that mirror the sequence of zeros and ones (starting from the least-significant bit) in the binary representation of the integer input **number**. For example, if **number** equals the decimal integer 48, then the **Boolean array** output will be [F, F, F, F, T, T, F, F], since $48_{10} = 00110000_2$. **Index Array** can then be used to ascertain the value of a particular element in this array. **Serial Poll**'s While Loop will continue to iterate until **Which Bit?** becomes TRUE.

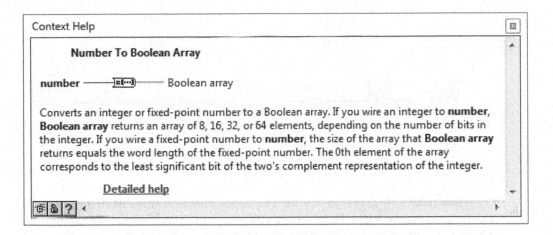

Run Serial Poll under **Highlight Execution** (which is activated by clicking the ☐ in the toolbar) and, through your observations, gain a better understanding of its operation. Remember to input values for **VISA resource name** and **Which Bit?** on the front panel. When run in this isolated manner, the VI will most likely never be able to exit the While Loop, so you'll have to stop it using the **Abort Execution** button in the toolbar. Also, turn off **Highlight Execution** by clicking the ☐.

We're finally ready to write **VISA Query (Serial Poll)**. This top-level program implements serial polling to synchronize the interface bus activities necessary in acquiring 1500 voltage samples using a 34410A multimeter.

Open **Simple VISA Query (Long Delay)**, and then use **Save As...** to create **VISA Query (Serial Poll)**. The front panel can remain unchanged. Type the following command into the **Command** control, much of which may already be there by default; be sure to include the *OPC (on the older 34401A model DMM, use *15* rather than *1500*).

CONF:VOLT:DC<Space>10,0.00001;:SAMP:COUN<Space>
*1500;:INIT;*OPC;:FETC?*

Switch to the block diagram and modify it as shown below, with **Status Config (Serial Poll)** and **Serial Poll** used as subVIs.

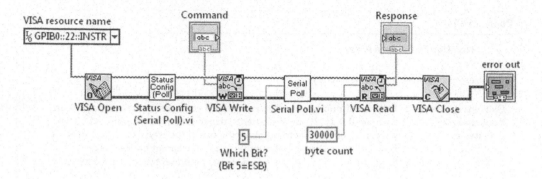

Here is how this diagram works: The concatenated command string is sent to the instrument by **VISA Write**. After configuring the multimeter for the desired DC Voltage measurement function, the acquisition process is begun by the *INIT* command. The succession of 1500 samples is acquired and temporarily stored in the multimeter's internal memory. After the 1500th sample is obtained, **OPC* instructs the instrument to set its SESR's OPC bit (which, in turn, sets the SBR's ESB), and then *FETC?* loads the contents of the internal memory into the instrument's output buffer. At that point, **Serial Poll** detects the setting of ESB, which then triggers the instrument's output buffer to be read by **VISA Read**. One might be tempted to write the concatenated command with **OPC* after *FETC?*, rather than sandwiched between *INIT* and *FETC?*, as above. It is best, however, to avoid sending **OPC* after a query (a query is a command like *FETC?* that ends in a question mark) because such commands cause a message to be loaded into an instrument's output buffer. If the message exceeds the finite size of the output buffer, the query must be immediately followed by **VISA Read** as the program executes in order to read the long message string over the bus successfully.

Return to the front panel, save your work, and then run **VISA Query (Serial Poll)**. Does the VI obtain the requested 1500 DC voltage samples successfully? If so, try running it again with **Highlight Execution** activated for both **VISA Query (Serial Poll)** and its subVI **Serial Poll**. This exercise will illustrate the weakness of the Serial Poll Method, namely, the large volume of interface bus traffic required by this technique. During the 5 seconds while the 1500 data samples are being gathered, the instrument is polled numerous times by the interface bus so that its status can be continuously monitored. Although effective, the Serial Poll Method is rather inefficient because of its excessive use of the interface bus and processor time.

14.15 MEASUREMENT VI BASED ON THE SERVICE REQUEST METHOD

The Service Request Method provides status reporting with a minimum of interface bus activity. To configure an IEEE 488.2 compliant instrument for status reporting using the Service Request Method, open **Status Config (Serial Poll)**, and then create **Status Config (SRQ)** using **Save As...** and save it in **YourName\Chapter 14**. The front panel and terminal assignments can remain as they are, but the icon should be redesigned as shown here.

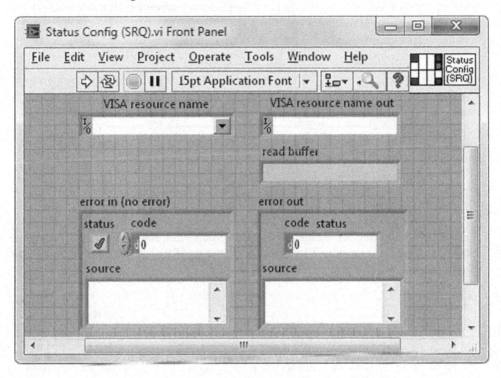

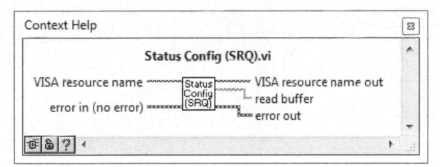

Only two modifications of the block diagram are needed. First, by changing *SRE 0* to *SRE 32* in the command string sent to the instrument, the instrument will assert the SRQ line when the SBR's fifth (ESB) bit is set. The already present *ESE 1* command configures the instrument to set the ESB in response to the setting of the SESR's OPC (Operation Complete) bit. Second, for VISA icons to detect service request (SRQ) events during this VISA session, **VISA Enable Event**, with **Service Request** wired to its **event type** input, must be included in the diagram as shown. **VISA Enable Event** is found in **Functions>>Instrument I/O>>VISA>>VISA Advanced>>Event Handling**.

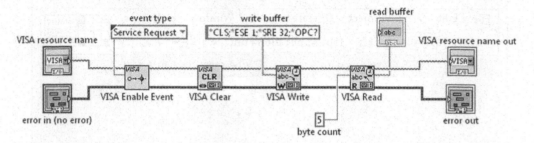

Save your work as you close this VI.

Now, let's write a program to acquire 1500 voltage samples from a 34410A DMM using the Service Request Method. Open **VISA Query (Serial Poll)**, then use **Save As...** to create a new VI named **VISA Query (SRQ)**, and store it in **YourName\ Chapter 14**. The front panel is fine as it is.

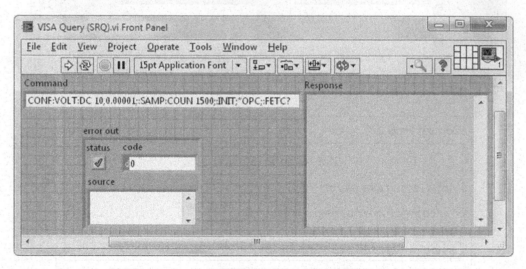

Switch to the block diagram and modify it as shown next.

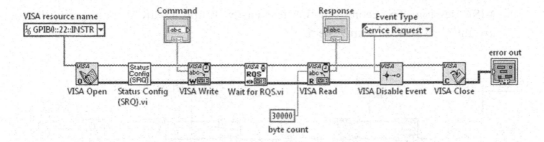

Here, **VISA Disable Event** (found in **Functions>>Instrument I/O>>VISA>>VISA Advanced>>Event Handling**) must be included to disable VISA servicing of SRQ events before the VISA session is closed.

Wait for RQS.vi—also found in **Functions>>Instrument I/O>>VISA>>VISA Advanced>>Event Handling** (Help Window shown below)—sits idly until the instrument denoted by **VISA resource name** asserts a SRQ. However, there is a limit to the patience of this icon. It will only wait up to a total time of **timeout**, with a default value of 25000 ms = 25 seconds. Because our measurement requires only 5 seconds, we can go with this default value by keeping the **timeout** input unwired, as shown in the above diagram.

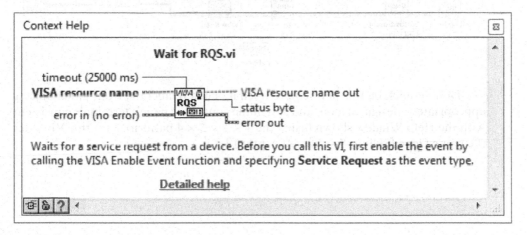

Save your work and then run **VISA Query (SRQ)**. Does it successfully acquire the requested 1500 DC voltage samples from your DMM? Do you understand the operation of this program and how the Service Request Method manages to work with a minimum of interface bus activity?

To simplify the block diagram of **VISA Query (SRQ)**, you might consider packaging **VISA Disable Event** and **VISA Close** together in a subVI called **Close (SRQ)**, since both of these icons are involved in closing down the service request-based

VISA session. To accomplish this feat easily, simply create a highlighting box around the two icons using the ⬆.

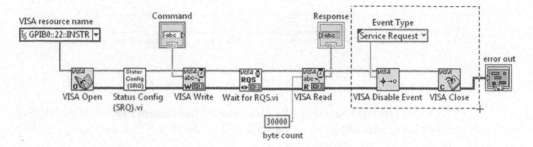

Then select **Edit>>Create SubVI**. A new subVI icon will appear wired on your diagram.

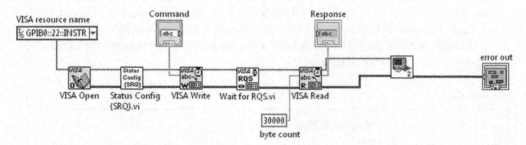

Double-click on this new icon to open it. Then relabel the front-panel objects appropriately, design an icon, and assign the connector pane's terminals consistent with the Help Window shown (using the $4 \times 2 \times 2 \times 4$ pattern). Save this VI under the name **Close (SRQ)** in **YourName\Chapter 14**.

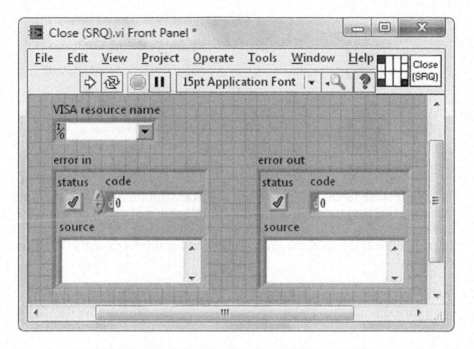

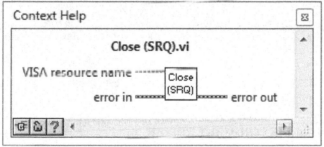

Switch to the block diagram of **Close (SRQ)**. It should appear as follows.

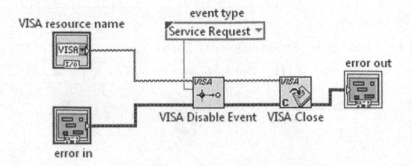

Close **Close (SRQ)**, and return to the block diagram of **VISA Query (SRQ)**. You may have to delete the originally created subVI and load a new copy of **Close (SRQ)** in its place, using **Functions>>Select a VI....** After that, the finished block of **VISA Query (SRQ)** will appear as shown next. Try running this VI to verify that it functions correctly.

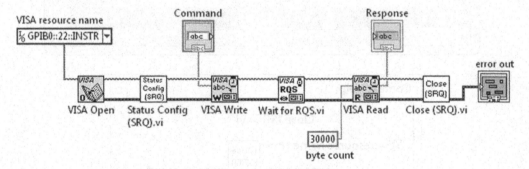

14.16 CREATING AN INSTRUMENT DRIVER

An instrument driver is a collection of modular software routines that perform the operations required in the computer control of a programmable instrument. These operations include configuring, triggering, status checking, sending commands to, and receiving data from, the instrument. Above, **Status Config (Serial Poll)** and **Status Config (SRQ)** are examples of configuration VIs that would be useful to include as part of the 34410A DMM instrument driver. We will now write another configuration VI, this time one that prepares the multimeter for taking a desired measurement.

The 34410A DMM is capable of implementing 14 types of measurement functions, including DC and AC voltage, DC and AC current, 2- and 4-wire resistance (2-wire is the "normal" method for measuring resistance; the more involved 4-wire technique is necessary only when measuring very small resistance samples), frequency and period of an AC signal, continuity, diode check, capacitance, and temperature. To gain experience with some of the LabVIEW tools available for developing instrument drivers, let's write a driver that offers the choice of configuring the 34410A DMM for a DC voltage, AC voltage, or 2-wire resistance measurement. You, of course, can be more ambitious and write your VI to control up to all 14 possible measurement functions.

Referring to the 34410A DMM user manual, we find that our driver must allow a user to select one of the following three possible commands to configure the instrument for the desired measurement function:

CONFigure:VOLTage:DC <Space> <Range>, <Resolution>
CONFigure:VOLTage:AC <Space> <Range>, <Resolution>
CONFigure:RESistance <Space> <Range>, <Resolution>

Here, the possible values of *<Range>* for both the DC and AC voltage measurements are 0.1, 1, 10, and 100 V (we ignore the highest range because it's different—1000 vs. 750—for DC and AC voltage). For the resistance measurement, the allowed *<Range>* values are 100, 1 k, 10 k, 100 k, 1 M, 10 M, and 100 M ohms (here, we ignore the 1 Gohm range because it is not available on the older 34401A model). In all cases, the measurement precision may be 5 1/2 or 6 1/2 digits, which corresponds to *<Resolution>* being 10^{-5} or 10^{-6} times the *<Range>* value, respectively.

We will write two programs called **Range and Resolution Decoder** and **Command String**, which will allow a user to construct the desired command string using front-panel controls. On **Range and Resolution Decoder**, given range and resolution choices from a user-friendly front-panel listing of the multimeter's available offerings, the program will convert these choices to the double-precision floating-point numeric format needed in **Command String**. **Command String** will construct the appropriate ASCII command string to be sent to the multimeter, based on selections made on its front-panel controls.

Create a new VI named **Range and Resolution Decoder** and save it in **YourName\ Chapter 14**. Place four **Enum** controls (found in **Controls>>Modern>>Ring & Enum**) on the front panel and label them **Function, Voltage Range, Resistance Range**, and **Resolution**, respectively. Pop up each **Enum**, select **Edit Items...**, and then program it with the items, in the given order, shown in the following list.

> **Function:** *DC Voltage, AC Voltage, Resistance*
> **Voltage Range:** *100 mV, 1 V, 10 V, 100 V*
> **Resistance Range:** *100 ohm, 1 kohm, 10 kohm, 100 kohm, 1 Mohm, 10 Mohm, 100 Mohm*
> **Resolution:** *5 1/2 Digits, 6 1/2 Digits*

Then place these four **Enum** controls in a cluster shell (found in **Controls>> Modern>>Array, Matrix & Cluster**) labeled **Function Parameters** as shown next.

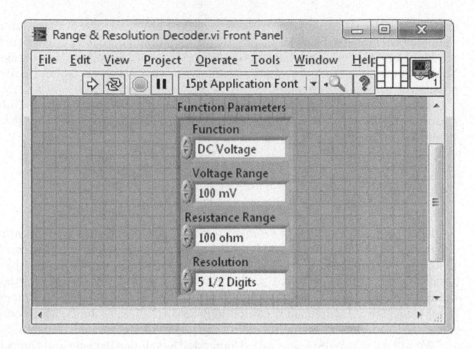

Switch to the block diagram, place a **Case Structure** there, and complete the code shown. Pop up on the Case Structure's border and select **Add Case for Every Value** and then verify that it has three cases labeled **DC Voltage**, **AC Voltage**, and **Resistance**.

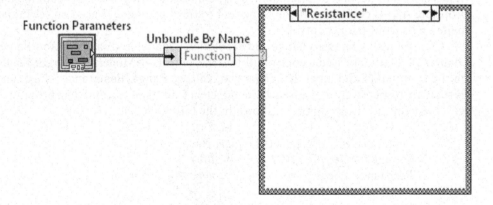

Select the **DC Voltage** case, and then place an **Index Array** icon within it. Pop up on Index Array's **n-dimension array** input and select **Create>>Constant** to create an **Array Constant** and label it **Voltage Ranges**. Next, program the index-0 through index-3 elements of this **Array Constant** as *0.1*, *1.0*, *10.0*, and *100.0*, respectively. The **Array Constant** then will serve as a *look-up table* of the multimeter's allowed voltage ranges, given as double-precision floating-point numbers. Complete the code for the **DC Voltage** case shown below. Here, the integer associated with a selected **Voltage Range** on the front-panel Enum control provides the index of the desired look-up table element. This element is then output by **Index Array**.

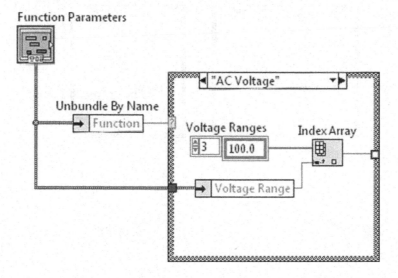

Clone the **Voltage Ranges** Array Constant (mouse-click while holding down *<Ctrl>*), and place the copy somewhere on the block diagram. Then switch to the **AC Voltage** case and (using your cloned **Voltage Ranges**) write the code shown.

Finally, switch to the **Resistance** case, and program it as shown. Here, the index-0 through index-6 elements of the **Resistance Ranges** Array Constant are *1.0E2, 1.0E3, 1.0E4, 1.0E5, 1.0E6, 1.0E7*, and *1.0E8*, respectively.

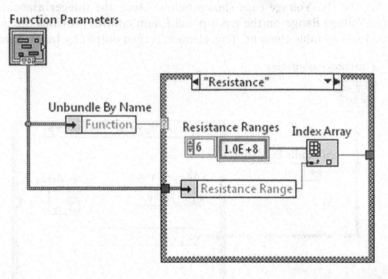

Add a second **Case Structure** and complete the diagram as shown next. The **Range & Resolution** indicator cluster is created by popping up on **Bundle** and using **Create>>Indicator**.

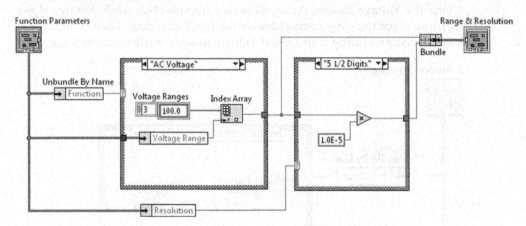

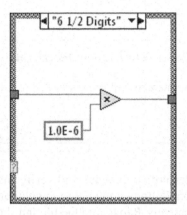

Return to the front panel. Within the **Range & Resolution** indicator cluster, name the owned labels of the top and bottom arrays **Numeric Indicator** as **Range** and **Resolution**, respectively, by popping up on each array's index display and selecting **Visible Items>>Label** (do not carry out the labeling using the [A] as this will create free, rather than owned, labels). Design an icon and assign the connector pane's terminals as shown. Save your work.

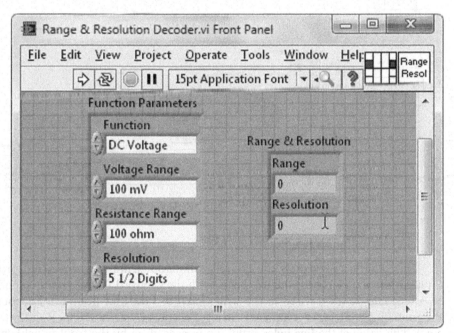

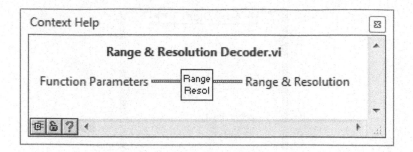

Run **Range and Resolution Decoder** and verify that it functions properly. For example, with **Function**, **Voltage Range**, and **Resolution** equal to *DC Voltage*, *10 V*, and *6 1/2 Digits*, respectively, **Range** and **Resolution** should equal *10.0* and *0.00001*.

Next, open a blank VI and save it under the name **Command String** in **YourName\Chapter 14**. Switch to the block diagram and write the following code, which constructs the desired ASCII command string. The **Function Parameters** control cluster and **output string** string indicator are made using the autocreation feature in the pop-up menus.

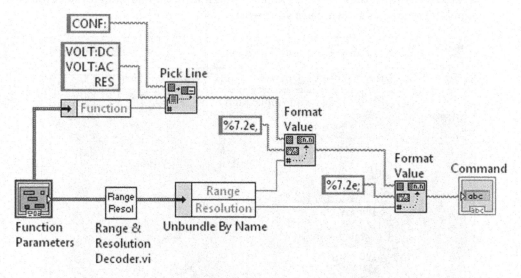

This diagram constructs the desired command string in a three-step process. First, all three possible commands begin with the keyword *CONF:*, so this sequence of ASCII characters is wired to the **string** input of **Pick Line** (found in **Functions>> Programming>>String>>Additional String Functions** with its Help Window given next). The value of the **line index** input (an integer given by the front-panel **Function Enum** control) then selects which of the three possible lines programmed into the **String Constant** wired to **multi-line string** is to be appended to *CONF:*. Create the three lines in this **String Constant** by the following sequence of keystrokes: *VOLT:DC<Space>* *<Enter>VOLT:AC<Space> <Enter>RES<Space>*. Be sure to include the *<Space>* character at the end of each command string. You can make the invisible space and line feed characters visible by popping up on the **String Constant** and selecting '\' **Codes Display**. The correct entry will then appear as *VOLT:DC\s\nVOLT:AC\s \nRES\s*, where \s and \n are backslash codes for space and line feed, respectively.

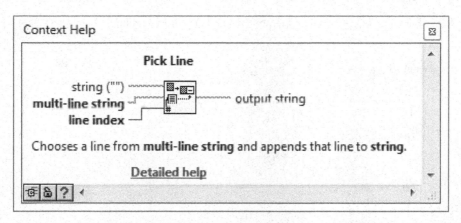

Format Value, from **Functions>>Programming>>String>>String/Number Conversion** (Help Window shown next), then is used to attach two more string fragments, each with embedded ASCII-coded numbers that program the *<Range>* and *<Resolution>* settings of the multimeter. The **Format Value** icon takes the number at its **value** input and converts it to an ASCII string representation with the format defined at its **format string** input. This ASCII string is appended to **string** and presented at **output string**. In the above diagram, the scientific notation format *%7.2e* (see Section 6.6) is used for both *<Range>* and *<Resolution>* parameters. Note that a comma (,) and semicolon (;) follow *<Range>* and *<Resolution>*, respectively.

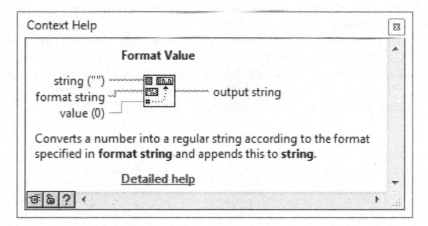

Switch to the front panel and change the label of the String Indicator from **output string** to **Command**. Run the VI with a given choice of the controls within **Function Parameters** and verify that the correct command string appears in the **Command** indicator. Save your work.

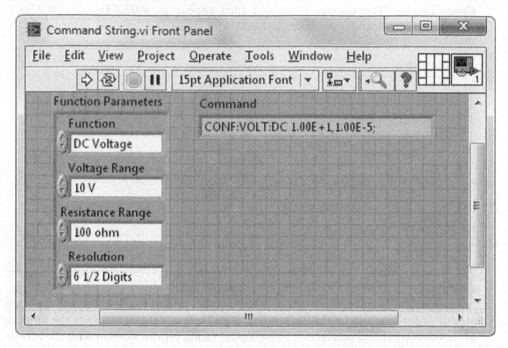

We will next add the ability to control the multimeter's autozeroing feature and to program the desired number of data samples to be taken. Add a **Push Button** (found in **Controls>>Modern>>Boolean**) and a **Numeric Control** to the front panel and label them **Autozero** and **Sample Count**, respectively. Change the representation of **Sample Count** to **U16**.

Switch to the block diagram and then include the autozero and sample count code shown below. The *%5d* format in the *SAMPle:COUNt* command specifies a five-place decimal integer because the maximum allowed value for *SAMPle:COUNt* (according to the 34410A DMM user manual) is 50000. The format string entry for this command should be *:SAMP:COUN<Space>%5d*.

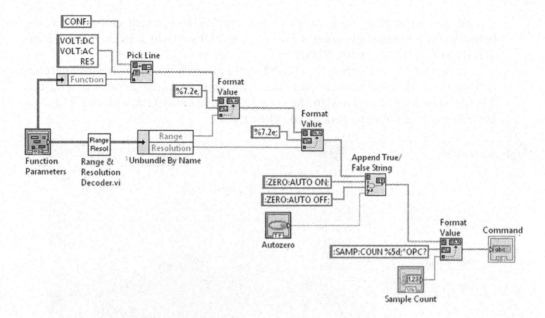

Autozero can either be turned on or off with the following commands:

$$:ZERO:AUTO<Space>ON$$
$$:ZERO:AUTO<Space>OFF$$

Append True/False String (found in **Functions>>Programming>>String Additional String Functions**), whose Help Window follows, provides an easy way to choose which of these two choices is concatenated to the command string. Remember to include the leading colon and final semicolon in the **false string** and **true string** entries to assure proper command concatenation.

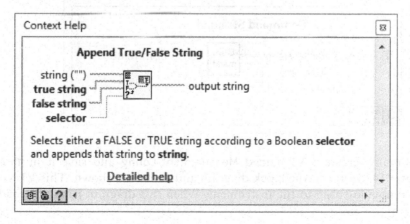

To guarantee that the instrument fully processes the sent command string before exiting the configuration VI (which you will write in a moment), the command string concludes with *OPC?*.

Return to the front panel. Run the VI with a given choice of the front-panel controls to verify that the correct command string appears in the **Command** indicator. Then design an icon and assign the connector pane consistent with the following Help Window. Save your work as you close the VI.

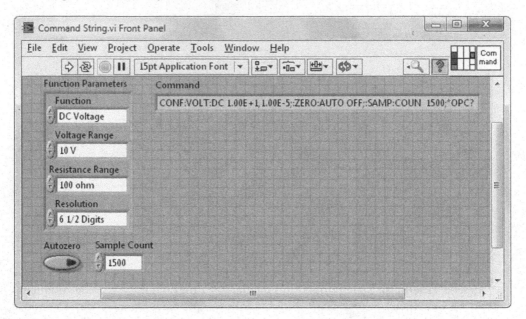

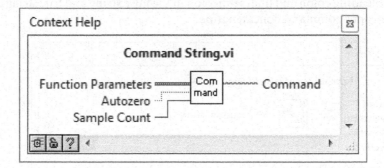

Finally, create a VI named **Measurement Config** and save it in **YourName\ Chapter 14**. Switch to the block diagram and code it as shown. This VI will write the command string to the instrument. When this diagram runs, the **read buffer**

indicator will display an ASCII "1" if the command string was successfully read by the instrument.

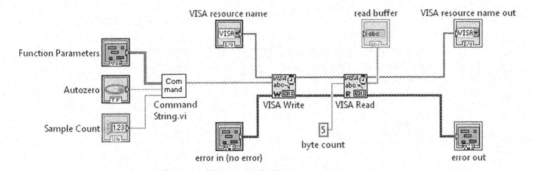

Switch to the front panel and arrange the objects there as you wish. Then design an icon and assign the connector pane's terminals consistent with the Help Window shown below. Make the default value for the **Sample Count** control equal to *1* using **Edit>>Make Current Values Default**. Save your work.

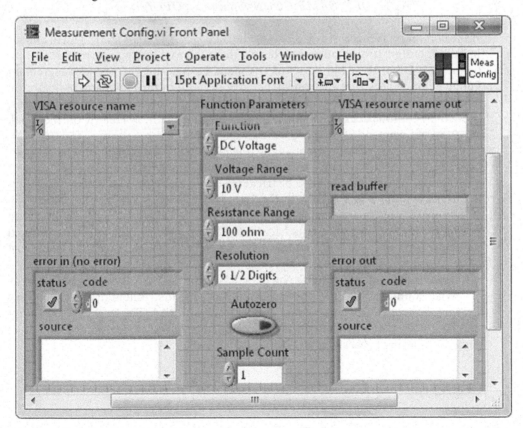

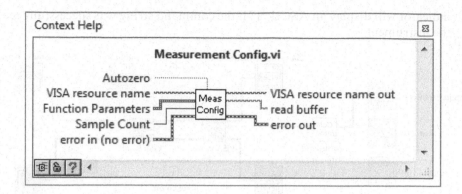

Input the VISA resource name for your instrument into the VISA resource name control, and then run Measurement Config with front-panel control settings shown above so that the following command is sent to the instrument:

$$CONF{:}VOLT{:}DC{<}Space{>}1.00E{+}1,1.00E{-}5{;}{:}ZERO{:}AUTO{<}Space{>}$$
$$OFF{;}{:}SAMP{:}COUN{<}Space{>}\ 1{;}{*}OPC?$$

If the command is successfully sent over the interface bus, an ASCII "1" will appear in output buffer. If the multimeter beeps, there is most likely an error in the sent command. Open the front panel of Command String, and then run Measurement Config again. Check that the concatenated command in the Command indicator on Command String's front panel has a form given above; make sure all of the colons, semicolons, and spaces are included. If there is an error, correct it on the block diagram of Command String.

After running Measurement Config, the multimeter will be left in *remote mode*. You can switch to *local mode* by pressing the instrument's front-panel Shift/Local key. The 34410A DMM can then be triggered (equivalent to sending the INIT command over the interface bus) with the Trigger button. A star (*) annunciator will blink on the instrument's front-panel display as it acquires voltage samples. Run your VI with Sample Count equal to *1500* (*15*, for 34401A owners), then in local mode press the Trigger button. Does the annuciator blink the expected amount of time after the Trigger button is depressed?

Save Measurement Config as you close it.

Write a final modular VI for your 34401A DMM instrument driver called Take Data as shown below, and save it in YourName\Chapter 14. The leading *CLS command assures that all bits in the SESR and SBR register are set to zero, prior to each data-taking process.

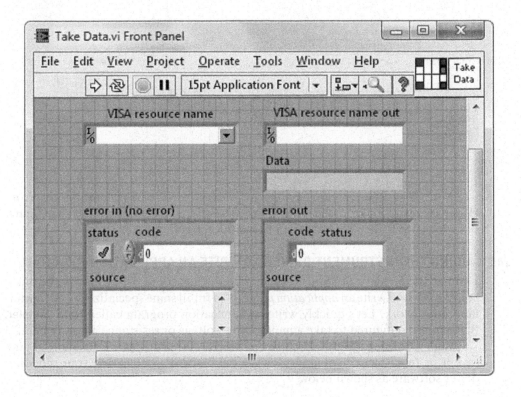

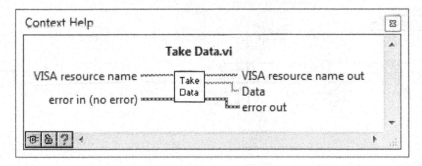

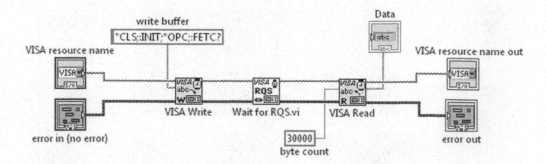

Note: **Take Data** cannot be run independently without generating an error. However, if you first run one of the other VIs that you have written (can you figure out which one?), then **Take Data** can be run successfully.

14.17 USING THE INSTRUMENT DRIVER TO WRITE AN APPLICATION PROGRAM

Ultimately, the merit of an instrument driver is measured by the ease with which you can use it to write an *application program* to fulfill some specialized need in your laboratory work. Let's quickly write an application program called **Data Sampler** that can be configured to take a multisample voltage or resistance measurement.

With **VISA Query (SRQ)** open, use **File>>Save As...** to create **Data Sampler** and save it in **Your Name\Chapter 14**. Rewrite the block diagram using your modular driver software as shown below.

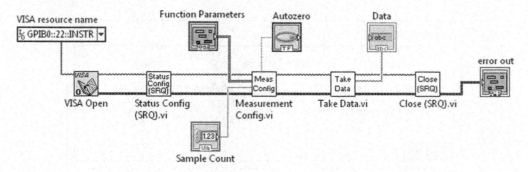

Return to the front panel and arrange the objects there as desired. Save your work.

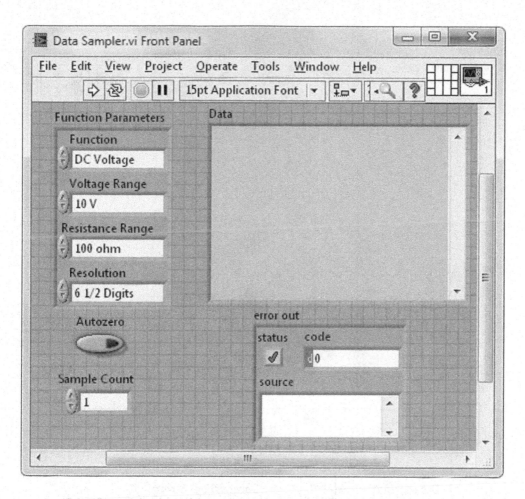

Run **Data Sampler** with various choices of front-panel settings (or interface buses!), and then pat yourself on the back for a job well done.

Finally, a useful 34410A DMM instrument driver utility would perform the following task: Take the instrument's **Data** string (data samples delimited by commas and the string terminated by a LF character) and convert it to a numeric array and a spreadsheet format. One manifestation of that utility called **Reformat Data String** is shown next. On the front panel, the string control and string indicator have been resized and scrollbars have been activated by selecting **Visible Items>>Vertical Scrollbar** in the pop-up menus. The **Single Numeric Sample** indicator is included for convenience (converting the array format to a single DBL numeric) when **Sample Count** is equal to *1*.

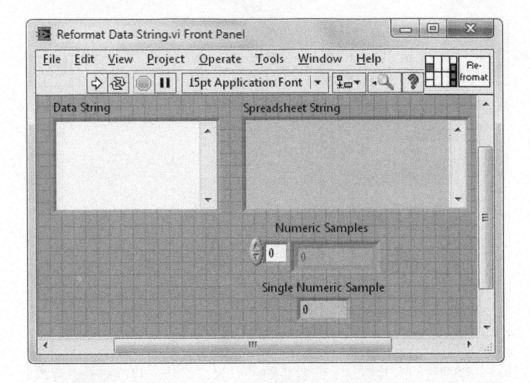

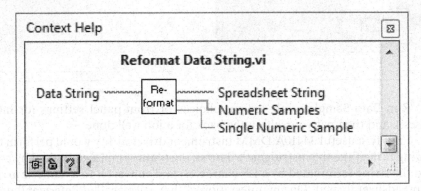

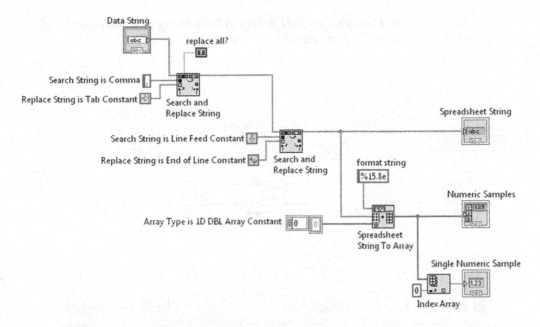

Write **Reformat Data String**. This VI implements the **Search and Replace String** icon found in **Functions>>Programming>>String** (Help Window shown below) to coerce the original **Data** string into the spreadsheet format (by replacing comma delimiters and the LF terminator with tabs and an EOL, respectively). Do you understand how it works?

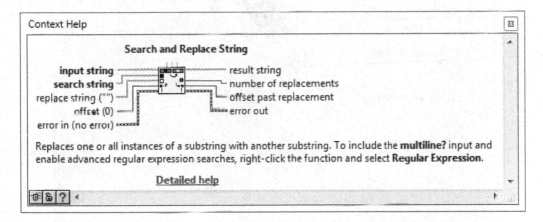

Return to **Data Sampler**. Include **Reformat Data String** as a subVI on its block diagram and modify the front panel as shown.

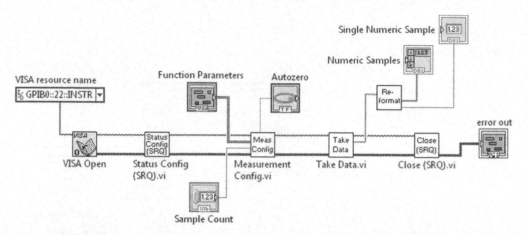

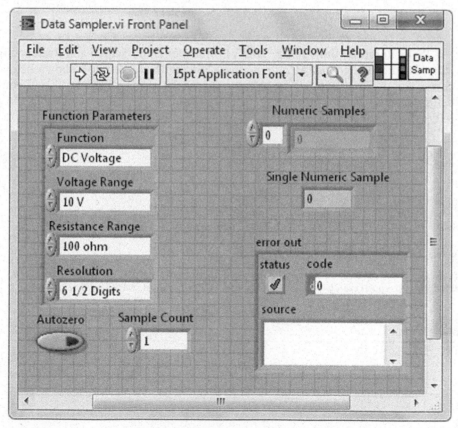

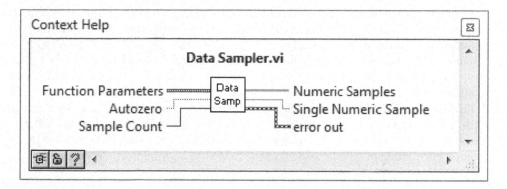

Once written, run **Data Sampler** and watch it perform its magic.

Now that you know some of what goes into writing an instrument driver, here's some very good news. In many cases, the LabVIEW instrument driver you will need for a particular instrument in your laboratory has already been written and is available for your use free of charge. National Instruments provides an extensive library of downloadable instrument drivers at *www.ni.com/downloads/ instrument-drivers*. You can also access this resource within LabVIEW by selecting **Tools>>Instrumentation>>Find Instrument Drivers…**. Most of these drivers are written using VISA icons and so, using the interface-appropriate VISA resource name for your instrument, can be used to communicate over various interface buses—RS-232, GPIB, Ethernet, and USB. If interested, try using **Tools>>Instrumentation>>Find Instrument Drivers…** to download the 34410A DMM instrument driver. After exiting LabVIEW and then restarting it, you should find the driver in a subpalette within **Functions>>Instrument I/O>>Instrument Drivers**. Take a look at some of the icons in this palette and see if you can decipher them.

DO IT YOURSELF

Time Evolution of X.vi Assume that you have a widget in your laboratory that is providing you with some interesting information about X, where X might be the position of an object or the intensity of a light source. Additionally, say, the widget provides this information about X in the form of a "voltage code," that is, it produces an output voltage V that is some known function of X. Then, with a 34410A DMM and an appropriate application VI, you can monitor X (via measurement of V) as a function of time.

Using your 34410A DMM instrument driver programs as subVIs, write the application VI. When run, this top-level VI continuously obtains a single DC voltage sample after each time step Δt, where Δt is given by the value of the front-panel control **Wait Time (second)**, until the front-panel **Stop Button** is pressed. While running, the VI provides real-time graphing of *Voltage* vs. *Time* data on a **Waveform**

Chart with the Chart's *Time* axis properly calibrated. The front panel also provides the option of storing all of the accumulated data in a spreadsheet file with the *Time* and *Voltage* data in the spreadsheet's first and second columns, respectively. Define *Time* = 0 at the moment that the first voltage sample is acquired.

The front panel of your VI should appear as shown below. All needed parameters without a front-panel control should be input on the block diagram. After building this VI, run it to observe a time-dependent voltage input (e.g., from a function generator) to the multimeter and save the resulting *Voltage* vs. *Time* data in a spreadsheet. Determine the minimum value allowed for **Wait Time (second)** using your knowledge of how long it takes the DMM to acquire a single voltage sample.

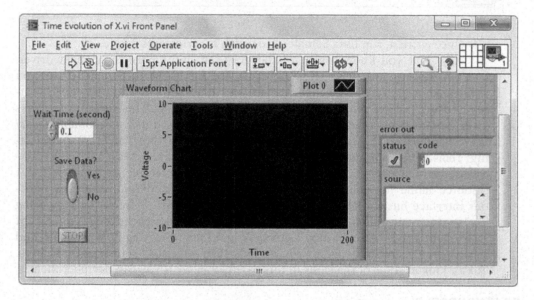

A helpful tip: The *Time* axis of the Waveform Chart can be calibrated using a Property Node. Pop up on the Waveform Chart's icon terminal and select **Create>>Property Node>>X Scale>>Offset and Multiplier>>Multiplier**. Then set the **Multiplier** property appropriately.

USE IT!

Arduino-Based Voltage Measurement Arduino USB boards are low-cost data acquisition devices that communicate with a computer over the USB interface. These boards are popular with hobbyists, are widely used in education, and have proved useful for simple research laboratory tasks. Although Arduinos are typically programmed using a language based on C and C++, these boards can be easily controlled using LabVIEW with the aid of an open source add-on called LINX. LINX can be obtained free of charge by following the instructions on the webpage "*Getting*

Started with LINX." Once the instructions are successfully implemented, the Functions Palette will have a subpalette named **Functions>>MakerHub>>LINX**, which contains a collection of icons that can control the data acquisition functions of an Arduino board.

The following program implements icons from **Functions>>MakerHub>>LINX** to acquire a voltage reading from an Arduino Uno board's Analog In Channel A0 four times per second. Within a LINX icon, if you open its subVIs (by double-clicking on them) down to the lowest-level subVI, you will find that VISA icons are used to send and receive messages over the USB interface connecting the Arduino and computer. The voltage difference being measured is connected at the Arduino's A0 and GND pins and it must fall within the range from 0 V to 5 V.

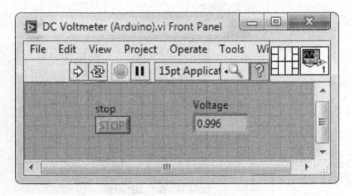

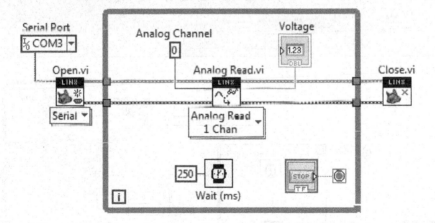

The next program demonstrates the utility of using LabVIEW to program an Arduino. An Arduino Uno board does not have hardware-timed analog input capabilities. That is, this board is not designed to acquire N voltage samples at a given sampling rate, a task required to build a computer-based digital oscilloscope.

However, using LabVIEW's software timing and triggering icons—**Wait (ms).vi** and **Basic Level Trigger Detection.vi**, respectively—the Arduino Uno can be programmed as a digital oscilloscope with analog triggering capabilities as shown below. This oscilloscope has a sampling rate of 100 samples per second and works well for signals of frequency up to about 10 Hz. The lowest level of the measured signal must be at least 0 V, while the highest level must be at most 5 V.

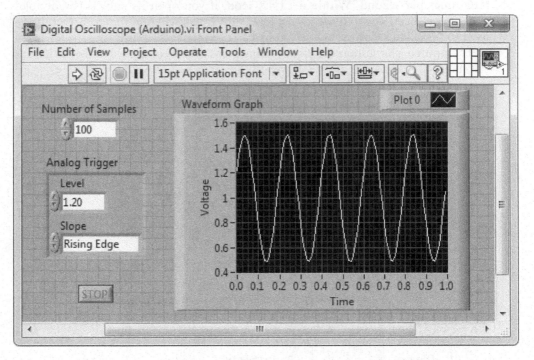

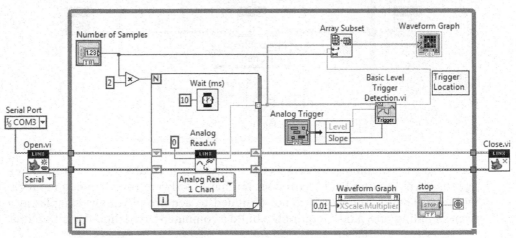

PROBLEMS

Each problem statement begins with a suggested descriptive name (including the **.vi** extension) for the program that you will write. Suggested icons for use in the VI can be found with the aid of **Quick Drop**.

1. **Thermocouple Thermometer (VISA).vi** Thermocouples are widely used as temperature sensors. A thermocouple is constructed by joining the ends of two dissimilar metals, for example, a copper and a constantan wire for a type T thermocouple. This junction produces a millivolt-level voltage that has a well-documented temperature dependence, where the temperature is measured relative to a "cold junction" reference temperature. Conveniently, this cold junction can be provided by a compact electronic device called a Cold Junction Compensator (CJC), which effectively makes the reference temperature equal to 0 °C.

   On its most sensitive scale, the 34410A multimeter can directly measure thermocouple voltages accurately enough to resolve small temperature changes. Connect a thermocouple to a CJC (e.g., Omega MCJ Series device) and then connect the plus and minus output of the CJC to the HI and LO Voltage Inputs of the 34410A multimeter. Then write a program that, every 250 ms until a front-panel Stop Button is pressed, reads the thermocouple voltage, converts this value to the corresponding temperature in Celsius, and then displays this temperature in a front-panel indicator.

   To convert the thermocouple voltage to its corresponding temperature, use the **Convert Thermocouple Reading.vi** icon with its **CJC Voltage** input wired to *0* (the **CJC Sensor** and **Type of Excitation** inputs can be left unwired). Program **Thermocouple Type** for your particular type of thermocouple (e.g., T).

   Run your VI and use it to measure room temperature as well as the temperature of your skin.

2. **Serial Poll with Timeout.vi** As written in this chapter, **Serial Poll** is flawed in that, if the bit being monitored in the Status Byte Register is never set, this VI will loop endlessly. With **Serial Poll** open, use **Save As...** to create a new VI. Then modify its block diagram so that, if the bit being monitored is not set within 10 seconds, the While Loop is stopped.

3. **Time Evolution of X (Built-In Driver).vi** Use **Tools>>Instrumentation>>Find Instrument Drivers...** to download the 34410A DMM instrument driver. After closing and restarting LabVIEW, this driver software will be found in a sub-palette within **Functions>>Instrument I/O>>Instrument Drivers**. Use the icons from this "built-in" driver to write a program that carries out the task described in this chapter's Do It Yourself project. The icon's VI Tree gives a helpful overview of the "built-in" driver.

4. **Take Data (Accurate Resolution).vi** Regardless of the chosen resolution, the 34410A DMM always reports data sample values with eight digits to the right of the decimal point. Thus, some of these decimal-place values are not significant. With **Take Data** open, use **File>>Save As...** to create a new VI. Add a **Function Parameters** front-panel control to this new VI (so that the selected resolution setting can be input), and then modify the block diagram so that the **data** output reports values with the actual resolution selected (e.g., 5 1/2 digits if that is the selected resolution).

5. **Simple VISA Query (Express).vi** Use the **Instrument I/O Assistant** Express VI to query the 34410A multimeter. Place an **Instrument I/O Assistant** icon on the block diagram of your VI. When this Express VI's dialog window opens, select the desired instrument, and then click on **Add Step**. In the **Add Step** dialog window that appears, double-click on **Query and Parse**. In the Enter a command box, type

    CONF:VOLT:DC 10,0.00001;:INIT;:FETC?

    and then click **Run this step**. The command will be sent to the 34410A DMM, and its string response will be displayed. Click the **Auto parse** button to convert the response string to numeric format and then close the dialog box by clicking the **OK** button. When returned to the block diagram, simply create an indicator for the icon's **token** output terminal.

    Run your VI and demonstrate that it successfully obtains a DC Voltage sample from the 34410A multimeter.

Formula Node Programming for Chapter 4

A.1 FORMULA NODE BASICS

The *Formula Node* is a built-in function of LabVIEW, which is found in **Functions>> Programming>>Structures**. Just like the MathScript Node, the Formula Node is a resizable box that can be used to enter text-based commands directly on the block diagram. In comparison to the MathScript Node, however, the Formula Node is fairly primitive in its capabilities. As shown in its Help Window below, the Formula Node can be programmed to carry out mathematical operations such as evaluating trigonometric and logarithmic functions as well as to perform simple Boolean logic, comparisons, and conditional statements.

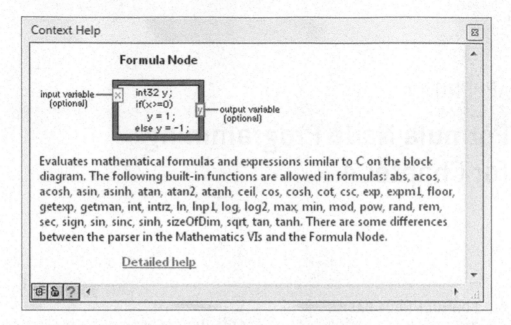

Evaluates mathematical formulas and expressions similar to C on the block diagram. The following built-in functions are allowed in formulas: abs, acos, acosh, asin, asinh, atan, atan2, atanh, ceil, cos, cosh, cot, csc, exp, expm1, floor, getexp, getman, int, intrz, ln, lnp1, log, log2, max, min, mod, pow, rand, rem, sec, sign, sin, sinc, sinh, sizeOfDim, sqrt, tan, tanh. There are some differences between the parser in the Mathematics VIs and the Formula Node.

Detailed help

A complete listing of the Formula Node's capabilities can be found by clicking on the **Detailed help** hypertext in its Help Window. There, you will find hyperlinks to a complete listing of the Formula Node's built-in functions, its syntax, and allowed operators (e.g., ** and = = are the power and logical equivalence operators, respectively). For example, the equation $y = 3x^2 + 2x + 1$ is programmed using the Formula Node as shown next.

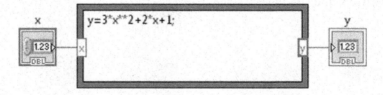

The capacity of a particular Formula Node is not limited to a single equation. The structure can be resized to contain as many equations as you require with each equation terminated by a semicolon (;).

A.2   QUICK FORMULA NODE EXAMPLE: SINE-WAVE PLOT (SECTION 4.2)

As an introduction to implementing it in a program, let's write the **Sine Wave Graph** program developed in Section 4.2 using a Formula Node (rather than a MathScript

Node). Our goal is to write a VI that will evaluate $\sin(x)$ at 100 evenly spaced x-values in the range from $x = 0$ to $x = 19.8$ and plot the result.

Open a new VI and then place a **Waveform Graph** on its front panel as shown below. Using the 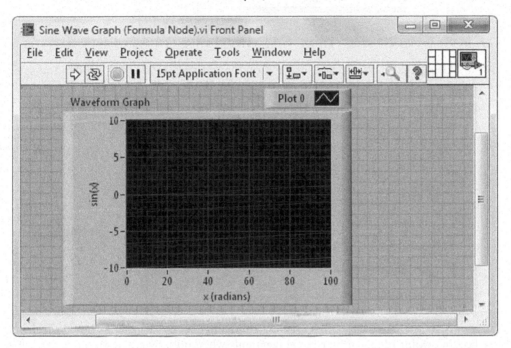, label the x- and y-axes as **x (radians)** and **sin(x)**, respectively. By default, autoscaling will be activated on both of the axes. Using **File>>Save**, save this VI under the name **Sine Wave Graph (Formula Node)**.

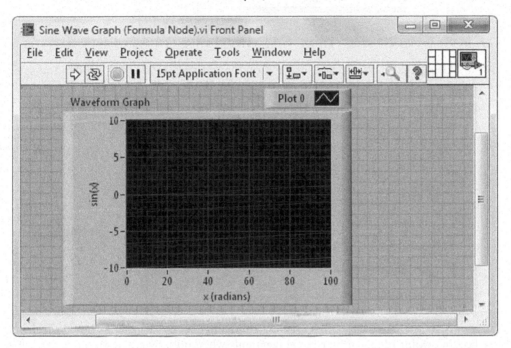

In Chapter 4, the command $x=start:step:stop$ is used within a MathScript Node to generate an N-element array of x-values in two programs: **Sine Wave Graph (Math-Script Node)** and **Waveform Simulator**. In the first program, the array size N is a fixed constant; in the second program, N is a variable specified during the program's runtime through the appropriate choice of values for $start$, $step$, and $stop$. Defining the size of an array is more difficult within the Formula Node (especially when done during runtime), so it is easiest to use a For Loop to do this job as follows.

Switch to the block diagram and place a **For Loop** there. Pop up on the count terminal [N], select **Create>>Constant**, and then enter *100* for the value of this constant. Next, place a **Formula Node** (found in **Functions>>Programming>>Structures**) within the For Loop.

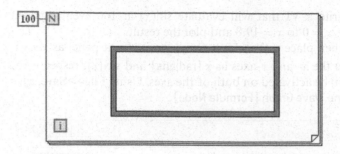

Waveform Graph

Using the ⬉, pop up at a location on the Formula Node's left border and select **Add Input**. In the highlighted input box that appears, enter the name *i*, then wire the For Loop's iteration terminal to this input. In a similar way, create two outputs (by popping up, then selecting **Add Output**) called *x* and *y* on the Formula Node's right border.

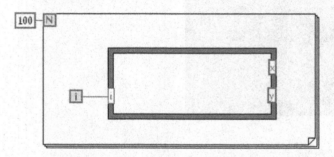

Waveform Graph

Note that the default data type for input and output variable is double-precision, floating-point (denoted by orange terminals when initially created). However, when the **I32** integer-formatted is wired to the input terminal labeled *i*, the input terminal changes to the integer representation (denoted by its blue color). If, in the future, you want to change the data type of an output terminal, create an input terminal with exactly the same name as the output terminal and wire a constant of the desired data type to that input terminal. Doing so also provides a default value for the terminal. Alternatively, you also can use a command within the Formula Node to define an output variable's data type. For example, the command *int32 y;* changes the data type of the output terminal *y* to a 32-bit integer. For our present program, the default data type **DBL** is a good choice for both the *x* and *y* output variables.

The two outputs created above define the *x* and *y* variables we will use in our coding of the Formula Node. Using the 🖑, enter the following two equations into the interior of the Formula Node. Note that all equations within a Formula Node must be terminated by a semicolon.

$$x = i * (1 / 5);$$
$$y = \sin(x);$$

Then wire the y output to the border of the For Loop so that the loop's auto-indexing feature will construct the desired 100-element array of sine-wave values. Note, each variable used within the Formula Node must be defined by either an input or output terminal. Thus, the x output terminal is necessary to define the x variable used within the Formula Node. Because the x variable is not needed further (e.g., for display), its terminal is not wired to anything.

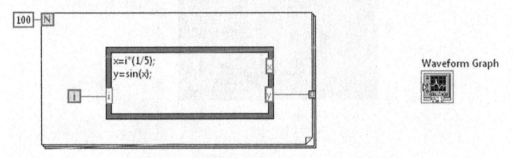

If an error occurs while coding a Formula Node, click the broken **Run** button to display the **Error List** window. LabVIEW marks the error with a # symbol.

Finally, complete the block diagram as shown below (and described in Section 4.2).

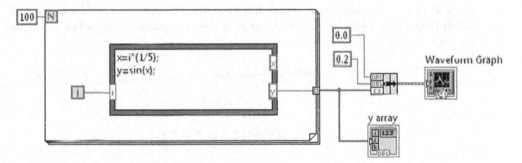

Save your work. Then return to the front panel and run your VI, verifying that it fulfills our sine-wave plotting goal.

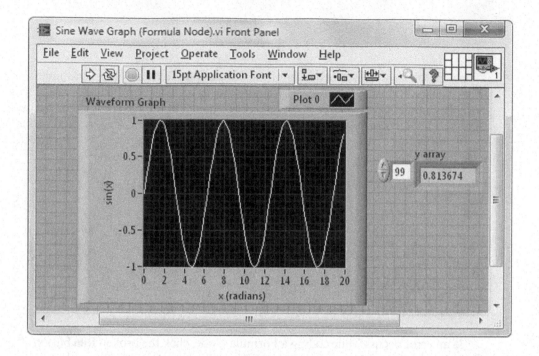

A.3 FORMULA NODE-BASED WAVEFORM SIMULATOR (SECTIONS 4.3–4.4)

For the Formula Node version of the **Waveform Simulator** program built in Sections 4.3–4.4, create the same front panel described there. Then, on the block diagram, rather than using a MathScript Node, place a Formula Node within a For Loop and program it with the following code:

$$delta\_t = 1 \: / \: f_s;$$

$$x = i * delta\_t;$$
$$y = A * \sin(2 * \text{pi} * f * t);$$

When complete, the block diagram will appear as follows. Remember to include semicolons at the end of each equation and also to create the **delta_t** output (which is necessary for defining this variable within the Formula Node).

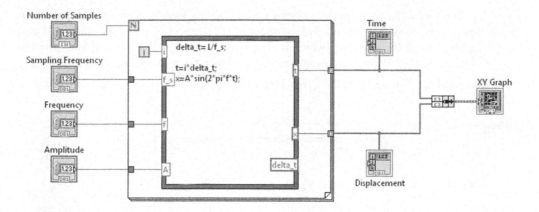

A.4 FORMULA NODE-BASED WAVEFORM SIMULATOR (SECTION 4.8)

The Formula Node has an *if* and an *if-else* conditional statement, but not an *if-elseif* statement. Thus, the Formula Node-based code for **Waveform Simulator** is written as follows.

$$delta\_t = 1 / f\_s;$$

$$t = i * delta\_t;$$

$$if\,(s == 0)\quad x = A * \sin(2 * pi * f * t);$$

$$if\,(s == 1)\quad x = A * \cos(2 * pi * f * t);$$

$$if\,(s == 2)\quad x = A;$$

$$if\,(s == 3)$$

$$\quad if\,\big(\mathrm{mod}\,(floor(2 * f * t), 2) == 0\big)\ x = 0;$$

$$\quad else\ x = A;$$

$$if\,(s == 4)\quad x = 4.0 * \sin(2 * pi * 100 * t) + 6.0 * \cos(2 * pi * 200 * t);$$

Here is how the square wave of period $T = 1/f$ is created: First, note that the time for a half-cycle is given by $\dfrac{T}{2} = \dfrac{1}{2f}$ so that $floor\left(\dfrac{t}{T/2}\right) = floor\,(2 * f * t)$ determines how many integer half-cycles have been completed after time t. Then, since the $\mathrm{mod}\,(x, y)$ function determines the remainder when x is divided by y, $\mathrm{mod}\,(floor(2 * f * t), 2)$ equals 0 and 1 during even and odd integer half-cycles, respectively. This fact is used to create a square wave that toggles between the values of 0 and A.

The block diagram for the Formula Node-based **Waveform Simulator** program is shown next. Remember to include semicolons at the end of each equation and also to create the **delta_t** output (which is necessary for defining this variable within the Formula Node). If an error occurs while coding a Formula Node, click the broken **Run** button to display the **Error List** window. LabVIEW marks the error with a # symbol.

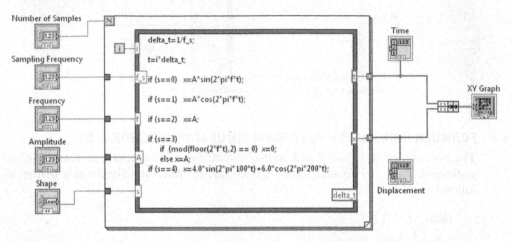

A.5   FORMULA NODE-BASED WAVEFORM SIMULATOR (SECTION 4.10)

After including control and indicator clusters on the front panel, the Formula Node-based block diagram will appear as shown below.

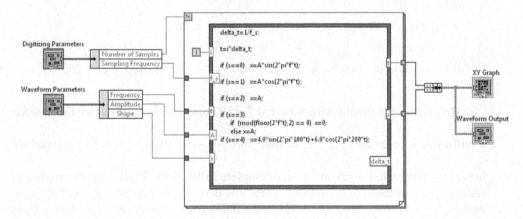

APPENDIX B

Mathematics of Leakage and Windowing

B.1 ANALYTIC DESCRIPTION OF LEAKAGE

Assume that an analog waveform $x(t)$ is discretely sampled at N equally spaced times $t_j = j\Delta t$, where $j = 0, 1, 2, \ldots, N-1$ and the sampling frequency $f_s = 1/\Delta t$. Then, as discussed in Chapter 12, the discrete Fourier transform is defined as

$$X_k \equiv \sum_{j=0}^{N-1} x(t_j) e^{-i2\pi jk/N} \qquad k = -\frac{N}{2} + 1, \ldots, 0, \ldots, +\frac{N}{2} \qquad \text{[B.1]}$$

where each complex-valued X_k is associated with the discrete frequency f_k given by

$$f_k = k\left(\frac{f_s}{N}\right) \equiv k\Delta f \qquad k = -\frac{N}{2} + 1, \ldots, 0, \ldots, +\frac{N}{2} \qquad \text{[B.2]}$$

To obtain an analytic expression that describes the leakage phenomenon we observed in Chapter 12, let's explore the discrete Fourier transform of a complex-exponential waveform.

Consider an oscillatory waveform with frequency f given by $x(t) = A \exp(i2\pi ft)$, where A is a constant. Inserting this choice for $x(t)$ into Eq. [B.1], the values of X_k are

$$X_k = \sum_{j=0}^{N-1} A e^{i2\pi f(j\Delta t)} e^{-2\pi jk/N} \qquad \text{[B.3]}$$

661

which can be rewritten as

$$X_k = A \sum_{j=0}^{N-1} \left[e^{i2\pi(f\Delta t - k/N)} \right]^j \qquad \text{[B.4]}$$

This series is the well-known finite geometric series, whose form and summation value are given by

$$\sum_{j=0}^{N-1} x^j = \frac{1 - x^N}{1 - x} \qquad \text{[B.5]}$$

After using Eq. [B.5] to evaluate the sum in Eq. [B.4], a few lines of algebra and trigonometric relations yield the following expression for the magnitude of the discrete Fourier transform values X_k:

$$|X_k| = \left| A \frac{\sin\left[\pi N (f\Delta t - k/N)\right]}{\sin\left[\pi(f\Delta t - k/N)\right]} \right| \qquad \text{[B.6]}$$

This equation then describes the quantity determined by **FFT (Magnitude Only)**.

Let's investigate the meaning of Eq. [B.6]. First, consider the case when the input frequency f exactly equals one of the frequencies f_k in Eq. [B.2], say, $f_{k'}$. Setting $f = f_{k'} = k'(f_s/N) = k'(1/N\Delta t)$, where we used the fact that $f_s = 1/\Delta t$, Eq. [B.6] becomes

$$|X_k| = \left| A \frac{\sin\left[\pi(k' - k)\right]}{\sin\left[\pi(k' - k)/N\right]} \right| \qquad \text{[B.7]}$$

For $k \neq k'$, the difference $(k' - k)$ will be an integer less than N, making the numerator and denominator of Eq. [B.7] zero and nonzero, respectively. Thus, $|X_k| = 0$ for $k \neq k'$. However, for $k = k'$, both the numerator and denominator of Eq. [B.7] are zero and l'Hôpital's rule gives $|X_{k'}| = AN$. So the resulting frequency spectrum is a delta function with a single spike at frequency $f_{k'}$ of height $|A_{k'}| = |X_{k'}|/N = A$. This prediction is consistent with what we observed when we input the oscillations at $f = f_{128} = 250$ Hz and $f = f_{256} = 500$ Hz inputs to the FFT algorithm.

Second, consider the case when the input frequency f is not equal to one of the frequencies f_k. We can simplify the appearance of Eqn. [B.6] by noting from Eqn. [B.2], along with the fact that $f_s = 1/\Delta t$, that

$$f \Delta t - \frac{k}{N} = \frac{f}{f_s} - \frac{f_k}{f_s} = \frac{f - f_k}{f_s} \qquad [\text{B.8}]$$

Then, putting Eq. [B.8] into Eq. [B.6], we find

$$\left| X_k \right| = \left| A \frac{\sin\left[\pi N\left(\dfrac{f - f_k}{f_s}\right)\right]}{\sin\left[\pi\left(\dfrac{f - f_k}{f_s}\right)\right]} \right| \qquad [\text{B.9}]$$

and, since $\Delta f = f_s/N$,

$$\left| X_k \right| = \left| A \frac{\sin\left[\pi\left(\dfrac{f - f_k}{\Delta f}\right)\right]}{\sin\left[\pi\left(\dfrac{f - f_k}{f_s}\right)\right]} \right| \qquad [\text{B.10}]$$

Inspecting Eq. [B.10], we see that, under our assumption that f does not equal one of the f_k, none of the $|X_k|$ will be zero. However, the frequencies f_k that fall closest to f will make the denominator of Eq. [B.10] the smallest. Thus, the $|X_k|$ associated with the f_k neighboring f will take on the largest values. To better visualize the meaning of Eq. [B.10], let's use it to plot the magnitude of the complex-amplitude $|A_k| = |X_k|/N$ vs. f_k with the following familiar choice of parameters: $f_s = 2000$ Hz, $N = 1024$, $A = 4.0$, and $f = 249$ Hz. Then $\Delta f = f_s/N = 2000/1024$ and $f_k = k\Delta f$, where $k = -511, \ldots, 0, \ldots, +512$. This plot, which is shown in Figure B.1, predicts the output of FFT (Magnitude Only) when a 249 Hz cosine function of amplitude 4 is input.

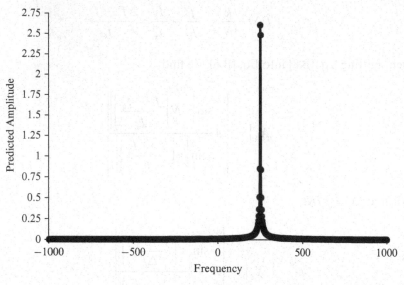

FIG. B.1

Zooming in on the peak to get a closer look, we see in Figure B.2 that Eq. [B.10] provides a perfect theoretical prediction of the spectral leakage we observed in Section 12.10 with a 249 Hz input to FFT (Magnitude Only).

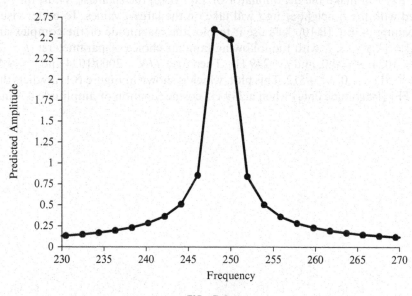

FIG. B.2

B.2 DESCRIPTION OF LEAKAGE USING THE CONVOLUTION THEOREM

The *Convolution Theorem*, a powerful result from higher mathematics, provides another vantage point from which to understand the problem of leakage. From this point of view, the situation appears as follows: When we acquire a finite number N of discretely sampled points for FFT spectral evaluation, we are in effect observing an infinite set of data d_j (where $j = -\infty, \ldots, -1, 0, +1, \ldots +\infty$) through a rectangular viewing window in time. Mathematically, we can define the rectangular window function $w(t_j)$ to be zero at all times $t_j = j\Delta t$, except during the "data-viewing" time interval from $j = 0$ to $j = N - 1$, when it is equal to 1. Then our finite set of N sampled data points x_j is given by the product $x_j = d_j w_j$. This idea is illustrated in Figure B.3.

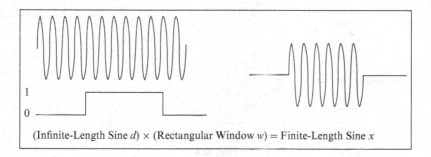

(Infinite-Length Sine d) $\times$ (Rectangular Window w) = Finite-Length Sine x

FIG. B.3

Let's say that the Fourier transforms of d_j and w_j are D_k and W_k, respectively. The question relevant to our finite-length data set then becomes, "What happens when one takes the Fourier transform of the product $x_j = d_j w_j$?" Well, according to the famous Convolution Theorem, the Fourier transform of the product of two functions $d_j w_j$ is equal to the *convolution* of the two Fourier transforms D_k and W_k. The convolution for continuous functions, denoted by $D * W$, is defined by

$$D(f) * W(f) = \int_{-\infty}^{+\infty} D(\phi) W(f - \phi) d\phi \qquad \text{[B.11]}$$

In the discrete case, this definition becomes

$$(D * W)_k = \sum_{m=-N/2+1}^{N/2} D_m W_{k-m} \qquad \text{[B.12]}$$

The convolution of two functions, although complicated in general to determine, is simple to ascertain in the following important case. Let D_m be a unit-amplitude delta function located at the frequency f_n, that is, $D_m = 0$ for all m, except $D_{m=n} = 1$. Then, from Eq. [B.12], we find that $(D * W)_k = W_{k-n}$, meaning that the convolution is just the function W displaced so that it is now centered at f_n rather than $f = 0$. This idea is illustrated in Figure B.4 using continuous functions.

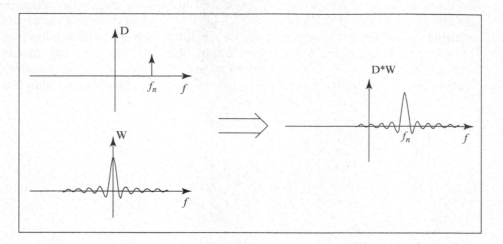

FIG. B.4

This example provides new insight into the leakage phenomenon. Consider the case of a finite-length complex-exponential input of the form $x\left(t_j\right) = A\exp\left(i2\pi f t_j\right)$, which can be described as the product of an infinite-length complex-exponential $d(t_j)$ and a rectangular window function $w(t_j)$. The Fourier transform D of the infinite complex-exponential is, of course, a delta function of height A located at frequency $+f$, while the discrete Fourier transform of the rectangular window (apart from a unity-amplitude phase factor) is easily shown to be

$$\left(W_{rectangle}\right)_k = \frac{\sin\left[\dfrac{\pi f_k}{\Delta f}\right]}{\sin\left[\dfrac{\pi f_k}{f_s}\right]} \qquad [B.13]$$

A plot of Eq. [B.13] (treating frequency as a continuous variable) with $f_s = 2000$ Hz and $N = 1024$ is shown in Figure B.5. Note the substantial amplitudes of $W_{rectangle}$ at high frequencies, which we know qualitatively result from the sharp turn-on and turn-off at the edges of the rectangular window (recall from the Fourier analysis of a square wave, its sharp edges are produced by the presence of high-frequency harmonics). The values of $W_{rectangle}$ at the frequencies f_k, that is, $(W_{rectangle})_k$, are given by the dots.

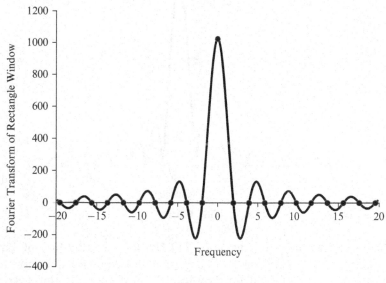

FIG. B.5

The magnitude of the finite-length complex-exponential's Fourier transform X is formed from the convolution of $W_{rectangle}$ with the delta function of height A at frequency $+f$. This convolution is simply $W_{rectangle}$, shifted along the frequency axis so that it is centered at frequency $+f$, and then multiplied by A. Thus,

$$|X_k| = |(D * W)_k| - \left| A \frac{\sin\left[\dfrac{\pi\left(f_k - f\right)}{\Delta f}\right]}{\sin\left[\dfrac{\pi\left(f_k - f\right)}{f_s}\right]} \right|$$ [B.14]

Note that Eq. [B.14] is equivalent to the "leakage description" given by Eq. [B.10].

Next, Figure B.6 shows a plot of Eq. [B.14] for $f_s = 2000$ Hz and $N = 1024$ with f equal to one of the f_k, namely, $f - f_{128} = 250$ Hz, where $A = 1$ is assumed for simplicity. Note that all of the values of $|X_k|$, which are given by the dots, fall on the zero-crossings of Eq. [B.14], except $|X_{128}|$ at $f_{128} = 250$. Thus, the discrete Fourier spectrum will be a delta function in this case.

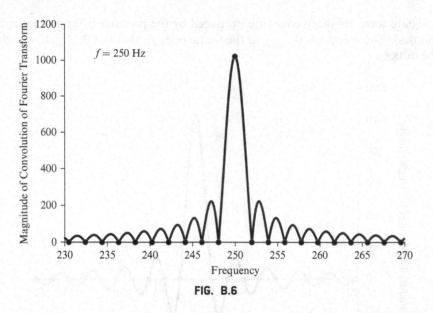

FIG. B.6

Changing f to the non-f_k value of 249 Hz, the resulting plot of Eq. [B.14] is shown in Figure B.7. Note that the $|X_k|$, again indicated by the dots, now fall at non-zero locations on the curve defined by Eq. [B.14], resulting in the spectral leakage phenomenon.

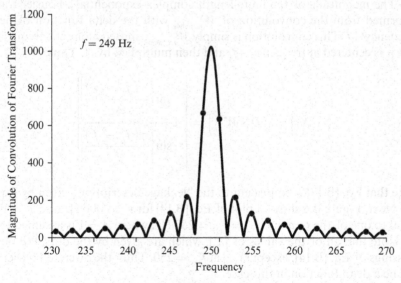

FIG. B.7

The Convolution Theorem then gives us the following insight: It is the wide-ranging frequency content of the rectangular window function (as given in Eq. [B.13] and shown in the previous plot) that produces leakage of spectral amplitude to frequencies f_k far removed from the input frequency f.

APPENDIX C

PID Temperature Control Project

C.1 PROJECT DESCRIPTION

In this project, you will apply your LabVIEW programming skills in the construction of a feedback-based system that controls the temperature of an aluminum block with great precision. This feat will be accomplished by controlling the current through a *thermoelectric (TE) device*, which performs as a "heat pump" between the aluminum block and a large temperature reservoir. When current flows in one direction through the TE device, the device pumps heat from the block to the reservoir, hence cooling the block. When current flows in the other direction, the TE device pumps heat from the reservoir to the block, heating the block. Your job will be to write a top-level program called **PID Temperature Controller** that uses the *Proportion-Integral-Derivative (PID) control algorithm* to produce the current flow required to maintain the aluminum block at a desired "*set point*" *temperature* $T_{set\text{-}point}$. If your program is written correctly, **PID Temperature Controller** should be able to maintain the block's temperature at $T_{set\text{-}point}$ with fluctuations less than about 0.05° C.

C.2 VOLTAGE-CONTROLLED BIDIRECTIONAL CURRENT DRIVER FOR THERMOELECTRIC DEVICE

Build the circuit in Figure C.1. It is designed to provide the rather large bidirectional current flow necessary to power both the heating and cooling capabilities of the TE device. The voltage level V_{in} serves as a "selector code word" that dictates the magnitude and direction of the current flow through the TE module. If the power source that supplies the ± 10 V in your circuit has built-in ammeters, use them to monitor the current flow through the TE device, eliminating the need to insert the ammeter shown in the diagram. See Section C.5 for a discussion of important practical issues regarding this circuit, especially the need (and method) for dissipating the prodigious amount of heat it produces.

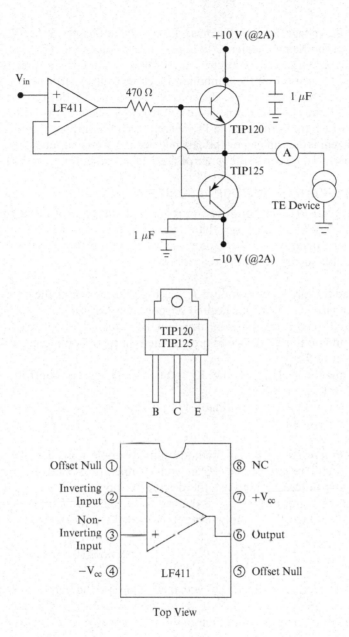

FIG. C.1 Voltage-controlled bidirectional current driver

Use **DC Voltage Source (Express)** in **YourName\Chapter 5** or **DC Voltage Source (DAQmx)** in **YourName\Chapter 12** to apply various voltages at V_{in}. First, test that when V_{in} is positive (negative), an electric current flows in the proper direction through the TE device to heat (cool) the aluminum block in your experimental setup. If a positive V_{in} cools the block, reverse the connections to the TE device within your circuit. Second, find the positive and negative voltages V_{sat}, where V_{sat} is the minimum voltage at V_{in} for which the current to the TE device becomes saturated. The value of this saturation current is determined by the capability of the power supplies attached to the TIP transistors. We then define the range of acceptable values for V_{in} to be between $\pm V_{sat}$.

C.3   PID TEMPERATURE CONTROL ALGORITHM

Build a digital PID temperature controller that can control the temperature of an aluminum block to within less than 0.05° C of a given set-point temperature. The set-point temperature can be chosen anywhere within the range of 0 to 40° C. An algorithm for such a controller is as follows:

1. Read the block's temperature T_{block} using a thermistor as the temperature sensor.
2. Compare T_{block} with the desired set-point temperature $T_{set\text{-}point}$.
3. Based on this comparison, decide what value of V_{in} will most appropriately command the TE device to provide the heating or cooling needed to bring T_{block} closer to $T_{set\text{-}point}$.
4. Apply this voltage at the V_{in} input of the voltage-controlled bidirectional current driver circuit.
5. Repeat this process continuously to obtain the desired temperature control of the aluminum block.

The *Proportional Control* method offers a simple procedure for deciding what voltage should be applied at V_{in} in your temperature-control algorithm. In this method, the difference between the desired set-point temperature $T_{set\text{-}point}$ and the actual sample temperature T_{block} is defined to be the error $E \equiv T_{set\text{-}point} - T_{block}$. Then the control voltage V_{in} is simply taken to be directly proportional to E:

$$V_{in} = A E \qquad \text{(Proportional Control)} \qquad \text{[C.1]}$$

where A is a constant called the *gain*. The value for the gain is chosen empirically such that, when $T_{set\text{-}point}$ is changed, the optimal value for A will cause the system to ramp to the new set-point and then stabilize near there (see the following discussion) quickly.

Here is a practical consideration you will confront when trying to implement Eq. [C.1]: When the sample temperature is far from the set-point, the error E may become large enough to generate a calculated value for V_{in} that falls outside its acceptable range of $\pm V_{sat}$. The solution to this problem is to truncate the expression given in Eq. [C.1] so that the magnitude of V_{in} is never allowed to be greater than V_{sat}. The graphical representation for the Proportional Control algorithm then is as shown in Figure C.2.

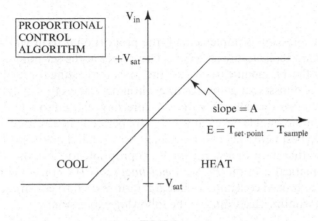

FIG. C.2

Although pleasingly simple, the Proportional Control method is intrinsically flawed. Here's why. Assume the sample is initially at room temperature and that you select $T_{set\text{-}point}$ to be above room temperature. Turning on the Proportional Control algorithm, the initial error E will be positive, resulting in a command to the TE device to heat the sample. So far, so good. As time goes on, the algorithm will issue the proper heating instructions to bring the sample closer and closer to the desired set-point temperature. When T_{block} nears $T_{set\text{-}point}$, the positive error E will become small, causing the Proportional Control to ease off on the applied heating power so that $T_{set\text{-}point}$ is gently approached. Then, at the decisive moment when T_{block} equals $T_{set\text{-}point}$, the error E becomes zero and the Proportional Control turns off power to the TE device. There is the flaw. Unfortunately, when at an elevated temperature, the sample will constantly be losing heat to its surrounding (room-temperature) environment through the heat-transferring processes of conduction, convection, and radiation. Thus, to maintain a sample at a set-point above room temperature, heat must constantly flow into the sample to counteract the heat it is losing to its environment. Since the Proportional Control turns off the heating when T_{block} equals $T_{set\text{-}point}$, the sample will never be able to stabilize at the desired temperature. Rather, the sample will stabilize at some equilibrium temperature $T_{eq} < T_{set\text{-}point}$. The positive error E produced at the magic temperature T_{eq} commands just the right amount of heating by the TE device to cancel out the heat being lost by the sample to the environment at that temperature. By similar reasoning, a set-point below room temperature will result in the sample stabilizing at an equilibrium temperature $T_{eq} > T_{set\text{-}point}$.

Luckily, there is an easy remedy to the above-described defect in the Proportional Control algorithm—simply include a constant term V_0 on the right-hand side of Eq. [C.1]:

$$V_{in} = A\,E + V_0 \qquad\qquad\qquad [C.2]$$

When this expression is implemented, the proportional term will ramp the sample to the set-point temperature as before. Then, once at the set-point (where $E = 0$), V_0 will instruct the TE module to provide the constant heating (or cooling) necessary to counteract heat losses (or gains) to the environment, thereby stabilizing the sample at $T_{set\text{-}point}$. Of course, the value of V_0 must be precisely chosen so that the environmental influences on the sample, when at $T_{set\text{-}point}$, are perfectly neutralized.

Now the best news: There is an elegant way to build intelligence into the control algorithm so that the proper value for V_0, appropriate to the chosen set-point, will be found automatically. In the *Proportional-Integral (PI) Control* method, rather than defining V_0 as a fixed constant, an integral term is used to construct the correct constant during runtime according to the following expression:

$$V_{in} = A\,E + B \int E\,dt \qquad\text{(PI Control)}\qquad [C.3]$$

Here, the integral keeps a running sum of all the error values that occur over the entire execution time of the algorithm. During the times when T_{block} is below $T_{set\text{-}point}$, a positive contribution will be made to the sum. When T_{block} is above $T_{set\text{-}point}$, a negative contribution will be made. Because of the self-correcting manner in which these contributions are made, the second term in Eq. [C.3] will eventually converge to the constant V_0 that allows the sample to stabilize at the set-point $T_{set\text{-}point}$. From that point on, the error E will equal zero and the value of the integral (and thus constant V_0) will no longer change.

Inclusion of a derivative term to damp out oscillations provides one further refinement to the control algorithm. The expression for this so-called *Proportional-Integral-Derivative (PID) Control* algorithm then is given by

$$V_{in} = A\,E + B \int E\,dt + C\frac{dE}{dt} \qquad\text{(PID Control)}\qquad [C.4]$$

where A, B, and C are constants. For the experimental situation of discrete data sampling, where the error E is determined every Δt seconds, the value of the control voltage V_{in} after the nth sampling can be approximated by

$$V_{in} = A\,E_n + B\,\Delta t \sum_{m=0}^{n} E_m + \frac{C}{\Delta t}\{E_n - E_{n-1}\} \quad\text{(Discrete-Sampling PID)}\quad [C.5]$$

where the summation $\sum_{m=0}^{n} E_m$ is over *all* of the error values determined since the

control algorithm was turned on.

C.4 PID TEMPERATURE CONTROL SYSTEM

Start construction of your digital temperature-control system by rebuilding the circuit of the thermistor-based digital thermometer described in Chapter 11's Do It Yourself project. The thermistor, which is embedded in the aluminum block, will be used to monitor T_{block}. If needed, add the 1 µF capacitor as shown in Figure C.3 to shunt high-frequency noise signals to ground and the unity-gain buffer to prevent the voltage-sensing circuitry from loading the thermistor circuit at low temperatures (when the thermistor resistance is large). Then connect V_{out} and GND to the inputs of an analog input channel of your LabVIEW system.

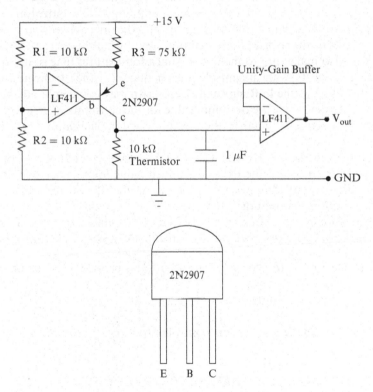

FIG. C.3 Hardware circuit for a digital thermometer. A constant current of 0.1 mA flows through thermistor.

Next, create a VI called **PID Temperature Controller**. Code **Temperature Controller** according to the following guidelines:

A. **Block Temperature Determination:** Noise-free temperature measurements of the block's temperature (assumed equal to the embedded thermistor's temperature) will be a "must" for your temperature-control algorithm to function properly. To average out random fluctuations when reading the thermistor temperature, you might try the following: At each determination of the temperature, quickly sample the thermistor temperature N times, and then analyze these data to determine the mean. Given the Steinhart–Hart coefficients for your thermistor, the **DAQ Assistant** or **DAQmx** icons make it simple to acquire the array of N temperature samples. To do the job of finding the mean from this data set, look at the VIs available in **Functions>>Mathematics>>Probability & Statistics**.

B. **Control Algorithm:** To determine the proper heating or cooling needed by the TE device, implement the Discrete-Sampling PID Control algorithm in the software. Remember that Δt in Eq. [C.5] is the time between successive determinations of the error E; think carefully about what this time is in your particular VI. Also make sure to include a truncation feature; that is, if Eq. [C.5] yields a value of V_{in} with magnitude greater than V_{sat}, then truncate the magnitude of V_{in} to V_{sat} (the **In Range and Coerce** icon may come in handy here). You will have to determine the optimum value for the constants A, B, and C empirically. If your setup is similar to the one described in Section C.5, try $A = 7$, $B = 1$, and $C = 0.5$.

C. **Output to the TE Device:** Use **DAQ Assistant** or **DAQmx** VIs on your block diagram to output the PID-determined value for V_{in} over one of LabVIEW's analog output channels to the input of the TE device's voltage-controlled bidirectional current driver circuit.

D. **Front Panel:** Use your creativity here. Remember you have a wide array of controls, indicators, and graphs available. A front-panel control that resets the summation $\sum_{m=0}^{n} E_m$ to zero when a button is pressed is an optional feature that you might find handy.

When you get the temperature controller to work, show it off to your friends and instructor!

C.5 CONSTRUCTION OF TEMPERATURE CONTROL SYSTEM

To perform this project, you will need access to an apparatus that controls the temperature of a small object through the use of a thermoelectric (TE) device. There is, of course, a multitude of ways to construct the required gadget. This section offers a design (at a total cost of approximately $100) that has worked successfully.

First, use a rectangular aluminum block of dimensions 2 in. × 1.5 in. × 5/16 in. (1 in. = 2.54 cm) as the object whose temperature is to be controlled. Hereafter, this object will simply be called "the block." So that accurate temperature measurements can be taken on the block, a small hole of diameter 3/16 in. is drilled into one of its sides to a depth of approximately 1/2 in. A thermistor (preferably coated with thermal grease; see below) can then be inserted into this hole and held securely by the addition of a 1/4 in. long #6-32 nylon set screw. For the thermistor, an inexpensive 10 kW model from Epcos (Model B57863S0103F040, available from Digikey P/N 495-2149-ND, www.digikey.com, $3) is used with soldered-on 12 in. lead wires for easy connection to a constant-current circuit. Finally, two #8-32 clearance holes are drilled (using a #19 drill bit) into the block as shown in Figure C.4. The 1.450 in. spacing is chosen so that a 30 cm wide TE module will fit between these holes.

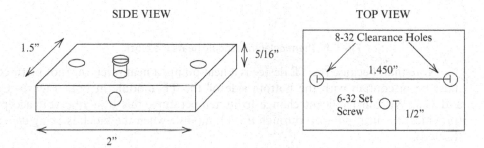

FIG. C.4 Aluminum block design

The required heating or cooling of the block is facilitated by placing it in contact with one side of a thermoelectric module. A TE module is a compact solid-state device that, via the Peltier Effect, acts as a heat pump. When current is caused to flow through the TE device in one direction, heat will be absorbed from the contacting block, cooling it, as shown in Figure C.5.

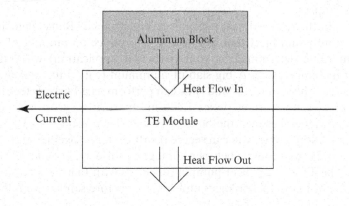

FIG. C.5 Thermoelectric module cools the block

When current is made to flow through the TE module in the opposite direction, heat is pumped into the block, heating it as in Figure C.6.

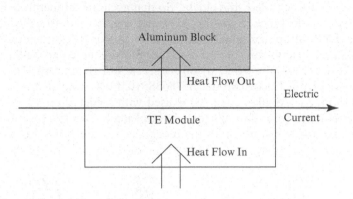

FIG. C.6 Thermoelectric module heats the block

Note that, because the TE device is a heat pump, a heat reservoir (not pictured) must be in contact with the bottom side of the TE module in both Figures C.5 and C.6. With insignificant change to its temperature, this heat reservoir accepts (provides) the heat pumped through the TE module when the block is being cooled (heated).

Laird Technologies (http://www.lairdtech.com) manufactures over 200 standard models of TE devices. The choice of a particular TE model is dictated by the heat-pumping capacity necessary and the electric power supply limitations in your application. For this project, use a Laird Technologies Model CP 1.0–127–05L1 TE device (available from Digikey P/N 926-1019-ND, $26 each). Within the relatively small range of current flow between ±2 Amps (A), this TE module has enough heat-pumping ability to change the temperature of a 2 in. × 1.5 in. × 5/16 in. aluminum block significantly away from room temperature (including cooling it to below 0° C and icing it up!).

How can the required heat reservoir be constructed? Remember that a heat reservoir is simply an object that can accept or produce an amount of heat (up to a maximum value determined by your particular application) without significantly changing its temperature. A big slab of aluminum (e.g., 5 in. × 3 in. × 3 in.) possesses a large "thermal mass" and thus can perform as a fairly decent heat reservoir. By tapping two #8-32 screw holes 1.450 in. apart into such a slab, a TE module can be sandwiched between the block and slab. Then using two 3/4 in. long #8-32 (insulating) nylon screws, you can secure the three items together and, in the process, obtain good thermal contact between contacting surfaces. To guarantee efficient heat flow, coat the TE device's heat-pumping surfaces with thermal grease (available, for instance, from Laird Technologies and Jameco) before sandwiching the three items.

This method for realizing a heat reservoir is inexpensive, but imperfect. If the TE module is run over moderately long time periods, the temperature of the aluminum slab will begin to significantly change from room temperature because of its rather poor efficiency at transferring heat with the surrounding room.

A much more stable heat reservoir can be constructed using a finned aluminum heat sink. Laird Technologies and Aavid Thermalloy (http://www.aavid.com) offer a line of products called *extrusions*, which are plate-like pieces of aluminum with many seamlessly attached fins. Because of the large surface area of its fins, an extrusion will very efficiently exchange heat with its surrounding room air, imbuing it with the desired properties of a heat reservoir. Good success has been found using 4 in. to 6 in. lengths of Aavid Thermalloy Extrusion (e.g., Newark P/N 93H1727, $23, http://www.newark.com) as the heat reservoir in the temperature-control system. Tapping two #8-32 screw threads 1.450 in. apart into the plate-like surface of the extrusion, the TE module with the block atop can be mounted and then secured with two #8-32 (insulating) nylon screws. Again, the use of thermal grease will enhance the conductance of heat between the block and the heat reservoir. A final improvement results by mounting the extrusion on top of a small fan (e.g., 120 VAC 4 in. Cooling Fan, Radio Shack P/N 273-238, $20), which forces air flow through the extrusion's array of fins. Legs for the fan and small angle brackets to connect it securely to the extrusion can be fashioned from 1/16 in. aluminum strips. Attach these strips to the fan using the fan's mounting holes. A photograph of the fan with its homemade legs and brackets is shown in Figure C.7.

FIG. C.7 Cooling fan assembly

Photos (from two different angles) of the entirely assembled temperature-control system are provided in Figures C.8 and C.9. A small aluminum bar attached to the plate region of the extrusion acts as a strain relief for the TE module wires.

FIG. C.8 Fully assembled temperature-control system hardware

FIG. C.9 Fully assembled temperature-control system hardware

Alternatively, a complete system similar to the one shown in Figure C.9 can be purchased from Laird Technologies (see, e.g., Model DA-014-12-02, available from Mouser, P/N 739-DA-014-12-020000, $180, www.mouser.com).

The bidirectional current driver circuit for the TE module is shown in Figure C.10. For simplicity, the TIP transistors are represented in this diagram as simple bipolar transistors but are, in actual fact, Darlington transistors. With regard to external connections, a Darlington transistor behaves exactly the same as a single bipolar transistor with a very large current gain β (hence, the acceptability of the simplified representation). The Darlington's large β (on the order of 1000) results from its two-transistor internal construction, in which the collector current of one transistor provides the base current to the other. In its ON state, the Darlington's base-emitter voltage is "two diode-drops" (≈ 1.2 V). Because of the coupling between its two internal transistors, the Darlington is especially susceptible to thermal instabilities, necessitating proper heat-sinking of these components for reliable service (see below).

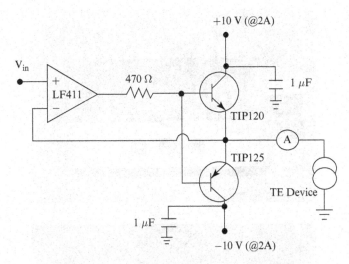

FIG. C.10 Voltage-controlled bidirectional current driver

The ± 10 V power sources attached to the TIP transistors' collectors must be capable of providing up to about 2 Amps (the ground of these power sources must be the same ground as the rest of the circuit (i.e., these grounds must be connected by, for example, cabling). The Laboratory DC Power Supplies sold by several electronic equipment manufacturers (e.g., BK Precision 1760A), which contain two identical DC supplies, are ideal instruments for providing both the negative and the positive voltage levels. These Laboratory DC Supplies typically have built-in ammeters that, in the above circuit, can be used to monitor the current flow through the TE device. The 1 μF capacitors in Figure C.10 remove any high-frequency noise present in the power supply voltage levels.

When this circuit is in operation, the TIP power transistors generate A LOT of heat. These hot transistors present the following problems: (1) If you touch them, they will hurt your fingers (an avoidable danger, once you're aware of it); (2) if the circuit is set up on a solderless breadboard, they will melt the breadboard's plastic (this is the voice of experience talking); and (3) if the transistors are allowed to get too hot, they will become thermally unstable and possibly destroy themselves (again, the voice of experience). Proper heat-sinking of the transistors will avoid these problems.

Because this process is somewhat involved, I provide students with a widget that contains the two transistors mounted on a heat sink. The noise-suppressing 1 μF capacitors are included as well. Top and bottom views of the widget (total cost approximately $30) are shown in Figure C.11.

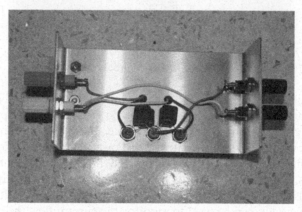

FIG. C.11 Top and bottom views of widget

The two TIP transistors have TO-220 packaging; however, I mount them on a TO-3 Heat Sink (Wakefield Engineering Model 401K, available from Newark P/N 58F502, $15). The pattern of mounting holes (appropriate for a single TO-3 packaged power transistor) on this finned heat sink allows the two TIP transistors to be affixed side by side and also provides convenient openings through which to pass wiring. A TO-220 heat sink mounting kit (Jameco P/N 34121, http://www.jameco .com, $4 per five-piece kit) must be used in mounting each TIP transistor to insulate

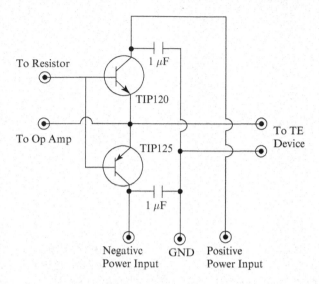

FIG. C.12 Widget circuit

its collector from the heat sink. The heat sink is then attached to a homemade chassis, fashioned from 1/16 in. aluminum sheet metal. Additionally, to facilitate necessary electrical connections, banana jacks and binding posts are mounted on the chassis.

The widget is then hardwired according to Figure C.12. All of the parts in this diagram may be purchased from Newark Electronics or from most any other electronics parts distributors.

Index